Springer Climate

Series Editor

John Dodson, Institute of Earth Environment, Chinese Academy of Sciences, Xian, China

Springer Climate is an interdisciplinary book series dedicated to climate research. This includes climatology, climate change impacts, climate change management, climate change policy, regional climate studies, climate monitoring and modeling, palaeoclimatology etc. The series publishes high quality research for scientists, researchers, students and policy makers. An author/editor questionnaire, instructions for authors and a book proposal form can be obtained from the Publishing Editor.

Now indexed in Scopus® !

Pritam Ghosh · Uday Chatterjee · Edris Alam
Editors

Climate Crisis and Changing Population Dynamics

Emerging Issues and Sustainable Solutions

Springer

Editors
Pritam Ghosh
Department of Geography
Hijli College
Kharagpur, West Bengal, India

Uday Chatterjee
Department of Geography
Bhatter College Dantan (Autonomous)
Kharagpur, West Bengal, India

Edris Alam
Faculty of Resilience, Integrated
Emergency Management
Rabdan Academy
Abu Dhabi, United Arab Emirates

ISSN 2352-0698 ISSN 2352-0701 (electronic)
Springer Climate
ISBN 978-3-032-24532-8 ISBN 978-3-032-24533-5 (eBook)
https://doi.org/10.1007/978-3-032-24533-5

This Springer imprint is published by the registered company Springer Nature Switzerland AG
The registered company address is: Gewerbestrasse 11, 6330 Cham, Switzerland

Dedicated to academicians, social scientists, environmentalists, planners, policy makers, Health and Development Practitioners, whose critical inquiry, ethical commitment, and action-oriented research continue to illuminate the relationship between climate change and changing population dynamics, and looking for sustainable solutions and effective policies related to climate change, migration, public health, and disaster response.

Preface

Climate change is not a distant threat; it is a present reality reshaping the population dynamics of the world. Anthropogenic stressors—including rising temperatures, prolonged drought, sea-level rise, coastline change, increasing frequency and severity of cyclones, and floods are not just environmental phenomena but also linked with human ecology, social and health crises. These climatic extremes intensify the prevalence of different communicable and non-communicable diseases, exacerbate physical health vulnerabilities, impact psychological health, elevate mortality rates and force various socioeconomically marginalized communities to migrate. Climate-induced displacement is threatening human rights, increasing food insecurity, loss of lives and poverty, and creating conflicts between various communities. Increasing climate-sensitive diseases like heat stress, undernutrition, vector-borne diseases, and diseases spread by air-water-soil pollution are ultimately increasing the additional burden of global mortality day by day. Understanding these dynamics is crucial for developing climate-smart, effective adaptation, resilience, and mitigation strategies. This edited volume offers an essential exploration into how extreme weather events, climate crises, and shifting climatic patterns influence health outcomes, increase mortality rates and drive migration worldwide.

The conceptual rationale and intellectual motivation of this edited volume stem from the growing recognition that climate change is not only an environmental crisis but also profoundly connected to socioeconomic and demographic processes that reshape various aspects of population dynamics through complex pathways. While existing scholarship has extensively documented climate hazards and their environmental effects, immediate impacts, and loss and mitigation strategies, less attention has been paid to how the climate crisis intersects with health challenges, morbidity and mortality, and human displacement across different parts of the world. Through interdisciplinary approaches, the book's chapters try to bridge climate science and climate change studies with population geography, social sciences, public health, sociology, demography, migration studies, and policy analysis.

The book comprises three distinct parts. **Part I** deals with the nexus of climate change and population dynamics, providing a contemporary overview of how climatic extremes impact human populations in diverse social and spatial contexts.

Chapter 1 of this part explores how climatic extreme events, such as flooding, intensify the burden of diseases among vulnerable populations or marginalized communities in different parts of the world. It demonstrates the pathways of disease transmission due to flooding, public health preparedness with special emphasis on equity-oriented, climate-resilient interventions. Chapter 2 critically revisits the "Gendered" nature of climatic impact, by demonstrating how the patriarchal structure of the society and unequal access to resources make women disproportionally susceptible to climatic shocks and systematic instigation of gender mainstreaming in climate policy. Chapter 3 analyses the paradox of urban expansion and climate footprint in Colombo, Sri Lanka. Utilizing high-resolution data, it shows how rapid demographic growth complicates Greenhouse Gas (GHG) emissions, sea-level rise (SLR), and how rapid urbanization amplifies climate risks among marginalized populations, informal settlements, and critical infrastructure in this region. The last chapter of this part visualizes the global research landscape on climate-induced mental health challenges through bibliometric analysis, outlining the thematic evolution, interdisciplinary linkage, and persistent geographic and collaborative gaps in the literature.

Part II outlines how climatic extremes influence the patterns of morbidity, mortality, and mental health distress across populations, highlighting the disproportionate burden of climate crises on vulnerable, marginalized populations in different parts of the world and underscoring the urgent need for climate-resilient interventions. The part comprises six chapters. The contributors in the part highlighted how climatic extreme events escalate the risk of different non-communicable diseases among different sections of society. Chapter 8 explores the impact of different types of climatic disasters on mental health, stress, trauma, and PTSD among vulnerable populations and outlines evidence-based mental health interventions within the disaster response and recovery framework. The next chapter mapped the impact of climate-induced drought on human morbidity and mortality, highlighting diverse pathways, regional disparities, and significant evidence gaps. Based on the micro-level empirical evidence, the following chapter explores the impact of climate disaster on the psychological health status and coping strategies of vulnerable, marginalized agricultural communities of eastern India. The volume then engages with outlining the growing incidents of heatwaves and the heatwave-related mortality trend in India. The last chapter of this part examines the post-disaster mental health crises based on South Asian case studies, advocating for trauma-informed, community-based psychological resilience as a core component of equitable climate recovery.

Part III of the volume, containing a total of eight chapters, illustrates how climate change and climatic extreme events drive human migration and displacement in different parts of the world, and the role of the policy framework for sustainable responses. The part starts with exploring the climate-induced economic losses and migration among marginalized communities in India. The next chapter highlights how the loss of agricultural land and livelihoods due to cyclones and salinity forces permanent migration in the coastal parts of Bangladesh. Utilizing field-based data and satellite imagery, the next two chapters outlined the empirical evidence of climate-induced flooding and residential displacement in different parts of India and Bangladesh. Adopting a multi-scalar, experiential design, the next chapter analyses the 2024 Brazil floods to explain how disasters remake "borderscapes" and administrative jurisdictions. In a related vein, the contributors outline how recurrent floods and riverbank erosion drive survival-oriented migration among char dwellers in northern Bangladesh. Attention then turns to systematic reviews on global policy strategies for protecting public health from climate extremes, identifying gaps in equity, governance, and anticipatory planning, particularly in low- and middle-income countries. The book ends with the chapter that reviews climate-induced migration in coastal regions worldwide, assessing legal and policy responses and advocating for coordinated, climate-resilient, and rights-based frameworks to protect displaced populations.

The chapters included in the volume represent a rich diversity of scholarly contributions from various disciplines, geographical contexts, and methodological approaches to outline the impact of the climate crisis on different aspects of changing population dynamics, such as population growth, human health, morbidity, mortality, psychological health, and displacement in different parts of the world. Collectively the book integrates the chapters on conceptual review, critical synthesis, systematic reviews, empirical investigation and policy analysis drawing from various subject areas such as population geography, public health, epidemiology, gender studies, environmental science, psychology and developmental policies etc. methodologically the contributions employ a broad spectrum of approaches including bibliometric, systematic, and scoping reviews, narrative and critical reviews, quantitative approaches, application of GIS and remote sensing, households and community-level survey based data analysis, mixed method framework and qualitative experimental inquiry. This methodological plurality underscores the inherently interdisciplinary nature of the climate crisis and population research. Bringing together macro-level global assessment along with micro-level community-level studies, the book offers a multi-scalar understanding of the vulnerability, resilience, and adaptive strategies in the face of accelerating climatic extremes.

An interdisciplinary perspective and comprehensive approach of this book ensure valuable insights for developing new research hypotheses and methodologies to the academician scholars and students in climate change, environmental science, geography, disaster management, public health, epidemiology, sociology, population studies, migration studies, and policy analysis. For policymakers and government agencies, this volume offers innovative solutions and evidence-based research crucial for formulating effective policies related to climate change, migration, public health, and disaster response. Public health officials can utilize the book's insights to improve preparedness and response plans, ensuring communities are better equipped to handle health crises induced by climate change. Disaster management professionals will find the volume's analysis of extreme weather events and their impacts on human populations particularly useful, and how more effective disaster response frameworks can be developed. By combining the synthesis of global literature with regionally grounded empirical research from the Global South and beyond, the volume serves as both a reference work and a teaching resource, offering conceptual clarity, methodological guidance, and practical insights for advancing equitable and sustainable responses to the challenges posed by climate-induced changes in population dynamics.

Kharagpur, India — Pritam Ghosh
Kharagpur, India — Uday Chatterjee
Abu Dhabi, United Arab Emirates — Edris Alam

Acknowledgments

Climate Crisis and Changing Population Dynamics: Emerging Issues and Sustainable Solutions is an outcome of collaborative intellectual efforts shaped by interdisciplinary investigation and scholarly dialogues to address one of the most important issues of current times. We are sincerely grateful to the authors whose original research works, critical insights, and methodological diversity have enriched this volume and advanced this academic dialogue outlining the intersections of climate crises and different aspects of population dynamics. We also acknowledge the anonymous reviewer for their insightful and constructive evaluation have advanced the standard, quality, coherence, and analytical depth of these chapters.

Contents

Editors and Contributors

About the Editors

Dr. Pritam Ghosh is an Assistant Professor and HOD at the Department of Geography, Hijli College, affiliated with Vidyasagar University, West Bengal, India and a Human Geographer with a Doctoral Degree from the University of Calcutta, Kolkata, India. His area of interest covers Medical and Health Geography, Epidemiology, and Public Health. As a dedicated researcher specializing in Health Geography and Social Sciences, his work is centred around investigating the geospatial, sociocultural, and environmental dimensions of public health. His research critically examines issues like childhood malnutrition, infant feeding practices, access to water, sanitation, and hygiene, and the role of allied public policies in India. He also contributes to emerging discourses on climate change and health, particularly in South Asia. With several peer-reviewed international journal articles and book chapters to his name, Dr. Ghosh is a prolific scholar and thought leader in his field. He currently serves as an Academic Editor for *PLOS ONE*, *Discover Social Science*, and *Health Journal* and reviews for more than 35 international journals spanning geography, public health, and the social sciences. He is also the editor of four upcoming Scopus-indexed books on health geography, under contract with a leading international publisher.

Dr. Uday Chatterjee is an Assistant Professor at the Department of Geography, Bhatter College, Dantan, affiliated with Vidyasagar University, West Bengal, India. He earned his Doctoral Degree in Applied Geography from Ravenshaw University, Cuttack, Odisha. His research interests encompass Urban Planning, Social and Human Geography, Applied Geomorphology, Indigenous knowledge, Hazards and Disasters, Risk Assessment and Resilience, Community Response and Environmental Issues. Dr. Chatterjee has been actively involved in editing scholarly works, including the Special Issue on Urbanism, Smart Cities, and Modelling for *Geojournal*, Springer, Special issue in Global Social Welfare, Springer and *Frontiers Environmental Science Journal*, *Geological Journal* Wiley, Editorial Board Member of *Scientific Report* (Nature Portfolio) journal and serving as a Book Series Editor for *Developments in Environmental Science*, Elsevier and *Sustainability Solutions-An Interdisciplinary and Transdisciplinary Nexus*, Springer Nature. His extensive research and editorial contributions highlight his commitment to advancing geographical knowledge and its applications. His research work has been funded by the West Bengal Pollution Control Board (WBPCB), Government of West Bengal, India, Indian Council of Social Science Research (ICSSR), Government of India. He has published research papers in reputed journals, edited books with reputed publishers like Springer, Elsevier, CRC Press, and authored books, book chapters, and conference proceedings. Currently, Dr. Chatterjee is doing an international project, in collaboration with Indonesia, Malaysia, and Japan, funded by APN (Asia Pacific-Global Change Research, Grant DOI https://doi.org/10.30852/p.22128), and visited Saudi Arabia for an international collaboration and conference session chair and Keio University, Tokyo, Japan, as a Visiting Professor.

Dr. Edris Alam is a Professor of Integrated Emergency Management in the Faculty of Resilience at Rabdan Academy, Abu Dhabi, United Arab Emirates. Professor Edris Alam is ranked among the world's top 2% of scientists based on a citation-based assessment conducted by Stanford University in collaboration with Elsevier. Dr. Edris Alam has studied and worked in Bangladesh, Australia, the UAE, and the United Kingdom, and earned his Ph.D. in Disaster Risk Reduction from the University of New South Wales (UNSW), Australia. Professor Alam has successfully secured competitive research funding from governmental organizations, non-governmental organizations, international NGOs, universities, and international development partners, including UNDP and DFID. In addition, Professor Alam has served as a consultant to NGOs, the Government of Bangladesh, international NGOs, and United Nations agencies. Professor Alam has an extensive record of high-quality research outputs in the fields of natural hazards, vulnerability, and disaster risk management. He has published more than a hundred peer-reviewed journal articles as a lead author in major international outlets. Professor Edris Alam is also the author of a book on disaster coping and adaptation and has contributed chapters to edited volumes published by leading international academic publishers. The findings of his research have been widely used to inform disaster risk reduction and development policy formulation in Bangladesh and other contexts. Driven by a strong commitment to reducing avoidable disaster-related deaths and strengthening societal resilience, Professor Alam founded the Disaster and Development Organization in 2008. Established as a trust and now registered as a national NGO with the Government of the People's Republic of Bangladesh, the organization operates through a multidisciplinary network spanning more than 20 countries across multiple continents.

Contributors

Fadzilah Adibah Abdul Majid Institute of Climate Adaptation and Marine Biotechnology (ICAMB), Universiti Malaysia Terengganu (UMT), Kuala Nerus, Terengganu, Malaysia

Maisarah Abdul Mutalib School of Graduate Studies, Postgraduate Centre, Management and Science University (MSU), University Drive, Shah Alam, Selangor, Malaysia;
International Medical School, Management and Science University (MSU), University Drive, Shah Alam, Selangor, Malaysia

Nurul Hafiza Ab Razak Institute for Environment and Development (LESTARI), Universiti Kebangsaan Malaysia (UKM), Bangi, Selangor, Malaysia

T. Adarsh Department of Geography, Sree Sankaracharya University of Sanskrit, Kalady, Kerala, India

Azyyati Ridha Alfian Institute for Environment and Development (LESTARI), Universiti Kebangsaan Malaysia (UKM), Bangi, Selangor, Malaysia;
Environmental and Occupational Health Department, Faculty of Public Health, Andalas University, Padang, Indonesia

Mohammad Aves Department of CSE, Graphic Era (Deemed to Be University), Clement Town, Dehradun, India

Tanmoy Basu Department of Geography, Katwa College, Katwa, Purba Bardhaman, West Bengal, India

Abubakar Abubakar Bello Department of Geography, Nigerian Defence Academy, Kaduna, Nigeria

Bibhore Kumar Deo United Nations Development Programme (UNDP), New Delhi, India

Nor Eliani Ezani Department of Environmental and Occupational Health, Faculty of Medicine and Health Sciences, Universiti Putra Malaysia (UPM), Serdang, Selangor, Malaysia

Mohamed Hassan Fathima Nuskiya Department of Geography, South Eastern University of Sri Lanka, University Park, Oluvil, Sri Lanka

S. Gayathri Department of Transport Planning, School of Planning and Architecture, Bhopal, India

Puranjani Ghosh School of Oceanographic Studies, Jadavpur University, Kolkata, India

Tuhin Ghosh School of Oceanographic Studies, Jadavpur University, Kolkata, India

Daisy Gohain School of Social Sciences and Languages, Vellore Institute of Technology, Chennai, India

Tiny Tanushree Gohain Department of Management, Brainware University, Barasat, West Bengal, India

K. P. Haneena Sulthana Department of Psychology, Sree Sankaracharya University of Sanskrit, Kalady, Kerala, India

Mallik Akram Hossain Department of Geography and Environment, Faculty of Life and Earth Sciences, Jagannath University, Dhaka, Bangladesh

Md. Abdul Khaleque Department of Disaster Management & Resilience, Bangladesh University of Professionals (BUP), Dhaka, Bangladesh

Ejaz Ahmad Khan Department of Public Health, Health Services Academy, Islamabad, Pakistan

Rahul Kumar Department of Microbiology, Graphic Era (Deemed to Be University), Clement Town, Dehradun, India

T. S. Lancelet Department of Geography, Sree Sankaracharya University of Sanskrit, Kalady, Kerala, India

Khai Ern Lee Institute for Environment and Development (LESTARI), Universiti Kebangsaan Malaysia (UKM), Bangi, Selangor, Malaysia

Monika Makwana School of Public Policy, Indian Institute of Technology, Delhi, India

Bittu Mandal School of Humanities and Social Sciences, Indian Institute of Technology Indore, Indore, India

Md. Mohiuddin Department of Geography and Environment, Faculty of Life and Earth Sciences, Jagannath University, Dhaka, Bangladesh

Biraj Kanti Mondal Department of Geography, Netaji Subhas Open University, Kolkata, India

Nurfashareena Muhamad Institute for Environment and Development (LESTARI), Universiti Kebangsaan Malaysia (UKM), Bangi, Selangor, Malaysia

Jadab Munda Department of Migration and Urban Studies, International Institute for Population Sciences, Mumbai, India;
Department of Geography, Magrahat College, University of Calcutta, Kolkata, India

Kamal Kumar Murari Centre for Climate Change and Sustainability Studies, School of Habitat Studies, Tata Institute of Social Sciences, Mumbai, India

Haninah Fadiah Mustafa Department of Environmental and Occupational Health, Faculty of Medicine and Health Sciences, Universiti Putra Malaysia (UPM), Serdang, Selangor, Malaysia

Nurulain Mustafa Udin School of Chemistry and Environment, Faculty of Applied Sciences, Universiti Teknologi MARA, Shah Alam, Selangor, Malaysia

Samshad Nowreen Department of Tourism and Hospitality Management, University of Dhaka, Dhaka, Bangladesh

Armania Nurdin Department of Biomedical Sciences, Faculty of Medicine and Health Sciences, Universiti Putra Malaysia (UPM), Serdang, Selangor, Malaysia; Laboratory of UPM-MAKNA Cancer Research (CANRES), Institute of Bioscience, Universiti Putra Malaysia (UPM), Serdang, Selangor, Malaysia

Unmesh Patnaik Centre for Climate Change and Sustainability Studies, School of Habitat Studies, Tata Institute of Social Sciences, Mumbai, India

Urmi Pattanayak ICMR-Regional Medical Research Centre, Bhubaneswar, Odisha, India

David Paul Department of English, Kristu Jayanti (Deemed to be University), Bengaluru, Karnataka, India

Mohd Roslan Rahmat Environmental Health Programme, Health Ministry Training Institute (ILKKM), Sungai Buloh, Selangor, Malaysia

Humaira Tabassum Raiyan Department of Geography and Environment, Jagannath University, Dhaka, Bangladesh

A. Revathi Department of Humanities and Social Sciences, BITS-Pilani K. K. Birla, Goa Campus, Sancoale, Goa, India

Raihan Riaz Department of Geography and Environment, Faculty of Life and Earth Sciences, Jagannath University, Dhaka, Bangladesh

Muhammad Naim Rosli School of Graduate Studies, Postgraduate Centre, Management and Science University (MSU), University Drive, Shah Alam, Selangor, Malaysia

Saima Ikbal Safi Department of Geography and Environment, Faculty of Life and Earth Sciences, Jagannath University, Dhaka, Bangladesh

Amiya Saha Department of Public Health and Mortality Studies, International Institute for Population Sciences (IIPS), Mumbai, Maharashtra, India

C. A. Sandhya Aravind Department of Psychology, Sree Sankaracharya University of Sanskrit, Kalady, Kerala, India

Sayantani Sarkar Department of Humanities and Social Sciences, BITS-Pilani K. K. Birla, Goa Campus, Sancoale, Goa, India

Sawjanya Sathyaseelan Department of Geography, University of Sri Jayewardenepura, Gangodawila, Nugegoda, Sri Lanka

Aida Soraya Shamsuddin Institute for Environment and Development (LESTARI), Universiti Kebangsaan Malaysia (UKM), Bangi, Selangor, Malaysia

Neegar Sultana Department of Geography and Environment, Jagannath University, Dhaka, Bangladesh

Sharifah Norkhadijah Syed Ismail Department of Environmental and Occupational Health, Faculty of Medicine and Health Sciences, Universiti Putra Malaysia (UPM), Serdang, Selangor, Malaysia

Gilbert Tayebwa The International Crane Foundation (Uganda Office), Kampala, Uganda

Tengku Nilam Baizura Tengku Ibrahim Department of Environmental Health, Faculty of Health Sciences, Universiti Teknologi MARA (UiTM), Bertam Campus, Kepala Batas, Pulau Pinang, Malaysia

Shruti Thakur School of Oceanographic Studies, Jadavpur University, Kolkata, India

Mumtahina Tisa Department of Geography and Environment, Faculty of Life and Earth Sciences, Jagannath University, Dhaka, Bangladesh

Benson Turyasingura Department of Environment and Natural Resources, Faculty of Agriculture and Environmental Sciences, Kabale University, Kabale, Uganda

Mesbah Uddin ADHVAN Inc, Shibuya City, Tokyo, Japan

Roberto Rodolfo Georg Uebel Department of International Relations, Graduate Program in Administration (PPGA), Escola Superior de Propaganda e Marketing (ESPM), São Paulo, SP, Brazil

Nor Hafizah Zakaria Institute of Climate Adaptation and Marine Biotechnology (ICAMB), Universiti Malaysia Terengganu (UMT), Kuala Nerus, Terengganu, Malaysia

Aziemah Zulkifli Institute for Environment and Development (LESTARI), Universiti Kebangsaan Malaysia (UKM), Bangi, Selangor, Malaysia

Part I
The Nexus of Climate Change and Population Dynamics

Chapter 1
Climate-Induced Flooding, Disease Burden, and Population Vulnerability: Health Risks for At-Risk Communities

Aida Soraya Shamsuddin, Aziemah Zulkifli, Nurfashareena Muhamad, Tengku Nilam Baizura Tengku Ibrahim, Maisarah Abdul Mutalib, Khai Ern Lee, Muhammad Naim Rosli, Nor Hafizah Zakaria, and Nurul Hafiza Ab Razak

Abstract Climate-induced flooding is increasingly recognised as a critical pathway through which environmental change affects infectious disease risks, especially in socially and infrastructurally vulnerable settings. Although the links between flooding and disease outbreaks are widely recognised, existing research remains

A. S. Shamsuddin (✉) · A. Zulkifli · N. Muhamad · K. E. Lee · N. H. Ab Razak
Institute for Environment and Development (LESTARI), Universiti Kebangsaan Malaysia (UKM), Bangi, Selangor, Malaysia
e-mail: aidasoraya@ukm.edu.my

A. Zulkifli
e-mail: aziemah_zul@ukm.edu.my

N. Muhamad
e-mail: fasha@ukm.edu.my

K. E. Lee
e-mail: khaiernlee@ukm.edu.my

N. H. Ab Razak
e-mail: hafiza@ukm.edu.my

T. N. B. Tengku Ibrahim
Department of Environmental Health, Faculty of Health Sciences, Universiti Teknologi MARA (UiTM), Bertam Campus, Kepala Batas, Pulau Pinang, Malaysia
e-mail: tengkunilam@uitm.edu.my

M. Abdul Mutalib · M. N. Rosli
School of Graduate Studies, Postgraduate Centre, Management and Science University (MSU), University Drive, Shah Alam, Selangor, Malaysia
e-mail: maisarah_abdulmutalib@msu.edu.my

M. N. Rosli
e-mail: muhd_rosli@msu.edu.my

M. Abdul Mutalib
International Medical School, Management and Science University (MSU), University Drive, Shah Alam, Selangor, Malaysia

P. Ghosh et al. (eds.), *Climate Crisis and Changing Population Dynamics*, Springer Climate, https://doi.org/10.1007/978-3-032-24533-5_1

fragmented, often focusing on single diseases or short-term outcomes while offering limited integration of transmission pathways, population vulnerability, and public health preparedness. What remains insufficiently addressed is how these factors interact to shape unequal health outcomes in flood-prone regions. This narrative review addresses this gap by synthesising multidisciplinary data on the pathways linking flooding to infectious disease transmission, patterns of community vulnerability, and the effectiveness of public health interventions. The published studies from Scopus, Web of Science (WOS), PubMed, Google Scholar and grey literatures form organisations such as World Health Organisation (WHO), Intergovernmental Panel on Climate Change (IPCC), World Bank Group and others were used to examine contributions of contaminated floodwaters, vector proliferation, sanitation failures, and also healthcare disruptions to the emergence of waterborne and vector-borne disease outbreaks. The findings show that the combination of inadequate WASH (Water, Sanitation and Hygiene) systems, limited healthcare access, and insufficient preparedness correlates with increased risks. This chapter discusses the interrelationship between pathways, vulnerabilities, and response strategies. It also emphasises the equity in the protective provision to the vulnerable population in flood-prone areas by describing early warning systems, climate-resilient WASH infrastructure, mobile health services, and community surveillance.

Keywords Climate resilience · Flooding · Infectious diseases · Public health · Vulnerability · WASH systems

Introduction

The extent and occurrences associated with flood events have risen due to the impact of climate change. In addition, the occurrence of storm events has also become more unpredictable and intense. According to Gül and Kayaalp (2024) and Promakhova et al. (2019), it has been demonstrated that the rise in global temperatures translates to the ability of the atmosphere to hold more moisture. This enhances floods associated with extreme precipitation. In contrast, the rise associated with global temperatures also enhances the rise associated with the level of the ocean. This has continued to threaten deltas and small island nations (Ali et al. 2024; Ke et al. 2021). Consequently, the impacts associated with climate-driven changes have affected vulnerable populations differently (Intergovernmental Panel on Climate Change, as cited in Carmona et al. 2022).

Flooding is one of the most common and catastrophic effects of climate change in the world. It has been reported that flooding has affected over 2.5 billion people

N. H. Zakaria
Institute of Climate Adaptation and Marine Biotechnology (ICAMB), Universiti Malaysia Terengganu (UMT), Kuala Nerus, Terengganu, Malaysia
e-mail: hafizah.zakaria@umt.edu.my

globally and that flooding represents over 44% of all climate change-related disasters over the past 50 years (WMO 2025; CRED 2015). Apart from destroying property and forcing people to move from one place to another in search of a new habitat, flooding can contribute to the spread of infectious diseases. This is mostly the case where there are no health facilities to cater to the needs of the local population. The most vulnerable regions to flooding are low- and middle-income countries that often have high urbanisation, but with inadequate sewage systems (Li et al. 2025; World Bank Group 2022; Sinharoy et al. 2019).

Previous literature has concluded that flooding is associated with increased water-borne, vector-borne, and environmentally induced infectious illnesses. For instance, floods carried Vibrio cholerae and Leptospira spp. that increased the risk of cholera, leptospirosis and hepatitis A (Carazo Gallego et al. 2025; Acosta-España et al. 2024; Shokri et al. 2020). Furthermore, the emergency situation during flooding, such as congested evacuation centres, malfunctioning sanitation systems, and limited clean water availability, would facilitate the progress of respiratory infections, gastrointestinal diseases, and dermatological problems among the flood victims (Carazo Gallego et al. 2025; Shokri et al. 2020). These condition is worsen by delayed detection, treatment, and containment of outbreaks due to disordered health facilities and monitoring systems during flooding (Dotse-Gborgbortsi et al. 2022; Nukavarapu and Durbha 2020).

Vulnerability and inadequate preparation affect health during flooding were insufficiently addressed in the previous literature, while most of them explicitly concluded that flooding affects health, especially waterborne and vector-borne diseases. The literature poses a noteworthy gap of limited data on vulnerable populations, public health management, and response mechanisms regarding climate-driven floods. Vulnerable populations such as those living in informal areas, children, the elderly, pregnant women, and people with disabilities are acknowledged; however, the inter-related factors (social, physical, and governmental variables) that elevate their health risk are often overlooked. Furthermore, the authority also normally discussed the preparedness and public health measures without considering their capacity to face the various flood-prone environments.

The research questions of this chapter are: (1) how does flooding caused by climate change impact the disease spread?; (2) which population is most vulnerable and what variables influence their susceptibility; and (3) which preparedness strategy, public initiatives, and policy approaches have been shown to reduce the negative health impacts of flooding. For this reason, this chapter aims to describe: (1) evidence on disease transmission pathways related to climate-driven flooding; (2) outlines of vulnerable population and their health discrepancies in flood-prone environments; and (3) preparedness approaches, public health interventions, and policy suggestions. This study contributes to the literature by providing insight into the variability of health impacts due to flooding driven by climate change based on different ways people are exposed and their susceptibility, as well as the responses of public health to these events. The government of flood-prone regions can use this review to develop new policies, improve community resilience and preparedness for the future.

Materials and Methods

This chapter utilised a narrative review approach to provide an overview of what is currently known about the impacts of flooding on human health caused by climate change. Various types of peer-reviewed articles, expert reports and case studies were reviewed to create this synthesis. The data was found using some of the most popular academic databases: Scopus, Web of Science, PubMed and Google Scholar. Additionally, grey literature provided by the WHO, World Bank Group, IPCC and others has been included. Search terms included: "flooding", "disease transmission", "public health", "vulnerability", and "WASH systems". The literature that was reviewed for inclusion in this chapter was primarily published between 2010 and 2025 focusing on works relevant to disease transmission pathways and social vulnerabilities associated with flooding, as well as public health interventions for flood-affected populations. Case studies from Malaysia and other regions of the world that experience frequent flooding were included to illustrate region-specific barriers and challenges (Fig. 1.1).

Results and Discussion

Pathways of Disease Transmission During Flooding

The altered ecological, infrastructural, and social systems due to floods have significantly increased the risk of infectious diseases. During and after the flood, the disease spread via interconnected pathways that were shaped by the interaction between human exposure, poor sanitation, and environmental contamination. Other than that, stagnant water also plays an important role in infections by providing ample conditions for the growth of waterborne pathogens and vector-borne diseases, and due to direct contact with the contaminated stagnant flood water (Acosta-España et al. 2024; Basaria et al. 2023). Ingestion of water-borne pathogens due to limited clean water access and crowded evacuation centres made the spread of skin and respiratory diseases, among others, pathways (Acosta-España et al. 2024; Basaria et al. 2023). The epidemiological effect after a flood event would be better predicted with consideration of these factors.

Role of Contaminated Floodwaters

Floodwaters are considered an important route of exposure to the infectious agents because of their capability to mobilise, retain, and spread the infectious agents across a wide geographic area. The floodwaters often contain untreated sewage, industry effluents, farm runoff, and solid waste from the community settlement, industrial zone and agricultural area. Therefore, the floodwater brings together pathogenic

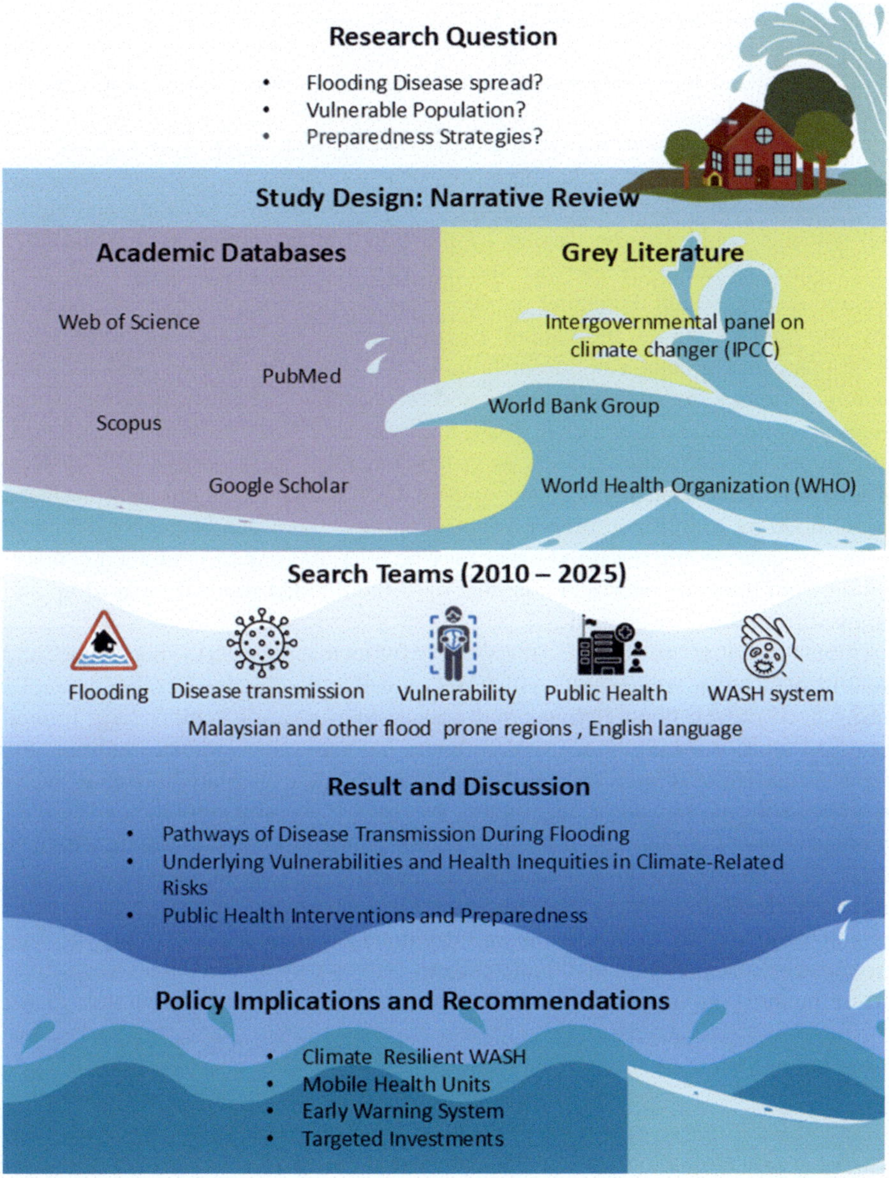

Fig. 1.1 Methodological flow

microorganisms, including bacteria, viruses, and protozoa, into water sources and surface areas. Previous studies have shown that faecal contamination from overloaded sewer systems is the main source of elevated *Escherichia coli* and *Enterococcus* levels in flooded areas (LeChevallier et al. 2024; Mukherjee et al. 2024). This route of exposure is prominent in highly crowded areas, such as in evacuation centres where there is insufficient sanitation infrastructure.

Other than that, chemical pollutants such as pesticides, heavy metals, and industrial solvents could also be carried by the floodwaters. These compounds have the potential to indirectly increase the risk of infectious diseases by weakening immune responses and damaging the ecosystem. These compounds may find their way into humans via the food chain. Environmental factors, such as the presence of various contaminants, can diminish the body's defensive abilities, which consequently increases the spread of illness and infection as a result of flooding within affected communities (Wang et al. 2024; Shetty et al. 2023). Flood events will typically impact groundwater aquifers as groundwater is accessed via a well system. Groundwater wells typically become compromised, allowing contaminants produced during flooding to infiltrate the wells and remain in the groundwater for extended periods of time. While the flood event may be over, the risk of continued contamination from flooded water will continue until the flooded water is cleaned up and removed from the environment.

Human activities are often also responsible for increasing both the persistence and establishment of environmental contamination within the floodwaters. For example, during flood events, flood victims are displaced to evacuation centres, have limited access to clean drinking water, and a functional sanitation system. Additionally, displaced individuals who are unable to access bottled or treated drinking water are more likely to consume contaminated floodwaters, significantly increasing their cumulative risk. Likewise, individuals working or walking in flood waters will have directly exposed their dermis to pathogens and infectious agents (i.e., *Leptospira*) originating from floodwaters. Hence, it is confirmed that contaminated floodwaters are a dynamic disease transmission pathway that connects the relationship between environmental degradation and human health.

Waterborne Diseases

Waterborne diseases represent one of the most direct and devastating effects of flood-associated contamination routes. Pathogens such as *Vibrio cholerae, Salmonella typhi, Shigella spp.*, and *Giardia lamblia* thrive in water habitats filled with organic matter and faecal pollutants introduced during floods (Shafii et al. 2023; Shayo et al. 2023). Cholera is a recurring post-flood outbreak in many low- and middle-income countries due to the bacteria's ability to grow quickly in contaminated water sources and spread efficiently through faecal-oral transmission. The ingestion of even small amounts of tainted water or food can cause serious gastrointestinal diseases, adding to dehydration, malnutrition, and increased mortality, especially among children and immunocompromised people (Abdullahi et al. 2025; Asantewaa et al. 2024).

Leptospirosis is a zoonotic bacterial disease caused by *Leptospira spp.*, widely spread through water contaminated with the urine of sick rats and other animals (Wang and Dunn 2024). Flood conditions aid transmission by displacing rodents from burrows, raising their contact with people, and spreading urine-borne germs widely in standing water. Human infection usually happens when *Leptospira* bacteria enter the body through cuts, abrasions, or mucous membranes during wading, swimming, or working in floods (Wang and Dunn 2024; Bradley and Lockaby, 2023). Symptoms range from mild febrile sickness to serious, life-threatening conditions such as Weil's disease, which includes kidney failure, jaundice, and haemorrhage (CDC 2025). Ziliotto et al. (2024) have reported a large leptospirosis outbreak with 725 cases in the state of Rio Grande do Sul in southern Brazil, due to an extreme flood event. The leptospirosis outbreak in May 2024 has claimed 26 lives and 96.2% of them were directly linked to the flood exposure.

Management of water-borne disease during and after a flood event is challenging due to interrupted sanitation infrastructure and limited clean water access. For example, when sewage water treatment facilities were malfunctioning, the sewage water could contaminate and create optimum conditions for pathogen reproduction and existence in a potable water system. Numerous cases reported that the open containers usage as clean water storage methods in evacuation centres has promoted bacterial growth and biofilm formation (Hayward et al. 2025; Meradji et al. 2025; Li et al. 2023). Due to this, the exposure risk is increased, and a cycle of reinfections is inevitable with a lack of broad-scale clean water and sanitation management. Prior incidents, for instance, the cholera outbreak in 2019 that occurred in Mozambique after Cyclone Idai, have proved the grave problem posed by waterborne pathogens when storms hit weak public health systems (Chen and Azman 2019; WHO 2019).

Moreover, the incidence of viral waterborne diseases such as noroviruses, rotaviruses, and hepatitis A is also elevated after flood events. These viruses have the ability to survive in extreme environments and require low infectious doses (Bleotu et al. 2024; Lanrewaju et al. 2022). In contrast to bacteria, these viruses are capable of surviving common water disinfection procedures and have weeks of longer life span (Shayo et al. 2023). Occurrences of viral gastroenteritis in evacuation centres underscore the necessity to identify viral persistence as an interrelated transmission pathway of health risk among the vulnerable population. Taken together, these factors show that flood-related waterborne diseases pose a broader range of pathogens that can spread over a long time after a flood disaster, rather than an acute bacterial outbreak.

Vector-Borne Diseases and Stagnant Water

Figure 1.2 shows pathways of disease transmission during flooding. Small stagnant pools or water puddles left after floods create ideal conditions for breeding grounds for mosquitoes, including Anopheles, Aedes, and Culex species. The condition allows the spread of vector-borne diseases such as malaria, dengue, chikungunya, and Japanese encephalitis (Sharma et al. 2022). The population of these mosquitoes

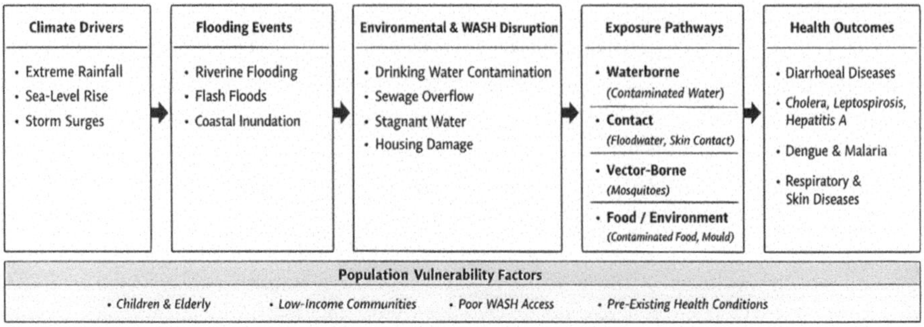

Fig. 1.2 Pathways of disease transmission during flooding

often increases in number after a flood event (Sharma et al. 2022; Coalson et al. 2021). The rapid growth of these insects population resulting in more contact between humans and mosquitoes, specifically in overcrowded evacuation centres that usually do not provide or have limited bed nets and mosquito repellents.

Other than that, changes in natural and human settlement landscapes also contribute to the spread of vector-borne diseases after a flood disaster. Furthermore, natural predators of mosquitoes, such as frogs, were most likely washed away by the floodwaters, consequently increasing the mosquito population (Yan et al. 2024; Chowell et al. 2019). The movement of flood victims to the edges of flood plain, wooded areas, or higher grounds may expose them to the mosquito species and zoonotic reservoirs they haven't encountered before (Zavaleta-Monestel et al. 2025; Fouque and Reeder 2019). This situation unlocks a new pathway for disease transmission that subsequently increases the health risk among the vulnerable population. Additionally, mice displaced due to the flood typically carry ectoparasites and act as vectors for zoonotic pathogens like hantaviruses, broadening the variety of flood-induced vector-borne diseases (Rulli et al. 2025; Shehata et al. 2025; Kim et al. 2024).

Insects such as blackflies, sandflies, and ticks thrive in moist environments and also act as vectors that spread onchocerciasis, leishmaniasis, and rickettsial diseases (Ratcliffe et al. 2024; Cecilio et al. 2022; Traub-Csekö et al. 2022). The disrupted solid waste management during flood events causes the accumulation of daily solid waste together with flood-related waste such as damaged household items, resulting in the reinforcement of flies and cockroach population growth that often act as vectors of intestinal pathogens. Remarkably, the time postponement between flood events and the peak in vector population density emphasises the vital and delayed impact of stagnant water on long-term disease risk.

Underlying Vulnerabilities and Health Inequities in Climate-Related Risks

The impacts of flood events are inequally distributed across society, and the significant influence was not limited to social, economic and political disparities. Various factors, including poverty, unstable housing, exclusion from essential services, and limited adaptive ability, may lead to the unequal health burden. According to analytical frameworks like the Pressure and Release (PAR) model and the Intergovernmental Panel on Climate Change (IPCC) Sixth Assessment Report, susceptibility is the point where exposure, sensitivity, and adaptation capacity converge (Carmona et al. 2022). While operational tools like Cutter's Social Vulnerability Index (SoVI) reveal that health risks are systematically shaped by age, gender, income, and access to healthcare. In summary, these standpoints underscore the fact that disasters are socially facilitated crises where exposure and outcome are determined by structural discrepancies rather than being equal-opportunity occurrences.

High-Risk Populations and Informal Settlements

Informal settlements are usually located along floodplains and riverbanks, hence making them the representative of the geography risk. Rapid urbanisation and limited affordable homes were forcing low-income families and refugees into these unwarranted settings. The poorly made homes made of plywood, ribbed metal, or salvaged wood offer little resistance against flood events, resulting in those living in the entire areas being displaced. Previous study has provide evident from Southeast Asia, Africa, and Latin America regions that informal settlements faced the biggest health risks during and after flood events due to fragile house structures, insufficient drainage infrastructure, and limited access to public services (Satterthwaite et al. 2020).

A detailed study on flood events in Malaysia from 2010 to 2022 has exposed that demographic and socioeconomic factors are directly associated with health risk. Low-income families generally have limited access to healthcare facilities and recovery programs (Isia et al. 2023). Besides, those living in the informal settlements are not protected legally and left out of mitigation and compensation plans. This situation led to cycles of repeated exposure to flood or other disasters and relocation (Mucherera and Spiegel, 2022; Hendricks and Van Zandt 2021). The Malaysian Social Vulnerability Index (SoVI) for Kajang district demonstrates underlying disadvantages, including advanced age, lower level of income and education, and limited access to healthcare facilities, which combine to reduce coping capacity (Idris et al. 2025). In this context, children, the elderly, and disable person living in informal settlements are at the peak of health risk. Children are mostly susceptible to malnutrition and infectious diseases. The elderly often have difficulty evacuating due to limited mobility and struggle to manage their chronic illness. Meanwhile, people with disabilities are particularly overlooked in relief plans. Moreover, women face

Table 1.1 High-risk groups and their disproportionate vulnerabilities during flooding, with selected examples from Malaysia and international contexts

Group	Key vulnerabilities	Examples/evidence
Children	More prone to diarrheal disease, malnutrition, dehydration, disrupted schooling, and long-term psychosocial stress	Diarrheal outbreaks tripled in poor districts post-flood in South Asia (Biswas et al. 2024); school closures in Malaysia after the 2014 Kelantan floods
Women (esp. pregnant/ lactating)	Loss of maternal healthcare access, higher malnutrition risk, and increased exposure to gender-based violence in shelters	Maternal health challenges during floods in South Asia (Meherali et al. 2024); women-headed households in Malaysia reported food insecurity (Isia et al. 2023)
Elderly	Reduced mobility in evacuation, higher vulnerability from chronic diseases, and social isolation	SoVI assessment in Kajang, Malaysia, identified older adults as highly vulnerable (Idris et al. 2025)
Persons with disabilities	Barriers to evacuation, overlooked in relief design, and difficulty accessing healthcare	Exclusion documented in the Bangladesh flood response (Rahman et al. 2025)
Informal settlement residents	Fragile housing, insecure tenure, poor drainage, no insurance, precarious livelihoods	Malaysian review of flood cases (Isia et al. 2023); limited adaptive capacity in Greater Kuala Lumpur (Kumaresen et al. 2025)

extra encounters of caregiving, loss of income that elevates their risks of food insecurity, and not to mention, gender-based violence in evacuation centres (Rahman et al. 2025; Nahin et al. 2024).

Livelihood uncertainty further compounds these weaknesses. Many informal settlers live on daily-wage work or small-scale businesses that can be wiped out in a single flood. Without funds or insurance, recovery becomes slow, if not impossible, continuing a loop where poverty and exposure support one another (Jha et al. 2012). Thus, informal villages show the intersections of physical marginalisation, social exclusion, and economic instability. Their disproportionate exposure is not incidental but the expected result of structural injustices that shape how tragedies occur. These shared weaknesses show that flooding does not affect all groups equally. As shown in Table 1.1, high-risk groups face separate but linked challenges that increase their health risks during and after flood events.

Health Inequities and Access to Clean Water and Healthcare

Unequal access to clean water and healthcare is a key determinant of health outcomes in disaster settings. Communities already reliant on dangerous water sources such as shallow wells, community taps, or polluted surface water are particularly at risk when these systems fail. For many, dependence on contaminated sources becomes a matter of life, changing chronic lack into a severe crisis. Previous studies across the South Asia region indicate that limited access to clean water is significantly associated with post-disaster diarrhoeal outbreaks (Biswas et al. 2024; Chyon et al. 2023). Studies done in Nigeria reported that children from low-income areas have a higher rate of waterborne illness compared to their counterparts (Abdullahi Tunde et al. 2025; Ugbede et al. 2025). Meanwhile, communities living in Greater Kuala Lumpur show low adaptability due to economic and political barriers, despite their high level of knowledge regarding environmental dangers (Kumaresen et al. 2025).

On top of that, healthcare inequity also follows this trend, which subsequently exacerbates the conditions. In rural areas where the healthcare facilities are often underequipped, the facilities may be totally shut down, while in urban informal settlements, fiscal struggles limit their access to the healthcare facilities and services (UN-Habitat 2020). Pregnant women face delayed maternal services during a flood event would increase their risk of pregnancy complications (Meherali et al. 2024). While children with lower immunity compared to adults are exposed to higher rates of malnutrition, dehydration and flood-related communicable diseases (Agabiirwe et al. 2022). Nevertheless, the situation in Malaysia has shown that social solidarity plays a moderating role in disaster events. Households surrounded by strong community networks often have better access to healthcare facilities compared to isolated households, showing that social solidarity creates resilience against disaster (Lugova and Haque 2024).

Another important factor is mental health inequities that are often unnoticed. The flood will cause displacement, job loss, insecurity, drive trauma, sadness, and post-traumatic stress among the victims. However, the flood response team rarely pays attention to the victim's mental health. Rahman et al. (2025) have reported that the displaced flood victims in Bangladesh show high levels of untreated sadness and post-traumatic stress disorder (PTSD). A similar situation was observed in Malaysia, where the flood response team prioritises the basic needs while psychological support is often underscored. Consequently, the situation widens the recovery gap where high-income families are supported by insurance and private healthcare, while low-income families are trapped in cycles of deprivation (Tasnim et al. 2023; Chan 2015).

Public Health Interventions and Preparedness

Flood disasters present complex and intensifying public health issues, particularly for those living in high-risk areas with existing social, economic, and infrastructural insufficiencies. The interruption of clean water sources, displacement of flood victims, and the failure of sanitation systems further promote ample conditions for the spread of flood-induced communicable diseases such as leptospirosis, cholera, and dengue. On top of that, the emergency situation disrupts chronic illness management that requires consistent ongoing care. An effective public health strategy in areas at high risk of flooding should implement a cohesive, preventive, and conditional approach that concurrently tackles the acute health threats and improves the long-term resilience of affected communities, particularly those in a vulnerable state.

Integrated Public Health Response

An integrated public health response to flood events involved intersectoral coordination across health, crisis management, environmental, and social safety agencies. This approach considers waterborne and vector-borne diseases, and injury-related health outcomes share similar aspects based on environmental circumstances and access to critical services. For example, the management of the leptospirosis outbreak during the 2014 Kelantan floods has profited from collaboration between the Ministry of Health Malaysia, local councils, and community health volunteers to recover drainage, ensure safe waste disposal, and conduct targeted rodent control operations (Mohd Yusof et al. 2025; Philip and Ahmed 2023).

In terms of preparedness and protection, measures may consist of mapping vulnerable groups, forward placement of medical supplies, and executing focused immunisation programs, especially for hepatitis A and typhoid diseases. During the reaction phase following flooding, implementing quick health needs assessments into broader emergency procedures can help assign resources effectively. Finally, the recovery phase should include "build back better" principles, such as elevating rural clinics above previous flood levels or adding green backup power for vaccine storage, ensuring continuity of critical health services (Hashim and Hashim 2016). On top of that, an integrated response should involve the community as one of the key components. The community could contribute their local knowledge, including traditional water safety practices and informal warning systems, which can improve the cultural relevance and support of interventions (Samsudin et al. 2024).

Early Warning and Disease Surveillance Systems

The early flood warning systems, depending on hydrometeorological parameters (rainfall and river levels data), can be upgraded by integrating the existing system with health monitoring in order to improve preparation and outbreak prevention. Connecting hazard forecasts with predictive epidemiological models aids public health authorities in anticipating disease risks and moving resources accordingly (WHO 2013). For example, heavy rainfall events in Southeast Asia regions have been linked with increases in diarrhoeal and dengue cases within weeks, showing the need for proactive interventions (Alied et al. 2022; Sugeno et al. 2023; Wang et al. 2022).

In this setting, digital health tools should be completely utilised. Mobile-assisted reporting tools for frontline health workforces can significantly lessen delays in disease reporting and improve spatial coverage (Samsudin et al. 2024). Trial projects using mobile reporting during floods aided the fast discovery of dengue hotspots and allowed focused vector control in Malaysia setting. Moreover, community-oriented monitoring networks, such as trained volunteers reporting odd disease tendencies, extend the range of formal systems into remote areas, connecting information gaps during disaster events (Samsudin et al. 2024).

Mobile Health Units and Emergency Healthcare Access

Flood events usually disturb or isolate the fixed healthcare facilities and services, consequently creating a barrier to the affected flood victims to access the critical services. Those in need of regular treatment for chronic conditions, including diabetes, hypertension, or kidney disease, are severely impacted (Aziz et al. 2016). Alternatives such as Mobile Health Units (MHUs) are capable of offering a flexible choice, providing on-site healthcare services in flood-affected areas where healthcare infrastructure is damaged or disrupted.

Aziz et al. (2016) have described that an MHU may range from basic cars containing essential medicines to fully prepared as a mobile hospital equipped with diagnostic tools, vaccine storage, and telemedicine capabilities. Advanced preparation is important, which includes route planning, volunteer recruitment, and service packages that should be carefully considered according to community risk profiles and anticipated health requirements. For example, during the East Coast Malaysia flood events, the MHUs filled with various teams with the capability to offer acute care, maternal and child health services, mental health support, and health education in evacuation centres.

The MHUs can also be the public health education spots where they can distribute hygiene kits, demonstrate basic water purification methods, and offer psychological support (Hashim and Hashim 2016). Coordination with emergency response teams is vital to avoid repetition, and the data collected during these activities should inform monitoring systems for improved preparedness planning in the future.

Hygiene and Sanitation Promotion Campaigns

Disruption of drinking water sources and sanitation systems due to flood events often increases the risk of waterborne disease transmission (Howard et al. 2020). Plans of cleaning and sanitation program should aim to decrease the risk of waterborne disease transmission via behavioural change, provision of cleaning resources, and environmental health gains (MacLeod et al. 2025).

An effective cleaning and sanitation program should combine both hardware equipment (latrines, water treatment kits) and software tools (health education, habit change communication). Previously, joint efforts between the Ministry of Health Malaysia and NGOs distributing clean water kits and water purification pills, while organising hygiene awareness workshops, have decreased diarrhoeal cases after the 2014 Kelantan floods (Mohd Radi et al. 2018). Furthermore, the effort shows that messages tailored according to regional settings, including using local languages and trusted messages, have improved uptake by the flood victims.

Besides, active community participation is also a vital key. Engaging religious organisations, women's groups, and youth leaders in supporting handwashing practices, safe clean water storage, and community clean-ups after flood events can improve both adoption and sustainability. Mass media efforts combined with direct communication, such as live examples of handwashing practices or chlorination of a clean water source, resulting an advance knowledge and practical skills. Finally, hygiene promotion integrated into disaster preparation shows potential that affected communities will keep the good practices all year-round and consequently lower baseline disease burden and refine their resistance against flood disasters.

Policy Implications and Recommendations

Building Climate-Resilient WASH Infrastructure

Capitalising on climate-resistant WASH infrastructure is vital for protecting public health in flood-prone areas due to the basic WASH services worldwide being affected by climate change uncertainties. The World Health Organization (WHO) stresses that climate change impacts, including higher flood levels, heavier rainfall, hotter temperatures, and unusual weather, present direct threats to clean water supply and functional sanitation systems. Floods can disrupt the pipe system, contaminate clean water sources, and overwhelm sewerage and treatment facilities, causing major postponements in both clean water and effective sanitation service delivery, and further exposing people to flood-related disease risks. Robust infrastructure, such as elevated clean water distribution points, effective emergency sanitation facilities, and systems, can keep critical WASH operational during and after flood events. These measures reflect the guidance in the Strategic Framework for WASH Climate-Resilient Development by the Global Water Partnership and UNICEF (GWP and UNICEF, 2022), which identifies the requirement to integrate climate considerations at all stages of WASH planning, including risk assessment, design standards, implementation, and monitoring. By integrating a climate-resilient WASH system, related government organisation can decrease infrastructure (healthcare, clean water and sanitation services) vulnerability against flood events and sustain continuous access to the crucial services. These measures serve as preventive investments that safeguard health systems, support long-term sustainability, and line with global development goals. Rather than just technical upgrades, the measures ensure equal access to crucial services in the era of climate change.

WASH as a Core Policy Instrument for Flood-Related Disease Prevention

WASH plays the key role in protecting public health during a disaster through its capability to intersect every crucial connection in the course of flood-induced communicable disease infection. The functional safe water source facility significantly decreases the exposure to diseases via ingestion, while effective sanitation systems avoid excreta from infecting the broader area. Simultaneously, the constant practice of hygiene habits, including handwashing practice and safe water conduct, further offers an important barrier that limits person-to-person disease transmission (MacLeod et al. 2025). A carefully designed WASH program has the ability to decrease more than 60% effect of diarrhoeal diseases, emphasising their crucial role in disease prevention during flood events (WHO 2025).

In Malaysia settings, the disrupted WASH during and post-flood period has been shown by the 2014 Kelantan floods, where a spatial–temporal study showed that leptospirosis frequency doubled in the disaster aftermath. The study reported the number of subdistricts with more than 100 cases per 100,000 people soaring from four to twelve subdistricts (Mohd Radi et al. 2018). This case reflects that floods are regularly followed by flood-induced communicable disease outbreaks, driven by deteriorating hygiene conditions, including limited clean drinking water access and overcapacity at evacuation centres (Shafii et al. 2023). These facts have given insight into the fragility of WASH system in high-risk areas of flooding and their key role as an effective barrier to curb the flood-induced communicable disease outbreaks.

During the December 2021 floods in Peninsular Malaysia, non-government organisation, namely the International Federation of Red Cross and Red Crescent Societies (IFRC) and Malaysian Red Crescent Society (MRCS) were taking initiative to adopt a few number of WASH interventions matched with Sphere standards. The interventions include a large-scale distribution of hygiene kits, fast and continuous supply of clean water, building of effective emergency sanitation facilities, and a comprehensive hygiene promotion programs (IFRC 2021). A preliminary assessment suggested that the integrative measures taken significantly curbed the extent of disease transmission when compared with the earlier flood event in 2014, where such integrative protective interventions were absent. A recent cluster randomised study conducted in rural Bangladesh showed that WASH enhancement during rainy seasons had a significant effect of lowering diarrhoea among children, particularly those from the low-income families (Ante-Testard et al. 2024).

Based on these evidence in terms of health and disaster policy, governments should apply WASH in flood or other disaster response plans as a key protection approach. The strategies that implement defensive hygiene promotion, guard clean water resources, and build resilient sanitary facilities have been proven to decrease the disease burden related to flood events. This indication shows that successful WASH actions in protecting communities during acute disease risks and shape adaptive capacity to face future health threats related to climate uncertainties (Ante-Testard et al. 2024).

Targeted Investments in Flood-Prone Communities

The application of environment-driven solutions has appeared as a vital support to adaptable WASH procedures while supporting existing standard engineering methods. Engineering methods such as constructed ponds, bioretention systems, green pathways, and permeable sidewalks contribute to better floodwater management and natural filtering of contaminants, thus improving water quality indirectly. These methods mutually reduce the risk of flood and expand ecosystem resilience, consequently offer complimentary benefits such as species protection and carbon sequestration (Ventura et al. 2024). In areas with fast urbanisation, such as Southeast Asia regions, cautious urban planning, such as optimisation natural drainage volume and the restoration of green infrastructure, offers sustainable and cost-effective route for reducing health risks related to flood disasters.

Comprehensive governance is another important key role of effective disaster management. Based on evidence, planning and resource distribution processes frequently show that flood risk and health costs unfairly fall on marginalised groups (Hendricks and Van Zandt 2021). Therefore, WASH investments must target flood or disaster risk and health costs by warranting open decision-making, subsidised access to clean water and sanitation, and the addition of social safety measures. For instance, participatory planning strategies that include women, local leaders, and informal settlement members significantly improve the uptake and sustainability of WASH services (Sisay et al. 2024). Incoporation of equity into WASH planning aids to address the structural disparities that intensify disaster effects and guarantees that efforts to develop resilience do not accidentally support cycles of vulnerability.

Finally, the construction of integrated tracking and risk-oriented surveillance systems is crucial to sustain long-term WASH resilience. Integrating hydrological models and public health data permits governments and related agencies to forecast infectious disease outbreaks related to floods, forward placement of resources, and guide focused necessary actions (Yarveisi et al. 2025). Emerging technologies, such as real-time river water levels and streamflow monitors, water quality monitors and mobile-assisted community reporting tools, expand the reach of monitoring systems into rural areas while decreasing delays in outbreak detection. A resilient WASH system is an integration of layered, linked system of protection, avoidance, and preparation, which includes technology innovation, ecological design, fair governance, and risk-driven monitoring, considered collectively that can secure public health during both acute and prolonged flood disasters.

Conclusion

Through a number of interconnected factors, such as water contamination, vector propagation, poor sanitation, and interruption of healthcare services, this narrative review shows how climate-driven floods intensely raise the risk of flood-linked infectious diseases. Due to limited access to clean water and healthcare facilities, vulnerable populations particularly those living in informal settlements and areas at high risk of flooding, consequently experience more severe health effects compared to those with more advantages. The results show that basic social, economic, and inefficient governance facilitate the flood-linked disease burdens in addition to environmental exposure. The assessment highlights the need to reconsider climate-driven floods as a health risk threat that demands integrated strategies from the viewpoints of public health and climate resilience. Strengthening climate-resistant WASH systems, enhancing disease monitoring, early warning systems, and accessible critical health services are vital to reduce the health risks associated with floods. Plans for preparing for and responding to catastrophes should also include community involvement. This increases the effectiveness of risk communication, monitoring, and intervention.

Future studies should examine the cumulative and increasing health effects of frequent floods in a changing environment, going beyond disease-specific or short-term studies. Determining the effectiveness of integrated public health treatments, governance systems, and policy frameworks in various social and environmental contexts should receive more attention. It will be crucial to coordinate public health preparation, infrastructure planning, and community-based approaches in order to address climate-related floods as a sign of health and to support inclusive, resilient health systems in places that are vulnerable to floods.

AI Disclosure Statement

The authors declare that artificial intelligence–based tools were applied during the preparation of this manuscript only to support language refinement and enhance manuscript clarity. All scientific content conclusions were independently produced by the authors, who assume full responsibility for the accuracy and integrity of the study.

Acknowledgements The authors thank all collaborators and institutions whose insights and resources contributed to the development of this work.

Authors' Contribution: All authors contributed equally to the conceptualisation, literature review, analysis, and writing of this work.

Funding:

Informed consent: The work was not supported by any funding.

Not applicable.

Conflict of Interest: The authors declare that there are no conflicts of interest related to this work.

References

Abdullahi Tunde A, Oche Joseph O, Abdulkareem Olakilekun A, Ogun Sedoten O, Grace Adeola A, Nenger Jerome A, Iyanuoluwa Oyedeji O, Adewunmi A, Godfred Yawson S, Barakat Olajumoke K, Joel JK (2025) Impact of climate change-induced flooding water related diseases and malnutrition in Borno state, Nigeria: a public health crisis. Environ Health Insights 19:11786302251321683. https://doi.org/10.1177/11786302251321683

Abdullahi IN, Dangana A, Usman MS, Miri N, Mohammed Y, Avong JC, Dansura ML, Samuel BE, Gagari VF, Gyang NM, Daudu OU, Nanbol HD, Ajao OI, Adejuyigbe ZA, Ugwu CN, Omoare AA (2025) Multifaceted epidemic preparedness against cholera outbreak by one health approach and genomic surveillance. Infect Genet Evol 133:105798. https://doi.org/10.1016/j.meegid.2025.105798

Acosta-España JD, Romero-Alvarez D, Luna C, Rodriguez-Morales AJ (2024) Infectious disease outbreaks in the wake of natural flood disasters: global patterns and local implications. Le infezioni in medicina 32(4):451–462. https://doi.org/10.53854/liim-3204-4

Agabiirwe CN, Dambach P, Methula TC, Phalkey RK (2022) Impact of floods on undernutrition among children under five years of age in low-and middle-income countries: a systematic review. Environ Health 21:98. https://doi.org/10.1186/s12940-022-00910-7

Ali T, Mortula MM, Gawai R (2024) Coastal vulnerability analysis of the UAE coast due to sea level rise. Adv Sci Technol Innov 749–752. https://doi.org/10.1007/978-3-031-51904-8_164

Alied M, Nguyen D, Aziz JMA, Vinch DP, Huy NT (2022) Dengue fever on the rise in Southeast Asia. Pathog Glob Health 117(1):1–2. https://doi.org/10.1080/20477724.2022.2116550

Ante-Testard PA, Rerolle F, Nguyen AT, Ashraf S, Parvez SM, Naser AM, Benmarhnia T, Rahman M, Luby SP, Benjamin-Chung J, Arnold BF (2024) WASH interventions and child diarrhea at the interface of climate and socioeconomic position in bangladesh. Nat Commun 15:1556. https://doi.org/10.1038/s41467-024-45624-1

Asantewaa AA, Odoom A, Owusu-Okyere G, Donkor ES (2024) Cholera outbreaks in low- and middle-income countries in the last decade: a systematic review and meta-analysis. Microorganisms 12(12):2504. https://doi.org/10.3390/microorganisms12122504

Aziz NA, Ismail N, Zaki RA (2016) Mobile health units in disaster response: the Malaysian experience. Prehosp Disaster Med 31(5):553–556. https://doi.org/10.1017/S1049023X16000753

Basaria AAA, Ahsan A, Nadeem A, Tariq R, Raufi N (2023) Infectious diseases following hydrometeorological disasters: current scenario, prevention, and control measures. Ann Med Surg 85(8):37782–43782. https://doi.org/10.1097/MS9.0000000000001056

Biswas B, Remesan R, Tiwari MK, Ghosh S (2024) Rapid assessment of flood inundation due to tropical cyclones in part of Sundarbans in Google Earth Engine environment. In: Mishra AP, Kaushik A, Pande CB (eds) Natural resource monitoring, planning and management based on advanced programming. Springer Nature Singapore, Singapore

Bleotu C, Matei L, Dragu LD, Necula LG, Pitica IM, Chivu-Economescu M, Diaconu CC (2024) Viruses in wastewater—A concern for public health and the environment. Microorganisms 12(7):1430. https://doi.org/10.3390/microorganisms12071430

Bradley E, Lockaby G (2023) Leptospirosis and the environment: a review and future directions. Pathogens 12(9):1167. https://doi.org/10.3390/pathogens12091167

Carazo Gallego B, Montesinos Sanchis E, Martínez Campos L, López Medina EM, Alegría Coll I, García López Hortelano M, Calvo Rey C (2025) Enfermedades infecciosas más frecuentes asociadas a inundaciones en niños. Anales de Pediatría 102(3):503777. https://doi.org/10.1016/j.anpedi.2025.503777

Carmona R, MacDonald JP, Dorough DS, Rai TB, Sanago GA, Thorsell S (2022) Recognising the contributions of indigenous peoples in global climate action?—An analysis of the IPCC report on impacts, adaptation and vulnerability. Iwgia Briefing Note. Retrieved from https://iwgia.org/en/resources/publications/4621-iwgia-briefing-analysing-recognition-contrubutions-indigenous-peoples-ipcc-report.html

CDC (2025) Clinical overview of leptospirosis. https://www.cdc.gov/leptospirosis/hcp/clinical-overview/index.html

Cecilio P, Cordeiro-da-Silva A, Oliveira F (2022) Sand flies: basic information on the vectors of leishmaniasis and their interactions with Leishmania parasites. Community Biol 5(1):305. https://doi.org/10.1038/s42003-022-03240-z

Centre for Research on the Epidemiology of Disasters (CRED) (2015) Retrieved from https://www.preventionweb.net/publication/human-cost-natural-disasters-global-perspective#:~:text=EM%2DDAT%20data%20also%20show,%24936%20billion%20in%20recorded%20damage

Chan NW (2015) Impacts of disasters and disaster risk management in Malaysia: the case of floods. In: Aldrich DP, Oum S, Sawada Y (eds) Resilience and recovery in Asian disasters: community ties, market mechanisms, and governance. Springer Japan, Tokyo

Chen WH, Azman AS (2019) Mitigating cholera in the aftermath of cyclone Idai. Am J Trop Med Hyg 101(5):960–962. https://doi.org/10.4269/ajtmh.19-0285

Chowell G, Mizumoto K, Banda JM, Poccia S, Perrings C (2019) Assessing the potential impact of vector-borne disease transmission following heavy rainfall events: a mathematical framework. Philos Trans R Soc B 374(1775):20180272. https://doi.org/10.1098/rstb.2018.0272

Chyon MSA, Biswas S, Mondal MS, Roy B, Rahman A (2023) Integrated assessment of flood risk in Arial Khan floodplain of Bangladesh under changing climate and socioeconomic conditions. J Flood Risk Manage 16:E12876. https://doi.org/10.1111/jfr3.12876

Coalson JE, Anderson EJ, Santos EM, Garcia VM, Romine JK, Luzingu JK, Domingues B, Richard DM, Little AC, Hayden MH, Ernst KC (2021) The complex epidemiological relationship between flooding events and human outbreaks of mosquito-borne diseases: a scoping review. Environ Health Perspect 129(9):096002. https://doi.org/10.1289/EHP8887

Dotse-Gborgbortsi W, Dwomoh D, Asamoah M, Gyimah FT, Dzodzomenyo M, Li C, Akowuah G, Ofosu A, Wright J (2022) Dam-mediated flooding impact on outpatient attendance and diarrhoea cases in northern Ghana: a mixed methods study. BMC Public Health 22(1). https://doi.org/10.1186/s12889-022-14568-w

Fouque F, Reeder JC (2019) Impact of past and on-going changes on climate and weather on vector-borne diseases transmission: a look at the evidence. Infect Dis Poverty 8:51. https://doi.org/10.1186/s40249-019-0565-1

Gül B, Kayaalp N (2024) Investigation of the flood event under global climate change with different analysis methods for both historical and future periods. J Water Clim Chang 15(8):3939–3965. https://doi.org/10.2166/wcc.2024.196

Hashim JH, Hashim Z (2016) Climate change, extreme weather events, and human health implications in the Asia Pacific region. Asia Pac J Public Health 28(2_suppl):8S–14S. https://doi.org/10.1177/1010539515599030

Hayward C, Ross KE, Brown MH, Bentham R, Nisar MA, Hinds J, Xi J, Whiley H (2025) Microbial risks in drinking water systems: persistence and public health implications of opportunistic premise plumbing pathogens. Front Microbiol 16:1575789. https://doi.org/10.3389/fmicb.2025.1575789

Hendricks MD, Van Zandt S (2021) Unequal protection revisited: planning for environmental justice, hazard vulnerability, and critical infrastructure in communities of color. Environ Justice 14:87–97. https://doi.org/10.1089/env.2020.0054

Howard G, Calow R, Macdonald A, Bartram J (2020) Climate change and water and sanitation: likely impacts and emerging trends for action. Annu Rev Environ Resour 45:253–283

Idris NDM, Zulkepli NA, Taha MR, Ashraf O, Muhamad N, Muhammad Y (2025) Quantifying community vulnerability to natural hazards and hydrometeorological hazards: application of the social vulnerability index in Kajang area. J Sustain Sci Manag 20:919–944. https://doi.org/10.46754/jssm.2025.05.003

IFRC (2021) Malaysian 2021 floods: In pictures. https://www.ifrc.org/article/malaysian-2021-floods-pictures

Isia I, Hadibarata T, Jusoh MN, Bhattacharjya RK, Shahedan NF, Fitriyani NL, Syafrudin M (2023) Identifying factors to develop and validate social vulnerability to floods in Malaysia: a systematic review study. Sustainability 15(17):12729. https://doi.org/10.3390/su151712729

Jha AK, Bloch R, Lamond J (2012) Cities and flooding: a guide to integrated urban flood risk management for the 21st Century. World Bank Publications

Ke Q, Yin J, Bricker JD, Savage N, Buonomo E, Ye Q, Visser P, Dong G, Wang S, Tian Z, Sun L, Toumi R, Jonkman SN (2021) An integrated framework of coastal flood modelling under the failures of sea dikes: a case study in Shanghai. Nat Hazards 109(1):671–703. https://doi.org/10.1007/s11069-021-04853-z

Kim B, Lee Y-J, Kwak D, Seo M-G (2024) Nationwide survey of vector-borne diseases in rodents and mites in Korea: Anaplasma, Ehrlichia, and Rickettsia. Animals 14(20):2950. https://doi.org/10.3390/ani14202950

Kumaresen M, Teo FY, Selvarajoo A, Sivapalan S, Falconer RA (2025) Assessing community perception, preparedness, and adaptation to urban flood risks in Malaysia. Water 17(15):2323. https://doi.org/10.3390/w17152323

Lanrewaju AA, Enitan-Folami AM, Sabiu S, Edokpayi JN, Swalaha FM (2022) Global public health implications of human exposure to viral contaminated water. Front Microbiol 13:981896. https://doi.org/10.3389/fmicb.2022.981896

LeChevallier MW, Prosser T, Stevens M (2024) Opportunistic pathogens in drinking water distribution–a review. Microorganisms 12(5):916. https://doi.org/10.3390/microorganisms12050916

Li P, Tong M, Wu J, Meng X, Fan X, Sun X, Li J, Xue T (2025) Flood exposure contributes to under-five mortality in low- and middle-income countries. One Earth 8(5):101305. https://doi.org/10.1016/j.oneear.2025.101305

Li Y, Qu Y, Yang H, Zgou X, Xiao P, Shao T (2023) Combatting biofilms in potable water systems: a comprehensive overview to ensuring industrial water safety. Environ Microbiol Rep 15(6):445–454. https://doi.org/10.1111/1758-2229.13207

Lugova H, Haque M (2024) Dynamics of community resilience in flood-prone areas of southeast Asia: a key component of the public health agenda. Adv Hum Biol 14(4):379–382. https://doi.org/10.4103/aihb.aihb_74_24

MacLeod C, Davies K, Mwenge MM, Chipungu J, Cumming O, Dreibelbis R (2025) Behaviour change interventions to improve household sanitation and hygiene practices in urban settings: a systematic scoping review. Int J Hyg Environ Health 264:114519. https://doi.org/10.1016/j.ijheh.2025.114519

Meherali S, Nisa S, Aynalem YA, Kennedy M, Salami B, Adjorlolo S, Ali P, Silva KL, Aziato L, Richter S, Lassi ZS (2024) Impact of climate change on maternal health outcomes: an evidence gap map review. Plos Global Public Health 4(8):e0003540. https://doi.org/10.1371/journal.pgph.0003540

Meradji S, Basher NS, Sassi A, Ibrahim NA, Idres T, Touati A (2025) The role of water as a reservoir for antibiotic-resistant bacteria. Antibiotics 14(8):763. https://doi.org/10.3390/antibiotics14080763

Mohd Radi MF, Hashim JH, Jaafar MH, Hod, Rozita Ahmad N, Mohammed Nawi A, Baloch GM, Ismail R, Ayub NIF (2018) Leptospirosis outbreak after the 2014 major flooding event in Kelantan, Malaysia: a spatial-temporal analysis. Am J Trop Med Hyg 98(5). https://doi.org/10.4269/ajtmh.16-0922

Mohd Yusof MS, Ahmed N, Bakhir SNI, Shiva Shanghari RV, Hashairi Fauzi M (2025) The prevalence of leptospirosis infections among humans in Malaysia: a systematic review and meta-analysis. Top Biomed 42(2):201–212. https://doi.org/10.47665/tb.42.2.013

Mucherera B, Spiegel S (2022) Forced displacement: critical lessons in the protracted aftermath of a flood disaster. GeoJournal 87:3855–3875. https://doi.org/10.1007/s10708-021-10471-w

Mukherjee M, Hossain MS, Boswell J, Zhang Y, Allen MS, LaMontagne MG, Gentry TJ (2024) Large, but short-term, increase in fecal indicator bacteria following extreme flooding from Hurricane Harvey in Houston TX. Front Water 6:1466377. https://doi.org/10.3389/frwa.2024.1466377

Nahin KTK, Nowreen S, Lima M (2024) Gendered vulnerability index for flood in Brahmaputra river floodplain. SSRN, p 5005685. https://doi.org/10.2139/ssrn.5005685

Nukavarapu N, Durbha S (2020) Interdependent healthcare critical infrastructure analysis in a spatiotemporal environment: a case study. ISPRS Int J Geo-Inf 9(6). https://doi.org/10.3390/ijgi9060387

Philip N, Ahmed K (2023) Leptospirosis in Malaysia: current status, insights, and prospects. J Physiol Anthropol 42:30. https://doi.org/10.1186/s40101-023-00347-y

Promakhova EV, Kuksina LV, Golosov VN (2019) Extreme erosion events and climate change. In: Springer proceedings in earth and environmental sciences, pp 118–120. https://doi.org/10.1007/978-3-030-03646-1_22

Rahman MM, Shobuj IA, Alam S, Sadia AM, Khanam S, Hossain MdT, Alam E (2025) The unforeseen tide: exploring mental health impacts of the 2024 flash flood in Bangladesh. BMC Public Health 25(1):2728. https://doi.org/10.1186/s12889-025-24118-9

Ratcliffe NA, Mello CB, Castro HC, Dyson P, Figueiredo M (2024) Immune reactions of vector insects to parasites and pathogens. Microorganisms 12(3):568. https://doi.org/10.3390/microorganisms12030568

Rulli MC, DÓdorico P, Galli N, John RS, Muylaert RL, Santini M, Hayman DTS (2025) Land use changes and infectious diseases emergence. Rev Geophys 63(2):e2022RG000785. https://doi.org/10.1029/2022RG000785

Samsudin NA, Karim N, Othman H, Naserrudin NA, Sahani M, Hod R, Siau CS, Harif MN, Abd Samad BH, Zaini ZI (2024) Exploring community behaviors and stakeholder challenges in engaging communities with dengue prevention behavior in Malaysia: implementation research for a qualitative study with a community-based participatory. Res Des 14(3):e074222. https://doi.org/10.1136/bmjopen-2023-074222

Satterthwaite D, Archer D, Colenbrander S, Dodman D, Hardoy J, Mitlin D, Patel S (2020) Building resilience to climate change in informal settlements. One Earth 2(2):143–156. https://doi.org/10.1016/j.oneear.2020.02.002

Shafii NZ, Mohd Saudi AS, Pang JC, Abu IF, Sapawe N, Kamarudin MKA, Mohamad MHN (2023) Association of flood risk patterns with waterborne bacterial diseases in Malaysia. Water 15(11):2121. https://doi.org/10.3390/w15112121

Sharma SN, Kumawat R, Singh SK (2022) Risk assessment and impact of floods on the transmission of vector borne diseases. J Commun Dis 54(3):15–21. https://doi.org/10.24321/0019.5138.202284

Shayo GM, Elimbinzi E, Shao GN, Fabian C (2023) Severity of waterborne diseases in developing countries and the effectiveness of ceramic filters for improving water quality. Bull Natl Res Cent 47:113. https://doi.org/10.1186/s42269-023-01088-9

Shehata AA, Parvin R, Tasnim S, Duarte PM, Rodriguez-Morales AJ, Basiouni S (2025) The hidden threat: roden-borne viruses and their impact on public health. Viruses 17(6):809. https://doi.org/10.3390/v17060809

Shetty SS, Deepthi D, Harshitta S, Sonkusare S, Naik PB, Suchetha Kumari N, Madhyastha H (2023) Environmental pollutants and their effects on human health. Heliyon 9(9):e19496. https://doi.org/10.1016/j.heliyon.2023.e19496

Shokri A, Sabzevari S, Hashemi SA (2020) Impacts of flood on health of Iranian population: infectious diseases with an emphasis on parasitic infections. Parasite Epidemiol Control 9:e00144. https://doi.org/10.1016/j.parepi.2020.e00144

Sinharoy SS, Pittluck R, Clasen T (2019) Reviews of drivers and barriers of water and sanitation policies for urban informal settlements in low-income and middle-income countries. Util Policy 60:100957. https://doi.org/10.1016/j.jup.2019.100957

Sisay SF, Gari SR, Ambelu A (2024) Fecal sludge management and sanitation safety: an assessment in Addis Ababa Ethiopia. Environ Health Insights 18:11786302241267188. https://doi.org/10.1177/11786302241267187

Sugeno M, Kawazu EC, Kim H, Banauvong V, Pehlivan N, Gilfillan D, Kim H, Kim Y (2023) Association between environmental factors and dengue incidence in Lao People's Democratic Republic: a nationwide time-series study. BMC Public Health 23:2348. https://doi.org/10.1186/s12889-023-17277-0

Tasnim S, Aripin S, Asif N (2023) A comparative analysis between experts and local people's perspective on challenges in creating flood resilient housing in Malaysia. Int J Bus Technol Manage 5:1–12. https://doi.org/10.55057/ijbtm.2023.5.3.1

Traub-Csekö Y, Ramalho-Ortigão M, Telleria EL (2022) Editorial: Sand flies and their interactions with Leishmania and other pathogens. Frontiers in Tropical Diseases 3:882512. https://doi.org/10.3389/fitd.2022.882512

Ugbede OJ, Okolo PN, Ochiaka DI, Eze PI, Ogbuyeme JN, Atuchi NM, Nwachukwu MC, Offor EC (2025) Public health implications of internal climate-induced migration: tracking disease patterns in IDP settlements in Northern Nigeria. Int J Educ Res Sci Dev 7(2):355–373. https://doi.org/10.59795/ijersd.v7i2.237

Un-Habitat 2020 (2015)World Cities Report 2020: The Value of Sustainable Urbanization, United Nations Office for Disaster Risk Reduction. Sendai Framework for Disaster Risk Reduction 2015–2030. https://www.undrr.org/publication/sendai-framework-disaster-risk-reduction-2015-2030

Ventura J-RS, Tulipan JU, Banawa A, Umali KDC, Villanueva JAL (2024) Advancements and challenges in decentralized wastewater treatment: a comprehensive review. Desalin Water Treat 320:100830. https://doi.org/10.1016/j.dwt.2024.100830

Wang F, Xiang L, Sze-Yin Leung K, Elsner M, Zhang Y, Guo Y, Pan B, Sun H, An T, Ying G, Brooks BW, Hou D, Helbling DE, Sun J, Qiu H, Vogel TM, Zhang W, Gao Y, Simpson MJ, Luo Y, Tiedje JM (2024) Emerging contaminants: a one health perspective. Innov (Camb (Mass)) 5(4):100612. https://doi.org/10.1016/j.xinn.2024.100612

Wang S, Dunn N (2024) Leptospirosis. StatPearls, Treasure Island (FL): StatPearls Publishing. https://www.ncbi.nlm.nih.gov/books/NBK441858/

Wang Y, Wei Y, Li K, Jiang X, Li C, Yue Q, Zee BC-Y, Chong KC (2022) Impact of extreme weather on dengue fever infection in four Asian countries: a modelling analysis. Environ Int 169:107518. https://doi.org/10.1016/j.envint.2022.107518

World Bank Group (2022) Policy Research Working Paper 10014: Rapid urban growth in flood zones. Retrieved from: https://documents1.worldbank.org/curated/en/099546404212214703/pdf/IDU0ef8bc63a022b304b7c08e7c04aac815d4d98.pdf

World Health Organization (WHO) (2013) Early warning surveillance and response in emergencies. Geneva: WHO. Retrieved from: https://www.who.int/publications/i/item/early-warning-surveillance-and-response-in-emergencies

World Health Organization (WHO) (2019) Tropical Cyclone Idai and Kenneth, Mozambique. Retrieved from: https://www.afro.who.int/sites/default/files/2019-05/WHOSitRep5Mozambique3May2019.pdf

World Meteorological Organization (WMO) (2025). Flood. Retrieved from https://wmo.int/topics/floods

Yan J, Mackay AJ, Stone CM (2024) Dynamics of invasive mosquitoes: introduction pathways, limiting factors, and their potential role in vector-borne pathogen transmission. Front Trop Dis 5:1503120. https://doi.org/10.3389/fitd.2024.1503120

Yarveisi F, Jafarzadegan K, Tripathy SS, Moftakhari H, Moradkhani H (2025) A data-driven framework for an efficient block-level coastal flood risk assessment. Int J Disaster Risk Reduct 122:105478. https://doi.org/10.1016/j.ijdrr.2025.105478

Zavaleta-Monestel E, Rojas-Chinchilla C, Molina-Sojo P, Murillo-Castro MF, Rojas-Molina JP, Martínez-Vargas E (2025) Impact of climate change on the global dynamics of vector-borne infectious diseases: a narrative review. Cureus 17(1):e77972. https://doi.org/10.7759/cureus.77972

Ziliotto M, Chies JAB, Ellwanger JH (2024) Extreme weather events and pathogen pollution fuel infectious diseases; The 2024 flood-related leptospirosis outbreak in Southern Brazil and other red lights. Pollutants 4(3):424–233. https://doi.org/10.3390/pollutants4030028

Chapter 2
Revisiting the Gender– Climate Change Nexus: A Critical Review

A. Revathi and Sayantani Sarkar

Abstract The discourse on climate change has emerged in international politics as a multidimensional challenge that requires both technical and interdisciplinary interventions. With the emergence of concepts such as differential vulnerability to climate change research, gender gained recognition as one of the significant axes determining vulnerability. The article explores existing research on gender and climate change by undertaking a critical analysis of selected works on gender and climate change to answer three questions: (a) Is climate change vulnerability gendered, (b) In comparison to men, are women more vulnerable to the impacts of climate change and (c) Should gender be considered as a significant factor in climate change policy? To address the objectives, a narrative review incorporating elements of critical analysis was employed. The process involved collecting literature, screening, preliminary reading, conceptual categorisation and critical analysis. The findings affirm that due to existing patriarchal structures, gendered roles and responsibilities and inequality of access to resources and information, vulnerability to climate change is gendered. Women are comparatively more vulnerable than men due to the climate-sensitive nature of their livelihood, greater vulnerability to poverty, comparatively more reliance on informal institutions and limited role in decision-making. Further, the study confirms that existing gender inequalities are reflected in climate policy-making, and there is a need for gender mainstreaming. This chapter revisits the existing scholarship on gender and climate change and provides an overview of how climate change impacts women and the caveats to consider in policy-making.

Keywords Climate change · Vulnerability · Gender · Climate policy · Women

A. Revathi (✉) · S. Sarkar
Department of Humanities and Social Sciences, BITS-Pilani K. K. Birla, Goa Campus, Sancoale, Goa, India
e-mail: p20200060@goa.bits-pilani.ac.in

S. Sarkar
e-mail: sayantanis@goa.bits-pilani.ac.in

P. Ghosh et al. (eds.), *Climate Crisis and Changing Population Dynamics*, Springer Climate, https://doi.org/10.1007/978-3-032-24533-5_2

Introduction

In the light of recent climatic events, it takes little effort to convey to people the gravity of the challenge that climate change has presented to humankind. According to the Intergovernmental Panel on Climate Change (IPCC), climate change is any discernible change in mean climate or variability in its properties that is usually noticed over a decade or longer (IPCC 2007). Climate change has become important in global politics over the years; once considered an issue of soft politics, it is now a significant security concern for nations.

With climate change emerging at the forefront of global politics, various nuances of its dimensions are being explored. New interdisciplinary approaches are implemented to comprehend the complexities of a problem of this magnitude. This has encouraged social scientists, including gender and development studies scholars, to conduct critical readings of its various aspects to understand its implications for society. The gender and climate change nexus has emerged in this particular context.

This nexus becomes important given the strong influence gender has on shaping opportunities accessible to individuals. The resources a person has access to, the influence they can have in decision-making and the role they play in society are all determined by gender among other factors. Understanding this nexus is essential for policy-making to ensure that existing inequality is accounted for in policy-making. If policy-making fails to do this there is a risk of widening the inequality. Therefore while designing sustainable solutions to deal with the challenges of climate change, gender becomes one of the most important axes to account for.

This chapter attempts to consolidate existing literature on gender and climate change. The preliminary reading of existing work on gender and climate change helped identify the key questions that are being raised by scholars working on the area. After identifying these pertinent questions, they were used as themes around which this review categorised the existing literature.

The objective of this study is therefore to address three distinct questions with the purpose of enhancing comprehension of the discourses pertaining to climate change and gender. The first question pertains to whether or not vulnerability to climate change is influenced by gender. The second inquiry aims to investigate the potential disparity in vulnerability associated with the consequences of climate change between genders, while the third question pertains to the importance of acknowledging differential gender impacts in the formulation of climate policy.

The chapter would first discuss the materials and methods followed to conduct the study. Subsequently, the chapter is structured into three distinct sections, followed by a discussion and conclusion. The initial segment concerning gender and climate change examines fundamental research on the subject and investigates the major themes associated with this discourse. The following segment on the vulnerability of women towards climate change focuses on the major themes associated with their increased vulnerability. Additionally, the third section, dedicated to gender debates within the context of climate policy, explores the emergence of gender as a significant actor in climate policy.

Materials and Methods

The study aimed to understand and critically analyse the literature on nexus of gender and climate change. A narrative review and critical analysis were conducted to explore the existing literature. In order to aid this exploration, three questions related to gender and climate change were identified:

1. Is climate change vulnerability gendered?
2. In comparison with men, are women more vulnerable to the impact of climate change?
3. Is gender a significant actor in climate policy?

This narrative review aimed to explore the association between climate change and gender. Therefore, the review first inquired into the existence of this nexus. Key debates and evidence regarding this association were identified in addressing this inquiry. Subsequently, by limiting gender to a binary framework, the review examines women's greater vulnerability. The review concludes by looking at the implications of this association for policy-making. These inquiries were selected based on an initial overview of the field acquired during the early phases of the review. These specific inquiries were considered relevant as they encapsulated the predominant themes that are currently significant within the gender and climate change discourse. Answering these questions essentially allows one to thematically organise existing works on gender and climate change and to identify existing gaps that could be addressed by future research.

The effort was to then seek relevant articles addressing these questions. The review thus had a thematic focus on these questions. While carrying out the literature selection, geographical limitations were not explicitly established. However, the scope was confined to studies within social sciences and gender studies.

The search strategy utilised the following databases: Google Scholar, Jstor, and Scopus. Keywords such as "GENDER AND CLIMATE CHANGE," "GENDER VULNERABILITY AND CLIMATE CHANGE," "WOMEN VULNERABILITY AND CLIMATE CHANGE," and "GENDER AND CLIMATE POLICY" were used to search the databases. The review included only the articles published in English, while excluding others. Journal articles, book chapters, reports, and working papers were considered for inclusion. Articles were screened based on the presence of keywords climate change, vulnerability, gender, women and climate policy. Some articles that were frequently cross-referenced were also included. Abstracts were reviewed for screening purposes, focusing on articles related to climate change and gender while excluding articles dealing with other aspects of climate change.

The selected articles were read thoroughly and critically analysed for the review. The material was organised conceptually to address the initial questions after a comprehensive reading. The review primarily serves as a narrative while critically engaging with the selected texts. Although the review includes significant works in the field, it does not adopt a systematic approach. It prioritises engaging with the most pertinent literature falling within the themes identified. A more systematic

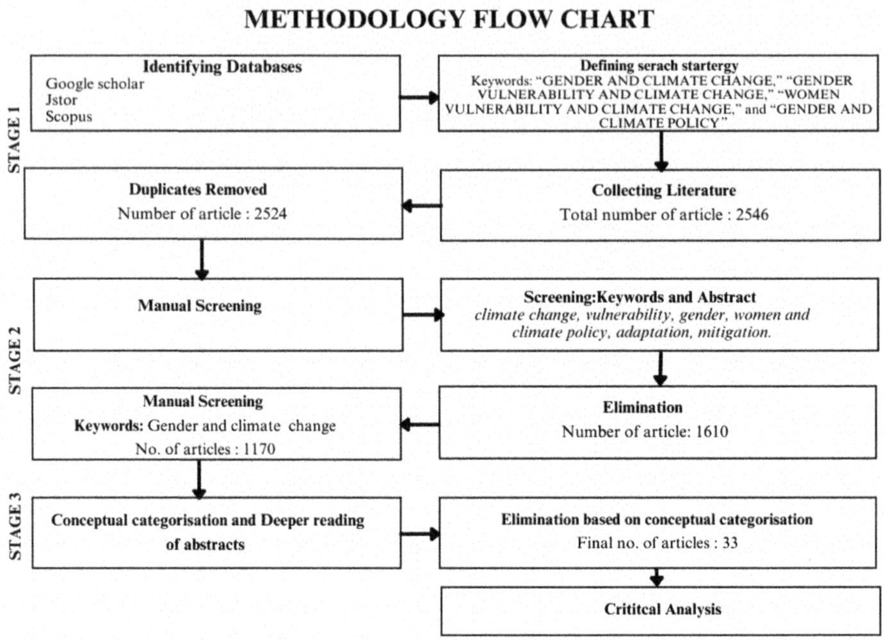

Fig. 2.1 Created by: authors

examination of existing works on gender and climate change, which extends beyond the themes this study deals with, is warranted, as this topic is receiving increasing scholarly attention, resulting in substantial literature production (Fig. 2.1).

Results and Discussions

Findings

Climate Change-Related Vulnerabilities and Gender

A. **Is climate change vulnerability gendered?**

The question of vulnerability has been dealt with by multiple researchers across fields. According to Kelly and Adger (2000), "the vulnerability of a population can be understood as the ability of individuals, households, social groups and communities to anticipate, cope with, resist and recover from or adapt to any shocks or external stress placed on their livelihood and well-being". Bohle et al. (1994) argue that vulnerability should be viewed as a context and not an outcome. Findings of existing studies indicate that vulnerability is not an inherent characteristic (Browne et al. 2016). Instead, it indicates contextual patterns of practices, power relations and processes that lead to differences in advantages and disadvantages enjoyed by certain groups and individuals (Enarson 1998). Vulnerability can thus be understood more as a product of the context and conditions within which the individual encounters the shock or stress than as an outcome of those.

Vulnerability is a multifaceted concept that should be understood more as a consequence of a complex interaction of varied circumstances, including exclusions and inequalities individuals face (Schofield and Gubbels 2019). It has political, socioeconomic and cultural aspects and is not determined by a single factor or indicator but by the interaction of various determinants.

Power shapes the context in which vulnerability operates. Hence, vulnerability is political. Power relations are vital in determining access to resources and producing inequalities and exclusions, contributing to vulnerability. Inequalities and differences existing within society lead to differential vulnerabilities.

Existing literature on climate change recognises differential vulnerability (Ahmed and Fajber 2009). Vulnerability of diverse individuals and groups to the effects of climate change varies according to multiple social factors. Thomas et al. (2018) suggest that differential vulnerability is a function of social rather than physical variables. Thus, existing research indicates that an individual's social location is one of the most important factors in deciding the level of their vulnerability.

Vulnerability to the impacts of climate change can be determined in part by gender. It was not until relatively recently that researchers began to make the association between gender and climate change. The technical approach towards climate change has predominated in discussions for a long time, while the political and social aspects of the issue have received less attention. The gender dimension of climate change is vital because it helps to understand the social dynamics of vulnerability (Eriksen et al. 2015). MacGregor (2010), in her study, argues that climate change has gender-differentiated causes and effects, which makes climate change vulnerability more gendered in nature.

The research literature has increased in the last decade. The causes and consequences have been discussed using recurring themes. Several theoretical and analytical frameworks have been used to approach the concept. Gender has emerged as an essential determinant with its effects on many climate change vulnerability indices. Yadav and Lal (2018) bring attention to the various determinants of climate change vulnerability, such as the roles and responsibilities of an individual, access to resources, and the rights they enjoy as individuals and by gender.

Gender differences are prevalent in the differential distribution of roles and related responsibilities among individuals. Within the household, women are traditionally assigned reproductive and care roles, and studies indicate that this frequently leads to greater vulnerability of women (Yadav and Lal 2018). Similarly, in some societies, men may be more vulnerable than women due to the nature of their livelihood activities. Another key factor contributing to vulnerability is a lack of resources to tackle the manifestations of climate change; resources are not equally accessible to all individuals. Gender is one of the elements that influences access to resources. Patt et al. (2009) emphasise the importance of gender in determining an individual's "ability to prepare for and respond to climate-related threats". Gender influences access to information, which is critical for building preparedness. Access to information differs based on the nature of the gender relations that exist in a society. Studies indicate that men in most societies have greater access to information than women, making them more aware and prepared. The preferred means for getting information also vary by gender, and these nuances must be addressed in policy-making for more effective climate policy. Furthermore, Patt et al. (2009) also emphasise the important role of gender roles in causing a difference on the impact on men and women.

In most societies, patriarchy is a dominating structure that governs gender relations. Gender inequalities are a manifestation of the patriarchal structures that most societies function within. As a result, the understanding of how patriarchal systems contribute to inequality and gendered vulnerability is important. Kantamanenei et al. (2018) focus on patriarchal structures that lead to unequal status and differential roles for men and women, resulting in differences in vulnerability between men and women.

In the matter of understanding gendered vulnerability, there are some essential caveats that one must account for. The first of these concerns the understanding of gender itself. Gender is often synonymous with 'women,' but such categorisations oversimplify the idea of gendered vulnerability and ignore its nuances. Understanding gender relations and how they influence roles and responsibilities is a more effective way to comprehend gendered vulnerability. MacGregor (2010) proposes viewing gender as a political relationship between individuals who identify as masculine or feminine.

The second caveat is that gendered vulnerability is contextual. According to Riverra-Ferra (2021), "climate variability and change occur in the framework of political, institutional, economic, and social structures and changes that interact dynamically with contextual circumstances". The social context in which the vulnerabilities manifest themselves can change over time. Climate change's gendered nature is time and location-sensitive (Patt et al. 2009). Therefore, as Nong et al. (2020) argue, "...the actual, social, economic and political circumstances in which vulnerability assessment takes place" must be considered.

The third caveat that this study seeks to emphasise is the subject of "intersectionality." Gender identity is an important predictor of vulnerability to climate change, but it does not work independently of other indicators. Understanding and recognising the intersectionalities that occur within a gender is critical. In this regard, Sara Ahmed and Fajber (2009) draw attention to how social stratification, including discrimination based on caste, class and religion, intersects gender identity. Aspects of social identity, including class, race, caste and sexuality, intersect with one's gender identity and need to be acknowledged while attempting to understand gendered vulnerability to climate change.

B. **In comparison with men, are women more vulnerable to the impact of climate change?**

As the association between gender and climate change became more widely accepted, various studies focusing on the differences in the impacts on men and women began to emerge. Along the way, a crucial realisation emerged about women's increased vulnerability to climate change. For instance, Goh (2012) claims in her research that "climate-related events affect men's and women's well-being and assets differently, and climate-related shocks affect women more negatively than men". Observations also indicate "women are particularly vulnerable to poverty, insecurity, and violence during and after climatic events"(Altson 2012).

The primary factor contributing to women's vulnerability is their standing as the poorest of the poor (MacGregor 2010). According to Onwutuebe's (2019) study of Nigerian society, a considerable proportion of women are poorer than men. The research highlights how the climate-sensitive nature of women's livelihood endeavours in these societies, combined with existing poverty, increases women's vulnerability (ibid). In addition, the study also critically analyses women's vulnerability within the context of the patriarchal structure that dictates gender relations (ibid). In their research on gender and climate change in Africa, Bob and Babugura (2014) conclude that "…due to social inequalities, ascribed social and economic roles, women are disproportionately affected by climate change impacts."

Bradshaw and Linncha (2014) address how women's vulnerability to climate change is influenced by access to resources, education, income, and knowledge. The finding that women rely on informal institutions and sources for information on climate change has significant policy implications. This also demonstrates the earlier argument on the gender difference in the medium of information access.

MacGregor (2010) identifies specific factors that make women more vulnerable compared to men. She contends that women's everyday care and provision roles will become harder due to the effects of climate change, with food shortages affecting women more than men. She also emphasises the lesser role women play in decision-making and the vulnerability of women to violence during climate-related events. Similar reasons for women's vulnerability have been advanced by Nelson et al. (2002) and Resurrecion (2012).

Climate change has resulted in unequally distributed costs and risks. According to Resurrecion (2012), the concurrent process of change and continuity that raises women's vulnerability should be recognised in policies on gender as well as climate change. Issues related to Climate change and related societal changes may force women to take on roles and related responsibilities traditionally assigned to male members while maintaining their care responsibilities (ibid).

Nelson et al. (2002) highlight an essential caution regarding assuming women as a homogeneous group. They argue that refugees, low-income people, people without housing, the elderly, disabled women, and recent migrant women are the major vulnerable groups of women. Women cannot be assumed to be a homogeneous group; differences exist among women along the lines of caste, class, race and sexuality.

Sara de Wit's (2021) work on how Masai women perceive climate change is another essential work to consider in discussing women's vulnerability. Wit (2021) makes several compelling arguments in her research against the "feminisation of climate change" and generalisations about women's vulnerability. She rightly warns us about the dangers of reducing women to "essentialising categories" and the policy advantage of such categorisation cannot be overlooked. Wit (2021) also opens new ways for integrating women into the climate change discourse as more active agents and highlights the various cultural aspects and perceptions requiring further exploration (ibid).

Aurora-Jonsson (2011) points out that the rhetoric on women and climate change is trapped within the binary of virtue and vulnerability. She contends that this emphasis can divert attention away from more pressing issues on the agenda, such as inequalities in decision-making. Jolly (2021) also draws attention to this issue about how the "voice of women" is often absent from the discourse.

This discourse on women and climate change is extensive and the paper will revisit its contextual nature before concluding. Women's increased vulnerability is determined by the context in which vulnerability is examined and by a confluence of variables, including but not limited to gender. Nong et al. (2020) research in Vietnam's Northern province, for example, found less difference in the Livelihood Vulnerability Index between men and women due to practices in Vietnam that support gender equality. Hence, it is imperative to consider the context in which the vulnerability arises.

C. **Should gender be a significant actor in climate policy?**

Since the recent acknowledgement of the climate-gender nexus, climate policy has also now begun to incorporate parameters of gender. The emphasis on gender in climate change policy was lacking as social issues were frequently overlooked in climate discourse due to a dearth of social scientists engaged in the topic (Terry 2009). Denton's (2002) research highlights how climate policy-making replicates the pre-existing inequalities, overlooking women and less developed countries, while men and developed nations dominate decision-making. MacGregor's (2010) research also aligns with this viewpoint, and she posits that there exists a tendency towards the masculinisation of policy-making related to climate change.

Gender constitutes a significant phenomenon that demands its place in climate policy formulation. Gender disparity is evidenced by inadequate representation of women in climate policy formulation (Patt et al. 2009). Kantamaneni et al. (2018) highlights that climate change policy cannot be regarded as a gender-neutral domain owing to its inherent gendered nature. The incorporation of gender in climate policy is essential, and it has the capacity to enhance the effectiveness of climate change policy. In contrast, the absence of gender perspectives may impede the actions to promote gender equity within the domain. Incorporating gendered perspectives and acknowledging the multifaceted nature of gender helps to identify underlying power structures leading to different experiences of vulnerability and adaptation related to climate change (Djoudi et al. 2016).

Facilitating equal representation of women in addressing climate change has its benefits. In their analysis of various European nations, Hemmati and Rohr (2009) observe that women are more inclined to endorse governmental efforts to address climate change. However, when integrating gender into climate policy, it is important to accept that merely having women present does not ensure a gender perspective. Effective participation and a conscious, consistent effort towards a gendered approach are necessary considerations.

Discussion

From the above findings, it becomes evident that vulnerability to climate change is gendered. Women are substantially more vulnerable to the impacts of climate change in comparison with men. It becomes important to consider this gendered dimension while examining the issues of climate change impacts and mitigation. Also, with the discussions on women's vulnerability and their resilience towards climate change, we should direct significant attention towards investigating women's representation and role in the decision-making bodies in the domain of climate change policy design, formulation and implementation. Their presence in leadership roles has a positive impact on addressing climate change and, therefore, should be encouraged. Analysis shows that social factors and social locations play significant roles in determining an individual's or a group's vulnerability to climate change. The gendered dimension of climate change-related vulnerabilities is unique as it is strongly dependent upon the existing power relations and dynamics of the society.

This study focuses on how patriarchy can be an added factor to the climate change-related vulnerabilities among genders. The study also points out that the existing and persisting poverty and inequalities among women add to their vulnerability towards climate change more than men. In extreme climatic situations like drought and floods, women face major challenges (Neumayer and Plümper 2007). Research points to the fact that the increase in water crisis (which is established as a climate change-driven phenomenon) has directly impacted women in rural areas where a piped water supply is not available (Chigusiwa et al. 2023). Women have to walk longer distances and carry heavier loads to ensure a water supply in their homes (Yadav and Lal 2018).

Despite the glaring differential vulnerabilities, the matter finds very little place in the climate change forum. There remains a serious gap in the recognition of gendered vulnerabilities in climate change policies. As pointed out earlier, this can be attributed to the dearth of adequate female presence in decision-making and policy formulation. This can be regarded as another variation of gender discrimination, popularly designated as the masculinisation of policy-making (Magnusdottir and Kronsell 2015). Another factor that emerges from the analysis is the lack of substantial academic attention towards the issue.

The analysis, further, brings out two major gaps existing in this field of research. Firstly, a significant portion of the scholarly literature pertaining to gender and climate change has concentrated on the rural areas of developing countries. Urban areas, including those in the Global South, need more focus. With urbanisation is rapidly occurring worldwide, leading to an increasing number of individuals moving into urban areas. Globally, urban areas have emerged the major residence for a significant population in many nations. The concentration of individuals in urban areas makes them salient in the context of climate change, as a greater number of people will be affected by its impacts in these locations. Informal settlements present in the Global South are characterised by the spontaneous occupation of land by individuals or groups who lack legal recognition of their property rights (Davis 2006). These settlements are often characterised by inadequate access to basic services and infrastructure such as sanitation, water, electricity and housing, and are subject to forced evictions. The emergence of informal settlements in the Global South is a complex challenge that is influenced by multiple factors, including unplanned urbanisation, poverty, and political instability. The discourse on climate change could potentially benefit from a focus on urban areas as these regions are characterised by a high degree of informality within their urban environments. Further investigation is required to thoroughly examine the particular nuances of gender-based vulnerability to climate change within informal settlements.

Secondly, studies examining the perspectives on climate change, particularly those of individuals who are more vulnerable to its effects, are important. The current research has primarily centred on the concept of climate change and gendered vulnerability; however, further investigation is required to explore how the historical evolution of gender relations has contributed to these vulnerabilities. The association between climate change and gender has been established, necessitating further qualitative research to investigate the intricacies of this relationship. There is also a need for an increase in participatory research alongside the prevalent impact-based studies. The investigation of how this nexus functions within diverse social structures and contexts is necessary to facilitate the framing and development of policies that are more attuned to gender sensitivity with the contextual nature of the problem at hand.

Conclusion

Intersection with gender has added a new direction in climate change research, with a substantial body of literature dedicated to its exploration. This study identifies three questions to explore the interplay between these two domains. The initial inquiry pertained to the gendered nature of vulnerability towards climate change. The research findings indicate that there are recurring trends in the discourse related to the elements that give rise to vulnerability based on gender. The analysis demonstrated how pre-existing gender-based disparities in roles and responsibilities, the availability of resources and knowledge, and the patriarchal structure of the community contribute to gender-based vulnerability. The analysis identifies certain precautions to be taken into account in future research on gender vulnerability. These precautions encompass the potential dangers of oversimplifying gender to exclusively refer to women and failing to recognise the intersectionality inherent in gender, and disregarding the contextual nature of gendered vulnerability.

The study investigated the issue of women's greater vulnerability to the impacts of climate change relative to men. The research findings indicate that within many societies, women are at a higher risk when compared to men. This is attributed to a variety of factors, such as their status as the comparatively marginalised in most societies and their limited access to available resources. The findings of the analysis indicate the importance of exercising caution in several areas, such as acknowledging the intersectionality of women and the potential risks associated with the 'feminisation of climate change'. Additionally, the study highlights the necessity for further research that amplifies the voices of women rather than reducing them to simplistic categories.

The third question on the significance of gender as an actor in climate policy helped shed light on the necessity of integrating gender considerations in designing climate policy. The interplay of gender and climate change has received increasing attention in climate policy research. Although there has been an increase in attention towards these areas in recent years, further efforts are required, particularly in the context of India.

The discourse has emphasised the concentration of research on the vulnerability of gender to climate change in rural areas. However, there is a need for additional investigation into the vulnerabilities and impacts in urban areas. The incorporation of gender considerations in climate policy is imperative from the local to the global scale. The analysis facilitated the identification of prevailing patterns in literature pertaining to gender and climate change and also identified areas that need further investigation.

The present study, nonetheless, is not without its limitations. Two specific limitations pertain to the scope of the questions examined and the methodology employed to execute the analysis. The study primarily centred on reviewing literature on climate change vulnerability to investigate the nexus between gender and climate change. However, it is important to focus on other climate change-related domains, such as adaptation and mitigation, to further explore this nexus. The employed methodology for the study is a comprehensive review of the available literature. There exists an opportunity for future research to utilise primary sources of data in order to address these inquiries.

AI Disclosure

This chapter was written without using AI assistance for conducting the review or analysis. However, AI has been employed for editing (Grammarly for grammar check) and improving readability.

References

Ahmed S, Fajber E (2009) Engendering adaptation to climate variability in Gujarat, India. Gend Dev 17(1):33–50. https://doi.org/10.1080/13552070802696896

Alston M (2012) Introducing gender and climate change: research, policy and action. In: Research, action and policy: addressing the gendered impacts of climate change, pp 3–14. https://doi.org/10.1007/978-94-007-5518-5_1

Arora-Jonsson S (2011) Virtue and vulnerability: discourses on women, gender and climate change. Glob Environ Chang 21(2):744–751. https://doi.org/10.1016/j.gloenvcha.2011.01.005

Bob U, Babugura A (2014) Contextualising and conceptualising gender and climate change in Africa. Agenda 28(3):3–15. https://doi.org/10.1080/10130950.2014.958907

Bohle HG, Downing TE, Watts MJ (1994) Climate change and social vulnerability. Glob Environ Chang 4(1):37–48. https://doi.org/10.1016/0959-3780(94)90020-5

Bradshaw S, Linneker B (2014) Contemporary debates on gender and climatic change. In: Gender and environmental change in the developing World. International Institute for Environment and Development, pp 11–19. http://www.jstor.org/stable/resrep18042.5

Browne G, Roberts J, Byrne C, Gafni A, Weir R, Majumdar B (2016) Translating research the costs and effects of addressing the needs of vulnerable populations: results of 10 years of research. Can J Nurs Res Arch 33(1). https://cjnr.archive.mcgill.ca/article/view/1621

Chigusiwa L, Kembo G, Kairiza T (2023) Drought and social conflict in rural Zimbabwe: Does the burden fall on women and girls? Rev Dev Econ 27(1):178–197. https://doi.org/10.1111/rode.12944

Davis M (2006) The planet of slums. Verso, New York

Denton F (2002) Climate change vulnerability, impacts, and adaptation: Why does gender matter? Gend Dev 10(2):10–20. https://doi.org/10.1080/13552070215903

de Wit S (2021) Gender and climate change as new development tropes of vulnerability for the global south: essentializing gender discourses in Maasailand, Tanzania. Tapuya: Lat Am Sci, Technol Soc 4(1). https://doi.org/10.1080/25729861.2021.1984638

Djoudi H, Locatelli B, Vaast C, Asher K, Brockhaus M, Basnett Sijapati B (2016) Beyond dichotomies: gender and intersecting inequalities in climate change studies. Ambio 45(S3):248–262. https://doi.org/10.1007/s13280-016-0825-2

Enarson E (1998) Through women's eyes: a gendered research agenda for disaster social science. Disasters 22(2):157–173. https://doi.org/10.1111/1467-7717.00083

Eriksen SH, Nightingale AJ, Eakin H (2015) Reframing adaptation: the political nature of climate change adaptation. Glob Environ Chang 35:523–533. https://doi.org/10.1016/j.gloenvcha.2015.09.014

Magnusdottir GL, Kronsell A (2015) The (In)Visibility of gender in scandinavian climate policy-making. Int Fem J Polit 17(2):308–326. https://doi.org/10.1080/14616742.2014.896661

Goh AHX(2012) A literature review of the gender-differentiated impacts of climate change on women's and men's assets and well-being in developing countries. CAPRi Working Paper No. 106. Washington, D.C.: International food policy research institute

Hemmati, M., & Röhr U (2009) Engendering the climate-change negotiations: Experiences, challenges, and steps forward. Gender & Development, 17(1), 19–32. https://doi.org/10.1080/13552070802696870

IPCC (2007) Climate Change 2007: impacts, adaptation and vulnerability. In: Contribution of working group ii to the fourth assessment report of the intergovernmental panel on climate change, Parry ML, Canziani OF, Palutikof JP, van der Linden PJ, Hanson CE (eds). Cambridge University Press, Cambridge, UK, p 976

Jolly, M. (2021). Gender and climate change. Int Encycl Anthropol 1–12. https://doi.org/10.1002/9781118924396.wbiea2419

Kantamaneni K, Phillips M, Thomas T, Jenkins R (2018) Assessing coastal vulnerability: development of a combined physical and economic index. Ocean Coast Manage 158:164–175. https://doi.org/10.1016/j.ocecoaman.2018.03.039

Kelly PM, Adger WN (2000) Theory and practice in assessing vulnerability to climate change and facilitating adaptation. Clim Chang 47(4):325–352. https://doi.org/10.1023/a:1005627828199

MacGregor S (2010) 'Gender and climate change': From impacts to discourses. J Indian Ocean Reg 6(2):223–238. https://doi.org/10.1080/19480881.2010.53666

Nelson V, Meadows K, Cannon T, Morton J, Martin A (2002) Uncertain predictions, invisible impacts, and the need to mainstream gender in climate change adaptations. Gend Dev 10(2):51–59. https://doi.org/10.1080/13552070215911

Neumayer E, Plümper T (2007) The gendered nature of natural disasters: the impact of catastrophic events on the gender gap in life expectancy, 1981–2002. Ann Assoc Am Geogr 97(3):551–566. https://doi.org/10.1111/j.1467-8306.2007.00563.x

Nong HT, Gan C, Hu B (2020) Climate change vulnerability and adaptation in Vietnam from a gender perspective: a case study of northern province of Vietnam. Int J Soc Econ 47(8):953–972. https://doi.org/10.1108/ijse-09-2019-0534

Onwutuebe CJ (2019) Patriarchy and women's vulnerability to adverse climate change in Nigeria. SAGE Open 9(1):215824401982591. https://doi.org/10.1177/2158244019825914

Patt AG, Dazé A, Suarez P (2009) Gender and climate change vulnerability: what's the problem, what's the solution?. In: Distributional impacts of climate change and disasters. https://doi.org/10.4337/9781849802338.00013

Resurreccion, P Bernadette (2012) The gender and climate debate: more of the same or new pathways of thinking and doing?. Clim Chang Southeast Asia 111–127. https://doi.org/10.4324/9780203107775-13

Rivera-Ferre M (2021) Climate change is not equal to all: the contribution of feminist studies to climate change research. Mètode Revista de Difusió de La Investigació (12). https://doi.org/10.7203/metode.12.20508

Schofield D, Gubbels F (2019) Informing notions of climate change adaptation: a case study of everyday gendered realities of climate change adaptation in an informal settlement in Dar es Salaam. Environ Urban 31(1):93–114. https://doi.org/10.1177/0956247819830074

Terry G (2009) No climate justice without gender justice: an overview of the issues. Gend Dev 17(1):5–18. https://doi.org/10.1080/13552070802696839

Thomas K, Hardy RD, Lazrus H, Mendez M, Orlove B, Rivera-Collazo I, Roberts JT, Rockman M, Warner BP, Winthrop R (2018) Explaining differential vulnerability to climate change: a social science review. WIREs Clim Chang 10(2). https://doi.org/10.1002/wcc.565

Yadav SS, Lal R (2018) Vulnerability of women to climate change in arid and semi-arid regions: the case of India and South Asia. J Arid Environ 149:4–17. https://doi.org/10.1016/j.jaridenv.2017.08.001

Chapter 3
Urban Expansion and Climate Footprint Paradox: An Integrated Assessment of Population Dynamics, GHG Emissions, and Sea-Level Rise in Colombo District, Sri Lanka

Mohamed Hassan Fathima Nuskiya and Sawjanya Sathyaseelan

Abstract Rapid urbanization in South Asian megacities fuels climate vulnerabilities, yet few studies comprehensively assess interactions between population growth, GHG emissions, and sea-level rise in Colombo District, Sri Lanka. It is a highly populated economic zone, which has experienced significant demographic and environmental changes in the past four decades. This chapter discusses these multi-scalar dynamics in three goals, which are analysing the demographic trends, quantifying the GHG emission patterns, and estimating the coastal threat due to SLR. The period 1981–2023 was designed on the basis of high-resolution population datasets (GHS), TROPOMI-derived satellite-based measurements of GHG concentrations, and projections of SLR by utilizing CoastalDEM. Correlation analyses were used to find relationships between population distribution, emissions, and coastal exposure. Findings show that the population of the study area grew by 45.9%, with the historic core having a clear outward growth. The CO_2 concentration was stabilized in the year 2010, but the levels of NO_2 were over 70 mol/m^2 with sustained transport congestion. The levels of CH_4 increased by 22% since 1993, with an uncontrolled waste system being its primary source of increase. Using SLR simulations, it is found that the coastline of Colombo has 19% of the total being highly exposed, which places a disproportionate risk on informal settlements and critical infrastructures. Results show the urban expansion paradox of the climate footprint increases risks, for which a suggested hybridized model measures can be implemented to reduce these combined effects and rebuild cities in the rapidly urbanizing Global South.

M. H. Fathima Nuskiya (✉)
Department of Geography, South Eastern University of Sri Lanka, University Park, Oluvil 32360, Sri Lanka
e-mail: nuskiyahassan@seu.ac.lk

S. Sathyaseelan
Department of Geography, University of Sri Jayewardenepura, Gangodawila, Nugegoda 10250, Sri Lanka

P. Ghosh et al. (eds.), *Climate Crisis and Changing Population Dynamics*, Springer Climate, https://doi.org/10.1007/978-3-032-24533-5_3

Keywords Urban sustainability · Climate footprint · GHG emissions · Population · Sea-level rise · Sri Lanka

Introduction

Climate change has become one of the most pressing international problems of the twenty-first century, having extensive impacts on ecosystems and human communities and economies (IPCC 2021). The cities, particularly in the developing world, are extremely susceptible because of the high growth rate of the population, unregulated urbanization, and the increased emission of greenhouse gases (GHG) (Revi et al. 2014). Such pressures are especially marked in the Global South in the coastal cities where the lack of infrastructure, overpopulation, and growing climatic risks meet. These dangers are exemplified by South Asian megacities, although the study of climate-population dynamics is usually quite disjointed, focusing on emissions or demographic change individually. Colombo District in Sri Lanka has been an example of this combination of city growth, migration, and climate exposure. Being the business and administrative centre of the country, the District has experienced significant changes in the environment and population over the last forty years (Wickramasinghe 2018). Population dynamics and GHG emission patterns are increasingly infused by climate change (IPCC 2023), and rapid urbanization and industrialization make most of the third-world situations more vulnerable. The main economic hub of the country, Colombo, is experiencing strong demographic changes and increasing emissions associated with urbanization and economic development (Census and Statistics Department, Sri Lanka [DCSSL] 2023). This chapter will look at how the climate change conditions affect the population trends and gaseous emissions in the Colombo District as a way of informing the long-term urban planning and mitigation risk.

The main research question that will be used in the proposed study is the lack of combined insights concerning how long-term population trends, GHG emissions, and sea-level rise (SLR) interplay to determine climate vulnerability in fast-urbanizing urban areas like Colombo. Despite the fact that the effects of climate in Sri Lanka are reported, their direct relations with the demographic change and the emission patterns across several decades are not properly studied. This discontinuity constrains planners and policymakers to come up with climate-resilient urban strategies. This matter is even more pressing, since the population and emissions of Colombo are growing, and climate risks are becoming more and more dangerous, especially coastal flooding, which is why it is necessary to plan it in an evidence-based way.

The relationship between urbanization and weather trade is complicated because urban areas contribute significantly to global GHG emissions, while simultaneously as they endure increased risks posed by climate-driven risks such as flooding, heat pressure, and sea-level rise (Grimm et al. 2008). Colombo is a city that is situated on the tropical coastline, and it has been facing the trend of increasing temperature, irregular rainfall, and frequently occurring extreme weather conditions over the study period (19,812,023) (Eriyagama et al. 2020; Nasar-u-Minallah et al. 2025). These climate changes have both direct and indirect impacts on the population alone, migration, and the infrastructure of a city, which not only exacerbate environmental degradation (De Silva and Kawasaki 2018). A person should understand those dynamics to come up with adaptive and mitigative guidelines that will guarantee sustainable city development. The necessity to find such trends results in an imperative need for an integrated climate model and methods of mitigation. It should also be noted that in accordance with the Nationally Determined Contributions (NDCs), the Sri Lankan government has pledged to cut its overall GHG emissions in the country by 14.5% by the year 2030 (United Nations Framework Convention on Climate Change [UNFCCC] 2023). This includes the policy changes like sustainable city shipping networks, energy-friendly infrastructure, and climate-resilient residential buildings that will minimise the emissions of Colombo as well as support its growing population (Ranasinghe et al. 2022). The untrained spaces can be improved, transport selling, and intense emission control can potentially cut down the carbon footprint of Colombo significantly (Gunathilaka et al. 2023).

Colombo District population structure is a result of a mixture of natural changes and urban–rural migrations, which are driven by economic opportunities and better living conditions (DCS 2021). Density in the district has rapidly increased, further increasing power consumption, vehicle emissions, and changes in land use, all of which have increased the amount of GHG emissions (Munasinghe et al. 2021; Zahir et al. 2024). It has been shown that urban sprawl and deforestation of the peri-city areas contributed to the decrease in the ability to absorb carbon and amplify the city's warm island effect (UHI), which has only contributed to the vulnerability of the areas to climate change (Wickramasinghe and Attanayake 2020; Nuskiya et al. 2025). Not only that, the use of fossil fuels to transport and do business has turned out to make Colombo a prolific producer of CO_2, CH_4, and N_2O, and a switch towards the low-carbon city-making strategies became inevitable (Weerakoon et al. 2019). The Colombo District has experienced massive urban growth, and the population has grown to approximately 2 as compared to 2.31 million in 2012. In 2022, fifty-five million, which is an annual growth of 1.03% (DCSSL 2023). This increase is pushed by rural-to-urban migration, economic possibilities, and improved infrastructure (World Bank 2023). However, weather-prompted activities, consisting of SLR and extreme climate, are influencing internal migration patterns, with a projected displacement of nearly 50,000 people by means of 2050 due to flooding and coastal erosion (Asian Development Bank [ADB] 2022). Rising temperatures and erratic rainfall patterns are also related to elevated incidences of vector-borne illnesses, adversely affecting public health (Ministry of Health, Sri Lanka 2023). Colombo contributes notably to Sri Lanka's typical GHG emissions, with CO_2 emissions from the shipping area alone accounting for 45% of the city's emissions (Ministry of Environment, Sri Lanka 2023). Energy consumption in Colombo grew by 4.7% annually between 2010 and 2022, broadly speaking, pushed by means of commercial and business sectors (Sri Lanka Sustainable Energy Authority 2023). The Intergovernmental Panel on Climate Change (IPCC) scenarios propose that underneath RCP 8.5 (high emissions scenario), Colombo should witness a temperature boom of 2.5–3.0 °C by

2050, exacerbating the urban heat island impact (IPCC 2023). If mitigation measures are carried out under RCP 4.5, emissions should stabilize, limiting temperature rise to beneath 2 °C (Eckstein et al. 2023).

The gaps in the existing literature, although there are several studies on urban risks of climate and population growth, are significant empirical and methodological gaps. To start with, long-term, multi-decadal demographic change satellite-based measurements of GHG are scarcely evaluated. Second, not many studies will measure the contribution of population emissions to the risk facing coastal climates in South Asian coastal cities. Third, there are still methodological gaps to integrate GHS population grids, TROPOMI GHG data, and CoastalDEM projections into one analytical approach. This chapter fills in these theoretical, empirical, and methodological gaps by coming up with a 43-year integrated assessment of Colombo District. The originality of the study is its combined framework connecting the dynamics of populations, trends of the sectoral GHG emissions, and exposure to SLR in four decades with its high-resolution spatial and satellite data. This method is not only progressive in terms of methodological innovation but also contributes to the support of climate-population interactions in the Global South. In line with that, the study aims to achieve three major goals: (1) to examine spatial and temporal changes in population distribution; (2) to estimate long-term GHG emission patterns; and (3) to evaluate the vulnerability of coastal areas to SLR. Placing Colombo district in the context of the global and regional climate-urbanization discussions and correlating the results with the central theme of the book, Climate Crisis and Changing Population Dynamics, this chapter can contribute to the creation of globally applicable knowledge and promote the creation of resilient urbanization plans in the face of climate change.

Materials and Methods

Study Area

The research was conducted in Colombo District, which is found along the western coast of Sri Lanka, with latitudes of 6° 54′ and longitudes of 79° 50′ and 80° 03′ (Fig. 3.1). Its area is about 699 km^2 and serves as the main commercial and administrative center of the country. In 2023, it was the densely populated district in the country; its a population density was 3548 persons/km^2 (DCS 2023). Based on the Geiger-Mueller classification of climate, this area is tropical rainforest (Af), with an annual mean rainfall of approximately 2400 mm and an average temperature of 26–31 °C (Peel et al. 2007). It was chosen as the study area because it is a unique area with a high level of urbanization, increasing GHG emissions, and exposure to SLR. Colombo is one of the embodiments of the Global South megacities, where population pressure and exposure to climate collide. That is why it is worth considering as one of the most important study sites to learn about long-term climate risks.

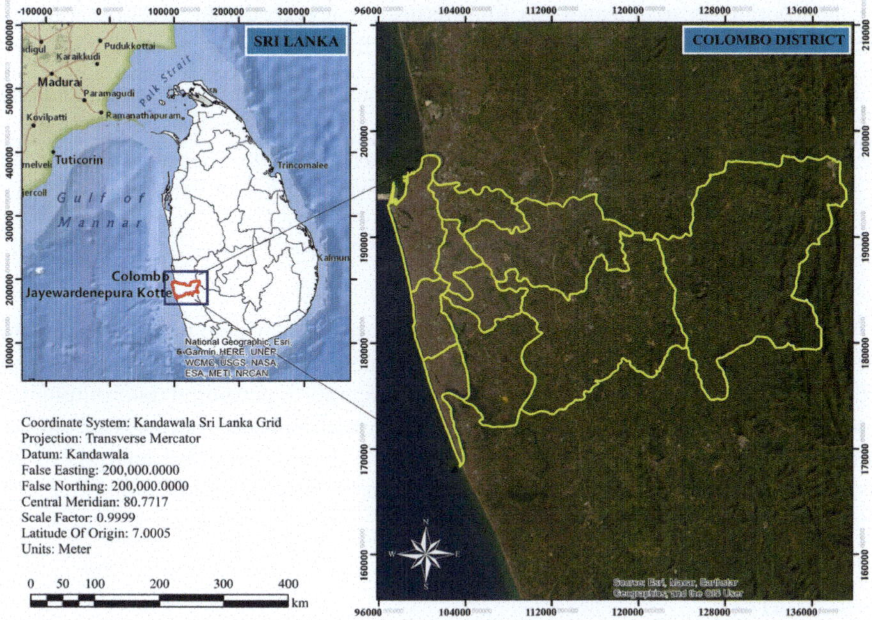

Fig. 3.1 Map of study area. *Source* Retrieved using ArcGIS 10.8

Colombo has witnessed high rates of industrialisation over the last 40 years, a notable rise in the service industry, and the expansion of the maritime trade in the huge port facilities. These changes have imposed more and more pressure on the resources at the cost of increasing the congestion of transport, and increased the factors of waste production that are closely associated with the increased GHG emissions. The coastal regions of the district are also low-lying areas, and they make the district very vulnerable to the effects of climate change, especially SLR. It is projected that, depending on the high-emission conditions, up to 19% of the shoreline of Colombo may be submerged by the end of the century, which puts the human livelihoods, economic infrastructure, and natural coastal systems at serious risk (IPCC 2021).

Research Design

The study was based on an integrative design that involved a combination of spatial, temporal, and statistical analysis, i.e., the interaction of population dynamics, GHG emissions, and sea-level rise risks in Colombo District, Sri Lanka, between the years 1981 and 2023. Three analytical parts were formulated as the methodological approach. The proposed study will follow the mixed-method integrative design that will incorporate geospatial analysis, long-term climate records, satellite observations, and statistical modelling. The model consists of three parts that are interconnected:

- **Population Dynamics Analysis**—The evaluation of population spatial changes between 1981 and 2023 based on census and gridded population data.
- **GHG Emission Trend Analysis**—The measurement of a shift in significant atmospheric contaminants (CO_2, NO_2, SO_2, CH_4) over time based on satellite-obtained data.
- **SLR and Coastal Flood Modelling**—Inundation risk forecasting in a variety of climatic conditions.

It was an integrative design selected to fill in the methodological gaps in the former studies that generally did not perform an analysis of the population, emissions, and SLR in a single framework. This kind of integration permitted determining the correlations between the indicators of urban growth and the indicators of climate vulnerability and enabled the possibility of doing a comprehensive analysis of the environment and socio-economic factors of the urban transition in Colombo.

Data Sources

The study used the high-resolution geospatial data, national census data, and satellite environmental data. Population distribution information in a fine grid has been retrieved by using the Global Human Settlement Layer (GHS) Population Grid in 1981 and 2023 (2023 and 1981) (Pesaresi et al. 2013; Schiavina et al. 2019), which presents the information on population distribution in a fine grid. The Sri Lanka Department of Census and Statistics (DCS) added these data sets with official census reports that included total population data and yearly growth rates to these data sets. Further details of the data source and its types are below (Table 3.1).

Data Processing and Analytical Methods

Pre-processing

Each dataset was pre-processed in a standardized workflow to have spatial consistency between sources. All raster and vector layers were clipped in ArcGIS 10.8 using the Colombo District administrative boundary. The GHS dataset created population grids, which were resampled to a uniform resolution of 100 m to enable a comparative study of the data over time. Google Earth Engine was used to create monthly TROPOMI composites to reduce atmospheric variability and retrieval. The elevation raster of the Coastal DEM was also adjusted to the coastline boundaries and improved to enhance the precision of the low-lying coastal elevations.

Table 3.1 Data sources utilized

Data	Types	Source
Population data	• Global Human Settlement (GHS) Population Grid (1981, 2023) • Sri Lanka Department of Census and Statistics (DCS) reports (1981–2023) • The GHS datasets (100–250 m resolution) provide spatially explicit population estimates suitable for long-term change detection, while DCS records provide administrative-level validation	GHS Population Grid: https://human-settlement.emergency.copernicus.eu DCS Sri Lanka: https://www.statistics.gov.lk
GHG emissions data	• TROPOspheric Monitoring Instrument (TROPOMI) accessed through Google Earth Engine (GEE) • Pollutants analysed: CO_2, NO_2, SO_2, CH_4 • Temporal resolution: daily to monthly composites • Spatial resolution: 3.5 × 7 km (improved resolution post-2018)	TROPOMI Explorer: https://showcase.earthengine.app
SLR and elevation data	• CoastalDEM (Version 2.1) • FloodMap.net SLR viewer • IPCC-based SLR scenarios (RCP 8.5, SSP2-4.5) • These datasets were selected for their ability to reduce vertical elevation errors common in SRTM-based models	FloodMap.net, CoastalDEM: https://floodmap.net
Climate variables	• CHIRPS rainfall (1981–2023) • MODIS surface temperature time series • GEE repository climate indices	Google Earth Engine: https://earthengine.google.com

Analytical Methods

Figure 3.2 explains the data analysis and procedures. The GHG concentration data of CO_2, NO_2, SO_2, and CH_4 were obtained by using the Tropospheric Monitoring Instrument (TROPOMI) on the Google Earth Engine platform, which provided daily to monthly average values both able to be utilized to determine changes in the trends with time (Veefkind et al. 2012). The modelling platforms of Coastal DEM and FloodMap.net were used to obtain SLR projections depending on both the IPCC RCP8.5 and SSP2-4.5 scenarios to consider a variety of climate futures. Additional temperature and precipitation data were also obtained using Google Earth Engine to give a climatic background of observed environmental changes.

ArcGIS 10.8 was used to create population distribution maps of 1981 and 2023 and then lined them with administrative boundary layers to facilitate spatial consistency. The growth of urban areas was measured by calculating the increase in the built-up land cover, and the build-up of the land cover on peripheries was observed, especially. To analyze GHG emissions, concentration data that were based on the satellites were clustered into three analytical periods that include 1981–2000, a period that is characterized by the early stage of urbanisation; 2000–2015, a period that is characterized by industrial growth; and 2015–2023, a period that is characterized by policy-based environmental intervention. Monthly composites of pollutants were created through

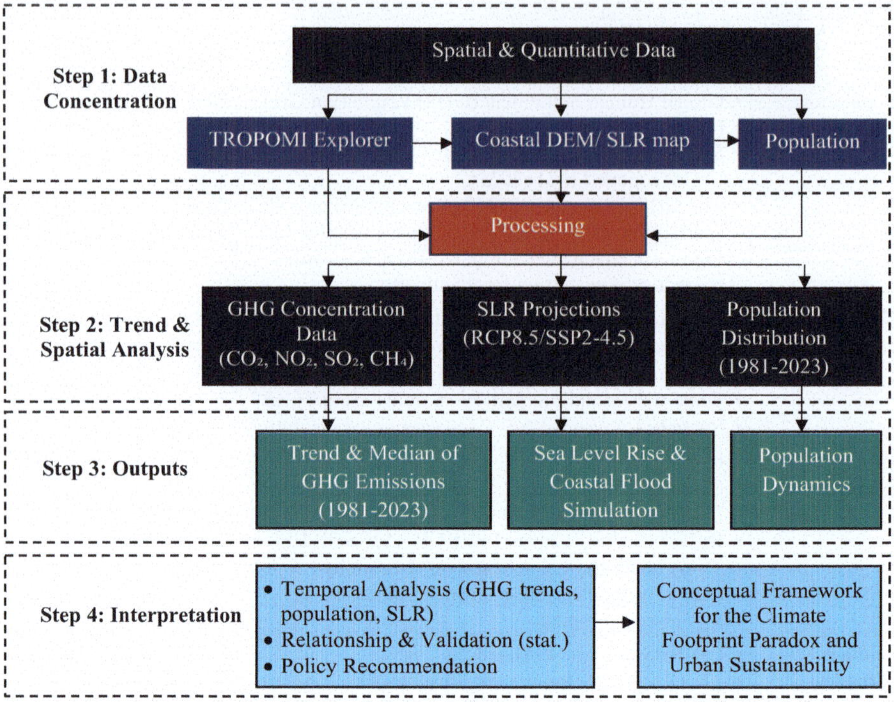

Fig. 3.2 Data analysis framework

the Earth Engine platform, and spatial means, medians, and interquartile ranges (IQR) were calculated.

Pearson correlation coefficients and linear regression models were used to statistically compare the emission trends and demographic trends to identify significant relationships. The risk assessments on the use of SLR were done by overlaying the projected inundation extents of the years 2030, 2050, and 2100 on the existing land use and infrastructure maps. The analysis of the SLR was performed by using Coastal DEM elevation layers, which were imported into QGIS and compared to the results of FloodMap.net. Inundation extents were generated by applying projections of + 0.5, + 1.0, and + 1.5 m of the increase in sea level to model projections in 2030, 2050, and 2100 scenarios using a static bathtub modelling technique. All datasets underwent cross-validation with national inventories from the Sri Lankan Ministry of Environment and peer-reviewed literature to minimize potential errors and enhance reliability.

Ethical Considerations

This study utilized publicly available datasets that do not contain any identifiable personal information. No ethical approval was required.

Result and Discussion

Population Dynamics in Colombo District (1981–2023)

The population spatial distribution maps of Colombo District over the years from 1981 to 2023 show a huge change in the settlement patterns. The population doubled in four decades, from about 1.7 million to 2.48 million, with the density rising from 2647 to 3548 peoples/km^2. The first map picture was of the central Colombo, still alive, though past its peak, but with the slowest population growth. Newly, about 68% of the new population has moved from the core to be assimilated into the suburban and peri-urban areas.

The demographic trends of the population flows within the Colombo District from 1981 to 2023 are signs of the radical transformations of the territory and a significant increase. The population increased from 1,699,241 people to 2,480,000 people, and the population density also rose to 3548 people/km^2 as compared to 2647 people/km^2. The greatly varying rates of urbanization are marked by the spatial map (Fig. 3.3): the urban centers have reached full development and, therefore, their development does not change significantly (less than 0.5% a year), but the peripheral areas absorbed 68% of the new population.

GHG Emission Trends in Colombo District (1981–2023)

CO_2 Emission Trends

The Colombo District showed different periods of CO_2 emission expansion from 1981 to 2023. Figure 3.4 shows that in the period 1981–2000, the concentration rose slowly (0.03–0.04 mol/m^2), due to early industrialization and urban growth as the manufacturing industry in the city grew. The 2000–2010 growth was much faster (peaking at 0.04 mol/m^2), with the rise in industrial activity and the boom in the number of people owning their own cars. Between the years 2010 and 2023, the emissions were either stagnant or experienced a minor downward trend, possibly because of the rise in renewable energy usage (especially solar power projects) and the tightening of fences on emissions policies by national climate-related policies. This stabilization indicates that there are early signs of decoupling economic growth and carbon emissions, but still, absolute levels are worrying.

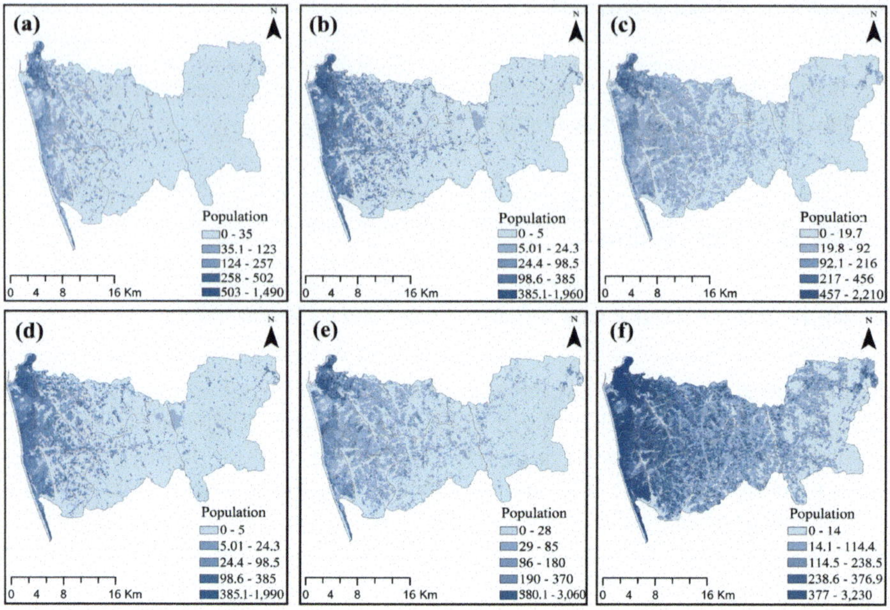

Fig. 3.3 Spatial distribution of population dynamics in Colombo District (1981–2023). *Source* Retrieved using ArcGIS 10.8

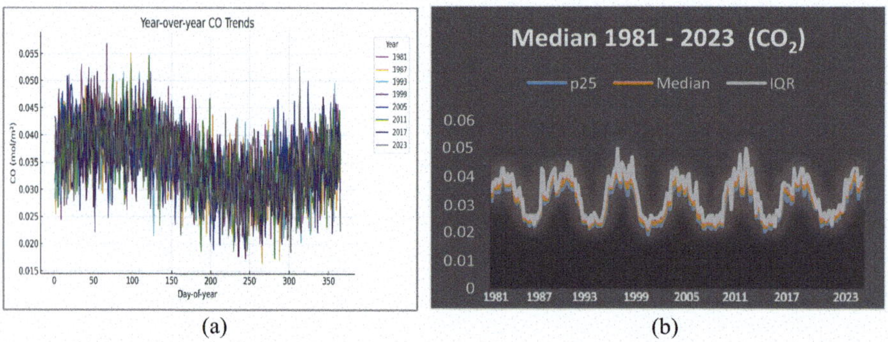

Fig. 3.4 **a** Trend of CO_2 emission (1981–2023); **b** median of CO_2 emission trends (1981–2023). *Source* Adopted from TROPOMI Explorer

SO_2 Emission Trends

The pattern of the SO_2 levels was unusual as it showed (Fig. 3.5) negative values (− 200 to 400 mol/m^2) during 1981–1987, which could be attributed to low levels of fossil fuel burnt and low industrialization in the early stages of Colombo development. This change was dramatic between 1995 and 2010, when the highest rate was 400 mol/m^2 with an increase in the use of coal in power generation and weak

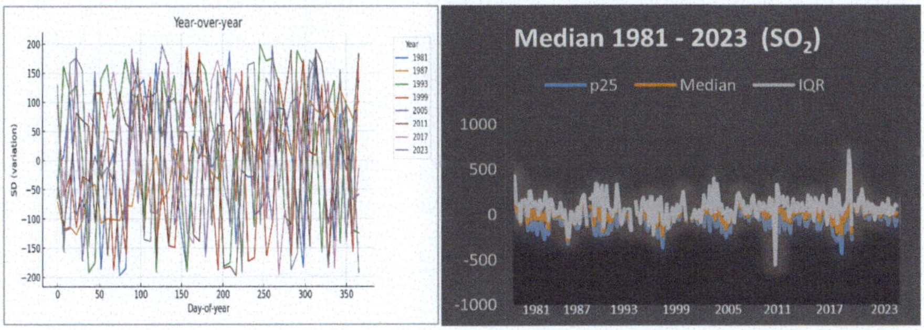

(a) (b)

Fig. 3.5 **a** Trend of SO_2 emission (1981–2023); **b** median of SO_2 emission trends (1981–2023). *Source* Adopted from TROPOMI Explorer

maritime fuel requirements. The period between 2010 and 2023 was at high levels, although there was a slight downturn due to continued difficulties in controlling the combustion of fuels in industries and in shipping, which was especially high around the Colombo Port. The slower response to SO_2 reduction than to CO_2 exposes the gaps in policy implementation in the sectors.

NO_2 Emission Trends

The pollution of NO_2 was also exhibiting disturbing progress, which is directly connected with the transport crisis in Colombo. This was observed (Fig. 3.6) in the 19,812,000 period (0.5–1.2 percent/year) when motorization started and in the 2000–2015 period (peak 4 percent/year), where vehicle ownership had risen exponentially in the absence of a proper public transport alternative. Most importantly, 2015 and 2023 figures surpassed 70 mol/m^2—one of the highest levels ever in South Asian cities—because of uncontrolled traffic jams and freight transportation based on diesel. Such a trend highlights a systemic ineffectiveness in urban mobility planning.

CH_4 Emission Trends

There were consistent trends of increasing emission of methane, and since 1993, there has been a steady increase in the emission due to agricultural growth and uncontrolled release of landfill gases. In the 1995–2010 period (Fig. 3.7), there was a significant increase (22%) because waste production exceeded the ability to manage it, and livestock farming became more common in the peri-urban regions. In contrast to other GHGs, CH_4 does not exhibit any stabilization, which indicates its continued use as an omission in climate action plans. The lack of methane capture in large landfills such as Bloemendhal is a key mitigation weakness, and the global warming

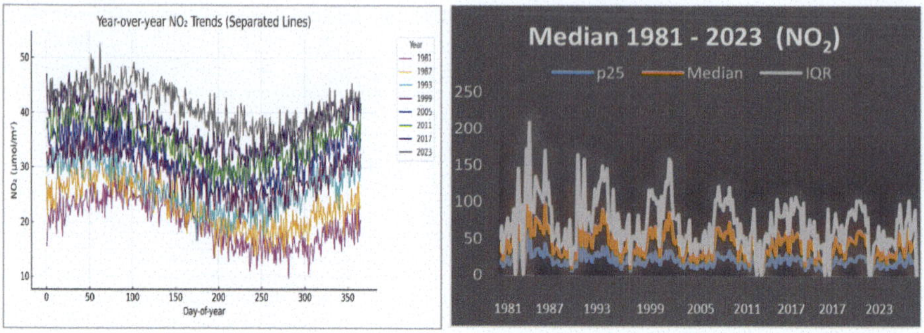

(a) (b)

Fig. 3.6 **a** Trend of NO_2 emission (1981–2023); **b** median of NO_2 emission trends (1981–2023). *Source* Adopted from TROPOMI Explorer

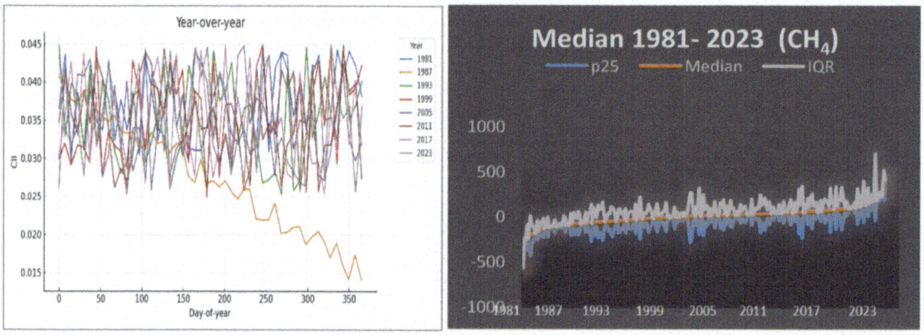

(a) (b)

Fig. 3.7 **a** Trend of CH_4 emission (1981–2023); **b** median of CH_4 emission trends (1981–2023). *Source* Adopted from TROPOMI Explorer

potential of CH_4, combined with its high priority target status in future interventions, identifies CH_4 as the most important target to find a solution.

Correlation Between Population Growth and GHG Emission

The correlation analysis (Fig. 3.8) indicates that the increase in population is also closely related to the increase in the concentration of GHG. The correlation coefficients of both CO_2 and NO_2 are very high ($r = 0.98$) (see Fig. 3.8a, b), which means that the growth of population is considerably correlated with the increased expenditure of energy, transportation, and industrial work. SO_2 has the greatest (see Fig. 3.8c) relationship ($r = 0.99$), which indicates the fast rate of development of industrial emissions with the increase of the population. CH_4 has a slightly less (see

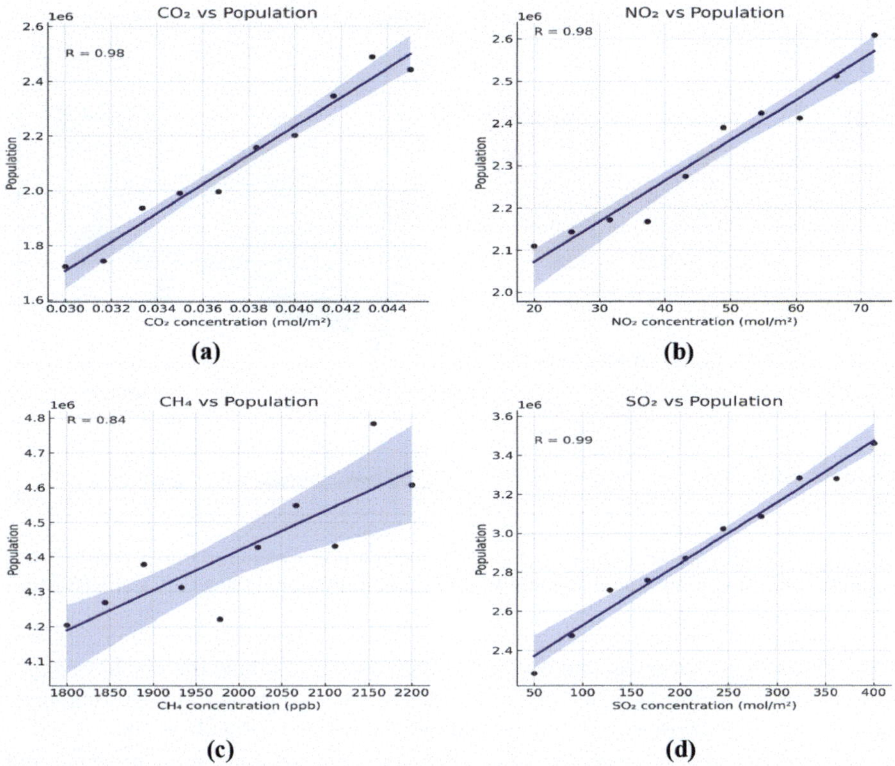

Fig. 3.8 **a** Correlation between CO_2 and population; **b** correlation between NO_2 and population; **c** correlation between CH_4 and population; **d** correlation between SO_2 and population. *Source* Adopted using TROPOMI Explorer data

Fig. 3.8d), but still significant correlation ($r = 0.84$), indicating that agricultural practices and the generation of waste materials also increase with the population, though not directly as other gases. Generally, it can be seen that the findings are highly indicative of the fact that population growth is a primary cause of high emissions of several GHGs.

Climate Variability: Temperature and Rainfall Trends (1981–2023)

The existing rainfall patterns (1981–2023) in the Colombo District show that there are no severe dry seasons, but a gradual reduction in rainfall, with urban areas witnessing frequent floods due to poor drainage management. Peculiar rainfall in the eastern and southern edges is now observed due to the changes in regional climates and changes in local land use, such as loss of wetlands and urban sprawl. Coastal regions

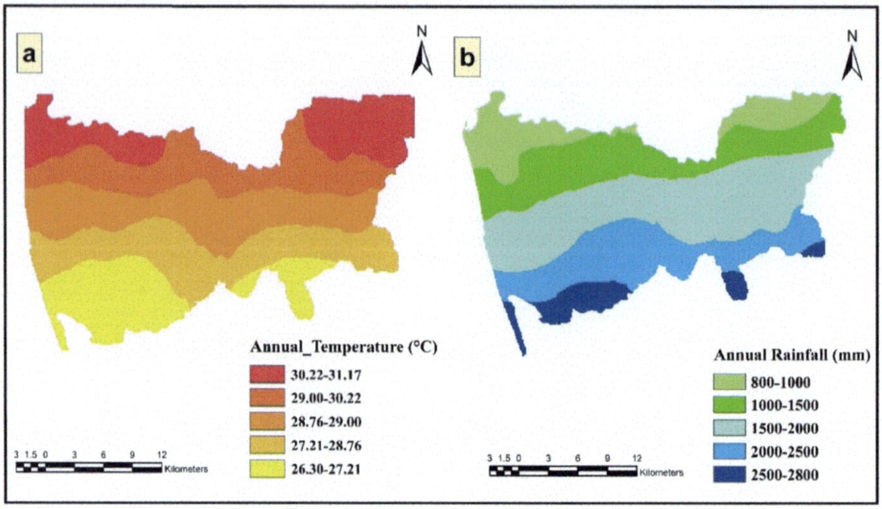

Fig. 3.9 **a** Spatial distribution of annual mean temperature in Colombo District (1981–2023); **b** spatial distribution of annual rainfall (1981–2023). *Source* Retrieved using GEE CHIRPS data

have alternating intense wet and dry years, and part of the inland regions have slight rainfall improvements due to localized convection and earlier environmental changes, representing space differences and broken rainfall patterns (Fig. 3.9a, b).

There is an evident increasing trend in surface temperature maps in urban and suburban regions. The temperatures are elevated in heart urban areas because of elevated building densities, scarce vegetation, urban heat islands, deforestation, and augmented automobile and industrial GHG emissions. According to satellite data, hotspots are correlated with fast development of infrastructure, emissions and population growth, validating non-uniform patterns of warming in the district (Figs. 3.9b, 3.10b). The combined effects of urbanization and climate change have a direct impact on the flood risk, water resources, and agriculture.

Sea-Level Rise (SLR) and Coastal Inundation Exposure

The GHG emissions over a long duration of time have led to the rise of SLR in Colombo District and urban and peri-urban low-lying areas (homes, industrial and residential) are the regions which are most vulnerable to the effects of the rise in sea level (Fig. 3.11). The overlap of demographic pressure, urban heat, and emissions in intensifying vulnerability can be seen by comparing Coastal DEM and satellite-derived elevation models, which show that densely populated zones are also the most exposed ones.

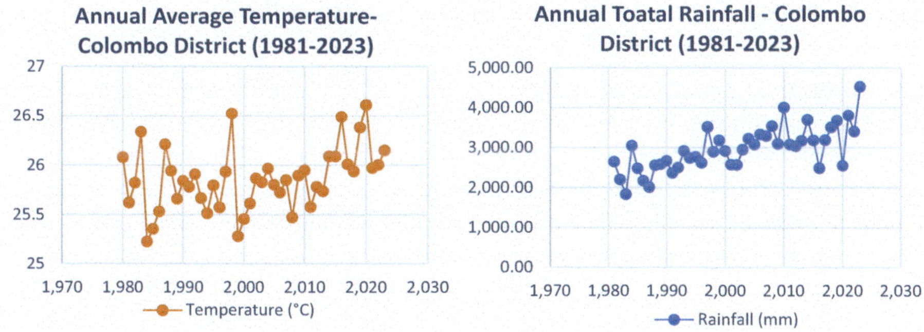

(a) (b)

Fig. 3.10 **a** Temporal changes of annual rainfall; **b** temporal changes of annual temperature. *Source* Retrieved using GEE CHIRPS data

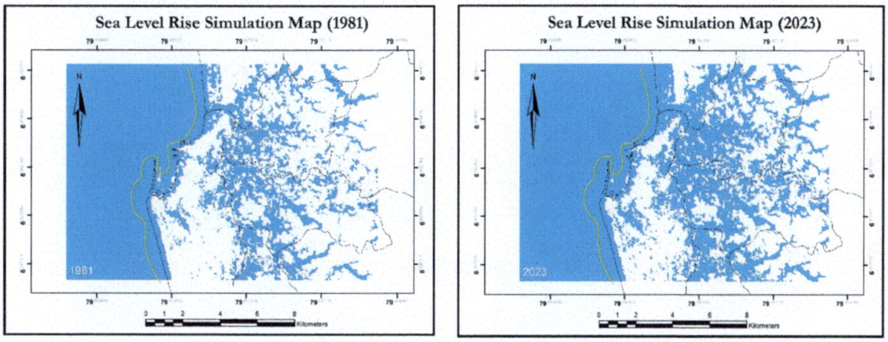

(a) (b)

Fig. 3.11 Sea-level rise simulation map (1981–2023). *Source* Retrieved using SLR data

Increase in CO_2, CH_4, NO_2, and SO_2, especially in times of industrialization (1995–2015), has increased global warming. Having a global warming potential of more than 25 times that of CO_2 within a period of one century, it is now rising as a result of poor waste management and the emission of landfills, and is contributing to the warming of the oceans, the expansion of thermal heat, and the melting of polar ice. The topography and high rate of urbanization of Colombo enhance such effects by reducing natural drainage and reducing the resilience of the shorelines. The case of SLR in the district indicates a compounded effect of past and present emissions, and not a solitary climate threat. Maps reveal that the ancient and current trends of GHG are actively defining the future geographical characteristics of the city near the coastline. Emission mitigation and adaptive urban design are necessary to avoid displacement and economic setbacks and additional enhancement of climate vulnerability.

Climate Footprint Paradox and Urban Sustainability

This section gives a conceptual framework (Fig. 3.12) indicating the multifaceted and contradictory interplay between environmental vulnerability, population dynamics, urban expansion and GHG emissions in Colombo, Sri Lanka. The current situation can be defined by the Climate Footprint Paradox, according to which urban development to enhance people's living standards leads to unintended GHG emissions, which contribute to climate change, and, therefore, conflicts with the objectives of development it is intended to accomplish (Ramadugu 2024; Allafta and Opp 2024).

This cycle starts with population growth, which leads to land conversion and rapid urbanization. This growth drives upward consumption of GHG-intensive industries, including transportation, energy, residential, and industrial production. Subsequently, the urbanization of the environment creates massive growth in the GHG emissions

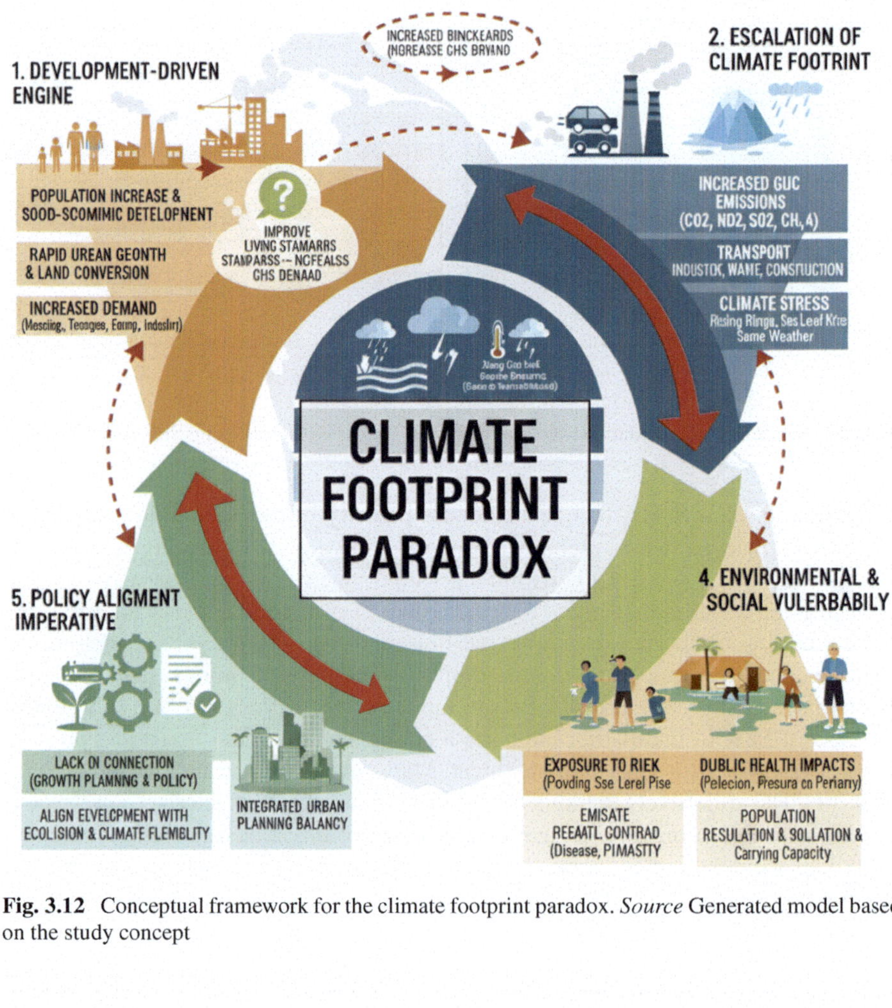

Fig. 3.12 Conceptual framework for the climate footprint paradox. *Source* Generated model based on the study concept

and puts the populations at a higher risk of climate-related impacts, such as sea-level rise, severe weather, and high temperatures. The framework is based on the fact that proactive policy alignment and strategic planning are necessary to balance between development goals and environmental sustainability to guarantee long-term resilience of cities and climate adaptation (Benna and Garba 2016; Satterthwaite 2009). The initial unit of the framework, which is called the Paradoxical Engine, determines socio-economic development and population growth as the key forces behind the urban growth and the intensive land conversion. The quest to have better living standards is, by itself, a boost in the demand for GHG, which makes the paradox of development creating environmental pressure obvious.

The second element is Escalation of the Climate Footprint, which describes how the growing urban activity increases CO_2, CH_4, NO_2, and SO_2 emissions. Even the transport sector becomes the greatest contributor in isolation, given that it uses fossil fuels, and additional methane emissions caused by poorly managed garbage in the ever-expanding suburbs further contribute to climate stress (Hammitt et al. 1996; Yaman 2024; Antal et al. 2020). These emissions are the direct causes of environmental risks such as the increasing temperatures, the increase of sea levels, and extreme weather variability.

The third component, Environmental and Social Vulnerability, points out the feedback mechanism of the climate stress that indirectly controls the population dynamics. The threat of flooding and other climate-related risks increases among the vulnerable populations of low-income earners, especially those living in informal settlements. This weakness produces demographic changes such as migration to fringe urban centers, thereby exerting further strain on infrastructure, housing and sanitation facilities. Moreover, the more the climate varies, the more the risks to the population's health increase, such as heatstroke, and water and vector-borne diseases, impacting the most vulnerable groups unequally (Amjad 2023; Gakaev et al. 2024). The last element is Policy Alignment Imperative, which determines the key intervention point of the cycle. The modern trend in urban development lacks the connection with environmental considerations. The solution to the paradoxical cycle is to have integrated urban planning and policies that are able to set the development goals alongside the ecological needs so as to ensure long-term sustainability of urban development and climate resilience.

Discussion

These findings indicate that the distribution of the population, GHG concentration, and exposure to the coastline have changed significantly over the last four decades in a manner that is interwoven in the Colombo District (De Costa and Dassanayake 2016). The growth of the population was forced out of the traditional urban center towards the peri-urban divisional secretariats, and signaled a change in Colombo city that was formerly a small city of the colonial days into a more polycentric metropolis. This external redistribution resembles the urbanization patterns of such Asian cities that are rapidly urbanizing, where congestion, decentralization of commercial activities and lack of availability of land push people to the urban periphery (Hugo 2019; Ranagalage 2021; Kawamura 1998). Such population changes are co-mortal with the large rise in atmospheric pollutants, particularly NO_2 and CH_4, that have continuously risen with the growth of vehicle fleets, expansion of the service sector, and the generation of solid waste. The correlation analysis also shows that there are positive yet significant positive correlations that exist between population growth and all the significant GHG concentrations, which also confirms the fact that demographic pressures do influence the atmospheric situation. The trend is also reflected in the climatic variables. Hydroclimatic variability and the increment in temperatures are comparable to the warming by anthropogenic activities in tropical cities worldwide. Within the framework of sea-level rise modelling, it is highlighted that much of the coastline of the low-lying region of Colombo is already vulnerable to flooding in all contingencies, and that the informal settlements and the most important infrastructures in the city are closely concentrated in the most prone areas (Patankar 2017; Fernando 2018). Together, they represent an intensifying and complex population-emissions-climate risks nexus that is changing. The environmental pressures found in Colombo are consistent with those in other cities in the Global South that are similarly coastally endowed, such as Mumbai, Dhaka, Lagos, and Manila (Pelling and Blackburn 2014; Anwar and Sur 2021; Ajibade 2022). The world literature demonstrates time and again that there exist many positive correlations between the demographic growth, rising NO_2 and CH_4 levels, and the creation of peri-urban growth corridors, and that the formation of Colombo is a part of a greater process of urbanisation and environmental transformations. One can also find analogues of locking of the atmospheric pollutants in the satellite-based observation of the rapidly developing tropical cities, which register a sudden increase in the atmospheric pollutants following the economic boosts, which is in tandem with the levels of this study (Yuen and Kong 2009; Fuchs et al. 2011);;. On the regional scale, South Asia is now generally recognized as a region of strong climate where the rise in sea levels and the hydrometeorological fluctuations are burning issues on the high-density coastline. Calculations of the inundation made concerning Colombo conform to the recent IPCC regional estimates which stress the point that even small increments in the rise in the sea level, will have a drastic effect on the infrastructure networks, residential and transport control systems of the regions of low elevated deltaic settlements and coastal settlements (Garschagen et al. 2021; Iturbide et al. 2020). The aspects that the research integrates the population and emissions and SLR variables into a single analysis tool directly respond to the growing academic expectations that complicated, multi-scalar analysis can be reduced to comprehend the intricacy of the city climate threat.

The chapter demonstrates how the changes in demographics are a benefit and are also modified by climate-related pressures within a coastal city. The findings have confirmed that, in fact, a climate-urbanization feedback exists with population growth causing higher emissions and changing land use that are themselves exerting pressure to further restrict the same spaces in which urban growth is occurring. The presented framework of the long-term analysis, which is holistic, contributes to the body of knowledge about this feedback mechanism by demonstrating how population change, atmospheric conditions, and coastal vulnerability interact within time and space. The Colombo case provides a practical illustration of how the future settlement patterns will be constrained by the climate risks, specifically an increase in sea level, forcing more individuals onto the land and imposing new demands on adaptation (Palamakumbure et al. 2020; Prasanna et al. 2023). The ability of the population trends to be correlated with atmospheric and coastal data in the chapter renders it an internationally applicable, scientifically reliable addition to the thematic congruence of the Climate Book and equips the insights that can be utilized in other cities that experience such changes as coastal cities.

Based on the results, it is possible to state that there are several policy priorities in the Colombo District and other urban areas. The strong dependence of the increasing population on the concentration of NO_2 in the atmosphere proves that there is a need to transition to the low-emission transport regime, namely, expansion of the public transport sector, promotion of the electrification of mobility, and optimal traffic management in congested transport zones. The gradual rise in the CH_4 levels indicates that there is a great need to revise the existing waste-management practices, improve landfills' operations, and popularize the technologies of capturing methane in the industrial and municipal sectors (Hayden and Mahin 2022; Krause 2011). The projection of the SLR deprives it of doubt that the adjustment of the coastal zone must be part of the center plan of the planning, and they must contain every such measure as strategic repositioning or elevation of the infrastructure, restoration of the mangroves, and limitation of the new development of the risky areas (Davar et al. 2021; Zhao and Liu 2020). Because the percentage of new population rise is shifting to the peri-urban divisions, climate-risk, sustainable drainage, and green-space networks would be added to the growth plans in the future to reduce the exposure and improve the resilience. Anybody reading this chapter of the research would find the price as presented by satellite-based surveillance and in-built data strategies to be very useful in terms of institutionalization of such systems to promote evidence-based city planning and early-warning systems.

The primary strength of the study lies in the design of the unified methodology, a mixture of multi-decadal population information, satellite-based measurements of GHG, and high-resolution coastal elevation models (Bove et al. 2020; Chaigneau et al. 2022). This will enable the study of urban and climatic interactions comprehensively in such an environment, which is not frequently carried out in South Asia. The fact that a number of the validation processes are incorporated also enhances the credibility and reliability of the findings. But several shortcomings have to be acknowledged. Nonetheless, uncertainties persist due to lower-resolution pre-2000 emissions data, potential underrepresentation of informal waste sites, and unmeasured local subsidence that may influence SLR estimates (Paolozzi et al. 2016; Zhao and Liu 2020. Despite these limitations, the results provide a scientifically rigorous foundation for climate-responsive urban planning and advance understanding of coupled population-climate dynamics in coastal megacities.

Recommendations

In Colombo, a complex array of forces of high-speed urbanization, growing emissions, and increasing vulnerability to rising sea levels needs a multidisciplinary and evidence-based policy intervention. The results of the study outline the urgent necessity to shift the current urban mobility system to a low-emission one, which is immediate by electrifying the public transport fleet, implementing better control over traffic, and increasing infrastructure that would make active and shared mobility possible. It is important to reduce the exceedingly high share of the transport sector in the concentration of NO_2 and reduce the overall air pollution in the city in order to reduce the air pollution in the area and bring Colombo into the series of emission-reduction paths, adopted worldwide (Carslaw and Beevers 2005). Simultaneously, the ongoing increase in CH_4 levels indicates the pressing need to modernize waste-management infrastructure by implementing technologies to capture methane, expanding waste-to-energy, and eliminating informal disposal locations that contribute to the reinforcement of spatial disparities in risk exposure (Yi et al. 2018; Moya et al. 2017; Stehlík 2009).

It is also essential that climate-risk analytics are included in the planning and regulation of coastal development. The discovery of large low-lying areas at risk of future flooding pressures requires policies that limit new development in risky locations, focus on ecosystem-based adaptation, and combine hybrid flood-defence structures that can absorb future SLR effects (Biraghi et al. 2024; Cengiz 2013);;. The peri-urban growth corridors need to incorporate green and blue infrastructure that can moderate levels of heat, positive hydrological operation, and the ecological resilience of growing settlement frontiers. These interventions need institutional reinforcement, which could be achieved by creating a centralized, data-driven urban climate agency that enhances the coordination of agencies and standardization of monitoring infrastructures and aligns the urban development decisions with the long-term adaptation strategies. In general, the current study leads to the recommendations that a transition to integrated climate governance, in which emission-reduction, risk-sensitive spatial planning, and coastal adaptation are seen as mutually supportive policy areas, should be taken. This would not only help to curb the short-term weaknesses that were established in the research but also set the pace towards a resilient urban pathway that will support the growing rate of climatic transformation in Sri Lanka and other coastal megacities of the Global South.

Conclusion

The chapter presents a comprehensive evaluation of population redistribution, greenhouse emissions, and exposure to SLR in the Colombo District over a forty-year period, as a classic example of how demographic change, environmental stressors, and climate risks interact in the fast-changing coastal areas. The analysis of the multi-decadal grid population data, satellite-inferred atmospheric data, and high-resolution coastal elevation model, the research paper develops an analytical framework that can discover long-term interplay between urban expansion and climatic danger (East et al. 2022; Woodley et al. 2006). The findings indicate that the trend of peri-urbanization in Colombo has been accompanied by sustained growth in NO_2 and CH_4, which supports the idea that the urbanization mechanisms in the most rapidly developed economies are closely connected with the destruction of the atmosphere (Plaia and Ruggieri 2011; Murena 2004). At the same time, using SLR simulations, we can demonstrate that extensive areas of the low-elevation coastline in the district are already vulnerable to the risk of inundation, which underlines the spatial incompatibility of the existing development patterns and the physical capacity of the coastal systems.

The scientific importance of this study is that its conception is multi-scalar, meaning that it adds to a global climate-urbanization literature by demonstrating that the joint use of demographic, atmospheric, and geophysical data can be used to depict complex human settlement-climate hazard interactions. This piece of work gives empirical support to the notion that rapidly urbanizing cities in the Global South, like Colombo, Mumbai, Lagos and Manila, have a structural vulnerability that is manifested in the fact that both population growth, emissions accumulation and growing SLR are converging (Davar et al. 2021; Zhao and Liu 2020). This methodological design also provides a model of transferability to climate-risk assessment in similar coastal megacities across the world to ensure that world urban development requirements are based on evidence-based planning in line with IPCC and UN-Habitat and priorities (Rosenzweig et al. 2018; McBean et al. 2017). The study reinforces the scientific community on climate-urban shift and plays a stronger role in defining international patterns of resilient metropolitan development by showing the interaction of atmospheric pollution, settlement development and exposure to the coast over time.

Direct policy implications are also possible in the findings of the study. The prevailing relationships among population growth, the concentration of pollutants, and coastal exposure reveal that a coherent policy domain should be implemented to incorporate emission-reduction measures, planning of land use, and adaptation in one policy area. This evidence can guide policymakers toward focusing on low-emission mobility systems, modernized waste-management infrastructure and specific coastal protection efforts on the most vulnerable areas (Kauhanen 2024; Lee et al. 2018). Spatial hotspots of risk identification can give a scientific foundation to zoning changes, infrastructure hardening, and ecosystem-based adaptation-specific investment (Hagenlocher et al. 2018; Jones et al. 2020). In a broader sense, the combined strategy illustrated here provides governments with a model that is replicable to incorporate climate-risk analytics into master planning, and in effect, increase resilience of the country over time and the efficacy of the national adaptation strategy.

Limitations

The study faced a number of methodological constraints. Older data on emissions were of lower spatial resolution, which limited the ability to identify intra-urban differences in GHG concentrations. Satellite sensors may not have captured all the activities in the informal waste sector, thus underestimating methane gas emissions. The results of SLR modelling were also determined subject to the assumptions of global climate models and access to local data of vertical land motion, which brought certain uncertainties. However, the application of several complementary datasets and methodological triangulation contributed to the strength of the overall results.

AI Disclosure

AI tools were used to generate the conceptual model based on the authors' own perceptions in this chapter. Therefore, the authors were the sole makers of all the analytical decisions, interpretations, data processing, and inferences.

Author Contributions Both authors have seen and agreed to the final version of the manuscript, and both have had equal contributions to this work.

Funding No specific grant was received to accomplish this research by funding agencies in the public, commercial, or non-profit sectors.

Potential Conflict of Interests As the corresponding author on behalf of all the authors declare that there are no prospective conflicts of interest regarding the research, authorship, and or publication of the article.

Data Availability Statement The datasets employed and analysed in the present study can be obtained from the corresponding author. Satellite and census datasets are publicly accessible via the website (open access).

References

Ajibade I (2022) The resilience fix to climate disasters: recursive and contested relations with equity and justice-based transformations in the Global South. Ann Am Assoc Geogr 112(8):2230–2247

Allafta H, Opp C (2024) Climate change paradox: the least responsible for it encounters the most of its implications. Climate 12(3):38

Amjad MA (2023) Moderating the role of social progress with greenhouse gases to determine the health vulnerability in developing countries. Environ Sci Pollut Res 30(40):92123–92134

Antal M, Plank B, Mokos J, Wiedenhofer D (2020) Is working less really good for the environment? A systematic review of the empirical evidence for resource use, greenhouse gas emissions and the ecological footprint. Environ Res Lett 16(1):013002

Anwar N, Sur M (2021) Climate change, urban futures, and the gendering of cities in South Asia

Asian Development Bank (2022) Climate risk country profile: Sri Lanka. https://www.adb.org/publications/climate-risk-country-profile-sri-lanka

Benna UG, Garba SB (eds) (2016) Population growth and rapid urbanization in the developing world. IGI Global

Biraghi CA, Mandarino FC, Masera G, Souza LG, Fraga JPR, Tadi M, Mohammad Zadeh MH (2024) Integrated and sustainable urban regeneration Milan—Rio—REMIRIO: urban diagnosis based on integrated modification methodology (IMM). In: International conference on sustainability in energy and buildings. Springer Nature Singapore, Singapore, pp 421–433

Bove G, Becker A, Sweeney B, Vousdoukas M, Kulp S (2020) A method for regional estimation of climate change exposure of coastal infrastructure: case of USVI and the influence of digital elevation models on assessments. Sci Total Environ 710:136162

Carslaw DC, Beevers SD (2005) Development of an urban inventory for road transport emissions of NO_2 and comparison with estimates derived from ambient measurements. Atmos Environ 39(11):2049–2059

Cengiz C (2013) Urban ecology. In: Advances in landscape architecture. IntechOpen

Chaigneau AA, Reffray G, Voldoire A, Melet A (2022) IBI-CCS: a regional high-resolution model to simulate sea level in western Europe. Geosci Model Dev 15(5):2035–2062

Davar L, Griggs G, Danehkar A, Salmanmahiny A, Azarnivand H, Naimi B (2021) A spatial integrated SLR adaptive management plan framework (SISAMP) toward sustainable coasts. Water 13(16):2263

De Costa G, Dassanayake W (2016) Estimation of impact of sea level rise on land and water resources when data is sparse-case of Colombo. Environ Ecol Res

De Silva M, Kawasaki A (2018) Socioeconomic vulnerability to disaster risk: a case study of flood impacts in Colombo, Sri Lanka. Int J Disaster Risk Reduct 31:194–205. https://doi.org/10.1016/j.ecolecon.2018.05.010

Department of Census and Statistics (DCS) (2021) Sri Lanka population and housing census 2021. Government of Sri Lanka

Department of Census and Statistics, Sri Lanka (2023) Population and housing data. http://www.statistics.gov.lk/

East JD, Henderson BH, Napelenok SL, Koplitz SN, Sarwar G, Gilliam R, Lenzen A, Tong DQ, Bradley Pierce R, Garcia-Menendez F (2022) Inferring and evaluating satellite-based constraints on NO_x emissions estimates in air quality simulations. Atmos Chem Phys Discuss 2022:1–31

Eckstein D, Künzel V, Schäfer L (2023) Global climate risk index 2023. Germanwatch. https://germanwatch.org/en/cri

Eriyagama N, Smakhtin V, Chandrapala L, Fernando K (2020) Impacts of climate change on water resources in Sri Lanka: a review. Hydrol Sci J 65(2):166–188

Fernando N (2018) Voluntary or involuntary relocation of underserved settlers in the city of Colombo as a flood risk reduction strategy: a case study of three relocation projects. Procedia Eng 212:1026–1033

Fuchs R, Conran M, Louis E (2011) Climate change and Asia's coastal urban cities: can they meet the challenge? Environ Urban ASIA 2(1):13–28

Gakaev R, Gatsaeva L, Eskiev M (2024) Impacts of greenhouse gas-induced climate change: risks, vulnerabilities, and adaptation strategies. In: E3S web of conferences, vol 537. EDP Sciences, p 03009

Garschagen M, Doshi D, Reith J, Hagenlocher M (2021) Global patterns of disaster and climate risk—an analysis of the consistency of leading index-based assessments and their results. Clim Change 169(1):11

Grimm NB, Faeth SH, Golubiewski NE, Redman CL, Wu J, Bai X, Briggs JM (2008) Global change and the ecology of cities. Science 319(5864):756–760. https://doi.org/10.1126/science.1150195

Gunathilaka R, Perera K, Wijesinghe A (2023) Urban sustainability and emissions in Sri Lanka. J Environ Policy 15(3):45–62

Hagenlocher M, Schneiderbauer S, Sebesvari Z, Bertram M, Renner K, Renaud F, Wiley H, Zebisch M (2018) Climate risk assessment for ecosystem-based adaptation–a guidebook for planners and practitioners

Hammitt JK, Jain AK, Adams JL, Wuebbles DJ (1996) A welfare-based index for assessing environmental effects of greenhouse-gas emissions. Nature 381(6580):301–303

Hayden FG, Mahin TA (2022) Integrating the concepts of zero greenhouse-gas emissions, the precautionary principle, and environmental impact statements for climate change policy mitigation. J Econ Issues 56(2):400–407

Hugo G (2019) Patterns and trends of urbanization and urban growth in Asia. In: Internal migration, urbanization and poverty in Asia: dynamics and interrelationships. Springer Singapore, Singapore, pp 13–45

Intergovernmental Panel on Climate Change (IPCC) (2021) Climate change 2021: the physical science basis. Cambridge University Press. https://www.ipcc.ch/report/ar6/wg1/

Intergovernmental Panel on Climate Change (2023) Sixth assessment report on climate change. https://www.ipcc.ch/assessment-report/ar6/

Iturbide M, Gutiérrez JM, Alves LM, Bedia J, Cerezo-Mota R, Cimadevilla E et al (2020) An update of IPCC climate reference regions for subcontinental analysis of climate model data: definition and aggregated datasets. Earth Syst Sci Data 12(4):2959–2970

Jones HP, Nickel B, Srebotnjak T, Turner W, Gonzalez-Roglich M, Zavaleta E, Hole DG (2020) Global hotspots for coastal ecosystem-based adaptation. PLoS ONE 15(5):e0233005

Kauhanen L (2024) Towards a carbon-neutral waste management: an assessment of the potential CO_2 emissions reductions from waste transport optimization

Kawamura M, Jayamanna S, Tsujiko Y, Sugiyama A (1998) Comparison of urbanization of four Asian cities using satellite data. Doboku Gakkai Ronbunshu 1998(608):97–105

Krause RM (2011) Symbolic or substantive policy? Measuring the extent of local commitment to climate protection. Environ Plann C Gov Policy 29(1):46–62

Lee CT, Mohammad Rozali NE, Van Fan Y, Klemeš JJ, Towprayoon S (2018) Low-carbon emission development in Asia: energy sector, waste management and environmental management system. Clean Technol Environ Policy 20(3):443–449

McBean G, Cooper R, Joakim E (2017) Coastal cities at risk (CCaR): building adaptive capacity for managing climate change in coastal megacities

Ministry of Environment, Sri Lanka (2023) National emission data report. http://www.environmentmin.gov.lk/

Ministry of Health, Sri Lanka (2023) Climate change and public health report. http://www.health.gov.lk/

Moya D, Aldás C, López G, Kaparaju P (2017) Municipal solid waste as a valuable renewable energy resource: a worldwide opportunity of energy recovery by using waste-to-energy technologies. Energy Procedia 134:286–295

Munasinghe M, Deraniyagala Y, Dassanayake N (2021) Sustainable urban development in Sri Lanka: challenges and opportunities. Sustainability 13(5):2567

Murena F (2004) Measuring air quality over large urban areas: development and application of an air pollution index at the urban area of Naples. Atmos Environ 38(36):6195–6202

Nasar-u-Minallah M, Jabeen M, Parveen N, Abdullah M, Nuskiya MHF (2025) Exploring the seasonal variability and nexus between urban air pollution and urban heat Islands in Lahore, Pakistan. Acta Geophys 1–21

Nuskiya MHF, Perera O (2025) Population growth and urban heat island effect in Batticaloa Municipal Council, Sri Lanka. In: Conference on geography and global sustainability (ICGGS)-2025, p 25

Palamakumbure L, Ratnayake AS, Premasiri HR, Ratnayake NP, Katupotha J, Dushyantha N, Weththasinghe S, Weerakoon WP (2020) Sea-level inundation and risk assessment along the south and southwest coasts of Sri Lanka. Geoenviron Disasters 7(1):17

Paolozzi A, Paris C, Pavlis EC, Sindoni G, Ciufolini I, Vendittozzi C (2016) Monitoring global climate change using SLR data from LARES and other geodetic satellites. In: Sensors and smart structures technologies for civil, mechanical, and aerospace systems 2016, vol 9803. SPIE, pp 1242–1250

Patankar AM (2017) Colombo: exposure, vulnerability, and ability to respond to floods. World Bank policy research working paper (8084)

Peel MC, Finlayson BL, McMahon TA (2007) Updated world map of the Köppen-Geiger climate classification. Hydrol. Earth Syst. Sci. 11(5):1633–1644

Pelling M, Blackburn S (2014) Case studies: governing social and environmental transformation in coastal megacities. In: Megacities and the coast. Routledge, pp 200–236

Pesaresi M, Huadong G, Blaes X, Ehrlich D, Ferri S, Gueguen L et al (2013) A global human settlement layer from optical HR/VHR RS data: concept and first results. IEEE J. Sel. Top. Appl. Earth Obs. Remote Sens. 6(5):2102–2131

Plaia A, Ruggieri M (2011) Air quality indices: a review. Rev Environ Sci Bio/technol 10(2):165–179

Prasanna HMI, Gunathilaka MK, Iz HB (2023) Sea level variability at Colombo, Sri Lanka, inferred from the conflation of satellite altimetry and tide gauge measurements. J Natl Sci Found Sri Lanka 51(2)

Ramadugu R (2024) Unraveling the paradox: green premium vs. climate risk premium in sustainable investing. ABS Int J Manag 12(2):71–89

Ranagalage M, Morimoto T, Simwanda M, Murayama Y (2021) Spatial analysis of urbanization patterns in four rapidly growing south Asian cities using Sentinel-2 data. Remote Sens 13(8):1531

Ranasinghe P, Fernando W, Silva T (2022) Climate adaptation in Sri Lanka: policy and practice. Environ Policy Rev 18(2):88–105

Revi A, Satterthwaite DE, Aragón-Durand F, Corfee-Morlot J, Kiunsi RBR, Pelling M, Roberts DC, Solecki W (2014) Urban areas. In: Climate change 2014: impacts, adaptation, and vulnerability. Cambridge University Press, pp 535–612

Rosenzweig C, Solecki WD, Romero-Lankao P, Mehrotra S, Dhakal S, Ibrahim SA (eds) (2018) Climate change and cities: second assessment report of the urban climate change research network. Cambridge University Press

Satterthwaite D (2009) The implications of population growth and urbanization for climate change. Environ Urban 21(2):545–567

Schiavina M, Melchiorri M, Corbane C, Florczyk AJ, Freire S, Pesaresi M, Kemper T (2019) Multi-scale estimation of land use efficiency (SDG 11.3. 1) across 25 years using global open and free data. Sustainability 11(20):5674

Sri Lanka Sustainable Energy Authority (2023) Annual energy consumption report. https://www.energy.gov.lk/

Stehlík P (2009) Contribution to advances in waste-to-energy technologies. J Clean Prod 17(10):919–931

United Nations Framework Convention on Climate Change (2023) Nationally determined contributions (NDCs) report. https://unfccc.int/ndc

Veefkind JP, Aben I, McMullan K, Förster H, De Vries J, Otter G, Claas J, Eskes HJ, De Haan JF, Kleipool Q, Van Weele M (2012) TROPOMI on the ESA Sentinel-5 Precursor: a GMES mission for global observations of the atmospheric composition for climate, air quality and ozone layer applications. Remote Sens. Environ. 120:70–83

Weerakoon KGMT, Gunawardena ERN, Sonnadara DUJ (2019) Greenhouse gas emissions from the transport sector in Sri Lanka: trends and mitigation strategies. Energy Policy 129:1283–1291

Wickramasinghe D, Attanayake J (2020) Land-use changes and urban heat island effects in Colombo Metropolitan region. J Environ Manage 265:110525. https://doi.org/10.1016/j.jenvman.2020.110525

Wickramasinghe D (2018) Climate change and urbanization in Sri Lanka: Case of Colombo. University of Colombo Press

Woodley WL, Rosenfeld D, Lynn B, Givati A (2006) Physical/statistical and modeling documentation of the effects of urban and industrial air pollution in California on precipitation and stream flows in mountainous terrain downwind

World Bank (2023) Urban development in Sri Lanka. https://www.worldbank.org/en/country/srilanka
Yaman C (2024) A review on the process of greenhouse gas inventory preparation and proposed mitigation measures for reducing carbon footprint. Gases 4(1):18–40
Yi S, Jang YC, An AK (2018) Potential for energy recovery and greenhouse gas reduction through waste-to-energy technologies. J Clean Prod 176:503–511
Yuen B, Kong L (2009) Climate change and urban planning in Southeast Asia. SAPI EN. S. Surveys and Perspectives Integrating Environment and Society (2.3)
Zahir ILM, Nuskiya MHF, Sangasumana VP, Iyoob AL, Ameer MLF (2024) Monitoring urban green space using remote sensing derived vegetation indices in Colombo District, Sri Lanka. Procedia Comput Sci 236:248–256
Zhao L, Liu F (2020) Land-use planning adaptation in response to SLR based on a vulnerability analysis. Ocean Coast Manag 196:105297

Chapter 4
Visualizing Global Research Patterns of Mental Health in the Context of Climate Change-Induced Extreme Weather Events

T. Adarsh, K. P. Haneena Sulthana, T. S. Lancelet, and C. A. Sandhya Aravind

Abstract Climate change-induced extreme weather events have emerged as a major global challenge, with growing evidence demonstrating their profound impacts on mental health and psychological wellbeing. Despite increasing scholarly attention, a comprehensive synthesis of global research trends at the intersection of climate change, extreme weather events, and mental health remains limited. This study addresses this gap through a bibliometric analysis of 2496 unique publications retrieved from the Scopus and Web of Science databases after data merging and deduplication, covering the period 2000–2025. Using bibliometric mapping techniques implemented through Bibliometrix in RStudio, the study examines publication growth patterns, leading authors and institutions, international collaboration structures, and thematic evolution of the field. The findings reveal a rapidly expanding and increasingly human-centered research domain, with a strong intellectual core linking climate change and mental health, supported by public health, psychological, and disaster-related research streams. The recent years show a marked shift toward impact-oriented themes, including mental health outcomes, risk perception, adaptation, and extreme events such as heat waves and floods. While the field demonstrates growing maturity and interdisciplinary integration, notable geographic and

T. Adarsh (✉) · T. S. Lancelet
Department of Geography, Sree Sankaracharya University of Sanskrit, Kalady, Kerala, India
e-mail: Adarshgeo@ssus.ac.in

T. S. Lancelet
e-mail: drlancley2011@ssus.ac.in

K. P. Haneena Sulthana · C. A. Sandhya Aravind
Department of Psychology, Sree Sankaracharya University of Sanskrit, Kalady, Kerala, India
e-mail: haneenasulthana@ssus.ac.in

C. A. Sandhya Aravind
e-mail: drsandhyaaravind@ssus.ac.in

P. Ghosh et al. (eds.), *Climate Crisis and Changing Population Dynamics*, Springer Climate, https://doi.org/10.1007/978-3-032-24533-5_4

collaborative imbalances persist. The study underscores the need for greater interdisciplinary convergence, regionally inclusive research, and evidence-based strategies to strengthen mental health resilience in the face of escalating climate crises.

Keywords Climate change · Climate crisis · Psychological distress · Eco-anxiety · Post-traumatic stress disorder (PTSD)

Introduction

Climate change has emerged as a significant risk factor for mental health, impacting individuals and communities across the globe through various direct and indirect pathways (Clayton 2020; Cunsolo and Ellis 2018). Climate change refers to long-term shifts in temperatures, weather patterns, and climate systems primarily driven by human activities particularly the burning of fossil fuels, deforestation, and industrial processes that increase greenhouse gas concentrations in the atmosphere (IPCC 2021). Mental health is a state of psychological, emotional, and social well-being in which individuals can cope with normal stresses, work productively, and contribute to their communities, reflecting more than the absence of mental disorders (World Health Organization [WHO] 2022).

As climate hazards such as floods, droughts, cyclones, and heatwaves intensify, they not only disrupt physical environments but also profoundly affect the psychological wellbeing of impacted communities (IPCC 2022; Palinkas and Wong 2020). Mental health consequences ranging from trauma and PTSD to eco-anxiety and chronic stress shape how population respond and adapt to climate pressures (Hayes et al. 2018). Psychological distress can influence decisions to migrate, alter community resilience, and shift vulnerability patterns, thereby contributing to changes in population distribution, mobility, and long-term demographic structures (Burke et al. 2018; Kelman 2020). Understanding these mental health outcomes is therefore essential for explaining how climate crises drive evolving population dynamics at local, national, and global levels (Berry et al. 2010).

Research indicates that the psychological repercussions of climate change may be more extensive than the physical health effects associated with ecological disruptions. Direct effects stem from extreme weather events, which can precipitate conditions such as Post-traumatic Stress Disorder (PTSD), anxiety, and depression (Hayes and Poland 2018; Shoib et al. 2023). The indirect effects include economic loss, displacement, and anxiety related to the ongoing threat of climate change. These causes form a multi-layered framework for understanding mental health outcomes linked to climate-related events (Hayes et al. 2018; Saltzman et al. 2023).

Climate change also yields cognitive consequences, as individuals grapple with the trauma of loss and displacement following environmental disasters. Neuroscientific studies have indicated that exposure to climate-related traumas can lead to alterations in brain function and can manifest as cognitive deficits, thereby challenging a person's ability to reconstruct post-disaster (Grennan et al. 2023). Heightened environmental stressors can induce a unique psychological condition often referred to as "eco-anxiety," leading to chronic distress about the long-term impacts of climate change on personal and planetary futures (Mishra and Ramanathan 2023).

Research indicates that groups such as children, the elderly, women, and marginalized communities, including Indigenous population, are particularly at risk of experiencing eco-anxiety and related conditions. For instance, Reyes et al. (2021) elucidate how Gen Z Filipinos experience anxiety directly linked to climate change, highlighting the disproportionate effects on youth living in regions most susceptible to climate disasters. Similarly, Léger-Goodes et al. (2022) note that children exposed to the realities of climate change exhibit significant emotional disturbances, including depression and anxiety.

Socio-economic factors significantly influence vulnerability to climate-induced hazards. Gislason et al. (2021) note that the interaction between social conditions and ecological crises creates compounded stressors for certain population, reducing their ability to cope with climate impacts. Without addressing these vulnerabilities, extreme weather events may deepen existing disadvantages and worsen mental health outcomes. To respond to these risks, strengthening community resilience is essential, particularly through social support systems and improved access to mental health care (Charlson et al. 2021; Anantapong et al. 2024). Policy recommendations also highlight the value of a stepped-care approach, tailoring mental health interventions to the severity of post-disaster needs (Power et al. 2022).

A significant knowledge gap exists in this field because there is no comprehensive or systematic understanding of how global research has evolved at the intersection of climate-induced extreme weather events and mental health. The existing literature is dispersed across different disciplines, geographic regions, and types of extreme events. This fragmentation makes it difficult for researchers to identify overall trends, dominant themes, existing knowledge gaps, and the broader evolution of scientific attention in this area. Incorporating art, literature, and spirituality into coping mechanisms may also facilitate emotional resilience, thus enabling communities to navigate the psychological distress related to climate change more effectively (Anantapong et al. 2024).

The research problem of this study is the lack of a structured global analysis that identifies which climate-induced extreme events are most frequently associated with specific mental health outcomes, and how research emphasis differs across countries, disciplines, and time. Understanding these patterns is crucial because public health initiatives play a vital role in monitoring and responding to mental health crises triggered by climate-related disasters, making robust evidence essential for developing effective response frameworks (Saltzman et al. 2023; Clemens et al. 2022).

Bibliometric analysis in this area helps to map the intellectual structure of specific research domains by quantitatively assessing the relationships and productivity of various scholarly works. This method allows researchers to identify influential articles, prominent authors, and adaptive trends in publication practices over time, which is crucial for understanding the evolution of a subject area (Kalsoom and Zámečník 2023; Wang et al. 2024). The methodological rigor of bibliometric analysis strengthens its applicability across various fields, allowing researchers to assess complex patterns and dynamics influencing the scientific community (Hamidah et al. 2021).

Objectives

- To analyze publication trends in the field of mental health in climate crisis.
- To understand the most influential contributions in this domain, including leading authors, highly cited publications, key institutions, and contributing countries.
- To know the patterns of scholarly collaboration through co-authorship networks at the author, institutional, and country levels.
- To map the intellectual and conceptual structure of climate change and mental health research using co-citation and co-word analyses, highlighting dominant themes and emerging research areas.

Rationale of the Study

The growing frequency of climate-induced extreme weather events has intensified their psychological consequences, making it essential to systematically examine existing research to understand trends, gaps, and emerging priorities in climate change and mental health scholarship (Charlson et al. 2021). Although research on climate change and mental health has expanded, the literature remains fragmented across disciplines, regions, and climate hazards, limiting a clear understanding of research trends and knowledge gaps. A bibliometric analysis is therefore needed to map the global research landscape, identify influential contributions, and support future research, policy development, and mental health interventions in the context of the climate crisis.

Method

This study used a bibliometric research design to quantitatively analyze the pattern of the field. Bibliometric analysis enables the mapping of a field's intellectual structure, research collaboration networks, and conceptual trends through statistical and network-based techniques (Adarsh and Sulthana 2025). To ensure methodological transparency, 2000 and 1571 data were retrieved from Scopus and Web of Science, respectively. The search covered the period 2000–2025, and the time period was selected to comprehensively capture the emergence, progression, and recent expansion of scholarly research on the mental health impacts of climate change, spanning from early scientific recognition in the 2000s to contemporary investigations addressing intensified climate extremes and evolving psychological concerns. After removing 827 duplicates, a total of 2450 documents were retained for analysis (Fig. 4.1).

All bibliometric visualizations were generated using Biblioshiny application in RStudio. Figures were exported directly from these tools using their built-in visualization functions, ensuring reproducibility and consistent graphical formatting.

The search query has been used for extracting documents from Scopus and web of science is listed below;

=((“climate change” OR “global warming” OR “climate crisis” OR “climate emergency”) AND (“climate-related disaster” OR “climate-induced disaster*” OR “extreme weather event*” OR “climate extreme event*” OR heatwave* OR “heat*

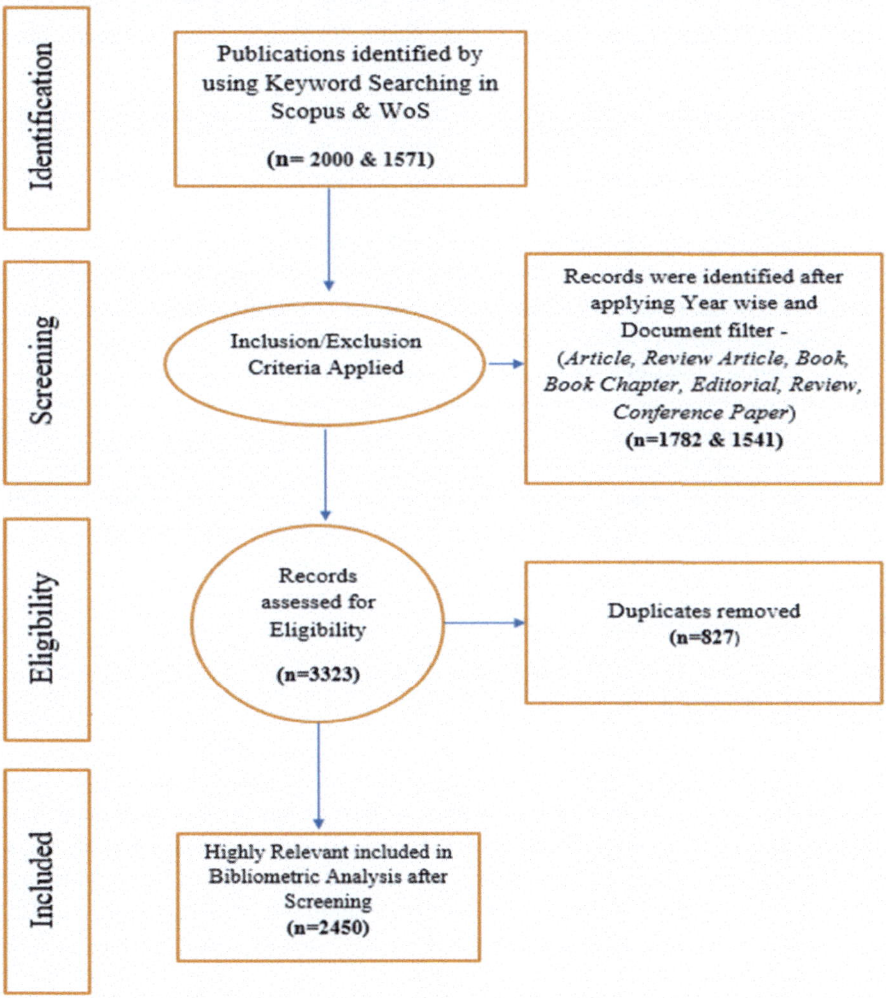

Fig. 4.1 Methodology flowchart

wave" OR flood* OR drought* OR cyclone* OR hurricane* OR typhoon* OR wildfire* OR "forest fire*" OR monsoon*) AND (psychology OR psychological* OR "mental health" OR "mental wellbeing" OR "mental disorder*" OR "psychological impact*" OR "psychological distress" OR PTSD OR "posttraumatic stress" OR trauma OR anxiety OR depression OR "eco-anxiety" OR solastalgia)) NOT (crop* OR crops).*

Data Analysis

The analysis was conducted using specialized bibliometric tools, Biblioshiny (an R-based bibliometric package) (Aria and Cuccurullo 2017). The following techniques were applied:

- Descriptive Analysis: This involved the identification of publication trends, prominent journals, influential authors, and citation metrics.
- Co-occurrence Analysis: Keywords were analyzed to visualize the thematic structure of the research field using network maps.
- Thematic Mapping: Themes were classified based on relevance and development degree using a two-dimensional analysis, categorizing topics into Basic Themes, Motor Themes, Niche Themes, and Emerging or Declining Themes.
- Thematic Evolution Analysis: Temporal changes in research focus were mapped using Sankey diagrams to trace the progression of topics over different time periods.
- Factorial Analysis: Applied to identify underlying structures within the data by performing multiple correspondence analysis, thereby revealing thematic clusters.
- Collaboration Network Analysis: Country-level collaboration networks were assessed to identify leading research hubs and international partnerships.

Result and Discussion

Main Information

The study spans 2000–2025 and examines 2450 documents published in 1026 sources, recording an annual growth rate of 20.73%, an average document age of 4.69 years, 28.27 average citations per document, and 0 references, with document contents comprising 8238 keywords plus and 5881 author keywords, authored by 8777 authors including 264 authors of single-authored works, resulting in 280 single-authored documents, an average of 4.67 co-authors per document, and 21.76% international co-authorship, and covering document types such as 1836 articles, 1 article–book chapter, 20 article–early access papers, 16 article–proceedings papers, 22 books, 152 book chapters, 53 conference papers, 2 editorial material–early access

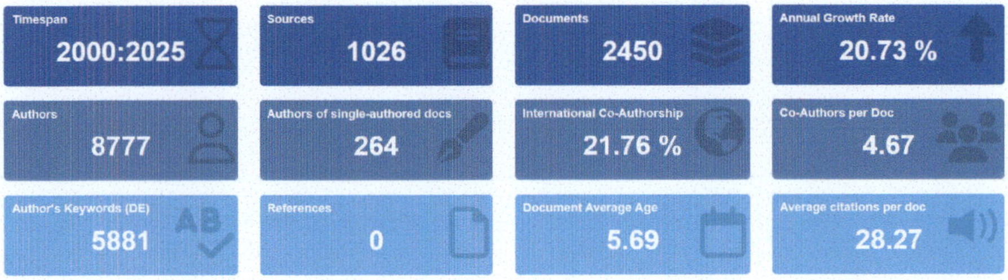

Fig. 4.2 Main information

items, 336 reviews, 4 review–book chapters, and 8 review–early access publications (Fig. 4.2).

Annual Scientific Production

The temporal distribution of publications reveals a *clear and sustained increase* in scholarly output on mental health in the context of climate change-induced extreme weather events over the study period (2000–2025). During the early phase (2000–2005), annual publication counts remained relatively low, ranging from 3 to 16 articles per year, reflecting the *nascent stage* of research where psychological dimensions of climate change were only beginning to receive academic attention (Fig. 4.3). A gradual upward trend becomes evident between 2006 and 2014, with annual outputs increasing from 23 articles in 2006 to 46 articles in 2014. This phase represents a period of *conceptual consolidation*, during which climate change began to be increasingly framed as a multidimensional risk with significant implications for mental health and psychological wellbeing.

From 2015 onwards, the field entered a phase of *accelerated growth.* Annual publications rose sharply from 55 articles in 2015 to 112 articles in 2018, indicating growing interdisciplinary engagement and heightened research interest. This expansion intensified after 2019, coinciding with increased global exposure to extreme weather events and broader recognition of climate-related psychological stressors. Publication output increased from 131 articles in 2019 to 207 in 2021, followed by 241 in 2022 and 249 in 2023. The most pronounced surge is observed in the most recent years, with 380 articles in 2024 and 444 articles in 2025, reflecting a *rapid escalation in research activity*. This sharp rise underscores the growing urgency of addressing mental health impacts within climate change discourse and highlights the emergence of this topic as a *central research priority* across disciplines.

The annual scientific production pattern demonstrates a transition from an *emerging research area* in the early 2000s to a *highly active and rapidly expanding field* in the 2020s. The observed trajectory confirms that mental health has become a critical component of climate change and extreme weather research, warranting sustained scholarly attention and systematic synthesis.

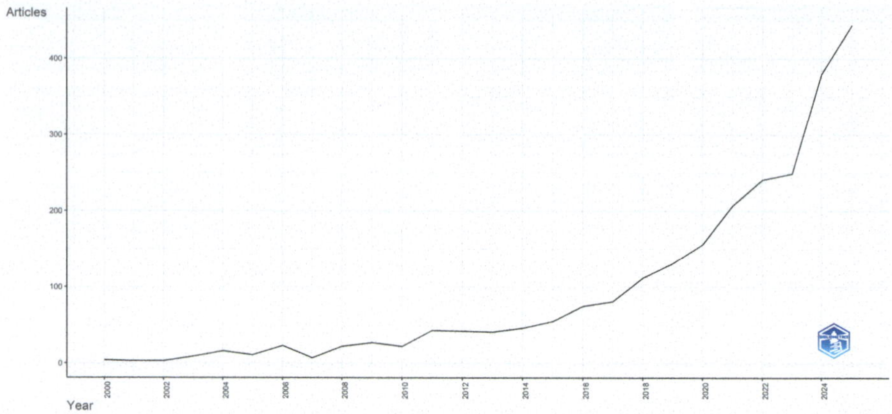

Fig. 4.3 Annual scientific production (all figures in this paper, except Fig. 4.7 has been prepared using Biblioshiny (Aria and Cuccurullo, 2017))

Average Citations per Year

The analysis of average citations per article reveals a clear temporal pattern in the scholarly impact of research on mental health in the context of climate change-induced extreme weather events. Earlier publications (2000–2002) exhibit notably high mean citation values, reflecting the influence of a small number of foundational studies that have accumulated citations over a longer period.

Between 2003 and 2015, average citation rates remain relatively stable, with several peaks indicating sustained academic attention to key contributions during the consolidation phase of the field. From 2016 onwards, a gradual decline in mean citations per article is observed, coinciding with a sharp increase in annual publication output.

The lowest average citation values occur in the most recent years (2022–2025), largely due to citation lag effects associated with newly published studies. The pattern reflects a transition from early high-impact contributions to a rapidly expanding body of recent literature that is still in the process of accumulating citations.

Most Relevant Sources

The analysis of source productivity identifies a set of core journals that have contributed substantially to the literature on mental health in the context of climate change-induced extreme weather events. The International Journal of Environmental Research and Public Health emerges as the most prolific outlet, with 96 publications, highlighting its central role in disseminating interdisciplinary research linking environmental change, public health, and psychological wellbeing. The International Journal of Disaster Risk Reduction follows with 47 publications, reflecting the strong emphasis on disaster-related mental health impacts within risk reduction and resilience frameworks. Science of the Total Environment (34 publications) and Sustainability (30 publications) further underscores the integration of psychological dimensions into broader environmental and sustainability-focused research.

Journals such as Climatic Change and PLOS ONE, each contributing 28 publications, indicate the growing recognition of mental health outcomes within mainstream climate change and multidisciplinary scientific discourse. Natural Hazards (25 publications) and BMC Public Health (22 publications) emphasize the relevance of psychological impacts in hazard-focused and population health research, respectively. The Scientific Reports (21 publications) and the Journal of Environmental Psychology (20 publications) represent key platforms for interdisciplinary and psychology-specific contributions. While environmental and public health journals dominate in terms of output, the presence of a dedicated environmental psychology journal among the most relevant sources reflects the increasing theoretical and empirical engagement of psychology with climate change and extreme weather research. The distribution of publications across these leading journals highlights the interdisciplinary dispersion of the field, with significant contributions from environmental sciences, disaster studies, public health, and psychological research domains.

Core Sources Identified Using Bradford's Law

The application of Bradford's Law of Scattering reveals a clearly defined set of *core journals* that account for a disproportionately large share of publications on mental health in the context of climate change-induced extreme weather events. As illustrated in the Bradford distribution, a small number of highly productive sources contribute the majority of articles, followed by a long tail of journals with progressively lower output (Bradford 1934). The core zone is dominated by interdisciplinary and applied journals, notably the *International Journal of Environmental Research and Public Health*, *International Journal of Disaster Risk Reduction*, *Science of the Total Environment*, *Sustainability*, and *Climatic Change* (Fig. 4.4). These sources collectively form the intellectual nucleus of the field, serving as primary platforms for disseminating research that integrates environmental change, disaster risk, public health, and psychological outcomes.

Beyond the core, the distribution shows a rapid decline in article counts across successive source ranks, consistent with Bradford's theoretical expectation. This pattern indicates that while a small set of journals functions as the *central communication channels*, a much larger number of peripheral journals contribute sporadically to the field. Such dispersion reflects the interdisciplinary breadth of climate–mental health research, spanning environmental sciences, psychology, public health, and hazard studies. The Bradford's Law analysis confirms that the literature is structured around a concentrated core of high-impact sources, accompanied by extensive

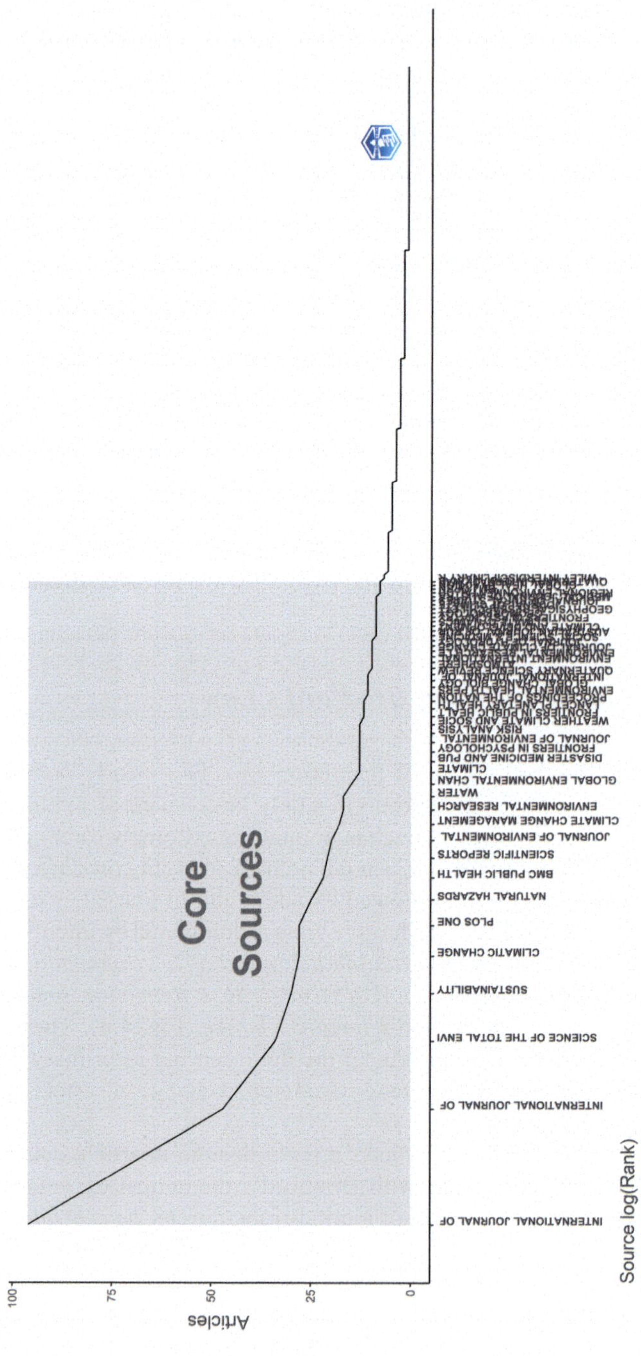

Fig. 4.4 Core sources identified by Bradford's law

peripheral scattering. This structure underscores both the consolidation of key publication venues and the wide disciplinary reach of research addressing psychological impacts of climate change-induced extreme weather events.

Most Relevant Authors

The analysis of author productivity highlights a group of key contributors who have played a significant role in shaping research on mental health in the context of climate change-induced extreme weather events. Zhang Y emerges as the most prolific author, with 35 publications, indicating sustained and high-level engagement with the field. When accounting for collaboration through fractionalized authorship, Zhang Y also maintains the highest adjusted contribution (5.05), reinforcing their central role within the research network.

Berry H follows with 21 publications and a substantial fractionalized count of 4.46, reflecting consistent scholarly influence, often within collaborative and interdisciplinary contexts. Wang Y (17 articles; 3.06 fractionalized) and Liu Y (15 articles; 2.76 fractionalized) further represent prominent contributors, particularly within climate–health and population-level mental health research.

Several authors with similar publication count Li Y, Li J, Li X, Liu J, and Zhang demonstrate the presence of *active research clusters*, especially among scholars working in closely related thematic areas. Their fractionalized contributions indicate varying degrees of collaboration, suggesting both leadership roles and participation in larger research teams. Ebi K, with 12 publications and a fractionalized count of 2.24, stands out as a key contributor linking climate science with health risk assessment and policy-relevant research. The distribution of prolific authors reflects a collaborative and internationally distributed authorship structure, where productivity is shared across interconnected research groups rather than dominated by a single individual. The use of fractionalized authorship provides a more nuanced understanding of individual contributions within this highly collaborative research domain.

Author Productivity Pattern Based on Lotka's Law

The distribution of author productivity in this study conforms closely to Lotka's Law (Lotka 1926), indicating a *highly skewed authorship structure* within research on mental health and climate change-induced extreme weather events. As illustrated by the Lotka distribution, the majority of authors have contributed only a single publication, while a progressively smaller proportion of authors account for multiple publications. The steep decline observed after the first publication category demonstrates that a large base of occasional contributors supports the field, whereas a small

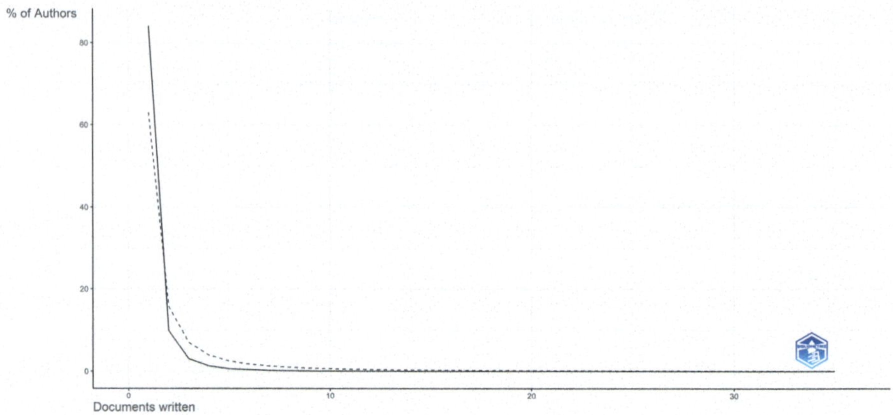

Fig. 4.5 Author productivity distribution following Lotka's law

core group of highly productive authors is responsible for a substantial share of the total scientific output (Fig. 4.5).

The close alignment between the observed productivity curve and the theoretical Lotka distribution suggests that author contribution patterns in this field follow established bibliometric regularities. This conformity indicates structural stability in knowledge production, despite the rapid growth of the literature. The Lotka's Law analysis confirms that research on climate change related extreme weather events and mental health is driven by a limited number of prolific authors, supported by a wide network of researchers contributing intermittently.

Most Relevant Affiliation

The analysis of institutional contributions reveals a concentration of research output among leading universities and national research systems, underscoring the strong academic foundation of studies on mental health in the context of climate change-induced extreme weather events. The University of California System emerges as the most productive affiliation, contributing 114 publications. The Chinese Academy of Sciences ranks second with 65 publications, highlighting China's growing prominence in climate–health research. The University of London (64 publications) and Harvard University (61 publications) further represent major hubs of research activity, demonstrating strong integration of climate science, public health, and psychological research.

Australian institutions feature prominently, with the Australian National University and Monash University each contributing 57 publications, followed closely by the University of Melbourne (56 publications) and the University of Sydney (54 publications). This strong representation underscores Australia's active research engagement, likely influenced by the country's exposure to climate extremes such as heatwaves, droughts, and bushfires. In the United States, additional leading contributors include the University of Washington (51 publications) and Stanford University (44 publications), both of which are recognized for interdisciplinary research at the intersection of environmental science, public health, and psychology. The distribution of affiliations indicates that research in this field is institutionally concentrated within globally influential research organizations, particularly in North America, Europe, Australia, and East Asia. This pattern reflects the availability of research infrastructure, funding capacity, and interdisciplinary expertise necessary to address the complex mental health implications of climate change-induced extreme weather events.

Countries' Scientific Production

The country-level analysis of scientific output reveals a geographically uneven distribution of research on mental health in the context of climate change-induced extreme weather events, with contributions concentrated in a limited number of high-income and research-intensive nations. The United States dominates the field, contributing 1690 publications, underscoring its central role in advancing interdisciplinary research linking climate science, public health, and psychology. Australia ranks second with 746 publications, reflecting strong national research engagement, likely influenced by recurrent exposure to climate extremes such as heatwaves, droughts, bushfires, and floods. China follows with 617 publications, indicating its rapidly expanding research capacity and growing emphasis on climate-related health and psychological outcomes.

European countries also make substantial contributions. The United Kingdom accounts for 433 publications, followed by Germany (324) and Canada (323), highlighting the prominence of these countries in climate–health and disaster-related mental health research. Italy (146), France (127), and Spain (110) further demonstrate active participation within the European research landscape. Among emerging economies, India contributes 158 publications, reflecting increasing scholarly attention to the mental health impacts of climate variability and extreme weather events in highly vulnerable socio-environmental contexts.

The spatial distribution of scientific production indicates that research in this domain is predominantly driven by countries with strong research infrastructure and funding capacity, while contributions from climate-vulnerable regions in the Global South remain comparatively limited. This imbalance highlights the need for greater inclusion of region-specific perspectives and locally grounded mental health research in future climate change studies.

Corresponding Author's Countries and International Collaboration Patterns

The analysis of corresponding authors' countries reveals pronounced differences in research leadership and collaboration structures within the field of *mental health in the context of climate change-induced extreme weather events*. The United States emerges as the dominant contributor, with 545 articles, clearly leading global research output. This is followed by Australia (261 articles) and China (209 articles), forming a second tier of highly productive countries (Fig. 4.6). Other major contributors include the United Kingdom (155 articles), Germany (115 articles), and Canada (109 articles), while India (80 articles) represents a notable contributor among emerging economies. The corresponding author productivity is concentrated in a small group of research-intensive nations.

International collaboration patterns, measured using the Multiple Country Publication (MCP) ratio, show substantial cross-national variation. Countries such as Switzerland (78.9%), Sweden (54.5%), Bangladesh (46.2%), and the Netherlands (37.1%) exhibit very high MCP ratios, indicating a strong reliance on international partnerships despite relatively lower total output. These countries function as *highly connected nodes* within the global research network. A second group which including the United Kingdom, China, Germany, France, Japan, Spain, and South Africa demonstrates moderate to high international engagement, with MCP ratios close to or above 30%, reflecting a balanced combination of domestic research capacity and global collaboration. In contrast, countries such as the USA (15.4%), Australia (19.9%), and Canada (20.2%) show lower MCP ratios, indicating that a substantial proportion of their research is conducted within national boundaries, supported by strong domestic research ecosystems.

Lower MCP ratios are observed for India, Brazil, Iran, Korea, and Pakistan, suggesting a comparatively greater emphasis on single-country publications. This

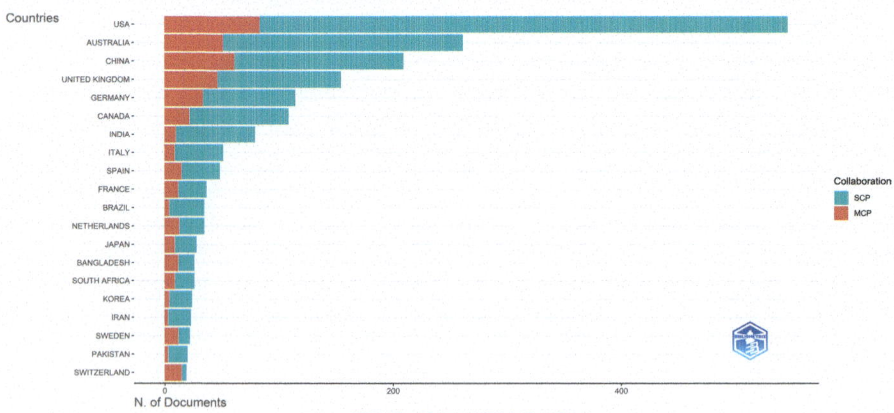

Fig. 4.6 Corresponding authors' countries

pattern may reflect national research priorities, developing international networks, or regionally focused collaboration structures. The findings indicate that research volume and international collaboration are not directly proportional. While countries such as the United States dominate in terms of output with a predominantly domestic orientation, others which is particularly smaller or highly specialized research systems demonstrate strong international connectivity. These patterns highlight diverse national strategies in knowledge production and underscore the importance of global collaboration for advancing research on the psychological impacts of climate change-induced extreme weather events.

Most Cited Countries

The analysis of citation performance at the country level highlights notable differences in research impact within the field of mental health in the context of climate change-induced extreme weather events. The United States records the highest total citations (18,818) (Fig. 4.7), reflecting both its large volume of publications and its sustained influence on the global research landscape. The average citation rate of 34.50 per article further indicates strong scholarly visibility. The United Kingdom and Australia demonstrate particularly high citation impact relative to output. The United Kingdom records 7721 total citations with the highest average citations per article (49.80), suggesting a concentration of highly influential studies. Australia, with 11,031 total citations and an average of 42.30 citations per article, similarly exhibits strong research visibility, likely driven by impactful work on climate extremes and mental health outcomes.

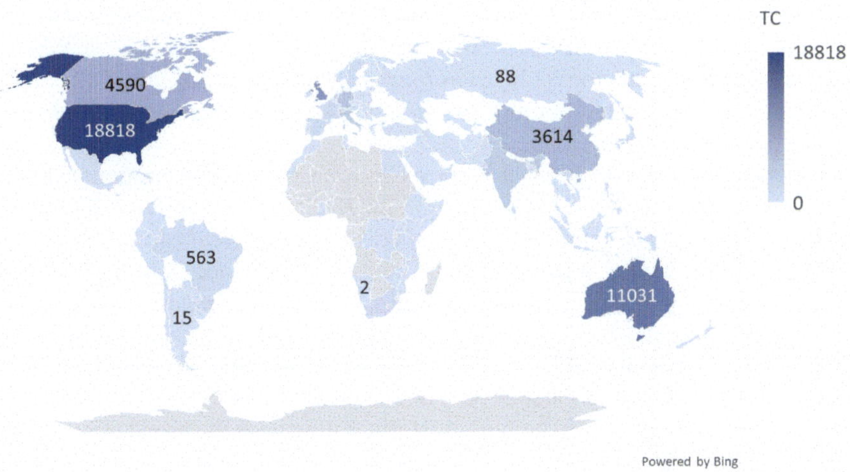

Fig. 4.7 Most cited countries

Canada (4590 total citations; 42.10 average citations) and Italy (2366 total citations; 45.50 average citations) also show high citation intensity, indicating that their contributions, while smaller in volume, have achieved substantial academic impact. In contrast, China, despite a relatively high publication output, records a lower average citation rate (17.30), suggesting more recent or less internationally cited contributions. Germany (26.50 average citations) and India (28.60 average citations) demonstrate moderate citation impact, reflecting steady but less concentrated influence.

Spain and the Netherlands exhibit comparatively lower total citation counts, though the Netherlands maintains a moderate average citation rate (29.30), indicating consistent scholarly relevance. The citation patterns reveal that research impact is not solely dependent on publication volume, but is strongly influenced by the visibility, maturity, and international reach of national research outputs. Countries such as the United Kingdom, Australia, Canada, and Italy demonstrate particularly high influence per article within the global climate–mental health research landscape.

Most Cited Papers

The citation analysis identifies a set of highly influential publications that have played a foundational role in shaping research on mental health and climate change-induced extreme weather events. The most cited paper is Reid C (2016) published in *Environmental Health Perspectives*, with 967 total citations, underscoring its seminal contribution to understanding climate-related health risks and their broader implications. Closely following is Goudie A (2001) in *Earth Science Reviews* (953 citations), representing an early and widely referenced work that provided critical conceptual grounding for climate–environment interactions. The strong citation performance of more recent publications, such as Cianconi P (2020) in *Frontiers in Psychiatry* (719 citations) and Clayton S (2020) in *Journal of Anxiety Disorders* (707 citations), reflects the growing integration of *psychological and psychiatric perspectives* into climate change research.

Papers by Cunsolo A (2018) in *Nature Climate Change* (718 citations) and Berry H (2010) in *International Journal of Public Health* (619 citations) highlight the emergence and consolidation of *climate-specific psychological constructs*, including ecological grief, mental wellbeing, and community resilience. Similarly, Rocque R (2021) in *BMJ Open* (685 citations) and Ebi K (2021) in *Annual Review of Public Health* (650 citations) emphasize the policy-relevant and public health dimensions of climate-related mental health impacts. The presence of interdisciplinary and authoritative sources, including Parsons K (2002) (668 citations) and Roxy M (2017) in *Nature Communications* (563 citations), further illustrates the broad scientific influence of works linking climate extremes, environmental stressors, and human health outcomes.

The most cited papers reveal that the intellectual foundation of this research field is shaped by interdisciplinary, high-impact studies published in leading environmental, public health, psychology, and climate science journals. These works have collectively advanced conceptual understanding, empirical evidence, and policy relevance, serving as key reference points for subsequent research on the mental health dimensions of climate change.

A Frequency Analysis of Key Words

The frequency analysis of keywords highlights the dominant thematic focus of the literature on *mental health in the context of climate change–induced extreme weather events*. The term *"climate change"* emerges as the most frequently occurring word (1683 occurrences), confirming its role as the *central conceptual anchor* of the research field (Fig. 4.8). The prominence of *"mental health"* (396 occurrences) underscores the explicit integration of psychological and psychiatric dimensions within climate-related research. This indicates a clear shift from purely environmental or physical health perspectives toward a more holistic understanding of climate impacts on human wellbeing.

Words such as *"human"* (306 occurrences) and *"humans"* (233 occurrences) emphasize the *anthropocentric orientation* of the literature, highlighting a strong focus on human exposure, experience, and response to climate-related stressors. The frequent appearance of *"drought"* (274 occurrences) reflects the significance of slow-onset and recurrent climate extremes as critical stressors influencing mental health outcomes. The occurrence of *"adaptation"* (243 occurrences) points to a growing emphasis on coping strategies, resilience-building, and policy-relevant responses to climate-related psychological risks. Similarly, the presence of *"vulnerability"* (197 occurrences) and *"risk"* (175 occurrences) indicates sustained attention to differential susceptibility across populations and contexts.

Fig. 4.8 Keywords

The repeated use of *"health"* (187 occurrences) further reinforces the interdisciplinary overlap between climate science, public health, and psychology. The term *"article"* (207 occurrences) reflects indexing characteristics rather than thematic content, but its frequency is typical in bibliographic datasets. The most frequent words reveal that the literature is strongly centered on climate change, mental health, human vulnerability, and adaptation, confirming the field's orientation toward understanding and addressing the psychological consequences of climate-related risks and extremes.

Collaboration Network of Keywords

The keyword co-occurrence network reveals a highly structured and thematically concentrated research domain, dominated by a small number of central concepts. The largest and most influential nodes are *climate change* and *mental health*, indicating both their high frequency and their central integrative role in the literature (Fig. 4.9). The strong linkage between these two concepts confirms that the *climate mental health nexus* constitutes the core analytical focus of the field. The prominence of the terms *human* and *humans* further underscores the anthropocentric orientation of the research, emphasizing human exposure, experience, and response to climate-related stressors. The overall structure of the network is modular yet highly interconnected, reflecting the interdisciplinary integration of psychology, public health, environmental science, and disaster research rather than isolated thematic silos.

Within this structure, three major thematic communities can be identified. The largest and most central community focuses on *climate change, mental health impacts, and adaptation*. This cluster captures the core intellectual foundation of the field by linking *psychological outcomes* such as *anxiety, depression, stress, posttraumatic stress disorder, health, and mortality* with *climate hazards* including *extreme weather, drought, heat waves, global warming, air pollution, temperature,*

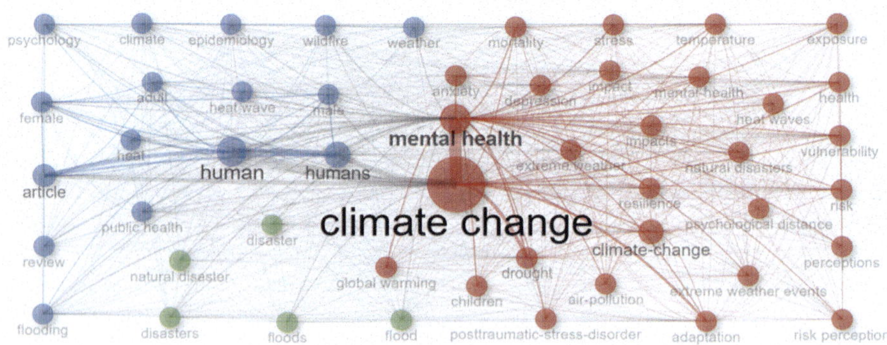

Fig. 4.9 Co-occurrence network of keywords

and natural disasters. It also incorporates concepts related to *vulnerability, resilience, adaptation, risk perception, exposure, and children*, highlighting how mental health outcomes are shaped by differential susceptibility, perception, and adaptive capacity. This community represents the primary research thrust examining how climate change and extreme events translate into psychological and health consequences.

A second thematic community centers on human dimensions and public health-oriented approaches. This cluster emphasizes *population-level analysis* through frequent use of demographic terms such as *human, humans, male, female, and adult*, alongside *disciplinary and methodological descriptors* including *public health, psychology, epidemiology, review, and article*. Environmental stressors such as *heat, heat wave, wildfire, weather, and flooding* are also prominent, indicating that this body of research applies public health and epidemiological frameworks to understand the distribution, prevalence, and risk factors of climate-related mental health impacts. This community complements the impact focused climate mental health cluster by providing methodological and population based perspectives.

The third and comparatively smaller community is focused on *disaster-related research*, with particular emphasis on *floods and flooding events*. Although more specialized in scope, this cluster maintains strong conceptual connections with the broader climate change and mental health literature, indicating that disaster specific studies contribute directly to the wider understanding of climate-related psychological impacts. The co-occurrence network demonstrates that research on climate change-induced extreme weather events and mental health is conceptually cohesive yet interdisciplinarily diverse, structured around a strong core linking *climate change* and *mental health*, and supported by interconnected public health and disaster focused research streams that together reflect a mature and integrated knowledge base.

Thematic Map

The strategic map provides a structured overview of the thematic organization of research on climate change and mental health by positioning themes according to their relevance degree (centrality) and development degree (density). Centrality reflects how strongly a theme is connected to other themes within the knowledge network, indicating its overall importance and integrative role, while density represents the internal coherence and maturity of research within that theme. Together, these dimensions divide the thematic space into motor, niche, basic, and emerging or declining themes, allowing a systematic interpretation of the intellectual structure of the field.

The cluster labeled *climate change mental health human* occupies the upper right quadrant, identifying it as a motor theme with both high centrality and high density. This positioning indicates that the intersection of climate change, mental health, and human dimensions represents the most influential and well-developed research area within the field. The strong internal cohesion of this cluster reflects a mature and actively evolving body of literature, while its high connectivity demonstrates that it serves as a central driver linking multiple research streams. This theme therefore

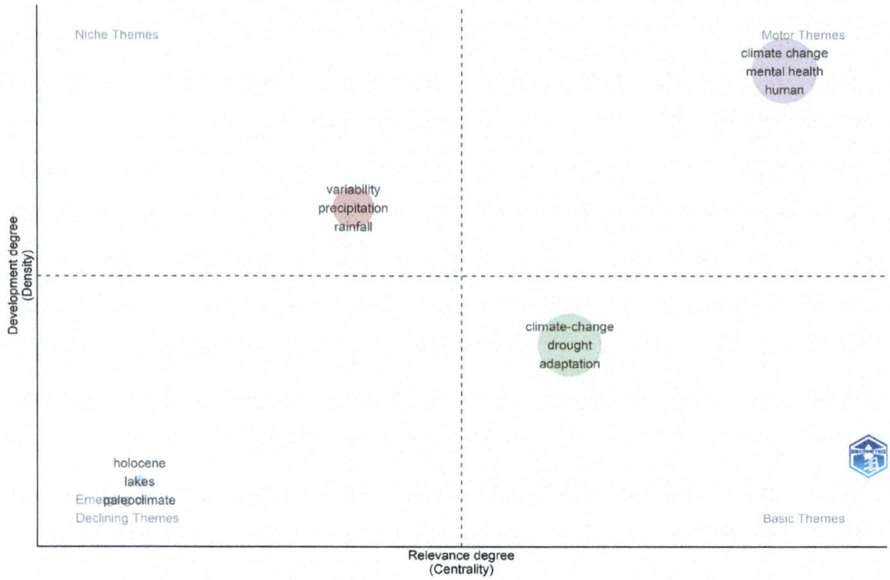

Fig. 4.10 Strategic visualization of the intellectual structure by plotting the main themes based on two key metrics: centrality and density

constitutes the primary intellectual core of the domain and plays a decisive role in shaping contemporary research directions (Fig. 4.10).

The cluster *variability precipitation rainfall* is located in the upper left quadrant, classifying it as a niche theme. Although this theme exhibits high internal density, indicating a well-developed and specialized research focus, its low centrality suggests limited integration with the broader thematic structure of the field. This pattern reflects a concentrated body of work addressing specific climatic processes and hydrometeorological variability, often pursued by specialized research communities. While methodologically and scientifically robust, this theme remains relatively peripheral to the dominant human and health-oriented research agenda.

The theme *climate change drought adaptation* appears in the lower right quadrant, corresponding to a basic theme with high centrality but low density. This indicates that it is a foundational and widely connected concept across the research landscape, underpinning multiple thematic areas. However, the relatively low internal cohesion suggests that research within this theme is diverse and fragmented, encompassing a broad range of perspectives and approaches rather than a tightly unified research front. As such, this theme functions as a conceptual backbone of the field, supporting both impacts focused and applied research without forming a highly specialized core itself.

The cluster *holocene lakes paleoclimate* is positioned in the lower left quadrant, identifying it as an emerging or declining theme. Its low centrality and low density indicate weak integration with other research themes and limited internal development. This pattern suggests that paleoclimatic and historical climate studies play a relatively marginal role within the current research landscape, which is increasingly oriented toward contemporary impacts, human vulnerability, and adaptation. While this theme may retain relevance for long-term climate understanding, it is not a major driver of present research activity within this domain.

The strategic map reveals a research landscape strongly driven by the human and mental health dimensions of climate change, with this motor theme anchoring the field both conceptually and structurally. Foundational themes related to climate change, drought, and adaptation provide essential contextual support, while specialized climatic variability research contributes depth without broad integration. Peripheral paleoclimate themes reflect either historical focus or declining relevance. Collectively, these patterns indicate a mature and human-centered research domain with clear priorities oriented toward health impacts, vulnerability, and adaptive responses to climate change.

Thematic Evolution

The thematic evolution analysis, based on merged author and index keywords using non stemmed single word terms, reveals a clear and systematic transformation in the intellectual structure of the research field over the period from 2000 to 2025. The Sankey-based topic flow demonstrates that *climate change* remains the most persistent and dominant theme across all periods, functioning as an overarching conceptual foundation. Over time, this broad framing increasingly branches into more specific and applied research directions. A notable trend is the progressive rise of *human* and *humans* as central concepts, particularly after 2010 (Fig. 4.11), indicating a strong shift toward anthropocentric perspectives that prioritize human exposure, vulnerability, and response to climate-related phenomena.

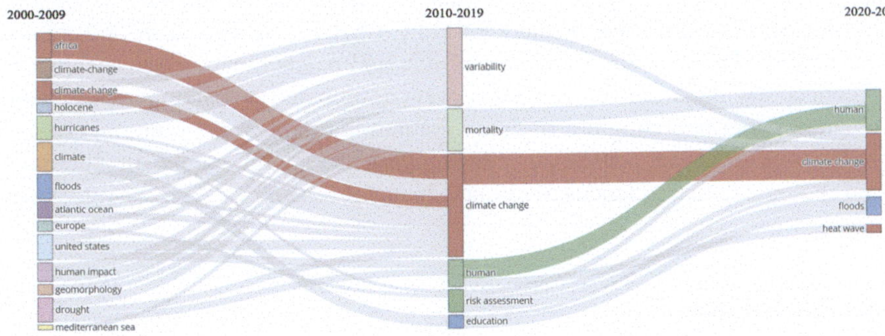

Fig. 4.11 The thematic evolution analysis provides a longitudinal view of research themes evolved over time

During the early phase from 2000 to 2009, the research landscape was characterized by a wide range of geographically and physically oriented themes. Motor themes in this period included *drought, flooding, storms, weather*, and *climate change*, often linked with broad regional or environmental contexts such as oceans and continental scales. At the same time, several niche themes related to geomorphology, paleoclimate, and atmospheric processes reflected a strong descriptive and Earth science orientation. Concepts such as *risk, global warming*, and *human impact* appeared as emerging themes, suggesting that human-centered and impact focused research was still in a formative stage.

The period from 2010 to 2019 marks a significant consolidation and reorientation of the field. Thematic maps show *human* and *humans* emerging as the most dominant motor themes, indicating that human dimensions had become the primary drivers of research development. *Climate change* transitions into a foundational theme with high centrality, while *adaptation* emerges as a key basic concept, reflecting growing concern with responses and coping strategies. Niche themes such as *mortality, floods, risk assessment*, and *policy* signal a movement toward applied, management oriented, and decision relevant research. In contrast, earlier broad physical descriptors such as hurricanes and global warming decline in prominence, suggesting their integration into more comprehensive climate impact frameworks.

In the most recent period from 2020 to 2025, the thematic structure becomes more focused and mature. *Human* and *humans* remain the strongest motor themes, firmly establishing the anthropocentric core of the field. *Climate change* continues as a highly central foundational concept, while *mental health* emerges as a major basic theme, marking a decisive shift toward psychological and wellbeing related impacts of climate change. The appearance of *heat wave* and *heat* as niche themes reflects intensified attention to extreme heat as a critical and specialized climate hazard, while the emergence of *China* indicates growing geographically focused research linked to high exposure and vulnerability contexts.

The thematic evolution demonstrates a clear progression from broadly descriptive and geographically oriented climate research toward a highly integrated, human centered, and impact driven knowledge structure. The increasing prominence of *adaptation, policy, risk assessment, heat wave*, and especially *mental health* indicates a maturing field that prioritizes societal relevance, health consequences, and actionable responses. The decline of earlier physical and paleoclimatic themes suggests either conceptual integration or reduced relevance compared to contemporary climate risks. These patterns highlight a research domain that has evolved into a cohesive and policy-relevant field focused on understanding and addressing the direct human and psychological consequences of climate change.

Factorial Analysis

The factorial analysis reveals a clearly structured intellectual landscape that is strongly shaped by two principal dimensions. *Dimension 1* explains an overwhelming

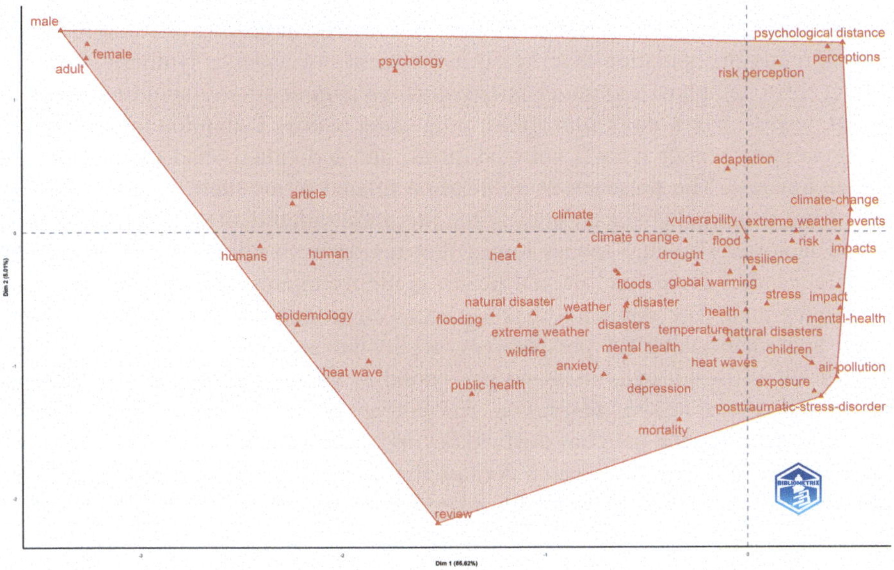

Fig. 4.12 The factorial analysis performed using Multiple Correspondence Analysis (MCA)

85.62 percent of the total variance, indicating that it represents the primary organizing axis of the research field. Along this dimension, a clear conceptual separation is evident between *environmental and climate-related phenomena* on the positive side and *human centered, health, and methodological perspectives* on the negative side (Fig. 4.12). This dominant axis therefore reflects the fundamental divide between climate drivers and impacts on one hand and human responses, experiences, and analytical approaches on the other. *Dimension 2* explains a smaller but meaningful 5.01% of the variance and provides a secondary layer of differentiation, refining the broader thematic separation established by Dimension 1.

On the climate-oriented side of the map, several closely connected clusters emerge that collectively define the environmental core of the field. A central cluster is dominated by *climate change, global warming, floods, flooding, heat, wildfire, weather, and temperature*, indicating a strong and cohesive focus on climate processes and extreme weather events. Closely linked to this cluster is a group of terms associated with *health outcomes and vulnerability*, including *mental health, stress, exposure, and mortality*, highlighting the direct pathways through which climatic stressors translate into adverse human health consequences. A further climate adjacent cluster emphasizes *risk perception, perceptions, adaptation, and risk*, reflecting growing attention to how societies understand, interpret, and respond to climate threats, and how adaptive strategies are conceptualized within the research literature.

On the human-centered side of the factorial map, the thematic structure reflects disciplinary and population focused approaches. One prominent cluster is oriented toward *public health and environmental epidemiology*, characterized by terms such as *epidemiology, heat wave, public health, and review*, indicating a strong methodological focus on population-level health assessment and evidence synthesis. Another distinct cluster emphasizes *human and psychological dimensions*, including *humans, human, female, psychology, and stress*, suggesting research attention to psychological processes, gender related vulnerabilities, and individual level experiences of climate impacts. The presence of publication related terms such as *article* further reflects the structural characteristics of the literature rather than substantive themes.

The factorial structure demonstrates that research on climate change-induced extreme weather events and mental health is highly interdisciplinary and conceptually integrated. The dominant axis separates environmental drivers from human health and methodological perspectives, while the secondary axis distinguishes between direct health outcomes and perceptual or adaptive responses. The strong presence of *mental health, adaptation, risk perception, public health, and epidemiology* confirms that the field has evolved beyond descriptive climate science toward a mature, human-centered research domain that integrates environmental processes with psychological, social, and health-related consequences.

Conclusion

This bibliometric analysis provides a comprehensive overview of the global research landscape on mental health in the context of climate change-induced extreme weather events, revealing a rapidly expanding, highly interdisciplinary, and increasingly human-centered field of inquiry. The temporal analysis demonstrates a clear transition from an early phase dominated by descriptive and geophysical perspectives toward a mature research domain focused on human vulnerability, psychological impacts, and adaptive responses to climate-related hazards. The sustained growth in scientific production, coupled with strong citation performance, highlights the growing academic and societal relevance of this research area.

The structural analyses based on Bradford Law, Lotka Law, co-occurrence networks, factorial analysis, and thematic evolution collectively indicate a well-defined intellectual core anchored by climate change and mental health. A small number of core journals, prolific authors, and leading institutions act as key knowledge hubs, while a broad periphery of contributors reflects the field's interdisciplinary reach. At the same time, notable geographic concentration in research output and citation impact underscores persistent regional imbalances, with high-income countries dominating knowledge production despite the disproportionate vulnerability of many regions in the Global South.

Thematic and network-based analyses reveal a decisive anthropocentric shift in research priorities. Human dimensions, public health perspectives, psychological outcomes, and adaptation strategies now occupy central positions within the knowledge structure, while earlier emphases on purely physical or historical climate processes have diminished or been conceptually integrated. The emergence of mental health as a foundational theme in recent years signals a critical evolution in climate research, positioning psychological wellbeing alongside physical health and environmental risk as a core concern. Increasing attention to extreme events such as heat waves and floods further reflects the field's responsiveness to contemporary climate realities.

Despite these advances, the findings also point to important gaps and future research needs. International collaboration remains uneven, and contributions from highly climate-vulnerable regions remain comparatively limited. Moreover, while adaptation and policy related themes are increasingly visible, stronger integration between empirical mental health research and actionable policy frameworks is still required. Addressing these gaps will be essential for translating scientific insights into effective interventions and equitable climate resilience strategies.

This study demonstrates that research on climate change-induced extreme weather events and mental health has evolved into a coherent and impactful scientific domain. By systematically mapping its growth, structure, and thematic evolution, the analysis provides valuable insights for researchers, policymakers, and funding agencies seeking to advance interdisciplinary, inclusive, and policy-relevant research at the intersection of climate change and mental health.

AI Declaration

Generative AI tools, including ChatGPT and Gemini, were used exclusively to assist in formulating and refining the keyword query structure for the bibliometric search, ensuring accuracy and completeness in identifying relevant terms. QuillBot was employed solely for grammar correction and language refinement during manuscript preparation. All methodological decisions, data extraction, bibliometric analyses, interpretations, and conclusions were independently conducted and validated by the authors. The authors take full responsibility for the integrity and accuracy of the final manuscript.

Acknowledgements This study is not funded by any organization or individual.

Funding This research did not receive any specific grant from funding agencies in the public, commercial, or not for profit sectors.

Conflict of Interest The authors declare that there are no conflicts of interest associated with this study.

Informed Consent Not applicable. This study is based exclusively on secondary data derived from published literature and publicly available bibliographic databases and does not involve human participants.

Author Credit Statement **Adarsh T:** Conceptualization, Methodology, Data curation, Formal analysis, Visualization, Writing—original draft, Writing—review & editing; **Haneena Sulthana K P:** Conceptualization, Methodology, Validation, Writing—review & editing; **Lancelet T. S:** Supervision, Methodology, Validation, Writing—review & editing; **Sandhya Aravind C. A:** Conceptualization, Interpretation, Validation, Writing—review & editing.

Data Availability Statement This study is based on bibliometric data retrieved from the Scopus and Web of Science databases. Due to licensing restrictions of the databases, the full datasets cannot be shared publicly; however, processed data and analytical outputs used in the study are available from the corresponding author upon reasonable request.

References

Adarsh T, Haneena Sulthana KP (2025) Trends and insights in nature-based solutions: bibliometric study. Int J Disaster Stud Clim Resil 1(1):22–34. https://doi.org/10.64866/j.ijdscr.2025.10003

Anantapong K, Moura H, Udomratn P, Persaud A, Javed A, Ramachandran P, Castaldelli-Maia JM, Torales J, Ventriglio A, Bhugra D (2024) Geopsychiatry: climate change, migration, and mental health. Ind Psychiatry J 33:S257–S260. https://doi.org/10.4103/ipj.ipj_33_24

Aria M, Cuccurullo C (2017) Bibliometrix: an R-tool for comprehensive science mapping analysis. J Infometrics 11(4):959–975. https://doi.org/10.1016/j.joi.2017.08.007

Berry HL, Bowen K, Kjellstrom T (2010) Climate change and mental health: a causal pathways framework. Int J Public Health 55(2):123–132. https://doi.org/10.1007/s00038-009-0112-0

Bradford SC (1934) Sources of information on specific subjects. Eng Illus Wkly J 137:85–86

Burke M, González F, Baylis P, Heft-Neal S, Baysan C, Basu S, Hsiang S (2018) Higher temperatures increase suicide rates in the United States and Mexico. Nat Clim Chang 8(8):723–729. https://doi.org/10.1038/s41558-018-0222-x

Charlson F, Ali S, Benmarhnia T, Pearl M, Massazza A, Augustinavicius J, Scott J (2021) Climate change and mental health: a scoping review. Int J Environ Res Public Health 18(9). https://doi.org/10.3390/ijerph18094486

Clayton S (2020) Climate anxiety: psychological responses to climate change. J Anxiety Disord 74. https://doi.org/10.1016/j.janxdis.2020.102263

Clemens V, von Hirschhausen E, Fegert JM (2022) Report of the intergovernmental panel on climate change: implications for the mental health policy of children and adolescents in Europe—a scoping review. Eur Child Adolesc Psychiatry 31:701–713. https://doi.org/10.1007/s00787-020-01615-3

Cunsolo A, Ellis NR (2018) Ecological grief as a mental health response to climate change-related loss. Nat Clim Chang 8(4):275–281. https://doi.org/10.1038/s41558-018-0092-2

Ebi KL, Vanos J, Baldwin JW, Bell JE, Hondula DM, Errett NA, Hayes K, Reid CE, Saha S, Spector J, Berry P (2021) Extreme weather and climate change: population health risks and responses. Annu Rev Public Health 42:293–315. https://doi.org/10.1146/annurev-publhealth-012420-105026

Gislason M, Kennedy A, Witham S (2021) The interplay between social and ecological determinants of mental health for children and youth in the climate crisis. Int J Environ Res Public Health 18(9). https://doi.org/10.3390/ijerph18094573

Goudie AS, Middleton NJ (2001) Saharan dust storms: nature and consequences. Earth Sci Rev 56(1–4):179–204. https://doi.org/10.1016/S0012-8252(01)00067-8

Grennan G, Withers M, Ramanathan D, Mishra J (2023) Differences in interference processing and frontal brain function with climate trauma from California's deadliest wildfire. PLoS Clim 2(1). https://doi.org/10.1371/journal.pclm.0000125

Hamidah I, Pawinanto R, Mulyanti B, Yunas J (2021) A bibliometric analysis of micro electro mechanical system energy harvester research. Heliyon 7(3). https://doi.org/10.1016/j.heliyon.2021.e06406

Hayes K, Poland B (2018) Addressing mental health in a changing climate: incorporating mental health indicators into climate change and health vulnerability and adaptation assessments. Int J Environ Res Public Health 15(9). https://doi.org/10.3390/ijerph15091806

Hayes K, Blashki G, Wiseman J, Burke S, Reifels L (2018) Climate change and mental health: risks, impacts and priority actions. Int J Ment Heal Syst 12(1). https://doi.org/10.1186/s13033-018-0210-6

IPCC (2021) Climate change 2021: the physical science basis. Intergovernmental Panel on Climate Change. Cambridge University Press

IPCC (2022) Climate change 2022: impacts, adaptation and vulnerability. Intergovernmental Panel on Climate Change. Cambridge University Press

Kalsoom G, Zámečník R (2023) Servant leadership knowledge structures: a bibliometric review. ECMLG 19(1):471–479. https://doi.org/10.34190/ecmlg.19.1.1674

Kelman I (2020) Disaster by choice: how our actions turn natural hazards into catastrophes. Oxford University Press

Léger-Goodes T, Malboeuf-Hurtubise C, Mastine T, Généreux M, Paradis P, Camden C (2022) Eco-anxiety in children: a scoping review of the mental health impacts of the awareness of climate change. Front Psychol 13. https://doi.org/10.3389/fpsyg.2022.872544

Lotka AJ (1926) The frequency distribution of scientific productivity. J Wash Acad Sci 16(12):317–323

Mishra J, Ramanathan V (2023) Harnessing the mental imprints of climate change for collective climate action. PLoS Clim 2(6). https://doi.org/10.1371/journal.pclm.0000233

Palinkas LA, Wong M (2020) Global climate change and mental health. Curr Opin Psychol 32:12–16. https://doi.org/10.1016/j.copsyc.2019.06.023

Parsons K (2002) Human thermal environments: the effects of hot, moderate, and cold environments on human health, comfort, and performance, 3rd edn. CRC Press. https://doi.org/10.1201/b16750

Power E, McCarthy N, Kelly I, Cannon M, Cotter D (2022) Climate change and mental health: time for action and advocacy. Ir J Psychol Med 40(1):6–8. https://doi.org/10.1017/ipm.2021.70

Reid CE, Brauer M, Johnston FH, Jerrett M, Balmes JR, Elliott CT (2016) Critical review of health impacts of wildfire smoke exposure. Environ Health Perspect 124(9):1334–1343. https://doi.org/10.1289/ehp.1409277

Reyes M, Carmen B, Luminarias M, Mangulabnan S, Ogunbode C (2021) An investigation into the relationship between climate change anxiety and mental health among Gen Z Filipinos. Curr Psychol 42(9):7448–7456. https://doi.org/10.1007/s12144-021-02099-3

Rocque RJ, Beaudoin C, Ndjaboue R, Cameron L, Poirier-Bergeron L, Poulin-Rheault RA, Fallon C, Tricco AC, Witteman HO (2021) Health effects of climate change: an overview of systematic reviews. BMJ Open 11(6). https://doi.org/10.1136/bmjopen-2020-046333

Roxy MK, Ghosh S, Pathak A, Athulya R, Mujumdar M, Murtugudde R, Terray P, Rajeevan M (2017) A threefold rise in widespread extreme rain events over Central India. Nat Commun 8. https://doi.org/10.1038/s41467-017-00744-9

Saltzman L, Hansel T, Ferreira R (2023) Climate change and health equity: expanding our reach using technology. PLoS Clim 2(7). https://doi.org/10.1371/journal.pclm.0000242

Shoib S, Hussaini S, Armiya'u A, Saeed F, Őri D, Roza T et al (2023) Prevention of suicides associated with global warming: perspectives from early career psychiatrists. Front Psych 14. https://doi.org/10.3389/fpsyt.2023.1251630

Wang J, Wang S, Li Q, Liu F, Wan Y, Liang H (2024) Bibliometric and visual analysis of single-cell multiomics in neurodegenerative disease arrest studies. Front Neurol 15. https://doi.org/10.3389/fneur.2024.1450663

World Health Organization (2022) World mental health report: transforming mental health for all. WHO Press

Part II
Health, Morbidity and Mortality in the Context of Climate Extremes

Chapter 5
The Double Burden: Climate Extremes and the Escalation of Non-communicable Diseases

Aida Soraya Shamsuddin, Maisarah Abdul Mutalib, Nor Hafizah Zakaria, Nurulain Mustafa Udin, Nor Eliani Ezani, Armania Nurdin, Haninah Fadiah Mustafa, Sharifah Norkhadijah Syed Ismail, Azyyati Ridha Alfian, Mohd Roslan Rahmat, and Fadzilah Adibah Abdul Majid

Abstract Climate change is increasingly recognised as a significant determinant of population health, yet its implications for non-communicable diseases (NCDs) remain unevenly conceptualised across disciplines and regions. While growing evidence links climate extremes to cardiovascular, respiratory, metabolic, and mental

---

A. S. Shamsuddin (✉) · A. R. Alfian
Institute for Environment and Development (LESTARI), Universiti Kebangsaan Malaysia (UKM), Bangi, Selangor, Malaysia
e-mail: aidasoraya@ukm.edu.my

A. R. Alfian
e-mail: azyyatiridhaalfian@ph.unand.ac.id

M. Abdul Mutalib
School of Graduate Studies, Management and Science University (MSU), Shah Alam, Selangor, Malaysia
e-mail: maisarah_abdulmutalib@msu.edu.my

N. H. Zakaria · F. A. Abdul Majid
Institute of Climate Adaptation and Marine Biotechnology (ICAMB), Universiti Malaysia Terengganu (UMT), Kuala Nerus, Terengganu, Malaysia
e-mail: hafizah.zakaria@umt.edu.my

F. A. Abdul Majid
e-mail: f.adibah@umt.edu.my

N. Mustafa Udin
School of Chemistry and Environment, Faculty of Applied Sciences, Universiti Teknologi MARA, Shah Alam, Selangor, Malaysia
e-mail: nurulainmustafa@uitm.edu.my

N. E. Ezani · H. F. Mustafa · S. N. Syed Ismail
Department of Environmental and Occupational Health, Faculty of Medicine and Health Sciences, Universiti Putra Malaysia (UPM), Serdang, Selangor, Malaysia
e-mail: elianiezani@upm.edu.my

P. Ghosh et al. (eds.), *Climate Crisis and Changing Population Dynamics*, Springer Climate, https://doi.org/10.1007/978-3-032-24533-5_5

health outcomes, existing research is often fragmented, hazard-specific, or disconnected from health system and equity considerations, particularly in low- and middle-income countries. This chapter addresses this gap through a narrative review synthesising global and regional evidence on how climate extremes, including heatwaves, air pollution, floods, droughts, and wildfires, interact with biological, social, and structural factors to shape NCD risks. The review draws on epidemiological evidence, health systems research, and selected international case studies to examine key exposure pathways, patterns of vulnerability, and the ability of healthcare systems to maintain continuity of chronic disease care during climate-related disruptions. A conceptual framework is presented to integrate these dimensions and illustrate how climate extremes act as systemic risk multipliers for NCDs. The findings underscore the need for climate-resilient, equity-centred strategies that integrate surveillance, urban planning, food system resilience, and governance reforms to strengthen sustainable NCD prevention and management in a changing climate.

Keywords Climate change · Climate extremes · Non-communicable diseases · Vulnerable populations · Health system · Resilience

Introduction

Climate change has emerged as one of the most pressing global health challenges of the twenty-first century, with far-reaching impacts that extend beyond infectious diseases to include non-communicable diseases (NCDs). The growing frequency,

H. F. Mustafa
e-mail: haninahfadiah@gmail.com

S. N. Syed Ismail
e-mail: norkhadijah@upm.edu.my

A. Nurdin
Department of Biomedical Sciences, Faculty of Medicine and Health Sciences, Universiti Putra Malaysia (UPM), Serdang, Selangor, Malaysia

Laboratory of UPM-MAKNA Cancer Research (CANRES), Institute of Bioscience, Universiti Putra Malaysia (UPM), Serdang, Selangor, Malaysia

A. Nurdin
e-mail: armania@upm.edu.my

A. R. Alfian
Environmental and Occupational Health Department, Faculty of Public Health, Andalas University, Padang, Indonesia

M. R. Rahmat
Environmental Health Programme, Health Ministry Training Institute (ILKKM), Sungai Buloh, Selangor, Malaysia
e-mail: p146230@siswa.ukm.edu.my

intensity, and duration of climate extremes, including heatwaves, floods, droughts, and wildfires, are disrupting environmental, social, and healthcare systems worldwide. These events influence air quality, food and water availability, and social determinants of health, thereby creating conditions that worsen chronic health conditions such as cardiovascular disease, chronic respiratory illness, diabetes, and mental health disorders (Ebi et al. 2021; Bolan et al. 2024). The World Health Organization (WHO) estimates that between 2030 and 2050, climate change will contribute to an additional 250,000 deaths annually from heat stress, malnutrition, and related health consequences (WHO 2023). The combined effects of temperature extremes, deteriorating air quality, and resource scarcity are particularly detrimental to vulnerable populations, including the elderly, individuals with pre-existing health conditions, and socioeconomically disadvantaged communities (UNFCCC 2018; Mkharia et al. 2023; Ngcamu 2023; USEPA 2025).

Recently, the global burden of NCDs has already been significant, accounting for over 75% of all deaths worldwide, or 43 million people in 2021, and is expected to rise due to demographic and lifestyle changes (WHO 2024a). Climate extremes act as an environmental multiplier of NCD risk through both direct and indirect pathways (WHO 2023). Direct impacts include physiological strain from heat exposure, which increases cardiovascular workload and impairs thermoregulation, as well as respiratory complications from inhalation of wildfire smoke and particulate matter (Chen et al. 2021; Münzel et al. 2021; WHO 2024b). Indirect pathways occur through climate-driven food and water insecurity, leading to nutritional deficiencies and metabolic imbalance, or through the disruption of healthcare services during extreme weather events, which interrupts ongoing management of chronic conditions (FAO 2015; Tong et al. 2022; Dewi et al. 2024; Martins et al. 2024).

Latest study by Singh et al. (2024) revealed that an increase of 2.1% in cardiovascular diseases (CVDs)-related mortality for every 1 °C rise in temperature (relative risk [RR], 1.021 [95% CI: 1.020–1.023]) and an increased risk of 11.7% in CVDs mortality associated with heat waves (RR, 1.117 [95% CI: 1.093–1.141]). In low- and middle-income countries, where health systems are often under-resourced, the capacity to adapt and respond to these compounded pressures is limited, further intensifying the health burden (Gkouliaveras et al. 2025). This convergence of climate extremes and NCDs constitutes a double burden that threatens to overwhelm public health systems.

Despite the growing body of literature linking climate extremes to NCDs, this evidence remains fragmented across disciplines and regions and is often insufficiently integrated into public health planning and policy frameworks, particularly in low- and middle-income countries where adaptive capacity and health system resilience are limited. Existing studies tend to focus on single hazards, outcomes, or populations, leaving gaps in understanding how climate extremes interact with biological, social, and structural determinants to shape chronic disease risks across diverse contexts.

Therefore, the objective of this chapter is to provide a narrative review of current global and regional evidence on the intersection between climate extremes and NCDs. This chapter synthesizes evidence on the pathways through which environmental stressors, including extreme heat, air pollution, flooding, and drought, affect health.

Of these, they contribute to the onset, progression, and exacerbation of chronic health conditions. It further examines how these risks are unevenly distributed across populations, with particular attention to older adults, individuals with pre-existing conditions, urban poor communities, and other marginalised groups. In addition, the chapter presents selected case studies to illustrate how these pathways and vulnerabilities manifest in real-world settings and discusses the implications for healthcare systems facing climate-related disruptions to chronic disease care. This chapter aims to advance understanding of the double burden of climate extremes and NCDs and to inform climate-resilient, equitable, and sustainable strategies for NCD prevention and management in line with global public health priorities.

Methodology

This chapter is based on a narrative literature review, synthesising and interpreting existing evidence on the links between climate extremes and the growing burden of non-communicable diseases (NCDs). A narrative approach was selected to allow flexibility in integrating findings across epidemiology, environmental science, public health, and policy, reflecting the interdisciplinary nature of climate-health research. The methodological framing was informed by established narrative review practices in climate and health research, including the approach described by Dalla Vecchia et al. (2024), and adapted to suit the broader conceptual scope of a book chapter.

The literature search was initiated on 20 April 2025 and complemented by manual searches conducted up to December 2025 to incorporate recently published studies and policy documents. Searches were undertaken across major bibliographic databases: PubMed, Google Scholar, and Scopus to identify peer-reviewed literature relevant to climate change, climate extremes, and chronic disease outcomes. Search terms were developed iteratively and included combinations of keywords related to climate change and climate extremes (e.g., climate change, global warming, heatwaves, wildfires, natural disasters) and health outcomes and systems (e.g., non-communicable diseases, vulnerable populations, health systems, resilience).

In addition to database searches, targeted Google searches were conducted to retrieve grey literature, including policy reports, technical documents, guidelines, and position statements published on the official websites of international organizations and professional bodies. These sources were included to capture health system perspectives, adaptation strategies, and policy responses that are often underrepresented in peer-reviewed journals but are highly relevant to climate-related NCD risks.

The review focused primarily on literature published from January 2018 onwards, reflecting the increasing frequency and intensity of climate extremes in recent years and their growing recognition as public health concerns. Commentaries, letters, duplicate publications, and isolated case reports were excluded to maintain analytical coherence. Rather than applying formal inclusion and exclusion criteria, sources were selected based on their relevance, conceptual contribution, and ability to inform understanding of climate-NCD pathways, vulnerability patterns, and health system impacts. Decisions regarding inclusion were refined through discussion to ensure consistency in thematic interpretation.

For each included source, descriptive information relating to publication type, thematic focus, geographical context, and relevance to climate-NCD linkages was extracted. The literature was then organised through thematic synthesis, allowing recurring pathways, vulnerable populations, and systemic challenges to be identified and interpreted across diverse contexts. The findings from this synthesis are presented and discussed in the following sections.

Result and Discussion

The narrative synthesis identified multiple interrelated pathways through which climate extremes contribute to the onset, progression, and exacerbation of non-communicable diseases. Across the literature, climate-related hazards, including extreme heat, air pollution, floods, droughts, and wildfires, were consistently linked to chronic disease risks through direct biological mechanisms, indirect environmental and social disruptions, and pressures on healthcare systems. The following sections present and discuss these pathways thematically, drawing together epidemiological evidence and interpretive insights to illustrate how climate extremes shape non-communicable disease burdens.

Figure 5.1 summarizes the key exposure pathways through which climate extremes influence non-communicable disease outcomes and illustrates how these pathways translate into differential vulnerability across population groups. Climate-related hazards influence non-communicable disease outcomes through multiple exposure pathways, including heat stress, air pollution, food and water insecurity, and psychological stress (Sects. 3.1–3.4). These pathways interact with biological, social, and structural factors to produce unequal health burdens across vulnerable population groups (Sect. 3.5).

Heat Stress and Cardiovascular Morbidity

Despite significant developments in public health that have reduced risk factors like tobacco use, dietary interventions, and physical activity, CVDs continue to be the world's leading cause of death, taking the lives of 18 million people annually (Al-Kindi et al. 2020). Nonetheless, there is growing evidence that environmental and climatic conditions, such as heat, have a negative impact on cardiovascular morbidity and death. The primary cause of weather-related fatalities is heat stress, which can also make pre-existing conditions like CVDs worsen (Liu et al. 2022). They highlight that an increase of 1 °C in temperature is significantly associated with an increase in 2.1% of morbidity related to CVD-related mortality, such as stroke, cardiac arrest,

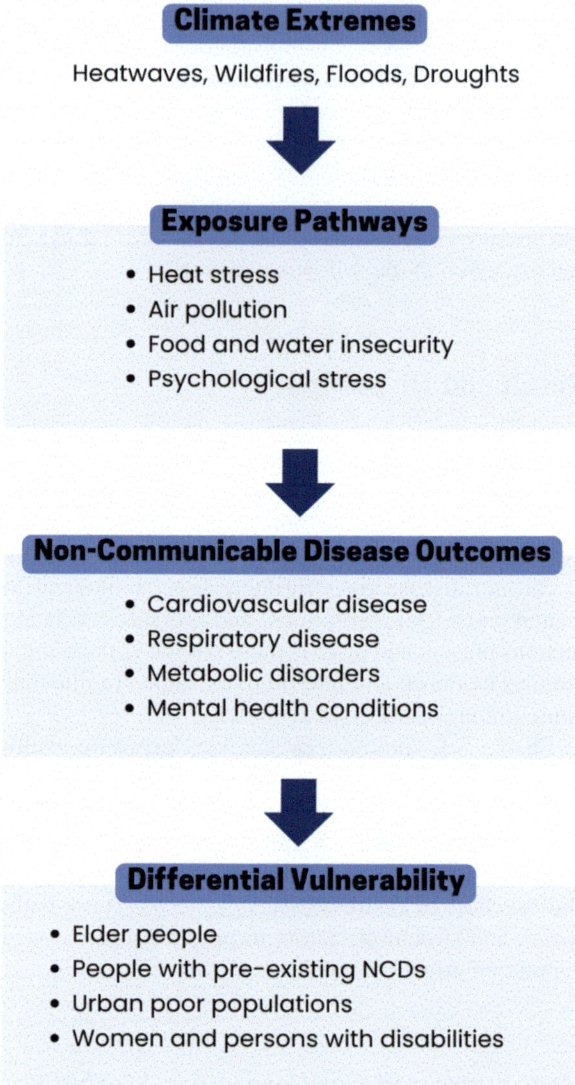

Fig. 5.1 Conceptual framework illustrating pathways linking climate extremes to non-communicable diseases (NCDs) and differential vulnerability. The figure was generated using the Canva online design platform (www.canva.com)

and coronary heart disease. The finding also aligns with Faurie et al. (2022), who reported that a 1 °C increase in temperature significantly increases the morbidity and mortality of heat-associated illnesses by 18%. A recent systematic review and meta-analysis of 266 studies also demonstrated that heatwaves increase mortality risk by 11.7%, with greater vulnerability observed among women, older adults, and populations in lower-middle-income tropical regions (Liu et al. 2022).

The link between heat and CVD mortality is thought to be mediated by several biological processes. Thermoregulatory processes promote vasodilation in response to high temperatures or extreme heat events, increasing blood flow to the skin to dissipate body heat through perspiration (Alahmad et al. 2020). Dehydration can result from thermoregulatory processes being worn down, raising core body temperature, and disrupting homeostasis. This can then cause haemoconcentration, a hypercoagulable state, electrolyte imbalances, hypotension, systemic inflammatory reactions, and a sudden rise in cardiac output (Lim et al. 2015).

Temperature-related changes in hydration status may be associated with elevated CVD mortality, according to a study by Lim et al. (2015) that examined the relationship between dehydration markers, temperature, and CVD mortality. They observed a significant association between temperature and several dehydration markers, including serum BUN/Cr, urine-specific gravity, plasma tonicity, and blood haematocrit, and further suggest that disturbance in hydration homeostasis may represent a key pathway linking extreme temperatures to adverse CVD outcomes.

Other epidemiological studies corroborated this conclusion, indicating that dehydration was the most likely biological mechanism because it causes haemoconcentration, or an increase in red blood cells, which in turn increases cardiac output and heart rate (Alahmad et al. 2020; Singh et al. 2024). Dehydration also activates the sympathetic nervous system. A number of related factors contribute to an increase in cardiac output: vasodilation lowers peripheral resistance, pulmonary pressure, and systemic pressure, and hypovolemia raises heart rate and systolic function. In vulnerable people, this increase in cardiac output can result in myocardial ischaemia, myocardial infarction, heart failure, and stroke, which can ultimately cause cardiac death (Li et al. 2019). Figure 5.2 shows the possible mechanisms by which heat stress affects the cardiovascular system.

Air Pollution, Respiratory Illness, and Chronic Exposure

Prolonged exposure to air pollution, exacerbated by climate change, disrupts the balance between oxygen supply and demand, thereby contributing to the global prevalence of NCDs (Alvarado-Calvo et al. 2025). Airborne pollutants, including ozone (O_3), particulate matter (PM), sulphur dioxide (SO_2), and nitrogen dioxide (NO_2), compromise pulmonary function by inducing airway inflammation, causing oxidative damage, and decreasing alveolar gas exchange, thus reducing systemic oxygen availability. Simultaneously, pollution-induced systemic inflammation and oxidative stress, exacerbated by climate-related stresses such as high temperatures, heightened metabolic demands, and increased respiratory exertion, increase oxygen requirements (Al-Kindi et al. 2020).

Air pollution is associated with 19% of cardiovascular deaths and 21% of stroke deaths, as well as chronic respiratory diseases such as chronic obstructive pulmonary disease (COPD) and lung cancer, which are estimated to cause 1.6 million and 500,000 annual deaths, respectively, as a result of the mismatch between oxygen delivery and utilisation (Schraufnagel et al. 2019). Moreover, the pathophysiological

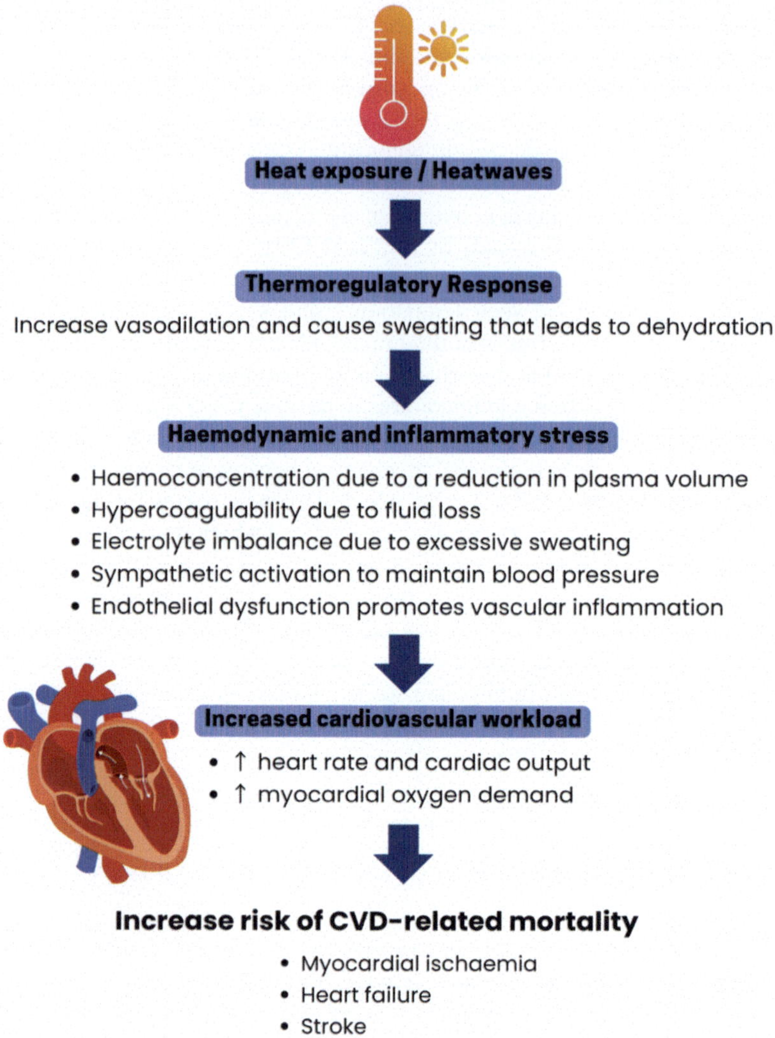

Fig. 5.2 Illustration of the biological pathways through which heat stress influences cardiovascular outcomes. Thermal exposure induces physiological responses that alter hydration status and vascular dynamics, contributing to inflammatory and endothelial changes that heighten the likelihood of acute cardiovascular events in vulnerable individuals. The figure was generated using the Canva online design platform (www.canva.com)

repercussions of an oxygen supply–demand imbalance in populations that are chronically exposed to polluted environments are underscored by the consistent association between acute elevations in PM and increases in emergency department visits for respiratory infections, COPD exacerbations, and asthma (Bell et al. 2018).

Food and Water Insecurity: Nutritional and Metabolic Risks

In addition to respiratory and cardiovascular impacts, climate change intensifies food and water insecurity, constituting another significant pathway connecting environmental change to the risk of NCDs. Restricted access to safe and nutritious food contributes to undernutrition, micronutrient deficiencies, and stunting, while concurrently encouraging dependence on low-cost, calorie-dense yet nutrient-deficient diets that heighten vulnerability to obesity and diet-related metabolic diseases (Hussain and Sharma 2025). Water insecurity exacerbates these risks by diminishing dietary diversity, hindering food preparation and hygiene, and increasing susceptibility to gastrointestinal illnesses that decrease nutritional absorption (Fanzo et al. 2022).

Chronic undernutrition and suboptimal food quality alter metabolic homeostasis, leading to impaired glucose control, dyslipidaemia, and heightened risk of metabolic syndrome, type 2 diabetes, and cardiovascular disease (Hussain and Sharma 2025). Chronic food and water insecurity induces physiological stress that triggers maladaptive neuroendocrine pathways, exacerbating oxidative stress and systemic inflammation, akin to the oxygen supply–demand imbalance observed in diseases associated with air pollution (Kuntić et al. 2024). Collectively, these strategies illustrate how climate change exacerbates nutritional, metabolic, respiratory, and cardiovascular risks via interrelated pathways of environmental stress and physiological disruption.

Psychological and Mental Health Effects of Climate Disasters

Climate-related disasters, such as floods, storms, droughts, wildfires, and heatwaves, have significant psychological and mental health effects that often persist beyond the immediate aftermath (Walinski et al. 2023). Survivors frequently endure acute stress, worry, and terror during and subsequent to the incident, with several individuals suffering chronic ailments such as post-traumatic stress disorder (PTSD), depression, generalised anxiety disorder, and substance use disorders. Displacement, destruction of residences, loss of livelihoods, and disruption of community support systems exacerbate this burden, placing vulnerable populations, including children, the elderly, women, and those with pre-existing psychiatric illnesses, at the most risk (Matthews et al. 2019).

In addition to acute trauma, the gradual impacts of climate change, such as droughts, rising sea levels, and ecosystem degradation, lead to chronic stress and the development of conditions like "eco-anxiety" and "climate grief" (Comtesse et al. 2021). These emotions are marked by fear, grief, and helplessness in anticipation of impending calamities and environmental degradation, especially among adolescents, many of whom express reduced hope and motivation for the future. Repeated exposure to disasters cumulatively diminishes resilience and heightens vulnerability to chronic psychiatric disorders, particularly in socially and economically disadvantaged areas. To tackle these challenges, it is essential to incorporate mental health services into disaster preparedness and recovery strategies, enhance psychosocial support networks, and develop climate-resilient health systems that can alleviate the significant psychological and mental health effects of climate change (Keya et al. 2023).

Vulnerability and Unequal Health Risks

Building on the exposure pathways outlined above, while the pathways described in Sects. 3.1–3.4 demonstrate how climate extremes influence non-communicable disease risks at the population level, their health consequences are not distributed evenly across societies. Instead, exposure to extreme heat, air pollution, food and water insecurity, and climate-related psychological stress is filtered through pre-existing biological susceptibility, social position, and structural conditions, resulting in marked inequalities in disease burden and health outcomes. Climate extremes, therefore, act not only as environmental stressors but also as risk multipliers, amplifying vulnerabilities shaped by age, health status, socioeconomic disadvantage, and access to healthcare. This section examines how these dynamics translate into disproportionate NCD risks among population groups that are consistently identified in the literature as most vulnerable to climate-related health impacts.

The health impacts of flood disasters are not distributed uniformly across populations but are mediated by structural, demographic, and socioeconomic determinants. Disparities in age, pre-existing health conditions, poverty, disability, and gender roles interact with fragile infrastructures and limited healthcare access to produce stratified risks of morbidity and mortality. Vulnerability to flood-related health risks should not be understood as an intrinsic attribute of individuals or communities, but as the outcome of cumulative interactions between biological susceptibility, socioeconomic constraints, and systemic limitations in healthcare and infrastructure. These processes shape differential patterns of exposure, protection, and recovery, producing stratified burdens of disease and mortality across population groups. This chapter examines the vulnerability within a framework that recognizes three critical axes: (1) analyses the intersection between biological senescence and systemic healthcare gaps, which together heighten the risks faced by older adults and those with pre-existing conditions; (2) considers urban poverty as both a spatial and socioeconomic phenomenon, where marginalization increases exposure to flood hazards and limits access to protective infrastructure; and (3) applies an intersectional lens to show how overlapping identities, including gender, disability, and socioeconomic status, compound risks and reinforce structural inequities.

The Elderly and People with Pre-existing Conditions

Older adults and people with chronic illnesses are among the most vulnerable during health crises. As people age, their immune system becomes weaker, a process known as immunosenescence. This includes a decline in the production of new immune cells and the presence of constant, low-level inflammation, sometimes referred to as “inflammaging” (Boren and Gershwin 2004). These changes make it more difficult for the body to fight infections and respond effectively to vaccines. This is why older adults are more likely to experience severe illness from diseases such as influenza or COVID-19 (Mueller et al. 2020).

Chronic conditions such as heart disease, diabetes, and respiratory problems add another layer of risk. During the COVID-19 pandemic, death rates were much higher in people with more than one chronic disease (Jordan et al. 2020). Infections can upset the fragile balance of the body, and in older patients, this often leads to dangerous immune reactions such as cytokine storms. The pre-existing inflammation that comes with aging makes this reaction more likely and more harmful.

However, vulnerability in this group is not only biological. Social and economic challenges also play a role. Reduced mobility, isolation, and limited financial resources often prevent older adults from accessing timely healthcare. Studies show that those in lower-income groups have fewer opportunities for preventive care, which widens health inequalities (The Lancet Healthy Longevity 2024). A life-course approach helps explain this, as disadvantages accumulated over a lifetime accumulate to create poorer health outcomes in old age.

The COVID-19 pandemic highlighted these issues most clearly in long-term care facilities. Many residents died, not only because of age or illness, but because of systemic problems such as underfunded institutions, poorly supported staff, and policies that focused on protecting hospitals rather than care homes. These failures demonstrate that vulnerability in older adults is not only the result of biology but is also produced by neglect within health and social systems.

The Urban Poor and Marginalized Communities

Urbanization, while often seen as a marker of progress, simultaneously creates concentrated zones of deprivation where health risks are magnified. Informal settlements, which house over one billion people globally, exemplify this paradox (UN-Habitat 2020). These communities are characterized by overcrowding, insecure housing, and poor water quality, sanitation, and hygiene (WASH) systems, which together create conditions ideal for the spread of infectious diseases (Ezeh et al. 2017). Geographic marginalization further compounds the risks, as many settlements are located on unsafe floodplains, drainage reserves, or industrial peripheries, where environmental hazards intersect with climate-related threats (Dodman et al. 2019).

Socioeconomic exclusion deepens these vulnerabilities. Employment in the informal sector typically lacks health insurance, social protections, and legal safeguards, while financial barriers and discrimination delay timely health-seeking behaviours (Kuddus et al. 2020). During the COVID-19 pandemic, these dynamics converged in urban slums, where limited access to healthcare and essential services resulted in disproportionately high infection and mortality rates (Wilkinson et al. 2020).

This clustering of multiple disadvantages is best understood through the concept of a syndemic, in which infectious diseases, malnutrition, and chronic stress interact with each other and are amplified by adverse social and economic conditions (Singer 2017). The COVID-19 syndemic in urban slums, where viral infection intersects with non-communicable diseases and structural poverty, is a clear example of how marginalization transforms health threats into systemic crises. Thus, vulnerability in urban poor populations is not an inherent condition but a socially and structurally produced reality. Infrastructural deficits, hazardous geography, and exclusionary governance interact to systematically generate and reinforce health inequities.

Gender, Disability, and Socioeconomic Inequalities

Health risks during disasters are not evenly distributed but are often magnified at the intersection of social identities such as gender, disability, and socioeconomic status. Intersectionality provides a framework for understanding how these overlapping conditions do not simply add to one another but interact to create compounded, often unique vulnerabilities (Crenshaw 1989). This approach is particularly relevant in disaster contexts, where pre-existing inequalities intensify the health burdens of marginalised populations.

Gender-based vulnerabilities are especially evident in flood situations. Women frequently shoulder caregiving responsibilities, placing them on the front line of exposure to unsafe water, inadequate sanitation, and food insecurity (Enarson and Chakrabarti 2009). Their mobility during evacuation is often constrained by cultural norms, caregiving duties, or pregnancy-related health needs, while the risk of gender-based violence increases significantly in displacement shelters (van Daalen et al. 2022). These factors highlight how disasters are not gender-neutral but reinforce existing inequities.

For people with disabilities, disasters multiply existing structural barriers. Physical impairments often limit access to evacuation routes, shelters, or healthcare facilities, which are rarely designed with accessibility in mind (Kett and Twigg 2007). Communication barriers, such as the absence of sign language interpretation or accessible emergency alerts, further exclude people with sensory disabilities. During the COVID-19 pandemic, individuals with intellectual disabilities experienced disproportionately higher mortality rates, partly due to systemic neglect and exclusion from health services (Turk et al. 2020).

Socioeconomic disadvantage intersects with gender and disability to deepen vulnerability. Poor households, particularly those headed by women or including disabled family members, face higher opportunity costs in accessing healthcare and relief resources. In many contexts, aid distribution systems are biased toward male household heads, further marginalizing women and disabled individuals (Gartrell et al. 2020). The clustering of poverty with other marginalized identities thus creates a layered vulnerability that cannot be addressed through single-axis interventions.

As illustrated in Fig. 5.3, these vulnerabilities can be understood not as separate issues but as interconnected domains. The Venn diagram shows the unique challenges within each category, such as inaccessible evacuation routes for people with disabilities, and more importantly, reveals the compounded risks at the intersections. For example, a woman with a disability from a low-income household falls at the center of these overlapping systems, facing the most acute risk: total exclusion from evacuation, healthcare, and relief resources.

Taken together, these dynamics illustrate that vulnerability during floods and related health crises cannot be reduced to biology or geography alone. It is actively produced at the intersection of social identities and structural inequalities. Addressing these challenges requires equity-centered approaches that recognize and respond to the compounded disadvantages faced by specific groups, ensuring that disaster preparedness and public health responses are inclusive.

The evidence underscores the urgent need for a paradigm shift in public health strategy, moving from targeting vulnerable groups to dismantling the structural processes that produce vulnerability. For the elderly, this requires moving beyond fragmented, disease-specific interventions towards integrated care models that explicitly address the social determinants of aging, such as isolation and economic insecurity. For the urban poor, addressing health demands a commitment to formalizing land tenure, upgrading housing and WASH infrastructure, and

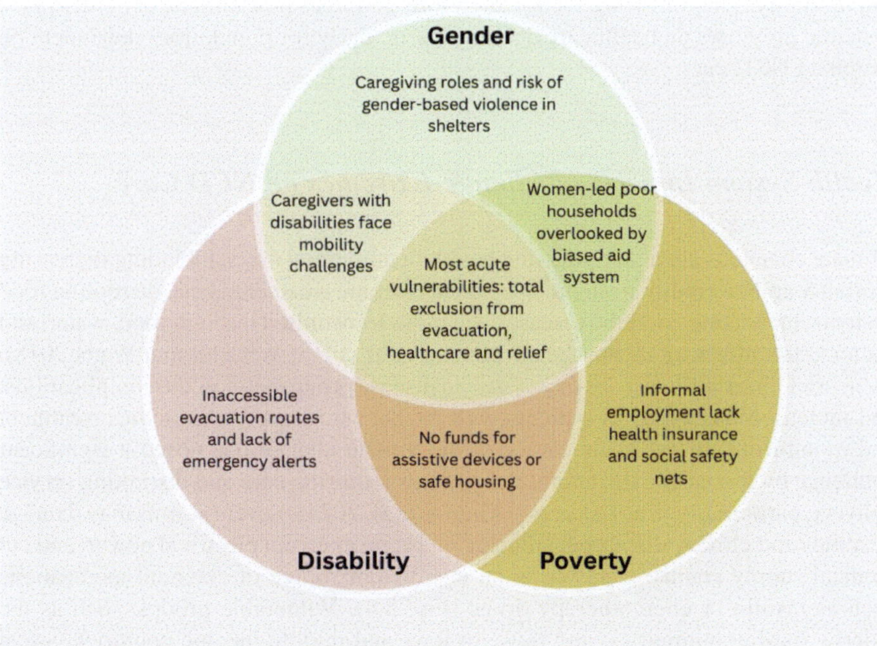

Fig. 5.3 Intersectional vulnerabilities in flood disasters. The figure was generated using the Canva online design platform (www.canva.com)

embedding universal health coverage within community-led primary care systems. This approach aligns with the World Health Organization's 'Health in All Policies' (HiAP) framework, which calls on all sectors to consider the health impacts of their decisions.

Finally, an intersectional lens must be embedded in all health policy to ensure that interventions are sensitive to the compounded risks faced by those marginalized by gender, disability, and class. Public health strategies that focus on individual behaviours while ignoring the structural inequities that constrain them are destined to fail. The most sustainable pathway toward health equity lies in addressing the political, social, and economic systems that distribute risk so unequally across society. Ultimately, achieving health equity requires a fundamental reorientation of public health, away from a focus on individual resilience and toward an unwavering commitment to dismantling the systems that produce vulnerability in the first place. Future research and advocacy must continue to unmask these processes and hold power to account.

These patterns of unequal vulnerability do not operate independently of health systems. Instead, they interact closely with the capacity, resilience, and responsiveness of healthcare infrastructure, shaping how effectively individuals with non-communicable diseases can maintain continuity of care during climate-related disruptions. In settings where health systems are fragile or overstretched, existing vulnerabilities are further amplified, translating exposure into preventable morbidity and mortality. The following section therefore examines how climate extremes place systemic pressures on healthcare delivery, particularly for populations dependent on sustained NCD care.

Health System Impacts of Climate Extremes on NCD Care

Climate change is currently affecting health in numerous ways, including increasing mortality and morbidity from more frequent extreme weather events, disrupting food systems, increasing zoonotic diseases and those transmitted through food, water, and vectors, and affecting mental health (Padhy et al. 2015; Wheeler and Watts 2018). Even brief interruptions in care can result in disease progression, acute complications, and increased mortality for patients with NCDs who require continuous treatment and monitoring. In the healthcare system, climate change has posed a significant challenge by increasing the number of patients requiring care and disrupting service delivery, particularly in at-risk areas (Martins et al. 2024). Severe storms may damage hospitals and clinics, displace healthcare workers, interrupt electrical power, and cut medical supply chains, resulting in an insufficient source of essential medications such as insulin or chemotherapy drugs (Fig. 5.4). Vulnerable groups, such as the elderly, rural communities, and those in low- and middle-income countries, are at an increased risk due to pre-existing constraints in health infrastructure and adaptive ability (Mkharia et al. 2023). These systemic disruptions reveal how climate

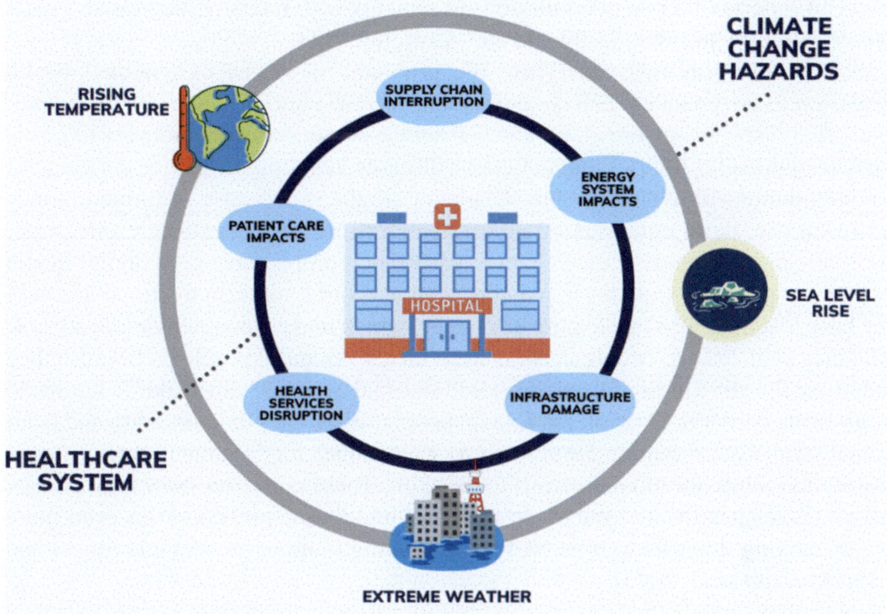

Fig. 5.4 Impacts of climate change on the healthcare system. The figure was generated using the Canva online design platform (www.canva.com)

extremes translate population-level vulnerability into concrete failures in chronic disease prevention, treatment, and long-term management.

Strained Infrastructure in Low- and Middle-Income Countries

Despite having a disproportionately high burden of NCDs, low- and middle-income countries are particularly vulnerable to the effects of climate change because their healthcare systems are persistently understaffed, underfunded, and unequally distributed. Due to their lack of climate-resilient infrastructure, many medical institutions are especially vulnerable to disruption during extreme weather events (Montero et al. 2025). Hospitals and clinics are frequently situated in flood-prone locations, rely on erratic electrical supply, or lack proper cooling systems.

Both governmental and private recovery efforts are hampered by diabetes and associated comorbidities, which are frequently made worse by disasters. People with diabetes had worse glycaemic control after disasters. Diabetic ketoacidosis and foot infections are among the consequences that put patients with diabetes at higher risk for morbidity and mortality, and ED visits among diabetics are rising. When disasters strike, some groups like the elderly, those from low socioeconomic backgrounds, those without health insurance, and various minorities become even more vulnerable and suffer from higher rates of morbidity and mortality. There have also been reports of a rise in admissions of patients with recently diagnosed Type 1 diabetes after the accident (Allweiss and Albright 2011).

Workforce constraints exacerbate the problem; shortages of qualified health professionals are prevalent in low- and middle-income nations, while climate disruptions often displace personnel, heighten absenteeism, or redirect resources to emergency responses for injuries and infectious diseases, resulting in service deficiencies in non-communicable disease care (Clarke et al. 2025). Because of limited money and resources, these countries often are unable to invest in healthcare infrastructure that can withstand climate change, emergency preparedness, or digital health systems. At the household level, patients cannot afford long-term treatment because they have lost their jobs due to climate change, which makes inequality more adverse (Nicholas et al. 2020). People in rural areas have even more problems because they cannot easily get to specialised services, and bad weather can make it harder to keep getting care. For example, in Kenya, people with high blood pressure and heart disease who live in remote areas have to wait longer for treatment during floods because the roads are too dangerous to drive on. These concerns show how climate change makes healthcare systems in low- and middle-income countries even more fragile, making it harder to treat NCDs and leading to more preventable illness and death (Muriuki et al. 2025).

Climate-Exacerbated Service Gaps for Chronic Disease Care

The management of chronic diseases is made more difficult by climate change, which exacerbates already-existing service shortages in healthcare systems. Long-term monitoring, essential medications, and diagnostic services must be continuously available to patients with NCDs. However, extreme climate events frequently disrupt this continuity. Landslides, floods, and storms can disrupt the cold chains required for temperature-sensitive drugs like insulin, destroy healthcare buildings, and cut off transportation routes, all of which can delay access to care (Ochi et al. 2014). Drug shortages brought on by supply chain failures during climate shocks frequently prevent patients from continuing life-sustaining medication. For instance, during the 2011 Great East Japan Earthquake, there were numerous reports of "drug refugees," and at least 283 people were said to have perished as a result of their pre-existing illnesses getting worse because they were unable to receive healthcare. Very little study has been done on the health effects of drug refugees (Ochi et al. 2013).

The elderly, low-income households, and rural areas are among the vulnerable groups that are disproportionately impacted because they already encounter obstacles while trying to obtain normal treatment. For example, during the 2010 floods in Pakistan, many people with diabetes and hypertension were hospitalised and experienced acute problems because they were unable to refill their prescriptions for weeks (Warraich et al. 2011). Likewise, heat waves in India have been linked to increases in heart attacks, and at these times, hospital overcrowding restricts the ability to provide continuous care for patients with long-term illnesses (de Bont et al. 2024). In addition to raising morbidity and mortality rates among NCD patients, these climate-exacerbated service gaps further exacerbate health disparities and put a sustained burden on already precarious healthcare systems.

Public Health Preparedness and Systemic Limitations

Global health systems have made strides in anticipating outbreaks of infectious diseases that are susceptible to climate change. However, the relationship between climate change and NCDs has received far less attention. This gap in preparedness exposes systemic limitations that leave millions of patients vulnerable during climate-related disasters. Several national health adaptation plans focus on urgent catastrophic events, such as vector-borne infections or food insecurity, but fail to incorporate cohesive measures for maintaining chronic disease management (Sutherst 2004). Emergency response strategies frequently neglect the requirements of patients dependent on routine drugs, dialysis, chemotherapy, or long-term monitoring, leading to avoidable complications and fatalities (Tomio and Sato 2014). The ability to predict and address climate—NCD interactions is further limited by structural flaws in surveillance, early warning systems, and intersectoral cooperation. Health systems in areas of Europe and South Asia, for instance, recorded increases in hospital admissions for respiratory and cardiovascular problems during extended heatwaves. However, they lacked adequate heat-health action plans to handle the spike in exacerbations of chronic diseases (Singh et al. 2024). Underfunded health budgets, disjointed governance, and inadequate integration of primary care and disaster response make these issues worse in many low- and middle-income nations.

Furthermore, insufficient investment in climate-resilient infrastructure and digital health systems impedes the provision of telemedicine or remote monitoring when physical access to care is compromised (Ansah et al. 2024). In addition to exposing systemic vulnerabilities, the lack of long-term planning causes health disparities by leaving vulnerable groups disproportionately unprotected. A paradigm shift will be necessary to improve readiness, going beyond models of infectious diseases to all-encompassing approaches that incorporate climate risk into the management of noncommunicable diseases, climate-proof health infrastructure, and guarantee continuous treatment of chronic illnesses both during and after climate shocks (Lokmic-Tomkins et al. 2023).

Case Studies on Climate Extremes and Non-communicable Disease (NCD)

Understanding the health consequences of climate extremes requires examining how these events affect populations across different contexts. Environmental conditions, socioeconomic structures, and the capacity of public health systems play a central role in shaping both the severity and the nature of health outcomes associated with climate-related hazards. In this context, case studies provide an opportunity to illustrate how the pathways and vulnerabilities discussed in the preceding sections manifest in real-world settings. This section presents selected case studies that highlight four key domains: the association between urban heat and non-communicable diseases, the health impacts of wildfire smoke on chronic illnesses, the consequences of flood-related sanitation disruptions, and the risks of drought-induced undernutrition and cardiometabolic disorders. Rather than providing an exhaustive review, these case studies are intended to contextualize the preceding analysis by illustrating how climate-driven pathways, vulnerabilities, and health system constraints converge in specific real-world settings.

Urban Heat and NCDs

Cities around the world are experiencing rapid urbanization, marked by both horizontal and vertical expansion, to accommodate growing populations. This transformation alters land use and land cover (LULC), leading to greater absorption of solar radiation due to the prevalence of artificial dark surfaces and construction materials with low albedo, high thermal capacity, and elevated conductivity compared to rural areas (Islam et al. 2024). As a result, urban environments experience the Urban Heat Island (UHI) effect, characterized by persistently higher near-surface air temperatures than surrounding non-urban areas, particularly at night (Saguansap et al. 2024). The "island" in UHI does not denote a geographical feature but metaphorically signifies concentrated heat zones within cities (Saguansap et al. 2024). The key drivers of UHI include the thermal properties of urban infrastructure, limited sky visibility, reduced relative humidity, diminished vegetation cover, restricted open spaces, and altered energy fluxes (Islam et al. 2024). These conditions not only increase thermal discomfort but also heighten health risks, including NCDs. Prolonged exposure to high ambient temperatures exacerbates cardiovascular strain, increases the risk of heat-related strokes, and worsens respiratory conditions such as chronic obstructive pulmonary disease (COPD) and asthma. For example, during the European heatwave of 2003, excess mortality reached nearly 70,000 deaths because of the June–August event, with cardiovascular and respiratory conditions being the leading contributors (WHO 2024b).

Several studies have shown a relationship between urban heat and NCDs. For example, a study by Chhetri et al. (2021) from Barasat City, India, demonstrated a high prevalence of heat cramps, strokes, and respiratory diseases in populations residing in high-heat zones. Similarly, evidence from other urban centres reinforced this trend, as evidenced in Bangkok, where elevated temperatures have been linked to increased cases of heat stress, heatstroke, cardiovascular and respiratory illnesses, and premature mortality, particularly among vulnerable groups such as the elderly, children, and individuals with pre-existing conditions (Saguansap et al. 2024). According to Saleem et al. (2024), Karachi alone reports around 1500 cases of heat stroke and several deaths daily due to the sharp rise in temperature, with the Edhi ambulance service noting that 30–40 bodies are brought to the mortuary every day. These alarming figures from an urban South Asian context underscore the severe toll of extreme heat on human health. Similarly, evidence from a longitudinal cohort study using the China Health and Retirement Longitudinal Study (CHARLS) further illustrated that the systemic risks posed by heat exposure, defined by a 92.5–97.5th percentile threshold over ≥2 days, were associated with 23–38% increased odds of rapid kidney function decline (OR 1.23–1.38 in fully adjusted models). The study, conducted among 6450 respondents aged 45 and above, covered 150 counties or districts and 450 villages or residential committees across 28 provinces in mainland China with follow-up data spanning from 2011 to 2015 (Zhang et al. 2025a). Together, these findings highlight that both acute mortality and long-term morbidity risks are magnified during prolonged periods of extreme heat, illustrating the multifaceted health burden of climate change.

Wildlife Smoke Exposure and Chronic Illnesses

Wildfires are becoming more frequent and intense due to climate change. Wildfire smoke contains several components known to be harmful to human health, including fine particulate matter ($PM_{2.5}$), carbon monoxide, ozone, and many hazardous air pollutants (HAPs) (Grant and Runkle 2022; Rizzo and Rizzo 2024). Exposure to wildfire smoke has been consistently associated with adverse health outcomes, such as respiratory infections, asthma, chronic obstructive pulmonary disease (COPD), and all-cause mortality (Grant and Runkle 2022). Among these pollutants, $PM_{2.5}$ is considered the most significant because it can penetrate deeply into the respiratory tract and bloodstream. These fine particles, with diameters of 2.5 μm or smaller, are composed of diverse chemical compounds that can exacerbate cardiovascular and respiratory diseases, impair lung function, and contribute to premature mortality. Evidence from large-scale studies underscores this burden. For example, Xu et al. (2025) investigated elderly populations aged between 65 and 100 years across the contiguous United States. The findings demonstrated that wildfire-related $PM_{2.5}$ was strongly associated with COPD mortality. A 1 $\mu g/m^3$ increase in $PM_{2.5}$ levels corresponded to a 9.2% increase in COPD mortality (95% CI: 8.8–9.7%). A 1 $\mu g/m^3$ increase in smoke $PM_{2.5}$ may increase deaths by 40.4, 9.6 and 3.9% in low, moderate, and high wildfire risk areas, respectively. Their study reveals that long-term exposure to smoke $PM_{2.5}$ significantly contributes to COPD mortality among elderly people.

Further evidence demonstrates that the adverse impacts of wildfire smoke extend beyond chronic mortality outcomes to include acute hospitalizations for a broad spectrum of respiratory conditions. A multicountry analysis conducted by Zhang et al. (2025b), which evaluated 1052 communities across Australia, Brazil, Canada, Chile, New Zealand, Vietnam, Thailand, and Taiwan, during 2000–2019, found that each 1 $\mu g/m^3$ increase in wildfire-specific $PM_{2.5}$ was associated with increased hospitalizations risks for all-cause respiratory, asthma, COPD, acute upper respiratory infection, influenza, and pneumonia by 0.36%, 0.48%, 0.38%, 0.42%, 0.79% and 0.36%, respectively. Higher risks were observed among the population ≤19 or ≥60 years old, from low-income or high non-wildfire $PM_{2.5}$ communities, and residing in Brazil, Thailand, Taiwan, and Vietnam. Occupational groups are also significantly affected, as highlighted by Ramos and Minghelli (2022), who reported that firefighters exposed to wildfire smoke developed restrictive respiratory patterns, with contributing factors including prolonged exposure, lack of protective equipment, and limited physical activity. These findings illustrate that wildfire smoke exposure is a growing global health concern, with both acute and long-term implications for respiratory morbidity and mortality, disproportionately affecting vulnerable populations and those in occupational roles with direct smoke exposure.

Floods, Sanitation Disruption, and Health Outcomes

Floods are among the most destructive climate-related hazards, with profound implications for sanitation systems and public health infrastructure. When sanitation facilities are inundated, sewage often contaminates floodwaters, creating a hazardous environment that fosters the spread of infectious and vector-borne diseases. While much of the literature has focused on outbreaks of cholera, leptospirosis, and diarrheal diseases, floods also exert substantial indirect impacts on non-communicable diseases (NCDs). Patients with chronic illnesses such as cardiovascular disease, diabetes, cancer, and respiratory conditions are particularly vulnerable during flood events, as disruptions to healthcare services and the supply of medications may compromise disease management (Ryan et al. 2016; Ghazanchaei et al. 2021). Evidence from Hurricane Katrina in the United States demonstrated how displaced individuals with chronic illnesses experienced rapid deterioration in health due to interrupted medical treatments, loss of essential medications, and reduced access to healthcare facilities (Bell et al. 2018). Similarly, sanitation disruptions disproportionately affect immunocompromised individuals, including cancer patients undergoing chemotherapy and transplant recipients, for whom access to clean water and hygienic conditions is critical. Those reliant on electricity-dependent therapies, such as dialysis patients, also face life-threatening challenges when infrastructure collapses (Ryan et al. 2016). Flooding not only displaces populations and disrupts sanitation infrastructure but also increases exposure to hazardous contaminants. In addition, according to Nshimiyimana et al. (2025), heavy metals, persistent organic pollutants, chemicals, and microplastics from agricultural runoff and leachates often mix with floodwaters, contaminating surface and groundwater sources. These contaminants have been reported to enhance endothelial dysfunction, oxidative stress, reactive oxygen species generation, apoptosis, and DNA damage, which are indirectly linked to NCDs.

Recent empirical studies highlight the broader NCD-related consequences of floods. Aggarwal et al. (2025) analyzed data on 72 major flood events in the United States between 2000 and 2016, along with more than 4.5 million hospitalizations, and found elevated risks for several non-communicable health outcomes. Specifically, flood exposure was associated with increased hospitalizations for skin diseases (3.1% [95% CI: 1.4–4.9]), nervous system diseases (2.5% [95% CI: 1.0–4.1]), musculoskeletal conditions (1.3% [95% CI: 0.3–2.3]), and injuries or poisoning (1.1% [95% CI: 0.2–2.0]). These findings illustrate how floods exacerbate health burdens beyond infectious disease pathways, impacting multiple physiological systems. Furthermore, the nutritional consequences of flooding are profound. Meanwhile, Hussain and Sharma (2025) investigated childhood health outcomes in India and reported significant increases in malnutrition rates in flood-affected regions. In areas experiencing three flood events, 38% of children were stunted, 20% were wasted, and 35.7% were underweight compared to non-flooded areas. The authors identified low birth weight, limited maternal education, inadequate sanitation facilities, and fragile household infrastructure as compounding risk factors. These studies underscore how floods, through sanitation disruption and systemic infrastructural breakdowns, intensify both chronic disease burdens and nutrition-related vulnerabilities, particularly among already marginalized populations.

Drought-Related Undernutrition and Cardiometabolic Risk

Drought represents one of the most severe slow-onset climate hazards, with significant health repercussions. Unlike acute disasters such as floods, drought unfolds gradually, exerting long-term pressures on food security, water availability, and public health. The impacts are multifaceted and vary according to the socio-economic status, adaptive capacity, and resilience of the affected populations (Bryan et al. 2020; Mani et al. 2024). In regions with poor governance structures, drought often acts as a threat multiplier, amplifying poverty, social inequalities, and even conflict over scarce resources (Mani et al. 2024). In many low- and middle-income countries, particularly in agrarian economies, droughts directly translate into reduced dietary diversity and caloric intake, leading to widespread undernutrition (FAO 2022). Children and women of reproductive age are especially vulnerable to stunting, wasting, and micronutrient deficiencies, which have long-term impacts on physical and cognitive development (Haron et al. 2023; WHO 2023). The cascading effects of drought on undernutrition extend beyond acute food shortages, as persistent nutritional deficiencies undermine immune function, increase susceptibility to infections, and impair recovery from disease (Fig. 5.5). Furthermore, chronic dehydration and hyperosmolarity have been linked to increased risks of obesity, diabetes, and metabolic syndrome (Mani et al. 2024).

Drought can exacerbate cardiometabolic risks through several pathways. These risks include increased exposure to air pollution and contaminated water sources, leading to higher rates of cardiovascular diseases (CVD). Studies have shown a synergistic effect between air pollution and drought, significantly increasing the risk of cardiovascular issues. For example, a study by Ling et al. (2025) found that individuals exposed to both high air pollution and severe drought had a much higher risk of CVD than those in lower-exposure groups. In addition, a study by Zhang et al. (2024) demonstrated that the simultaneous occurrence of drought and hot weather events was associated with an 18% increase in total daily stroke admissions and a 20% increase in ischemic stroke admissions, with attributable fractions of 23.40% and

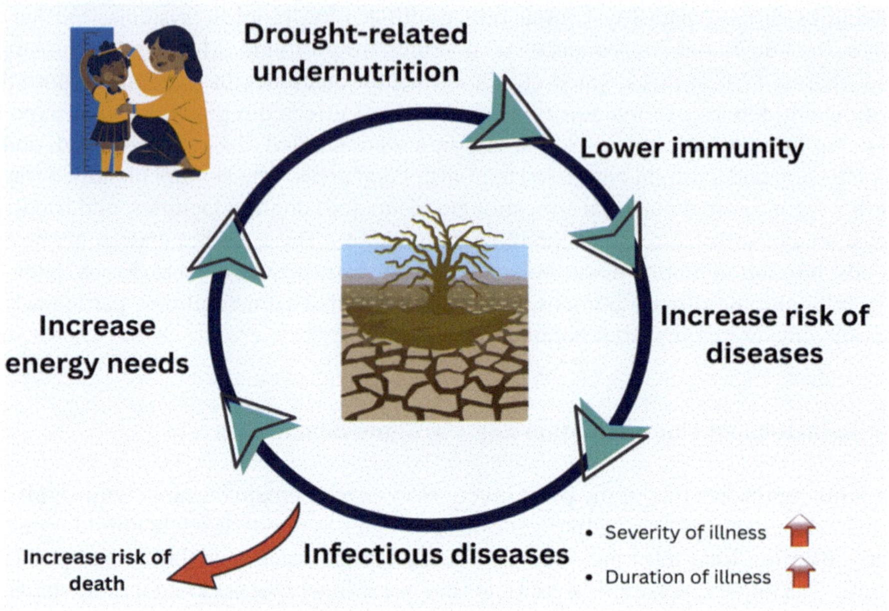

Fig. 5.5 The cascading effects of drought on undernutrition

24.13%, respectively. These findings illustrate how drought not only drives undernutrition but also amplifies cardiovascular and metabolic risks, reinforcing its role as a structural driver of chronic disease burdens.

Towards Climate-Resilient NCD Strategies

Climate change is progressively influencing NCDs through both direct and indirect pathways, including air pollution, food and water poverty, extreme weather events, and ecosystem degradation (Dewi et al. 2024). Developing climate-resilient solutions for NCD prevention and management is thus a top global health goal. Such solutions must go beyond traditional disease-centered approaches and incorporate climate adaptation, sustainable urban design, environmental health interventions, and equitable food system changes (Mosadeghrad et al. 2023). This section advances current understanding of climate–health interactions by synthesizing evidence across mechanistic pathways, population vulnerabilities, health system impacts, and contextual case studies to conceptualize climate extremes as systemic amplifiers of noncommunicable disease risk. By integrating biological, social, and structural perspectives, the review moves beyond single-hazard or disease-specific analyses and provides a comprehensive framework for understanding the double burden of climate extremes and chronic disease.

Integrating Climate Adaptation into NCDs Prevention

Conventional frameworks for non-communicable disease prevention prioritise lifestyle modifications, early identification, and clinical management strategies (Budreviciute et al. 2020). Climate change requires expanding this approach to include environmental determinants of health. Exposure to heatwaves elevates the risk of cardiovascular mortality, whereas prolonged air pollution is associated with chronic respiratory diseases, lung cancer, and metabolic disorders. Incorporating climate adaptation into non-communicable disease prevention involves integrating environmental risk reduction within public health strategies (Schraufnagel et al. 2019).

Aligning NCD prevention with climate adaptation policies at both national and local levels is a viable approach. This involves designing health action plans that explicitly consider climate hazards, including integrating heat-health warning systems with cardiovascular and respiratory disease management programs (Chandra and Lee 2025). Clinical guidelines must include recommendations regarding patient susceptibility to climate-related exposures, including medication modifications during heatwaves and air quality advisories for individuals with asthma and COPD. Incorporating education on climate-related risks into preventive healthcare, especially for vulnerable populations such as the elderly, children, and low-income groups, strengthens individual resilience and alleviates pressures on health systems (Mosadeghrad et al. 2023).

Urban Planning and Healthy Built Environments

NCDs are significantly influenced by the intersection of urbanisation and climate change (Siiba et al. 2024). Poorly designed urban environments exacerbate sedentary behaviour, unhealthy diets, and exposure to pollutants, while climate change heightens the risks of inundation, heat islands, and deteriorating air quality. Therefore, it is imperative to foster climate-resilient, healthful built environments in order to prevent NCDs.

Expanding access to verdant spaces and active transportation infrastructure are critical interventions. Greenery in urban areas not only promotes physical activity but also mitigates air pollution, lowers urban heat, and reduces stress, thereby simultaneously addressing multiple risk factors for NCDs (Nieuwenhuijsen 2016). For example, research indicates that cities with higher tree canopy cover have lower rates of cardiovascular and respiratory hospital admissions during heatwaves (Sinha et al. 2021). In the same way, the development of secure walking and cycling infrastructure reduces dependence on motorised transport, thereby promoting physical activity and reducing greenhouse gas emissions, a dual benefit for the prevention of NCDs and climate mitigation.

Strengthening Surveillance, Early Warning, and Air Quality Control

Effective NCD strategies in the context of climate change necessitate robust surveillance and early warning systems to identify, forecast, and address health risks influenced by climate factors. Traditional NCDs surveillance systems primarily emphasise disease prevalence, incidence, and risk factors, frequently neglecting environmental triggers and exposures. Incorporating climate-related variables, such as air quality, temperature extremes, and food or water safety, into surveillance systems would enable a more thorough assessment of health risks (Nugroho et al. 2025).

Integrating real-time air quality data with hospital admission records enables predictive modelling of respiratory and cardiovascular exacerbations, thereby allowing health systems to issue targeted warnings and allocate resources during high-risk periods (Ratcliff et al. 2024). Early warning systems for heatwaves have demonstrated effectiveness in decreasing mortality rates. These models can be adapted to predict pollution spikes, outbreaks of vector-borne diseases, or disruptions in food and water supply, all of which can indirectly affect non-communicable disease outcomes.

Air quality management is fundamental to climate-resilient non-communicable disease strategies. Regulatory frameworks should implement more stringent emissions standards across transportation, industry, and agriculture, while concurrently investing in renewable energy and clean mobility solutions. In addition to regulation, community-level air monitoring and public alert systems enable individuals to implement protective measures, such as refraining from outdoor activities or utilising filtration during pollution events (Nugroho et al. 2025).

Ensuring Resilient and Equitable Food Systems

Nutritional and metabolic risks are significantly impacted by food and water insecurity, which is further worsened by climate change. Food availability is disrupted by floods, droughts, and shifting agricultural yields, and low-income people are most affected by rising prices (Bhatnagar et al. 2024). Obesity, diabetes, and cardiovascular disease are accelerated by these stresses, which also lead to micronutrient shortages and an overreliance on inexpensive, calorically dense, nutritionally deficient meals (Ebi et al. 2021). Thus, initiatives to construct climate-resilient NCDs must centre on building resilient food systems.

To build food systems that can withstand natural disasters, farmers must use sustainable methods, diversify their crop yields, and invest in climate-smart farming technologies (Bhatnagar et al. 2024). To improve access to fresh produce in marginalised communities and reduce reliance on vulnerable global markets, it is recommended that local food supply chains be strengthened. Preventing contamination and foodborne infections, which can negatively impact nutritional health, is another benefit of water security measures such as safe storage and rainwater collection.

Policy and Governance for Inclusive Health Action

Vulnerable populations are disproportionately affected by environmental stressors like air pollution, extreme heat, and resource shortages, which exacerbate existing health conditions. Addressing these challenges requires policy and governance approaches that go beyond isolated, siloed efforts. A coordinated and inclusive approach, supported by responsive institutions, enabling policy structures, and active community participation, is necessary to reduce risks and build resilient health systems. This section outlines three key strategies: (1) explores multi-sectoral coordination in aligning health, environment, and urban policies; (2) focuses on community engagement and participatory governance; (3) focuses on the importance of advancing health equity by prioritizing at-risk groups and addressing the social determinants of health.

Multi-sectoral Coordination: Integrating Health, Environment, and Urban Planning

Climate change significantly affects human health by worsening key social and environmental determinants, contributing to the rising burden of NCDs (Dhimal et al. 2020). Rapid urbanization has intensified risks related to air pollution, inadequate housing, and limited access to healthcare (de Sa et al. 2021). Urban infrastructure plays a critical role, where access to green and walkable spaces supports health, while proximity to roads, industrial zones, and landfills increases disease risk (Zhang et al. 2023). Recognizing these challenges, world leaders pledged to reduce premature deaths related to NCDs by one-third by 2030 at the 2015 United Nations High-Level Political Forum (Sornpaisarn et al. 2024).

The National Multi-sectoral Action Plan (NMAP) 2017 outlines coordinated actions across non-health sectors to achieve the national target of reducing premature mortality from NCDs by 25% by 2025 (Tiwari et al. 2024). The WHO guiding framework on Multisectoral Action for NCD prevention outlines four key pillars for successful collaboration: (1) Governance and accountability through strong leadership to foster cross-sectoral engagement; (2) Leadership by leveraging governance structures for coordination; (3) Purposeful ways of working using tools and processes to operationalize actions; and (4) Resources and capabilities to ensure adequate human and financial support (Akselrod et al. 2024).

Rooted in the "whole-of-government" and "Health in All Policies" (HiAP) frameworks, the approach encourages all sectors to integrate health considerations and disease prevention into their policy-making processes. Effective implementation depends on sustained advocacy, capacity-building, and inter-agency knowledge exchange (Tiwari et al. 2024). The existing model employed by South Australia is the establishment of a HiAP unit to inform policymakers on the impact of decision-making on health. Meanwhile, European Countries like France, Finland, the UK, and

Norway established multisector committees within the government, participated by representatives from all relevant sectors.

WHO provides countries with governance frameworks, tools, and technical guidance to promote inclusive and healthy urban environments. Its Global Strategy on Health, Environment, and Climate Change sets a vision through 2030 to reduce environmental health risks by transforming systems of living, production, and governance (WHO 2020). Aligned with this vision, the HiAP approach embedded disease prevention and health promotion into urban and economic policymaking (Talebi et al. 2024). Besides, the WHO Housing and Health Guidelines are implemented at the regional, national, and local levels by providing evidence-based recommendations to minimize health risks from poor housing conditions. These comprehensive guidelines address issues such as overcrowding, indoor temperature extremes, home injury hazards, and housing accessibility, while also consolidating guidance on water and air quality, noise, and toxic substances such as tobacco smoke, lead, asbestos, and radon (de Sa et al. 2021).

A key challenge in sustainable development is equipping urban and regional leaders with the knowledge to incorporate health and well-being into planning processes. In response, WHO and UN-Habitat jointly developed the Sourcebook on Integrating Health in Urban and Territorial Planning, which provides practical guidance for governments, planners, health professionals, and civil society (De La Haye et al. 2024). The sourcebook outlines the importance of health in planning and offers frameworks, tools, and case studies to support implementation across governance levels and spatial scales. In addition, the Urban Health Initiative model (UHI) has been established to support healthier, climate-friendly cities, leveraging processes, knowledge, local data, and competencies in urban transformation. The model outlines six steps, including mapping current systems, applying context-specific health and economic tools, testing local scenarios, building engagement capacity, sustaining outreach, and monitoring to refine policy (de Sa et al. 2021).

Effective Multi-Sectoral Collaboration (MSC) relies on political commitment, shared vision across sectors, inclusive stakeholder involvement, effective coordination and communication, timely data sharing, adequate resources, and evaluation mechanisms (Sornpaisarn et al. 2024). However, barriers such as limited leadership, funding, accountability, and data availability impede the implementation of multisectoral action and reduce the effectiveness of integrated responses to complex health and environmental issues. Despite global commitments and resources, sectoral silos persist, particularly among health, climate, energy, mobility, housing, and urban planning, resulting in fragmented policies and missed opportunities for integration (McIntyre et al. 2024). These systemic disconnects contribute to inefficiencies and unsustainable practices, such as fossil-fuel dependence and car-centric urban design. Weak coordination across systems like water, energy, and food further compounds vulnerabilities. Limited collaboration among sectors and actors weakens governance and results in duplication of efforts, weak governance, and ineffective climate-resilient systems (Berry et al. 2018).

Community Engagement and Participatory Solutions

Inter-sectoral collaboration and community participation are essential strategies for achieving health equity and success in combating NCDs, especially among the most vulnerable communities, which are most exposed to extreme heat and other health-related challenges. Empowering communities is necessary for prevention, given that the prevalence of NCDs is closely linked to individual lifestyles and health behaviours, which are influenced by economic, political, cultural, social, behavioural, environmental, and biological factors (Talebi et al. 2024).

Participatory disaster preparedness strategies include engaging seminars, workshops, and awareness campaigns to enhance the understanding of local risks and encourage proactive behaviour. Medium-impact strategies like home visits and school programs promote early education and build household-level preparedness. Focus groups provide platforms for community dialogue and feedback for multi-level engagement in health resilience (Kangana et al. 2024).

Participatory disaster risk assessments are practical for identifying vulnerabilities and strengthening community capacities. Tools like the Vulnerability and Capacity Assessment (VCA) integrate community knowledge into development planning and expand methods such as mapping, transect walks, and seasonal calendars by incorporating household or neighbourhood vulnerability assessments (Schwerdtle et al. 2025). The process empowers communities to recognize risk and strengthen their capacities to natural hazards.

Local involvement in the design and dissemination of early warning systems, such as heat advisories and air quality alerts, ensures that these tools are accessible, timely, and contextually appropriate. Researchers highlighted the need for real-time, community-specific data and emphasized the need to consider local leaders' perspectives, available resources, and capacity for involvement in risk communication (McIntyre et al. 2024).

Advancing Health Equity in Climate and NCD Policies

Health equity is a core principle of public health aimed at ensuring that every individual has an equal opportunity to achieve optimal health, regardless of their economic, social, or demographic background. In addition to key social determinants such as economic stability, housing security, education, and environmental conditions, health equity recognizes that diverse populations encounter distinct barriers in accessing healthcare and related resources. Targeted interventions are necessary to address these disparities to reduce preventable diseases and improve life expectancy, particularly among marginalized groups (Saadati 2025). Despite global efforts to integrate equity into health strategies, systemic and institutional barriers persist and disproportionately affect vulnerable populations (Arifeen et al. 2024).

A growing body of locally focused research has explored strategies to reduce air quality and heat exposure, emphasizing community-engaged methods that address local priorities and aim to combine knowledge with action to reduce health disparities

(McIntyre et al. 2024). In this context, improving neighbourhood infrastructure is a global priority to address inequities. Therefore, identifying high-risk areas with strong evidence and prioritizing populations in poor conditions is essential. Following the past place-based interventions, targeted spatial strategies using urban design tools are urgently needed to support sustainable urban development (Zhang et al. 2023).

A comprehensive review of healthcare data sources, including electronic health records, surveys, and registries, is essential to address disparities (Anyanwu et al. 2024). Geospatial tools such as GIS and hotspot mapping further enhance understanding of variations in health outcomes, service access, and utilization across different demographic and geographic groups, by visualizing spatial inequities and informing targeted interventions. Health equity indices and environmental scans integrate social and environmental data to provide a more holistic view of disparities and guide equitable planning (Ilojianya et al. 2024).

Participatory needs assessments and collaboration with local leaders ensure interventions are context-specific and culturally responsive (Stockton et al. 2021). Multi-sectoral collaborations and health equity task forces provide coordination and accountability, while advocacy for systemic change addresses root causes such as poverty and racism (Babarinde et al. 2023). Reducing financial barriers is an important component of equity-focused health policy. Expanding insurance coverage, providing supportive funds, and offering cost-effective self-care tools enable access to health services (Tirado et al. 2022; Saadati 2025). For example, in Taiwan, insurance coverage for corticosteroids and inhalation therapies lowered lung cancer rates in women with COPD. In contrast, a study in Germany found that higher costs for services such as mammography screening reduced their use among individuals (Talebi et al. 2024).

Conclusion and Future Directions

This narrative review has shown that climate extremes substantially intensify the burden of non-communicable diseases (NCDs) through a constellation of interlinked biological, environmental, social, and health system pathways. Extreme heat, wildfire smoke, and deteriorating air quality impose direct physiological stress that elevates cardiovascular and respiratory morbidity and mortality, while floods and droughts disrupt food and water systems, contributing to undernutrition, metabolic disorders, and long-term health deterioration. Climate-related disasters further compromise continuity of care by disrupting healthcare infrastructure, medication supply chains, and access to essential services. Psychological and mental health effects, including stress, trauma, and anxiety, add a layer of risk that interacts with chronic disease vulnerability. Taken collectively, the evidence indicates that climate extremes function as risk multipliers, accelerating NCD onset, exacerbating disease progression, and increasing preventable mortality.

Beyond documenting these pathways, this review advances current understanding by explicitly integrating exposure mechanisms, population vulnerability, and health system constraints within a single conceptual framework. Rather than treating climate hazards, NCDs, and vulnerability as separate domains, the synthesis presented here demonstrates how their interaction produces uneven and escalating health burdens across settings. This integrative perspective helps to bridge disciplinary silos between climate science, environmental health, chronic disease epidemiology, and health systems research, offering a more comprehensive lens through which to interpret emerging evidence.

The findings also underscore that vulnerability to climate-related NCD impacts is not randomly distributed. Older adults, individuals with pre-existing conditions, socioeconomically disadvantaged populations, women, and persons with disabilities experience disproportionate harm due to the intersection of biological susceptibility, social position, and structural inequities in housing, infrastructure, and healthcare access. In many low- and middle-income countries, fragile health systems face compounded pressures as climate shocks disrupt service delivery, intensify workforce shortages, and weaken already constrained supply chains. These patterns highlight that vulnerability is socially produced and maintained through governance, planning, and resource allocation decisions, rather than being an inherent characteristic of individuals or communities.

From a public health and policy perspective, this review highlights the urgency of embedding climate resilience within NCD prevention and management strategies. Strengthening climate-informed surveillance and early warning systems, integrating environmental risk considerations into clinical guidelines, investing in climate-resilient health infrastructure, and aligning urban planning with health-promoting design principles are essential steps. Securing resilient food and water systems and advancing intersectoral action through approaches such as Health in All Policies are equally critical for addressing upstream drivers of risk.

As a narrative review, this chapter also points to several priorities for future research. There remains a need for longitudinal and comparative studies across regions to better quantify cumulative and synergistic effects of climate extremes on NCD trajectories. Greater attention is required to understudied populations, including informal urban communities and people with disabilities, as well as to mental health outcomes that intersect with chronic disease management. Methodologically, future work would benefit from integrating climate exposure data with health system performance metrics to better understand resilience and failure points during extreme events. In conclusion, addressing the double burden of climate extremes and NCDs requires moving beyond disease-specific or hazard-specific responses toward equity-centred, systems-oriented strategies. By reframing climate change as a central determinant of chronic disease risk, this review contributes to a growing body of evidence supporting climate justice–informed public health action. Protecting vulnerable populations and strengthening health systems against climate extremes will be essential for sustaining gains in NCD control and advancing long-term population health in a changing climate.

AI Disclosure Statement

The authors declare that artificial intelligence-based tools were applied during the preparation of this manuscript only to support language refinement and enhance manuscript clarity. All scientific content conclusions were independently produced by the authors, who assume full responsibility for the accuracy and integrity of the study.

Acknowledgements The authors thank all collaborators and institutions whose insights and resources contributed to the development of this work.

Authors' Contribution All authors contributed equally to the conceptualization, literature review, analysis, and writing of this work.

Funding The work was not supported by any funding.

Conflict of Interest The authors declare that there are no conflicts of interest related to this work.

Informed Consent Not applicable.

References

Aggarwal S, Hu JK, Sullivan JA, Parks RM, Nethery RC (2025) Severe flooding and cause-specific hospitalisation among older adults in the USA: a retrospective matched cohort analysis. Lancet Planet Health 9(7):101268. https://doi.org/10.1016/S2542-5196(25)00132-9

Akselrod S, Collins TE, Berlina D, De Pinho campos K, Fones G, de Sousa Neves D, Bashir F, Allen LN (2024) Multisectoral action to address noncommunicable diseases: lessons from three country case studies. Front Public Health 12:1303786. https://doi.org/10.3389/fpubh.2024.1303786

Alahmad B, Khraishah H, Shakarchi AF, Albaghdadi M, Rajagopalan S, Koutrakis P, Jaffer FA (2020) Cardiovascular mortality and exposure to heat in an inherently hot region: implications for climate change. Circulation 141(15):1271–1273. https://doi.org/10.1161/CIRCULATIONAHA.119.044860

Al-Kindi SG, Brook RD, Biswal S, Rajagopalan S (2020) Environmental determinants of cardiovascular disease: lessons learned from air pollution. Nat Rev Cardiol 17(10):656–672. https://doi.org/10.1038/s41569-020-0371-2

Allweiss P, Albright A (2011) Diabetes, disasters and decisions. Diab Manag 1(4):369–377. https://doi.org/10.2217/dmt.11.31

Alvarado-Calvo A, Alvarado-Rodríguez Y, Cruz-Mora K, Mora-Jiménez J, Arguedas-Chacón S, Zavaleta-Monestel E (2025) Trends and association of environmental exposure and climate change with non-communicable diseases in Latin America. Healthcare 13(14):1653. https://doi.org/10.3390/healthcare13141653

Ansah EW, Amoadu M, Obeng P, Sarfo JO (2024) Health systems response to climate change adaptation: a scoping review of global evidence. BMC Public Health 24(1):19459. https://doi.org/10.1186/s12889-024-19459-w

Anyanwu EC, Maduka CP, Ayo-Farai O, Okongwu CC, Daraojimba AI (2024) Maternal and child health policy: a global review of current practices and future directions. World J Adv Res Rev 21(2):1770–1781

Arifeen SE, Grove J, Hansen PM, Hargreaves J, Johnson HL, Johri M, Saville EA (2024) Evaluating global health initiatives to improve health equity. Bull World Health Organ 102(2):137–139. https://doi.org/10.2471/blt.23.290531

Babarinde AO, Ayo-Farai O, Maduka CP, Okongwu CC, Sodamade O (2023) Data analytics in public health, a USA perspective: a review. World J Adv Res Rev 20(3):211–224

Bell SA, Abir M, Choi H, Cooke C, Iwashyna T (2018) All-cause hospital admissions among older adults after a natural disaster. Ann Emerg Med 71(6):746–754. https://doi.org/10.1016/j.annemergmed.2017.06.042

Berry HL, Waite TD, Dear KBG, Capon AG, Murray V (2018) The case for systems thinking about climate change and mental health. Nat Clim Chang 8(4):282–290. https://doi.org/10.1038/s41558-018-0102-4

Bhatnagar S, Chaudhary R, Sharma S, Janjhua Y, Thakur P, Sharma P, Keprate A (2024) Exploring the dynamics of climate-smart agricultural practices for sustainable resilience in a changing climate. Environ Sustain Indicat 24:100535. https://doi.org/10.1016/j.indic.2024.100535

Bolan S, Padhye LP, Hasemizad T, Govarthanan M, Karmegam N, Wijesekara H, Amarasiri D, Hou D, Zhou P, Biswal BK, Balasubramanian R, Wang H, Sidddique KHM, Rinklebe J, Kirkham MB, Bolan N (2024) Impacts of climate change on the fate of contaminants through extreme weather events. Sci Total Environ 909:168388. https://doi.org/10.1016/j.scitotenv.2023.168388

Boren E, Gershwin ME (2004) Inflamm-aging: autoimmunity, and the immune-risk phenotype. Autoimmun Rev 3(5):401–406. https://doi.org/10.1016/j.autrev.2004.03.00

Bryan K, Ward S, Roberts L, White MP, Landeg O, Taylor T, McEwen L (2020) The health and well-being effects of drought: assessing multi-stakeholder perspectives through narratives from the UK. Clim Change 163:2073–2095. https://doi.org/10.1007/s10584-020-02916-x

Budreviciute A, Damiati S, Sabir DK, Onder K, Schuller-Goetzburg P, Plakys G, Katileviciute A, Khoja S, Kodzius R (2020) Management and prevention strategies for non-communicable diseases (NCDs) and their risk factors. Front Public Health 8:574111. https://doi.org/10.3389/fpubh.2020.574111

Chandra N, Lee JKW (2025) A systematic review of heat health warning systems: enhancing the framework towards effective health outcomes. Curr Environ Health Rep 12(1):31. https://doi.org/10.1007/s40572-025-00496-5

Chen H, Samet JM, Bromberg PA, Tong H (2021) Cardiovascular health impacts of wildfire smoke exposure. Part Fibre Toxicol 18:2. https://doi.org/10.1186/s12989-020-00394-8

Chhetri R, Mukherjee K, Dash P (2021) Potential effect of the changing urban climate on non-communicable diseases: a case study on Barasat City, India. IOP Conf Ser Earth Environ Sci 116491:012015

Clarke J, Blom IM, Campbell-Lendrum D, Kerry V (2025) Health workforce as a climate and health priority in nationally determined contributions. The Lancet Planetary Health 21:101288. https://doi.org/10.1016/j.lanplh.2025.101288

Comtesse H, Ertl V, Hengst SM, Rosner R, Smid GE (2021) Ecological grief as a response to environmental change: a mental health risk or functional response? Int J Environ Res Public Health 18(2):734. https://doi.org/10.3390/ijerph18020734

Crenshaw K (1989) Demarginalizing the intersection of race and sex: a black feminist critique of antidiscrimination doctrine. Univ Chic Leg Forum 1989(1):139–167

Dalla Vecchia C, Sadik A, Jonah L (2024) Impact of climate change in emergency medicine: a narrative review. J Public Health Emerg 8:27. https://doi.org/10.21037/jphe-23-87

de Bont J, Nori-Sarma A, Stafoggia M, Banerjee T, Ingole V, Jaganathan S, Mandal S, Rajiva A, Krishna B, Kloog I, Lane K, Mall RK, Tiwari A, Wei Y, Wellenius GA, Prabhakaran D, Schwartz J, Prabhakaran P, Ljungman P (2024) Impact of heatwaves on all-cause mortality in India: a comprehensive multi-city study. Environ Int 184:108461. https://doi.org/10.1016/j.envint.2024.108461

de Sa TH, Ludolph R, Hinsch M, Roebbel N (2021) Health at the heart of urban and territorial planning: the WHO's multi-sectoral leadership for human and planetary health. J Urban Health 98(4):438–441. https://doi.org/10.1007/s11524-021-00533-2

De La Haye A, Gow R, Lake AA, Ball C, van Herk S, Rofin-Serra M et al (2024) A thematic analysis of UK healthy planning frameworks and tools designed to support the inclusion of health in urban planning. Cit Health 8(5):956–972. https://doi.org/10.1080/23748834.2024.2353963

Dewi SP, Kasim R, Sutarsa N, Dykgraaf SH (2024) A scoping review of the impact of extreme weather events on health outcomes and healthcare utilization in rural and remote areas. BMC Health Serv Res 24(1):1333. https://doi.org/10.1186/s12913-024-11695-5

Dhimal M, Neupane T, Dhimal ML (2020) Understanding linkages between environmental risk factors and noncommunicable diseases—a review. FASEB BioAdv 16:119. https://doi.org/10.1096/fba.2020-00119

Dodman D, Archer D, Satterthwaite D (2019) Editorial: responding to climate change in contexts of urban poverty and informality. Environ Urban 31(1):3–12. https://doi.org/10.1177/0956247819830004

Ebi K, Vanos J, Baldwin JW, Bell JE, Hondula DM, Errett NA, Hayes K, Reid CE, Saha S, Spector J, Berry P (2021) Extreme weather and climate change: population health and health system implications. Annu Rev Public Health 42:293–315. https://doi.org/10.1146/annurev-publhealth-012420-105026

Enarson E, Chakrabarti PGD (2009) Women, gender and disaster: global issues and initiatives. Sage Publications India, New York

Ezeh A, Oyebode O, Satterthwaite D, Chen Y-F, Ndugwa R, Sartori J et al (2017) The history, geography, and sociology of slums and the health problems of people who live in slums. Lancet 389(10068):547–558. https://doi.org/10.1016/S0140-6736(16)31650-6

Fanzo J, Rudie C, Sigman I, Grinspoon S, Benton TG, Brown ME, Covic N, Fitch K, Golden CD, Grace D, Hivert MF, Huybers P, Jaacks LM, Masters WA, Nisbett N, Richardson RA, Singleton CR, Webb P, Willett WC (2022) Sustainable food systems and nutrition in the 21st century: a report from the 22nd annual Harvard nutrition obesity symposium. Am J Clin Nutr 115(1):18–33. https://doi.org/10.1093/ajcn/nqab315

Faurie C, Varghese BM, Liu J, Bi P (2022) Association between high temperature and heatwaves with heat-related illnesses: a systematic review and meta-analysis. Sci Total Environ 852:158332. https://doi.org/10.1016/j.scitotenv.2022.158332

Food and Agriculture Organization of the United Nations (FAO) (2015) Climate change and food security: risk and responses. https://openknowledge.fao.org/server/api/core/bitstreams/a4fd8ac5-4582-4a66-91b0-55abf642a400/content

Food and Agriculture Organization of the United Nations (FAO) (2022) The state of food security and nutrition in the world. https://openknowledge.fao.org/server/api/core/bitstreams/67b1e9c7-1a7f-4dc6-a19e-f6472a4ea83a/content

Gartrell A, Calgaro E, Goddard G, Saorath N (2020) Resilience for all? Critically examining disability and gender in disaster risk reduction policy and practice in Cambodia. Glob Environ Chang 64:102134. https://doi.org/10.1016/j.gloenvcha.2020.102134

Ghazanchaei E, Khorasani-Zavareh D, Aghazadeh-Attari J, Mohebbi I (2021) Identifying and describing impact of disasters on non-communicable diseases: a systematic review. Iran J Public Health 50(6):1143–1155

Gkouliaveras V, Kalogiannidis S, Kalfas D, Kontsas S (2025) Effects of climate change on health and health systems: a systematic review of preparedness, resilience, and challenges. Int J Environ Res Public Health 22(2):232. https://doi.org/10.3390/ijerph22020232

Grant E, Runkle JD (2022) Long-term health effects of wildfire exposure: a scoping review. J Clim Change Health 6:100110. https://doi.org/10.1016/j.joclim.2021.100110

Haron MZ, Rohama AJ, Hamid NAA, Omar MA, Abdullah NH (2023) Stunting and its associated factors among children below 5 years old on the east coast of Peninsular Malaysia: evidence from the national health and morbidity survey. Malaysian J Med Sci 30(5):155–168

Hussain D, Sharma S (2025) Flood exposure and childhood malnutrition in India using composite index of anthropometric failure (CIAF). Environ Res 276:121527. https://doi.org/10.1016/j.envres.2025.121527

Ilojianya VI, Usman FO, Ibekwe KI, Nwokediegwu ZQS, Umoh AA, Adefemi A (2024) Data-driven energy management: review of practices in Canada, USA, and Africa. Eng Sci Technol J 5(1):219–230

Islam S, Karipot A, Bhawar R, Sinha P, Kedia S, Khare M (2024) Urban heat island effect in India: a review of current status, impact and mitigation strategies. Discover Cit 1:34. https://doi.org/10.1007/s44327-024-00033-3

Jordan RE, Adab P, Cheng KK (2020) Covid-19: risk factors for severe disease and death. BMJ 368:m1198. https://doi.org/10.1136/bmj.m1198

Kangana N, Kankanamge N, De Silva C, Goonetilleke A, Mahamood R, Ranasinghe D (2024) Bridging community engagement and technological innovation for creating smart and resilient cities: a systematic literature review. Smart Cit 7(6):3823–3852. https://doi.org/10.3390/smartcities7060147

Kett M, Twigg J (2007) Disability and disasters: towards an inclusive approach. World Disaster Report 6:86–109

Keya TA, Leela A, Habib N, Rashid M, Bakthavatchalam P (2023) Mental health disorders due to disaster exposure: a systematic review and meta-analysis. Cureus 15(4):e37031. https://doi.org/10.7759/cureus.37031

Kuddus MA, Tynan E, McBryde E (2020) Urbanization: a problem for the rich and the poor? Public Health Rev 41(1):1. https://doi.org/10.1186/s40985-019-0116-0

Kuntić M, Hahad O, Münzel T, Daiber A (2024) Crosstalk between oxidative stress and inflammation caused by noise and air pollution—implications for neurodegenerative diseases. Antioxidants 13(3):266. https://doi.org/10.3390/antiox13030266

Li S, Peng Y, Wang X, Qian Y, Xiang P, Wade SW, Guo H, Lopez JA, Herzog CA, Handelsman Y (2019) Cardiovascular events and death after myocardial infarction or ischemic stroke in an older Medicare population. Clin Cardiol 42(3):391. https://doi.org/10.1002/clc.23160

Lim YH, Park MS, Kim Y, Kim H, Hong YC (2015) Effects of cold and hot temperature on dehydration: a mechanism of cardiovascular burden. Int J Biometeorol 59:1035–1043. https://doi.org/10.1007/s00484-014-0917-2

Ling C, Shen Q, Cui L, Yang M, Liu Z, Xu T, Sun W, He S (2025) Air pollution, drought, and their synergistic effects on cardiovascular disease risk in China: a nationwide study. Environ Pollut 384:126955. https://doi.org/10.1016/j.envpol.2025.126955

Liu J, Varghese BM, Hansen A, Zhang Y, Driscoll T, Morgan G, Dear K, Gourley M, Capon A, Bi P (2022) Heat exposure and cardiovascular health outcomes: a systematic review and meta-analysis. Lancet Planet Health 6(6):e484–e495. https://doi.org/10.1016/S2542-5196(22)00117-6

Lokmic-Tomkins Z, Bhandari D, Bain C, Borda A, Kariotis TC, Reser D (2023) Lessons learned from natural disasters around digital health technologies and delivering quality healthcare. Int J Environ Res Public Health 20(5):4542. https://doi.org/10.3390/ijerph20054542

Mani ZA, Khorram-Manesh A, Goniewicz K (2024) Global health emergencies of extreme drought events: historical impacts and future preparedness. Atmosphere 15(9):1137. https://doi.org/10.3390/atmos15091137

Martins FP, Paschoalotto MAC, Closs J, Bukowski M, Veras MM (2024) The double burden: climate change challenges for health systems. Environ Health Insights 18:789. https://doi.org/10.1177/11786302241298789

Matthews V, Longman J, Berry HL, Passey M, Bennett-Levy J, Morgan GG, Pit S, Rolfe M, Bailie RS (2019) Differential mental health impact six months after extensive river flooding in rural Australia: a cross-sectional analysis through an equity lens. Front Public Health 7:367. https://doi.org/10.3389/fpubh.2019.00367

McIntyre A, Heidari L, Hagen M, Bongiovanni R, Bowman BN, Fabian P, Kinney P, Scammel MK (2024) Extreme heat and air quality: community leader perspectives on information barriers and opportunities in two environmental justice communities. New Solut J Environ Occup Health Policy 34(4):256–267. https://doi.org/10.1177/10482911241290557

Mkharia G, Sadeghi A, Leddin D, Costello A (2023) Impacts of climate change on vulnerable populations. Gut 72(12):1195. https://doi.org/10.1136/gutjnl-2023-331195

Montero R, Chung I, Wright M (2025) Healthcare resilience in the climate crisis: do we have any? Fut Healthcare J 12(1):100241. https://doi.org/10.1016/j.fhj.2025.100241

Mosadeghrad AM, Isfahani P, Eslambolchi L, Zahmatkesh M, Afshari M (2023) Strategies to strengthen a climate-resilient health system: a scoping review. Glob Health 19(1):62. https://doi.org/10.1186/s12992-023-00965-2

Mueller AL, McNamara MS, Sinclair DA (2020) Why does COVID-19 disproportionately affect older people? Aging 12(10):9959–9981

Münzel T, Hadad O, Sørensen M, Lelieveld J, Duerr GD, Nieuwenhuijsen M, Daiber A (2021) Environmental risk factors and cardiovascular diseases: a comprehensive expert review. Cardiovasc Res 118(14):2880–2902. https://doi.org/10.1093/cvr/cvab316

Muriuki D, Ambrose M, Ahmed H, Foster M, Nguchu H, Goldberg L, Samia B (2025) Cardiovascular disease care beyond the cardiologist: an overview of the rollout of transthoracic echocardiography training and services in Kenya. Glob Heart 20(1):50. https://doi.org/10.5334/gh.1437

Ngcamu BS (2023) Climate change effects on vulnerable populations in the global south: a systematic review. Nat Hazards 118:977–991. https://doi.org/10.1007/s11069-023-06070-2

Nicholas PK, Breakey S, White BP, Brown MJ, Fanuele J, Starodub R, Ros AV (2020) Mental health impacts of climate change: perspectives for the ED clinician. J Emerg Nurs 46(5):590–599. https://doi.org/10.1016/j.jen.2020.05.014

Nieuwenhuijsen MJ (2016) Urban and transport planning, environmental exposures and health-new concepts, methods and tools to improve health in cities. Environ Health 15(Suppl 1):38. https://doi.org/10.1186/s12940-016-0108-1

Nshimiyimana C, Thakur N, Alalawy AI, Zheng Y, Kareri T, Salama E-S (2025) Low-water quality and non-communicable disease nexus: health impacts, mechanisms, and advanced water treatment approaches. J Environ Chem Eng 13(2):115744. https://doi.org/10.1016/j.jece.2025.115744

Nugroho HYSH, Basuki TM, Savitri E, Supangat AB, Putra PB, Wahyuningrum N, Adi RN, Setiawan O, Nandini R, Cahyono SA, Auliyani D, Nada FMH, Pratiwi D, Hasani M (2025) Advancing air quality monitoring systems towards sustainable green development: insight for metropolitan cities in Indonesia. Environ Sustain Indicat 26:100649. https://doi.org/10.1016/j.indic.2025.100649

Ochi S, Murray V, Hodgson S (2013) The great East Japan earthquake disaster: a compilation of published literature on health needs and relief activities, March 2011–september 2012. PLoS Curr 5:e765678. https://doi.org/10.1371/currents.dis.771beae7d8f41c31cd91e765678c005d

Ochi S, Hodgson S, Landeg O, Mayner L, Murray V (2014) Disaster-driven evacuation and medication loss: a systematic literature review. PLoS Curr 6:96. https://doi.org/10.1371/currents.dis.fa417630b566a0c7dfdbf945910edd96

Padhy SK, Sarkar S, Panigrahi M, Paul S (2015) Mental health effects of climate change. Indian J Occup Environ Med 19(1):3–7. http://doi.org/10.4103/0019-5278.156997

Ramos C, Minghelli B (2022) Prevalence and factors associated with poor respiratory function among firefighters exposed to wildfire smoke. Int J Environ Res Public Health 19(14):8492. https://doi.org/10.3390/ijerph19148492

Ratcliff GE, Matheny ME, Brown JR, Sullivan I, Richmond BW, Paulin LM, Conger AK, Davis SE (2024) Integrating clinical and air quality data to improve prediction of COPD exacerbations. AMIA Ann Symp Proc 2023:1209–1217

Rizzo LV, Rizzo MCFV (2024) Wildfire smoke and health impacts: a narrative review. J Pediatr 101:s56–s64. https://doi.org/10.1016/j.jped.2024.11.006

Ryan BJ, Franklin RC, Burkle FM Jr, Aitken P, Smith E, Watt K, Leggat P (2016) Reducing disaster exacerbated non-communicable diseases through public health infrastructure resilience: perspectives of Australian disasters service providers. PLoS Curr 8:266. https://doi.org/10.1371/currents.dis.d142f36b6f5eeca806d95266b20fed1f

Saadati SM (2025) The future of health equity: policy strategies to reduce disparities in public health. Foresight Health Govern 2(2):15–32

Saguansap P, Saguansap V, Mruksirisuk P, Thanvisitthpon NT (2024) Assessment of urban heat island vulnerability using sustainability-focused framework: a case study of Thailand's Bangkok metropolis. Curr Res Environ Sustain 8:100262. https://doi.org/10.1016/j.crsust.2024.100262

Saleem MS, Fatima SZ, Kamran H, Nouman A, Bibi K (2024) Impact of rising summer temperatures on government sector tertiary care emergency centers: addressing heatstroke and associated death rates in Karachi, Pakistan. Environ Health Insights 18:1–3. https://doi.org/10.1177/11786302241303584

Schraufnagel DE, Balmes JR, Cowl CT, De Matteis S, Jung SH, Mortimer K, Perez-Padilla R, Rice MB, Riojas-Rodriguez H, Sood A, Thurston GD, To T, Vanker A, Wuebbles DJ (2019) Air pollution and noncommunicable diseases: a review by the forum of international respiratory societies' environmental committee, part 2: air pollution and organ systems. Chest 155(2):417–426. https://doi.org/10.1016/j.chest.2018.10.041

Schwerdtle PN, Zidouemba DT, Dermbaye AR, Jobanputra K, McRae M, Tarabbo M, Njouonkou M, Madjissem M, Robert A, Haider Z (2025) Building climate resilience in health systems: a climate vulnerability and capacity assessment in a rural hospital in Chad. Ann Glob Health 91(1):50. https://doi.org/10.5334/aogh.4743

Siiba A, Kangmennaang J, Baatiema L, Luginaah I (2024) The relationship between climate change, globalization and non-communicable diseases in Africa: a systematic review. PLoS ONE 19(2):e0297393. https://doi.org/10.1371/journal.pone.0297393

Singer M (2017) Syndemics and the biosocial conception of health. Lancet 389(10072):941–950. https://doi.org/10.1016/S0140-6736(17)30003-X

Singh N, Areal AT, Breitner S, Zhang S, Agewall S, Schikowski T, Schneider A (2024) Heat and cardiovascular mortality: an epidemiological perspective. Circ Res 134(9):1098–1112. https://doi.org/10.1161/CIRCRESAHA.123.323615

Sinha P, Coville RC, Hirabayashi S, Lim B, Endreny TA, Nowak DJ (2021) Variation in estimates of heat-related mortality reduction due to tree cover in US cities. J Environ Manage 301:113751. https://doi.org/10.1016/j.jenvman.2021.113751

Sornpaisarn B, Chunharas S, Sornpaisarn S, Saonuam P, Nipun RF, Burtyee C, Samutachak B, Chandarasorn M, Supadulya N, Chunsuttiwat S, Singha S, Rojanapithayakorn W, Ungchusak K, Rehm J (2024) Enhancing multi-sectoral collaborations for the prevention and control of NCDs in Thailand with a new approach. Health Res Policy Syst 22:169. https://doi.org/10.1186/s12961-024-01262-z

Stockton DA, Fowler C, Debono D, Travaglia J (2021) World Health Organization building blocks in rural community health services: an integrative review. Health Sci Rep 4(2):e254. https://doi.org/10.1002/hsr2.254

Sutherst RW (2004) Global change and human vulnerability to vector-borne diseases. Clin Microbiol Rev 17(1):136–173. https://doi.org/10.1128/CMR.17.1.136-173.2004

Talebi E, Dabbaghi F, Mahmoudi G (2024) Proposing a model for engaging community participation in the prevention and control of non-communicable diseases. Int Arch Health Sci 11(4):242–248

The Lancet Healthy Longevity (2024) The decade of healthy ageing: progress and challenges ahead. Lancet Healthy Longevity 5(1):e1. https://doi.org/10.1016/S2666-7568(23)00271-4

Tirado MC, Vivero-Pol JL, Bezner Kerr R, Krishnamurthy K (2022) Feasibility and effectiveness assessment of multi-sectoral climate change adaptation for food security and nutrition. Curr Clim Change Rep 8(1):35–52. https://doi.org/10.1007/s40641-022-00181-x

Tiwari A, Singh SK, Manar M, Kanchan S, Kaur M (2024) A qualitative assessment of multi-sectoral action for non-communicable diseases (NCDs) in Uttar Pradesh, India. Discover Public Health 21(5):8

Tomio J, Sato H (2014) Emergency and disaster preparedness for chronically ill patients: a review of recommendations. Open Access Emerg Med OAEM 6:69–79. https://doi.org/10.2147/OAEM.S48532

Tong S, Banbrick H, Beggs PJ, Chen L, Hu Y, Ma W, Steffen W, Tan J (2022) Current and future threats to human health in the anthropocene. Environ Int 158:106892. https://doi.org/10.1016/j.envint.2021.106892

Turk MA, Landes SD, Formica MK, Goss KD (2020) Intellectual and developmental disability and COVID-19 case-fatality trends: TriNetX analysis. Disabil Health J 13(3):100942. https://doi.org/10.1016/j.dhjo.2020.100942

UN-Habitat (2020) World cities report 2020: the value of sustainable urbanization. UN-Habitat, Nairobi

United Nations Framework Convention on Climate Change (UNFCCC) (2018) Considerations regarding vulnerable groups, communities and ecosystems in the context of the national adaptation plan. https://unfccc.int/sites/default/files/resource/Considerations%20regarding%20vulnerable.pdf

United States of Environmental Protection Agency (USEPA) (2025) Climate change and the health of socially vulnerable people. https://www.epa.gov/climateimpacts/climate-change-and-health-socially-vulnerable-people

van Daalen KR, Kallesøe SS, Davey F, Dada S, Jung L, Singh L, Issa R, Emilian CA, Kuhn I, Keygnaert I, Nilsson M (2022) Extreme events and gender-based violence: a mixed-methods systematic review. Lancet Planet Health 6(6):e504–e523. https://doi.org/10.1016/S2542-5196(22)00088-2

Walinski A, Sander J, Gerlinger G, Clemens V, Meyer-Lindenberg A, Heinz A (2023) The effects of climate change on mental health. Deutsches Arzteblatt Int 120(8):117–124. https://doi.org/10.3238/arztebl.m2022.0403

Warraich H, Zaidi AK, Patel K (2011) Floods in Pakistan: a public health crisis. Bull World Health Organ 89(3):236–237. https://doi.org/10.2471/BLT.10.083386

Wheeler N, Watts N (2018) Climate change: from science to practice. Curr Environ Health Rep 5(1):170–178. https://doi.org/10.1007/s40572-018-0187-y

Wilkinson A, Ali H, Bedford J, Boonyabancha S, Connolly C, Conteh A (2020) Local response in health emergencies: key considerations for addressing the COVID-19 pandemic in informal urban settlements. Environ Urban 32(2):503–522

World Health Organization (WHO) (2023) Climate change. https://www.who.int/news-room/fact-sheets/detail/climate-change-and-health

World Health Organization (WHO) (2024a) Noncommunicable diseases. https://www.who.int/news-room/fact-sheets/detail/noncommunicable-diseases

World Health Organization (WHO) (2024b) Heat and health. https://www.who.int/news-room/fact-sheets/detail/climate-change-heat-and-health

Xu K, Hao H, Zhang D, Wang W, Li H, Deng Y, Ma T, Steenland K, Chang H, Liu Y (2025) Long-term exposure to smoke PM2.5 and COPD caused mortality for elderly people in the contiguous United States. Environ Int 199:109513. https://doi.org/10.1016/j.envint.2025.109513

Zhang Y, Liu N, Li Y, Long Y, Baumgartner J, Adamkiewicz G, Bhalla K, Rodriguez J, Gemmel E (2023) Neighborhood infrastructure-related risk factors and non-communicable diseases: a systematic meta-review. Environ Health 22(2):8. https://doi.org/10.1186/s12940-022-00955-8

Zhang H, Li X, Shang W, Wu T, Wang S, Ling L, Zhou W (2024) Risk and attributable fraction estimation for the impact of exposure to compound drought and hot events on daily stroke admissions. Environ Health Prevent Med 29:56. https://doi.org/10.1265/ehpm.24-00168

Zhang J, Wang B, Ye B, Zhu B, Liu Y (2025a) Association between heatwave exposure and rapid kidney function declines: a longitudinal cohort study from CHARLS. BMC Public Health 25:1596. https://doi.org/10.1186/s12889-025-22822-0

Zhang Y, Xu R, Huang W, Ye T, Yu P, Yu W, Wu Y, Liu Y, Yang Z, Wen B, Ju K, Song J, Abramson MJ, Johnson A, Capon A, Jalaudin B, Gree D, Lavigne E, Johnston FH, Morgan GG, Knibbs LD, Zhang Y, Marks G, Heyworth J, Arblaster J, Guo YL, Morawska L, Coelho MSZS, Saldiva PHN, Matus P, Bi P, Hales S, Hu W, Phung D, Guo Y, Li S (2025b) Respiratory risks from wildfire-specific $PM_{2.5}$ across multiple countries and territories. Nat Sustain 8:474–484. https://doi.org/10.1038/s41893-025-01533-9

Chapter 6
Mental Health Implications of Climate Disasters: Stress, Trauma, and Recovery

Rahul Kumar and Mohammad Aves

Abstract Over the past two decades, climate-related disasters have become increasingly common and more severe, which has greatly impacted the mental health of many different groups of people. According to the World Health Organization, between the years 2000 and 2020, over 1.5 billion people worldwide suffered as a direct result of climate disasters. It is estimated that out of those affected, ~20% of survivors will develop a mental illness, including post-traumatic stress disorder (PTSD), anxiety, and depression. Following Hurricane Katrina, it was reported that nearly 49% of all survivors would qualify as having a diagnosed mental illness and around 30% would still be experiencing long-term symptoms of PTSD. The wildfires in Australia during 2019–2020 resulted in a significant increase in mental health consultations among those who were impacted, with a 22% increase occurring in the areas concerned. The groups of people most vulnerable to the psychological effects of climate-related disasters, including children, the elderly, and individuals with pre-existing mental health conditions, typically bear the majority of the burden when it comes to dealing with mental illnesses resulting from a climate-related disaster. This chapter explores some of the psychological channels associated with how climate disasters affect an individual's stress and trauma, using both epidemiological and clinical research as the basis for our analysis. They examine both adaptive and maladaptive coping strategies as they relate to climate disasters, along with the influence of displacement, social and economic upheaval, and cultural/community resilience. For instance, in some parts of South Asia, up to 40% of all substance abusers experienced depressive symptoms at least 6 months following significant flooding. The evidence-based interventions such as psychological first aid (PFA), trauma-informed care, and policies that include mental health as part of disaster response efforts.

Keywords Climate disaster · Disaster response · Mental health · PTSD · Trauma · Resilience

R. Kumar (✉)
Department of Microbiology, Graphic Era (Deemed to Be University), Clement Town, Dehradun, India
e-mail: rahulkumar.phdmb@gmail.com

M. Aves
Department of CSE, Graphic Era (Deemed to Be University), Clement Town, Dehradun, India

P. Ghosh et al. (eds.), *Climate Crisis and Changing Population Dynamics*, Springer Climate, https://doi.org/10.1007/978-3-032-24533-5_6

Introduction

Climate disasters are among the most unimaginable experiences on Earth in the twenty-first century, touching infrastructure, ecosystems, and also seriously impacting human psychological well-being (David-Raj et al. 2024). Floods, cyclones, wildfires, drought, and prolonged extreme heat events are growing in frequency, scale, and unpredictability and even making climate change itself accelerate (Jha and Dev 2024). Science explains that these growing trends happen as climate change hastens, rising temperatures globally and changing patterns of precipitation combine uniquely with urbanization and unsustainable practices, widely giving rise to large-scale migrations and dislocation of livelihood and life. As more communities everywhere are exposed to consequential climate shocks cumulatively, starting to witness the psychological toll, it is a growing economic worry (Mishra and Michael 2024). The people that experience climate disasters quickly pay a mental health cost. Mental health problems can be caused when people are exposed to trauma, uncertainty, and catastrophic loss in climate disasters, leading to increased rates of conditions such as anxiety, depression, and post-traumatic stress disorder (PTSD) and protracted grief. Vulnerable populations, especially children and the elderly, low-income households, indigenous peoples, and those with existing mental health issues, pay the heaviest costs in terms of limited access to health-care and economic resources, and the latter are more at risk of environmental damage (Knapp and Wong 2020). There is a deepening phenomenon sometimes referred to as eco-anxiety, in which chronic worry and feelings of helplessness and fear over environmental damage and the threat of future climate disasters are reported as requiring mental health solution (Cianconi et al. 2020). These are, of course, mediated by complex social, cultural, and political contexts. In societies that use cultural beliefs and communal memories to achieve emotional equilibrium, as well as spiritual conceptions of being connected to one another through a higher power, two underlying influences failing government policies and systems and poverty can impede access to appropriate mental health-care services (Burns 2015). Therefore, learn about how disasters impact a community's mental wellness, using this information to develop comprehensive, culturally sensitive practices to build resilience against the worsening climate crisis which is important.

Rationale of the Study

Climate change disasters have a devastating impact on mental health, and very little research has been conducted into the psychological implications of climate change disasters. It is well documented that many people experience physical injury resulting from floods, hurricanes, severe heat storms (heatwave), and wildfires; however, the invisible psychological injuries (i.e., post-traumatic stress disorder (PTSD), anxiety, depression, protracted emotional distress) arising from these events have received much less research attention (De Rond and Lok 2016). Global data indicate that approximately 20% of those who survive a natural disaster develop a significant mental disorder; for many extreme weather events (Bromet et al. 2017) (such as Hurricane Katrina and the Australian bushfires) the percentage of people developing PTSD from climate change related events may be as high as 30–49%, depending upon the length of time exposed to the event (Patrick et al. 2023). Despite this alarming data, most disaster preparedness and response plans do not incorporate mental health. In most emergency management systems, physical safety, rebuilding critical infrastructure, and providing for an economic recovery are the priority, while providing psychosocial care has historically been relegated to the background, and often not available until days or weeks following the disaster (Bailey et al. 2025). The gap between the availability of resources for physical recovery and those for psychosocial recovery is even more pronounced in low-middle-income countries. These countries typically have a developed health-care system that is fragile; they spend minimal amounts on mental health; and there are very few trained mental-health professionals in these countries. Because of the disproportionate impact of climate change disasters on vulnerable populations, the fact that this study was needed is glaring (Smith et al. 2022). Vulnerable populations include children, the elderly, and people with disabilities, marginalized communities, and others with a pre-existing mental health condition. These groups are often at greater risk because they may have limited access to mental health care, be socially isolated, and have other systemic disadvantages; and many will endure prolonged displacement, loss of livelihood, or disruption of daily routine due to the impact of a climate change related disaster, which may be directly correlated with chronic psychological distress and suffering (Shareeff et al. 2024). Long-term psychosocial support/ISAS and IDA is significantly underdeveloped. Most of the support systems currently available are designed for short terms. Therefore, survivors do not receive long-term therapeutic interventions nor do they have means by which to reintegrate into the community (Ore and Foli 2020). As the effects of climate change continue to increase, emotional reactions such as ecological anxiety, solastalgia, and climate-related grief will be seen more frequently (Comtesse et al. 2021). This increased frequency of emotional responses related to climate change indicates that resilience building strategies and mental health policies need to be integrated into future efforts. The purpose of this study is to identify the psychological impact that climate-related disasters have on survivors, to identify gaps in current response systems; to advocate inclusive, culturally sensitized, and policy-based mental health interventions that can assist in building the resilience of the survivor's communities, particularly as climate-related threats increase (Table 6.1).

Materials and Methods

The information in this chapter is organized in a narrative literature review format. This type of review is used to gather all the available research regarding how climate-related disasters (CrDs) may affect mental health. The narrative review format was

Table 6.1 Dimensions of climate disasters and their mental health implications

Dimension	Description	Psychological impact	Most affected groups	Implications for recovery and policy	References
Sudden-onset climate events (e.g., flash floods, storm surges)	Rapid, unpredictable events causing mass displacement and destruction	Acute stress reactions, panic, shock, intrusive memories	Children, coastal communities, low-income families	Requires rapid deployment of psychological first aid and emergency mental health teams	(Akhter et al. 2015; Batra and Erbas 2025)
Slow-onset climate events (e.g., prolonged drought, sea-level rise)	Gradual environmental deterioration affecting livelihoods	Chronic anxiety, hopelessness, eco-grief, suicidal ideation	Farmers, indigenous populations, rural communities	Policy must support livelihood restoration and long-term psychosocial care	(Kõlves et al. 2025; Adams et al. 2016)
Loss of livelihood	Economic instability due to crop loss, destroyed property, halted tourism, etc	Depression, despair, heightened suicide risk, family stress	Farmers, informal workers, fishing communities	Integrate economic rehabilitation with mental health support	(Shaban et al. 2024; Nye et al. 2023)
Forced migration and displacement	Movement to temporary shelters or relocation zones	Identity loss, emotional numbness, fear, social withdrawal	Elderly, persons with disabilities, climate refugees	Need for safe shelter design, community support networks, culturally sensitive care	(Perez-Murcia 2021; Gutterman 2022)
Shelter overcrowding	Limited resources and privacy in relief camps	Irritability, sleep disturbance, interpersonal conflict	Women, children, marginalized groups	Trauma-informed shelter management and protection services	(Conzatti et al. 2022; Kahan et al. 2020)
Breakdown of social support systems	Disruption of family, community, and cultural bonds	Loss of belonging, isolation, emotional instability	Widowed elderly, single-parent households	Strengthening community-based resilience programs	(Cutrona 2000; Mampane 2020)
Disruption in health-care services	Inaccessibility of medication, therapy, or clinics during disasters	Relapse of mental illness, worsening symptoms	People with pre-existing psychiatric conditions	Disaster policy must prioritize continuity of mental health care	(Ritchie et al. 2019; Esterwood and Saeed 2020)

(continued)

Table 6.1 (continued)

Dimension	Description	Psychological impact	Most affected groups	Implications for recovery and policy	References
Exposure to traumatic visuals	Witnessing injury, death, destruction, or burnt landscapes	PTSD symptoms, recurring nightmares, avoidance behavior	First responders, adolescents, survivors of wildfires	Requires trauma-informed debriefing and long-term counseling	(Burback et al. 2024; Hartwell et al. 2023)
Environmental degradation	Destruction of forests, rivers, farmlands, habitats	Solastalgia, ecological grief, guilt, hopelessness	Indigenous groups, forest-dependent communities	Integrate cultural healing practices and ecological restoration	(Tschakert et al. 2024; Haq et al. 2023)
Heat-related psychological strain	Extreme temperature exposure and prolonged heatwaves	Irritability, aggression, cognitive fatigue	Urban poor, outdoor workers	Policies must combine heat management with mental health awareness	(Rony and Alamgir 2023; Lohrey et al. 2021)
Gender-based vulnerabilities	Increased risk of exploitation, harassment, and domestic violence	Trauma, hypervigilance, low self-esteem, depression	Women and girls in shelters	Need gender safe spaces, legal support, and counseling services	(Sabri and Granger 2018; Stark et al. 2022)
Impact on education systems	School closures and destruction disrupt learning and routines	Regression, attention difficulties, anxiety	Children and teenagers	Establish school-based trauma programs and emotional learning curriculums	(Colvin et al. 2022; Herrenkohl et al. 2019)
Food and water insecurity	Scarcity during disasters causing dependency and stress	Fear, helplessness, anger, chronic stress	Low-income households, children, pregnant women	Integrate nutrition support with psychosocial interventions	(Dasgupta and Basu 2023; Akter et al. 2020)
Cultural and spiritual disruption	Loss of sacred spaces, rituals, and community gatherings	Identity conflict, grief, emotional disorientation	Indigenous and tribal groups	Include cultural leaders in mental health recovery plans	(Counted et al. 2020; O'Keefe et al. 2021)

(continued)

Table 6.1 (continued)

Dimension	Description	Psychological impact	Most affected groups	Implications for recovery and policy	References
Misinformation and fear amplification	Spread of rumors about disasters and climate threats	Mass panic, catastrophic thinking, social distrust	Youth, digitally active populations	Strengthen communication strategies and mental health literacy	(Suryana et al. 2024; Tay et al. 2018)
Stigma around mental health	Cultural taboos limiting help-seeking behavior	Silent suffering, delayed treatment, worsening conditions	Rural communities, men, older adults	Need awareness campaigns and stigma reduction programs	(Dimitrov et al. 2022; Valentine et al. 2022)
Weak mental health infrastructure	Limited professionals, poor funding, lack of disaster protocols	Inadequate diagnosis, overburdened systems	LMICs and remote regions	Policy reforms must integrate mental health into disaster planning	(Ogbuagu et al. 2024; Shibesh and Nagabhatla 2025)
Barriers to telepsychiatry	Poor connectivity, low digital literacy	Lack of timely support, increased isolation	Remote rural areas, elderly	Invest in digital mental health platforms and training	(Sharma et al. 2025; Jackson et al. 2019)
Cascading disasters (multiple events in short time)	Repeated exposure to climate shocks	Compound trauma, exhaustion, emotional burnout	Communities in climate hotspots	Multi-hazard preparedness with continuous psychological support	(Alcántara-Ayala 2025; Setyawan et al. 2023)
Intergenerational trauma	Transmission of climate-related stress across generations	Fear of future, fatalism, climate anxiety	Youth in disaster-prone regions	Requires family-centered and community wide interventions	(Cuthrell 2025; Bessaha et al. 2022)

selected for this purpose due to the varied range of research methods used in CrD mental health studies, which include studies from many different fields (i.e., public health, psychology, climatology, sociology, and disaster management). Sources were identified from a comprehensive search of several large academic databases PubMed, Scopus, Web of Science, Google Scholar, and ScienceDirect as well as from a number of different gray literature sources, such as WHO Reports, Intergovernmental Panel on Climate Change (IPCC) Reports, United Nations Disaster Risk Reduction (UNDRR) Report, and other documents produced by Government Agencies and Non-Governmental Organizations (NGOs) about the effects of natural disasters on LarDb-mental health. The time frame of the literature review encompassed the period from 2000 to 2025 and reviewed current research relating to the climate crisis, increase in the frequency of natural disasters, and development of new models for understanding the psychological effects of disasters on people's mental health. Studies included in the literature review were only included if they addressed the psychological consequences of climate-related disasters (such as PTSD, anxiety, depression, and eco-anxiety; solastalgia; community resilience), and included particular vulnerable populations (e.g., children, the elderly, marginalized communities, individuals with pre-existing mental health conditions). The search criteria omitted all studies that were not directly related to climate-related disasters, studies that did not measure mental health outcomes, and studies that did not have a rigorous methodological framework. Keywords used in the search process included climate disaster, natural disaster, extreme weather event, climate change, mental health, PTSD, trauma, depression, anxiety, eco-anxiety, solastalgia, community resilience, coping mechanisms, disaster displacement, and vulnerable populations. For the case study, events were handpicked. The based on selections on a few criteria: severity mattered, documented psychological impact seemed vital, and then high-quality literature availability was key. This included the awful Hurricane Katrina; also, the 2019–2020 Australian bushfires had a large impact (Reiner et al. 2024). Moreover, monsoon floods in India came into play and drought-related crises, also, in sub-Saharan Africa made the list. These are just a few instances and will help us to obtain helpful observations, on diverse locations; society's involvement, plus many cultures, and different cultures are very important in many of the world's locations. This investigation provides clear ideas into various physical, sociable, and cultural settings. Generally, this system of working helped secure a through fact-based exploration of climate disasters and their impact on one's psyche all around the world. This ensured that all bases are properly covered.

Methodological and Database Utilization

To ensure transparency, reproducibility, and methodological precision, this chapter followed a structured methodological flow map for the identification, screening, selection, and synthesis of relevant literature examining the mental health impacts of climate-related disasters (Fig. 6.1). Major multidisciplinary and health-focused databases, including PubMed, Scopus, Web of Science, Google Scholar, and ScienceDirect, were systematically searched to capture a broad spectrum of peer-reviewed studies, while gray literature from authoritative sources such as the World Health Organization (WHO), Intergovernmental Panel on Climate Change (IPCC), United Nations Disaster Risk Reduction (UNDRR), and governmental and non-governmental organizations was consulted to reduce publication bias and incorporate policy relevant evidence. A structured search strategy using predefined keywords and Boolean operators covering terms related to climate disasters, extreme weather events, mental health outcomes, and resilience was applied across all databases, with results limited to English language publications from 2015 to 2025. Titles and abstracts were initially screened to remove duplicates and exclude studies not directly addressing climate-related disasters or mental health outcomes, followed by full text assessment based on predefined inclusion and exclusion criteria emphasizing methodological rigor and relevance to vulnerable populations. Relevant data were then systematically extracted and thematically organized according to disaster type, affected populations, psychological outcomes, mechanisms of stress and trauma, resilience processes, interventions, and policy implications. Finally, a narrative synthesis approach was employed to integrate findings across diverse study designs and geographical contexts, enabling a comprehensive and coherent understanding of the mental health consequences of climate-related disasters.

Results and Discussion

Understanding Climate Disasters

Climate disasters are extreme environmental events of the broadest kind, which are either caused or exacerbated by climate change. These are floods, droughts, wildfires, heatwaves, and cyclones that, while producing different impacts, are also sharing the interconnected drivers, like rising global temperatures, changing rainfall patterns, and ecosystem degradation (Loucks 2021). Floods are less rare than before due to heavy rains, melting of ice caps, and lack of efficient drainage structures, and they are bringing a large number of people to the state of being homeless and also destroying their properties. Droughts can be traced back to prolonged rainfall deficits and increasing evaporation rates, affecting agriculture, water availability, and food security most severely (Orimoloye 2022). Wildfires, which have been increased in frequency due to high temperatures and drying of the forest, are now also hitting areas that had not been previously threatened, leading to both ecological and economic injury of a great extent. Heatwaves have become longer, more frequent and intense; thus, human health is put at a severe risk, especially in urban heat islands. At the same time, as the warmer oceans give hurricanes and cyclones more fuel, they are turning into new powers and the cause of devastating wind and storm surge damages (Salarieh et al. 2023). Geographically, climate disasters portray different patterns in

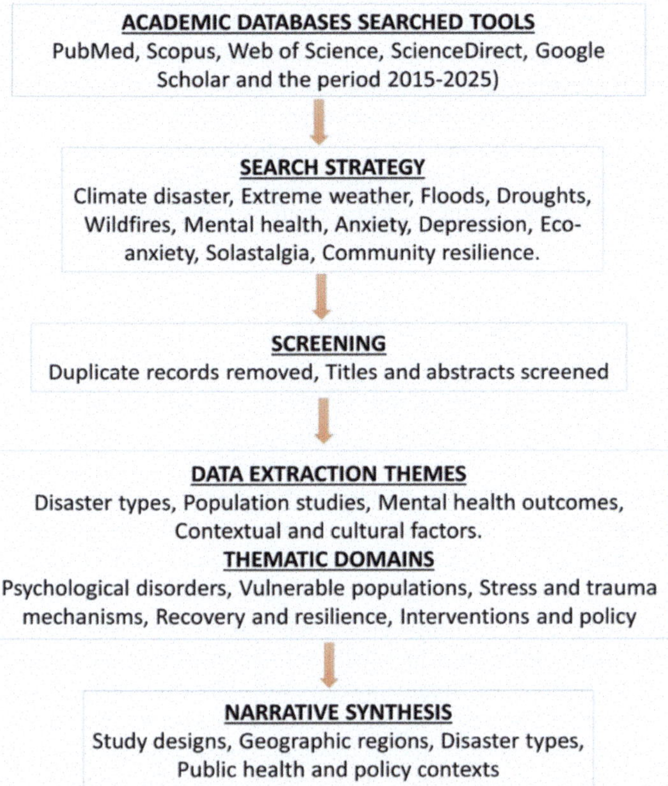

Fig. 6.1 Methodological flow map illustrating database selection, search strategy, screening, eligibility assessment, data extraction, and narrative synthesis used to examine the mental health impacts of climate-related disasters (2015–2025)

time and space where certain places experience more severe and frequent events. For example, South Asia is exposed to floods and cyclones on a regular basis, while Sub-Saharan Africa suffers from droughts, Australia and North America have to deal with wildfires, and Europe is getting used to heat waves that are unlike any it has ever known before (Lombe et al. 2024). A small island nation in the Pacific, for example, is going to lose its home due to the rising sea level and storm surges. Because of the continuing rise of global surface temperatures that are now even more than 1.2 °C above the level of pre-industrial times (Noor et al. 2021), these hazards have become more intense and are leading to increased human deaths and economic losses along with environmental changes irreversible (Smith et al. 2023). The severity of climate disasters hugely depends on risk factors of a socioeconomic and environmental nature that surround them. Badly built infrastructures, unplanned urbanization, deforestation, and building homes in areas where it is hazardous to do so are some of the factors that make people highly vulnerable. Poor communities are very often without the advantages of early warning systems, resilient housing, or easy access to emergency services, and what is more, these are they who suffer most in times of disaster. Even more, environmental degradation such as disappearing wetlands or soil erosion and ground cover is among the factors that make the earth defend less naturally against climate disasters (Chakraborty et al. 2023). To sum up, the origin of climate disasters lies in a complicated relationship between climate changes, human activities, and social vulnerabilities, which determine to what extent and in which way people are hit by disasters all over the world.

Psychological and Emotional Impact

Climate disasters, in particular, disrupt psychological wellbeing and mental health issues of a broad range and are not only limited to those individuals who suffer physically. Among the most researched mental health symptoms, one can find post-traumatic stress disorder (PTSD), anxiety, and depression, which appear as a result of a direct experience of a life-threatening event such as floods, cyclones, wildfires, or heatwaves (Negi et al. 2025). The people who have survived the tragedy often live through the traumatic event again and again in their heads, they are always on alert, have their sleep interrupted, are worried all the time, and even despair trying to make sense of the huge loss that overwhelms their power to understand it. Anxiety finds its origin in the unknown scenarios of new disasters, destroyed means of living, and the fear of getting hit again, whereas depression is very much associated with losing a loved one, going through financial crises, and the breaking of homes, communities, and cultural identities. The signs may reveal themselves right away after the occurrence of an event or emerge gradually as people cope with the long-term recovery challenges (Bonanno et al. 2010). In addition to traditional mental health issues, climate disasters can evoke various emotional responses, which in this case are called eco-grief and solastalgia (Cianconi et al. 2023). The term eco-grief is used for that type of grief or loss when one mourns the deterioration or death of the ecosystem, the disappearance of certain species, or the loss of the natural beauty (Pihkala 2024). Solastalgia is considered to be homesickness without leaving home, and it goes side by side with the feeling of loss of the familiar surroundings (Breth-Petersen et al. 2023) which have been totally changed or damaged by the impacts of the climate such as droughts, coastal erosion, deforestation, or wildfires. These emotional states arise most in case of Indigenous peoples; rural populations that feel their identities, culture, and ways of living are tightly bound up with nature and the natural environment. The psychological burden of the loss of one's ecological surroundings can be very demanding and as such can lead to the feeling of displacement, belonging, as well as existential anxiety (Rehling 2022). It is quite common that the first impacts of climate disasters on people include among others: shock, confusion, fear, acute stress, and emotional numbness. These are typical reactions which are usually immediate and with sufficient support, stability, and community resilience; these effects may be short-lived. In the meantime, the consequences of a long period after the event can still be felt all over the place for months or, even years after, especially in cases when the recovery has been slow or incomplete. Among the long-term effects are chronic PTSD, persistent depressive disorders, complicated grief, substance dependency, and anxiety about future climate emergencies. The psychological journey is not always the same for everyone, and very often it is determined by the people around them, access to care, as well as the person's own resources to regain a feeling of security and control. In the end, the emotional toll from climate disasters is not only a mirror of the event's intensity but also that of the social, cultural, and environmental factors that influence the recovery process.

Impact on Specific Groups

Climate crises can be quite unjust in their effects since they do not treat populations equally: some groups, from a developmental, physical, or psychological point of view, are at the risk of becoming extremely vulnerable. In this respect, children can be considered the most sensitive group when talking about the emotional and psychological impact of climate-related events. Limited coping skills, dependence on caregivers, and developing cognitive systems all contribute to children's vulnerability to fear, confusion, and traumatic stress (Zimmer-Gembeck and Skinner 2016). Experiencing disaster-related destruction, such as losing homes, being displaced, or witnessing injuries, children can develop nightmares, separation anxiety, behavioral regression, attention difficulties, and eventually the mental health disorders like PTSD or depression of a long-term nature. Moreover, through the disruption of schooling, social networks, and routines, children's resilience becomes even more fragile, and thus, early psychological support is of utmost importance for the prevention of chronic emotional disturbances. The old age population, on the other hand, is not psychologically indifferent to climate disasters either but is considerably less robust than it has been in the past (Rafiey et al. 2016). Physical limitations, chronic illnesses, reduced mobility, and social isolation, all associated with the aging process, increase the risk of elderly people during emergencies, and have a significant impact on their emotional states. A large number of elderly people say that helplessness, loss of independence, and fear of abandonment are among the most conspicuous feelings they undergo during chaotic disasters. Apart from that, the coming of disaster to the residential area, neighborhood, or cultural landscape of a person that has been there for a long time may be a source of grief, identity disruption, and elevated anxiety (Smid et al. 2018). Moreover, the old people with cognitive impairments such as dementia understand less; therefore, they become more distressed, confused, and, if supportive interventions are not available, psychologically declined in the long term. Besides that, without any support, they may experience distress, confusion, and psychological decline in the long run. Firstly, people having mental health problems before the onset of calamities face the hardest challenges during and after them. Treatment interruptions, medication losses, damage to health-care facilities, and difficulties in accessing usual care may all cause a relapse or aggravation of symptomatology. As a consequence, an individual prone to anxiety might experience it more intensively due to increased uncertainty, while a depressive person may feel more deeply melancholic due to more losses and a psychotic patient may considerably worsen in a chaotic environment (Bürgy 2018). In addition to stigma, poor accessibility to services further exacerbates that population's vulnerability. The very nature of disasters as a destabilizing factor elevates; thus, this group of persons is particularly in urgent need of continuous, coordinated, and accessible mental health support. Summarized, children, the elderly, and people with pre-existing mental illnesses should be the first to receive interventions that are geared toward their specific needs and are carried out during the time of disaster preparedness and recovery. Comprehending these groups' particular vulnerabilities is a prerequisite for good mental health care that is inclusive, equitable, and effective even when the climate is rapidly changing.

Mechanisms of Stress and Trauma

The psychological impact of climate disasters is shaped by multiple interconnected mechanisms that influence how individuals experience, process, and recover from traumatic events. At the biological level, stress responses are largely governed by the hypothalamic pituitary adrenal (HPA) axis, which regulates the body's reaction to perceived threats (Hinds and Sanchez 2022). During a climate disaster, the HPA axis is activated, leading to the release of cortisol, which is the primary stress hormone. While cortisol enhances immediate survival by sharpening alertness and mobilizing energy, prolonged elevation disrupts the neural pathways related to memory, mood regulation, and emotional control (Sitorus and Silitonga 2025). Chronic activation of the stress response, which is common in survivors of repeated disasters, prolonged displacement, or ongoing uncertainty, has been linked to the development of PTSD, anxiety disorders, depression, and cognitive impairment. The psychological pathways of trauma involve an individual's subjective experience, interpretation, and emotional processing of an event. Climate disasters often violate the fundamental assumptions of safety, stability, and predictability, leading to shock, fear, helplessness, and loss of control. Cognitive processes such as catastrophic thinking, intrusive memories, and negative belief systems shape the severity and duration of psychological distress (Sugiura and Sugiura 2016). Survivors may struggle with guilt, grief, and identity disruption, particularly when a disaster results in injury, loss of loved ones, or destruction of meaningful places. Displacement and loss of livelihood are major contributors to trauma, especially in communities dependent on natural resources. Forced relocation disrupts social networks, cultural ties, and daily routines, thereby leading to feelings of dislocation, uncertainty, and emotional exhaustion. Loss of livelihood due to drought, storms, or environmental degradation threatens an individual's sense of purpose and security, increasing the risk of depression, substance use, and suicidal behavior. Emotional consequences are often magnified when recovery is slow or when support systems are insufficient. The interaction between physical and mental health further complicates the trauma experience. Injury, chronic pain, heat-related illnesses, malnutrition, and infectious diseases associated with climatic disasters can exacerbate psychological distress (White et al. 2023). Conversely, poor mental health can weaken immune function, reduce treatment adherence, and heighten the perception of physical symptoms, thereby creating a cycle that hinders recovery. In the end, cultural influences shape how people react to trauma. Cultural beliefs, spiritual frameworks, and community rituals shape how individuals interpret the disaster, how individuals express the emotions, and how individuals seek the help (Sun et al. 2018). In the societies, coping and communal healing practices build the strength. The stigma around health can block the emotional expressions and can delay the treatment. Understanding the dimensions is making the interventions that are respectful work well and fit the situation. Together, these biological, psychological, social, and cultural mechanisms demonstrate that the trauma resulting from climate disasters is complex and multifaceted, requiring holistic and culturally informed approaches to support recovery.

Recovery and Resilience

Recovery from climate disasters is a multifaceted process involving individual, familial, cultural, and community level strengths. Coping mechanisms help survivors manage survivor distress after a disaster. Survivors often use resources like problem solving, controlling feelings, faith, and hope to face the unknown. Survivors also find that being in nature, keeping routines, paying attention, and making art can help survivors feel in control again and lower stress (Baker 2014). Coping strategies form the line of the defense. Coping strategies matter for the term health outcomes. Communities with networks, shared values, and effective communication have fewer long-term mental consequences. The collective actions such as volunteer support, community led rebuilding, shared resource distribution, and group decision-making give reassurance. Collective actions such as volunteer support, community-led rebuilding, shared resource distribution, and group decision-making promote a sense of solidarity. The local leaders, schools, faith-based organizations, and grass-roots groups build trust, cut isolation, and give survivors support during recovery. The community resilience restores stability, builds a sense of belonging, and gives people power to rebuild their lives (Kaluarachchi 2018). The cultural resilience provides meaning, identity, and continuity during changes. The cultural traditions, rituals, storytelling, and spiritual beliefs let people make sense of disaster experiences and process grief, in ways that fit their culture. Indigenous communities also draw on the connections to land and the collective memory to strengthen the recovery of Indigenous communities. Cultural continuity also builds the belief that recovery is possible, after a deep loss. Family support is a part of healing. Families give the safety, the comfort, and the practical help. Families also lower the feelings of fear and abandonment. Supportive family interactions communication, shared problem solving, empathy, and reassurance lower stress and raise coping skills in children, adults, and the elderly (Falloon et al. 2023). The stable family relationships help people rebuild their routine, their identity, and their normal life, after a disaster. The coping strategies, the community togetherness, the cultural identity, and the strong family bonds act as pillars of resilience. Strengthening these interconnected systems is essential for supporting long-term recovery and mitigating.

Interventions and Mental Health Support

The effective interventions and mental health support systems are important to lower the burden of the climate disasters and help the long-term recovery. The used of psychological first aid (PFA) as a useful plan that gives safety, comfort, and calm feelings to people, in crisis. Psychological first aid (PFA) focuses on listening checking what survivors need and linking survivors to help without making survivors tell the stuff again. After the help, trauma-informed care gives a plan, for long-term healing. The plan focuses on safety, empowerment, and trust. Trauma-informed care knows that trauma affects people. Trauma-informed care adds attention to every level of care. Trauma-informed care makes sure that services do not re-traumatize survivors. Governments and NGOs play a role, in putting these actions into practice. Governments and NGOs fund community-based programs. Governments and NGOs make sure that services are fair and reach everyone (Abiddin et al. 2022). Governments need to put health into the disaster management frameworks. The NGOs bring health, help to underserved communities, set up clinics community outreach, and running capacity building projects. The emerging technologies let telemental health give support, offer online counseling, and provide the digital self-help tools (Langarizadeh et al. 2017). The digital self-help tools work well in the areas and in the places cut off by disaster. This includes embedding psychosocial support into early warning systems, evacuation procedures, relief camp management, and long-term rehabilitation programs. By prioritizing mental health alongside physical safety and economic recovery, disaster planning can foster more resilient communities capable of withstanding the escalating impacts of climate change.

Policy, Planning, and Public Health

In the view, policy, planning, and public health frameworks help address the health consequences of the climate disasters. Policy, planning, and public health frameworks guide the fair and lasting responses. Global frameworks such as the Sendai Framework for Disaster Risk Reduction (2015–2030), the Paris Agreement, and the World Health Organization's guidelines on health and psychosocial support (MHPSS) stress that global frameworks must include well-being in the climate resilience strategies (Palinkas et al. 2020). Global frameworks link well-being to the climate resilience strategies. These frameworks push the nations to use the people focused approaches that make the early warning systems stronger, keep the people safe, and put health into the climate adaptation policies. The level the mental health policies are increasingly seeing climate change as a threat to public health, the countries are changing the policies to add the community health-care training, for health workers and tele-mental health services (Zangani et al. 2022). In middle-income countries, the financial constraints, the limited mental health infrastructure, and the stigma block progress. The strengthening policies need funding, inter-sectoral collaboration, and alignment with global guidelines. The goal is to keep well-being from being ignored during disaster management. The disaster risk reduction (DRR) is a part of health planning. Disaster risk reduction (DRR) works to cut the effects of climate hazards (Aitsi-Selmi and Murray 2016). Disaster risk reduction (DRR) does this through prevention through preparedness and through recovery. Putting health into disaster planning makes sure to meet health and emotional needs at every step of a disaster. They keep health services running during a crisis. They design buildings and evacuation routes that keep people safe and treat them with dignity. Public health systems have to see health as a part of climate safety. When mental health is part of the plan, the whole community recovers better. The aligning global frameworks, national policies, and DRR strategies, governments and public health institutions can build adaptive systems capable of safeguarding physical and psychological well-being in the face of increasing climate-related threats.

Case Studies

Real-world climate disasters show evidence of the psychological effects, on the people who live through them. Real-world climate disasters also show how the social, cultural, and environmental settings shape health outcomes. Hurricane Katrina in USA in 2005 stands as an example. The storm caused displacement, the collapse of infrastructure, and emergency responses that treated income and minority communities unfairly (Rubin and Henry Falk 2019). The combination of loss of home, loss of services, and the feeling that help was not given equally made the mental health impact even worse for income and minority communities. Survivors of Hurricane Katrina showed levels of PTSD. Survivors of Hurricane Katrina showed levels of depression (Lai et al. 2015). Survivors of Hurricane Katrina showed grief. Many survivors of Hurricane Katrina reported trauma years, after the storm. The slow recovery efforts make emotional distress worse. The broken social services make emotional distress worse. The unstable housing conditions make emotional distress worse. The situation shows the need, for disaster planning and long-term mental health support. The 2019–2020 bushfires called the Black Summer fires cause big mental trauma, for firefighters, for rural families, and for indigenous communities (Ahmed and Ledger 2023). The fire exposure, the wildlife loss, the land loss, and the smoke also trigger PTSD. The fire exposure, the wildlife loss, the land loss the smoke also trigger grief. The fire exposure, the wildlife loss, the land loss, and the smoke also trigger grief over loss of place as the landscape changes forever.

In India, recurring floods, in Kerala, Assam, and Bihar, show how the social and economic conditions shape the risk. The floods destroy the homes, the crops, and the livelihoods and cause stress for the people. The overcrowded relief camps, the uncertainty, and the lack of privacy increase the anxiety and the fear, among the women the children and the elderly (Kemei et al. 2023). The mental health stigma and the limited access to services keep people from getting better, culturally respectful, and gender considerate policies matter. In Sub-Saharan Africa, long droughts hit farm and herding communities hard. The long droughts cause long-term stress, depression, and higher risk of suicide because crops fail, livestock die, food is short, and water is scarce (Khatri et al. 2024). Strong ties to the land make grief and hopelessness worse when damage to the environment threatens identity and heritage. Weak mental health infrastructure and pervasive poverty further restrict access to care, making community-based and culturally grounded support essential. Collectively, these case studies demonstrate that, while climate disasters consistently create substantial psychological harm, the severity and nature of these effects vary according to local vulnerabilities and response capacities. They reinforce the need to integrate mental health into all stages of disaster management to reduce the long-term psychological consequences (Table 6.2).

Limitations of the Study

The use of data may limit the study from making conclusions or seeing the details of the context. Also the methods used in the cited studies are different. The difference in methods makes it hard to compare the studies directly or to combine the results. Research on slow onset disasters, like droughts or slow environmental changes, is still small. The small amount of research limits what we know about the effects of those events. Furthermore, mental health datasets from low and middle income countries (LMICs) are scarce, which constrains the generalizability of the findings across diverse global contexts. Finally, cultural variability in psychological reporting may influence how symptoms are expressed and recorded, potentially affecting the interpretation of mental-health outcomes.

Recommendations

These findings lead to recommendations to improve mental health outcomes in disaster-affected populations. Mental health and psychosocial support (MHPSS) must be integrated into the disaster planning, so that the response is timely and coordinated. Frontline workers must receive the training that helps frontline workers identify and address distress effectively. Community-based mental health systems can improve access to the care and keep the care sustainable. Interventions must fit the culture to match the beliefs and the practices. Strengthening the telepsychiatry

Table 6.2 Climate disaster case studies and mental health outcomes

Case study	Type of climate disaster	Environmental impacts	Psychological outcomes	Most affected groups	Major contributing factors to mental health burden	Recovery challenges	Lessons for future planning	References
Hurricane Katrina (USA 2005)	Category 5 hurricane, storm surge, flooding	Widespread flooding, infrastructure collapse, prolonged displacement	High rates of PTSD, depression, grief, anxiety, substance misuse	Low-income communities, African Americans, elderly, displaced families	Inequitable emergency response, loss of housing, economic instability, family separation	Slow rebuilding, housing shortages, service gaps, distrust in institutions	Need for equity focused disaster planning; long-term mental health programs; stronger housing security	(Baker and Cormier 2014; Bonanno et al. 2010)
Australian bushfires (2019–2020)	Intense wildfires fueled by heatwaves and drought	Loss of forests, wildlife deaths, community destruction, air pollution	PTSD, eco-grief, solastalgia, anxiety, burnout among firefighters	Rural families, indigenous communities, firefighters, wildlife-dependent livelihoods	Repeated exposure to fire, loss of cultural land, environmental devastation, smoke-related stress	Limited mental health access in rural areas; prolonged fire seasons; emotional fatigue	Provide specialized support for first responders; integrate cultural healing; invest in fire preparedness	(Van Eck et al. 2024; Australian 2024; Humphreys et al. 2022)
India floods (Kerala, Assam, Bihar)	Severe monsoon flooding	Crop loss, destruction of homes, displacement to relief camps	Anxiety, acute stress, depression, fear, sleep disorders	Women, children, elderly, farmers, people in relief shelters	Overcrowded camps, livelihood loss, poverty, waterborne diseases, social marginalization	Lack of privacy in camps; limited mental health services; stigma; delayed rehabilitation	Build gender sensitive shelters; expand community counseling; strengthen early warning and rescue systems	(Kunnathepeedikayil 2020; Ullah et al. 2025)

(continued)

Table 6.2 (continued)

Case study	Type of climate disaster	Environmental impacts	Psychological outcomes	Most affected groups	Major contributing factors to mental health burden	Recovery challenges	Lessons for future planning	References
Sub-Saharan droughts	Long-term drought, water scarcity, land degradation	Crop failure, livestock death, food insecurity, migration	Chronic stress, hopelessness, depression, suicide risk, eco-grief	Farmers, pastoralists, rural poor, nomadic tribes	Food scarcity, livelihood collapse, climate anxiety, strong dependence on land and livestock	Poor mental health infrastructure, extreme poverty, repeated drought cycles	Promote climate-resilient farming; integrate psychosocial support into aid programs; strengthen food security	(Adaawen et al. 2019; Mahbod and Parnian 2024)
Hurricane Katrina secondary impacts	Prolonged displacement following initial disaster	Long-term homelessness, economic losses	Survivor guilt, social isolation, family conflict	Single mothers, low-income households	Job loss, insurance barriers, disrupted education	Bureaucratic delays; unstable income sources	Improve insurance systems; provide mental health during relocation	(Fussell 2015; Carlson et al. 2022)
Australian bushfires—ecological losses	Large-scale destruction of biodiversity	Altered landscapes, habitat loss	Solastalgia due to environmental change	Indigenous groups, youth	Cultural identity tied to land; climate grief	Slow ecological regeneration	Engage communities in ecological restoration for healing	(Stanley et al. 2024; Burns et al. 2019)

(continued)

Table 6.2 (continued)

Case study	Type of climate disaster	Environmental impacts	Psychological outcomes	Most affected groups	Major contributing factors to mental health burden	Recovery challenges	Lessons for future planning	References
India floods rural impact	River overflow in farming communities	Soil erosion, crop destruction	Financial stress leading to anxiety, depression	Farmers, laborers	High dependency on agriculture	Debt cycles; lack of crop insurance	Introduce crop insurance + mental health support	(Shaban et al. 2024; Agarwal et al. 2022)
Sub-Saharan droughts displacement impact	Climate-induced migration	Loss of homeland, relocation stress	Trauma, identity loss	Women and children migrants	Insecurity, loss of cultural roots	Weak support systems	Build migrant-centered mental health services	(Eboreime et al. 2025; Miller et al. 2024)

infrastructure can expand to reach particularly in remote or resource-limited areas. More research is needed to understand the long-term psychological impact of disasters. Efforts to reduce stigma and raise awareness of the intersection of climate change and mental health are crucial for fostering resilient and informed communities.

Conclusions

To sum up, the study shows that mental health matters a lot in disasters. The study points out that mental health is especially important, in middle-income countries where resources are limited and data are scarce. The study says that research already gives ideas. Gaps still exist. The gaps include not knowing enough, about the long-term impact of slow onset disasters. The gaps also include not understanding the nuances of health reporting. Integrating mental health and psychosocial support into disaster preparedness, training frontline workers, developing community-based and culturally sensitive interventions, and leveraging telepsychiatry are essential strategies for mitigating psychological distress. Strengthening research, reducing stigma, and promoting climate mental health awareness will further enhance resilience and ensure that the affected populations receive comprehensive and effective support.

References

Abiddin NZ, Ibrahim I, Abdul Aziz SA (2022) Non-governmental organisations (NGOs) and their part towards sustainable community development. Sustainability 14(8):4386. https://doi.org/10.3390/su14084386

Adaawen S, Rademacher-Schulz C, Schraven B, Segadlo N (2019) Drought, migration, and conflict in sub-Saharan Africa: what are the links and policy options? Curr Direct Water Scarcity Res 2:15–31. https://doi.org/10.1016/B978-0-12-814820-4.00002-X

Adams C, Rodrigues ST, Calmon M, Kumar C (2016) Impacts of large-scale forest restoration on socioeconomic status and local livelihoods: what we know and do not know. Biotropica 48(6):731–744. https://doi.org/10.1111/btp.12385

Agarwal SK, Gupta S, Singh V, Sutar R, Nair D (2022) Insurance as an instrument of financial security in addressing mental illness among agricultural workers in the LMICs. medRxiv 6:5910. https://doi.org/10.1101/2022.06.09.22275910

Ahmed I, Ledger K (2023) Lessons from the 2019/2020 'black summer bushfires' in Australia. Int J Disast Risk Reduct 96:103947. https://doi.org/10.1016/j.ijdrr.2023.103947

Aitsi-Selmi A, Murray V (2016) Protecting the health and well-being of populations from disasters: health and health care in the Sendai framework for disaster risk reduction 2015–2030. Prehosp Disaster Med 31(1):74–78. https://doi.org/10.1017/S1049023X15005531

Akhter SR, Sarkar RK, Dutta M, Khanom R, Akter N, Chowdhury MR, Sultan M (2015) Issues with families and children in a disaster context: a qualitative perspective from rural Bangladesh. Int J Disast Risk Reduct 13:313–323. https://doi.org/10.1016/j.ijdrr.2015.07.011

Akter F, Rahman M, Pitchik HO, Winch PJ, Fernald LCH, Nurul Huda TM, Jahir T, Amin R, Das JB, Hossain K, Shoab AK, Khan R, Yeasmin F, Sultana J, Luby SP, Tofail F (2020) Adaptation and integration of psychosocial stimulation, maternal mental health and nutritional interventions for pregnant and lactating women in rural Bangladesh. Int J Environ Res Public Health 17(17):6233. https://doi.org/10.3390/ijerph17176233

Alcántara-Ayala I (2025) Cascading hazards and compound disasters. NPJ Nat Hazards 2(1):54. https://doi.org/10.1038/s44304-025-00111-5

Australian OQNSW (2024) As these events increase in frequency and intensity, recovery efforts will be more expensive, more prolonged and, most likely, the resources allocated for recovery will not adequately meet the needs of these communities. The rubber band of healthcare, which is stretched more and more, will eventually snap. What recent events tell us Australia's experience of climate-related events has largely been shaped by oscillations across extremes of heat, drought. In: Routledge handbook of climate change and health system sustainability, vol 10, p 452

Bailey K, Crawford T, Hodge S (2025) Recovery and resilience in healthcare emergency management. In: Emergency management for healthcare leaders. Productivity Press, pp 87–99. https://doi.org/10.4324/9781003529194

Baker LR, Cormier LA (2014) Disasters and vulnerable populations: evidence-based practice for the helping professions. Springer, Berlin

Baker BA (2014) Art speaks in healing survivors of war: the use of art therapy in treating trauma survivors. In: Trauma treatment techniques. Routledge, pp 183–198. https://doi.org/10.4324/9781315864518

Batra M, Erbas B (2025) Extreme weather, vulnerable populations, and mental health: the timely role of AI interventions. Int J Environ Res Public Health 22(4):602. https://doi.org/10.3390/ijerph22040602

Bessaha M, Hayward RA, Gatanas K (2022) A scoping review of youth and young adults' roles in natural disaster mitigation and response: considerations for youth wellbeing during a global ecological crisis. Child Adolesc Mental Health 27(1):14–21. https://doi.org/10.1111/camh.12517

Bonanno GA, Brewin CR, Kaniasty K, Greca AM (2010) Weighing the costs of disaster: consequences, risks, and resilience in individuals, families, and communities. Psychol Sci Public Interest A J Am Psychol Soc 11(1):1–49. https://doi.org/10.1177/1529100610387086

Breth-Petersen M, Garay J, Clancy K, Dickson M, Angelo C (2023) Homesickness at home: a scoping review of solastalgia experiences in Australia. Int J Environ Res Public Health 20(3):2541. https://doi.org/10.3390/ijerph20032541

Bromet EJ, Atwoli L, Kawakami N, Navarro-Mateu F, Piotrowski P, King AJ et al (2017) Posttraumatic stress disorder associated with natural and human-made disasters in the world mental health surveys. Psychol Med 47(2):227–241. https://doi.org/10.1017/S0033291716002026

Burback L, Brémault-Phillips S, Nijdam MJ, McFarlane A, Vermetten E (2024) Treatment of posttraumatic stress disorder: a state-of-the-art review. Curr Neuropharmacol 22(4):557–635. https://doi.org/10.2174/1570159X21666230428091433

Bürgy M (2018) Study of delusional depression: drive, dynamics, therapy. Neurol Psychiatry Brain Res 30:117–124. https://doi.org/10.1016/j.npbr.2018.08.003

Burns JK (2015) Poverty, inequality and a political economy of mental health. Epidemiol Psychiat Sci 24(2):107–113. https://doi.org/10.1017/S2045796015000086

Burns TJ, Boyd TW, Leslie CM (2019) Regenerative development and environmental ethics: healing the mismatch between culture and the environment in the third millennium. Regenerative urban development, climate change and the common good. Routledge, New York, pp 115–135

Carlson MJ, Wimer C, Haskins R (2022) Changing work, changing families, and public policies toward low-income families. RSF Russell Sage Found J Soc Sci 8(5):1. https://doi.org/10.7758/RSF.2022.8.5.01

Chakraborty SK, Sanyal P, Ray R (2023) Pollution, environmental perturbation and consequent loss of wetlands. In: Wetlands ecology: eco-biological uniqueness of a Ramsar site (East Kolkata Wetlands, India). Springer, Cham, pp 521–582. https://doi.org/10.1007/978-3-031-09253-4_8

Cianconi P, Betrò S, Janiri L (2020) The impact of climate change on mental health: a systematic descriptive review. Front Psych 11:490206. https://doi.org/10.3389/fpsyt.2020.00074

Cianconi P, Hanife B, Grillo F, Betro S, Lesmana CBJ, Janiri L (2023) Eco-emotions and psychoterratic syndromes: reshaping mental health assessment under climate change. Yale J Biol Med 96(2):211–226

Colvin MKM, Reesman J, Glen T (2022) The impact of COVID-19 related educational disruption on children and adolescents: an interim data summary and commentary on ten considerations for neuropsychological practice. Clin Neuropsychol 36(1):45–71. https://doi.org/10.1080/13854046.2021.1970230

Comtesse H, Ertl V, Hengst SM, Rosner R, Smid GE (2021) Ecological grief as a response to environmental change: a mental health risk or functional response? Int J Environ Res Public Health 18(2):734. https://doi.org/10.3390/ijerph18020734

Conzatti A, Kershaw T, Copping A, Coley D (2022) A review of the impact of shelter design on the health of displaced populations. J Int Hum Act 7(1):18. https://doi.org/10.1186/s41018-022-00123-0

Counted V, Neff MA, Captari LE, Cowden RG (2020) Transcending place attachment disruptions during a public health crisis: spiritual struggles, resilience, and transformation. J Psychol Christ 39(4):276–287

Cuthrell KM (2025) Climate psychiatry: the impacts of climate change intergenerational trauma's impact on reproductive, neuropsychiatric and neurodegenerative disorders. Int Neuropsych Dis J 22(3):93–112. https://doi.org/10.9734/indj/2025/v22i3483

Cutrona CE (2000) Social support principles for strengthening families. Fam Support Disadv Fam 12:103–122

Dasgupta R, Basu M (2023) Mental health and socio-psychological manifestations of cyclone-induced water insecurity in the Indian Sundarban delta. Int J Disast Risk Reduct 98:104085. https://doi.org/10.1016/j.ijdrr.2023.104085

David-Raj A, Padmapriya R, David-Raj A (2024) Climate crisis impact on ecosystem services and human well-being. In: Climate crisis, social responses and sustainability: socio-ecological study on global perspectives. Springer, Cham, pp 3–36. https://doi.org/10.1007/978-3-031-58261-5_1

De Rond M, Lok J (2016) Some things can never be unseen: the role of context in psychological injury at war. Acad Manag J 59(6):1965–1993. https://doi.org/10.5465/amj.2015.0681

Dimitrov R, Jelen A, L'Etang J (2022) Taboos in health communication: stigma, silence and voice. Public Relat Inquiry 11(1):3–35. https://doi.org/10.1177/2046147X211067002

Eboreime E, Anjorin O, Obi-Jeff C, Ojo TM, Hertelendy A (2025) From drought to displacement: assessing the impacts of climate change on conflict and forced migration in West Africa's Sahel Region. J Clim Change Health 23:100448. https://doi.org/10.1016/j.joclim.2025.100448

Esterwood E, Saeed SA (2020) Past epidemics, natural disasters, COVID19, and mental health: learning from history as we deal with the present and prepare for the future. Psych Quart 91(4):1121–1133. https://doi.org/10.1007/s11126-020-09808-4

Falloon IR, Laporta M, Fadden G, Graham-Hole V (2023) Managing stress in families: cognitive and behavioural strategies for enhancing coping skills. Routledge, New York

Fussell E (2015) The long term recovery of new orleans' population after Hurricane Katrina. Am Behav Sci 59(10):1231–1245. https://doi.org/10.1177/0002764215591181

Gutterman AS (2022) Older persons' rights to an adequate standard of living, social protection, participation and security. In: Gutterman A (ed) Older persons' rights to an adequate standard of living, social protection, participation and security. Older Persons' Rights Project, Oakland. https://doi.org/10.2139/ssrn.3893232

Haq SM, Pieroni A, Bussmann RW, Abd-ElGawad AM, El-Ansary HO (2023) Integrating traditional ecological knowledge into habitat restoration: implications for meeting forest restoration challenges. J Ethnobiol Ethnomed 19(1):33. https://doi.org/10.1186/s13002-023-00606-3

Hartwell C, Randazza JM, Bratman GN, Eisenman DP, Ellis B, Goodsell E et al (2023) Forest therapy as a trauma-informed approach to disaster recovery: insights from a wildfire-affected community. PLoS Climate 2(9):e0000096. https://doi.org/10.1371/journal.pclm.0000096

Herrenkohl TI, Hong S, Verbrugge B (2019) Trauma-informed programs based in schools: linking concepts to practices and assessing the evidence. Am J Community Psychol 64(3–4):373–388. https://doi.org/10.1002/ajcp.12362

Hinds JA, Sanchez ER (2022) The role of the hypothalamus-pituitary-adrenal (HPA) axis in test-induced anxiety: assessments, physiological responses, and molecular details. Stresses 2(1):146–155. https://doi.org/10.3390/stresses2010011

Humphreys A, Walker EG, Bratman GN, Errett NA (2022) What can we do when the smoke rolls in? An exploratory qualitative analysis of the impacts of rural wildfire smoke on mental health and wellbeing, and opportunities for adaptation. BMC Public Health 22(1):41. https://doi.org/10.1186/s12889-021-12411-2

Jackson K, Roberts R, McKay R (2019) Older people's mental health in rural areas: converting policy into service development, service access and a sustainable workforce. Aust J Rural Health 27(4):358–365. https://doi.org/10.1111/ajr.12529

Jha MK, Dev M (2024) Impacts of climate change. Smart internet of things for environment and healthcare. Springer, Cham, pp 139–159

Kahan D, Lamanna D, Rajakulendran T, Noble A, Stergiopoulos V (2020) Implementing a trauma-informed intervention for homeless female survivors of gender-based violence: lessons learned in a large Canadian urban centre. Health Soc Care Commun 28(3):823–832. https://doi.org/10.1111/hsc.12913

Kaluarachchi Y (2018) Building community resilience in the re-settlement of displaced communities. Proced Eng 212:443–450. https://doi.org/10.1016/j.proeng.2018.01.057

Kemei J, Salami B, Soboka M, Gommaa HIM, Okeke-Ihejirika P, Lavin T (2023) The forms and adverse effects of insecurities among internally displaced children in Ethiopia. BMC Public Health 23(1):200. https://doi.org/10.1186/s12889-023-15109-9

Khatri P, Kumar P, Shakya KS, Kirlas MC, Tiwari KK (2024) Understanding the intertwined nature of rising multiple risks in modern agriculture and food system. Environ Dev Sustain 26(9):24107–24150. https://doi.org/10.1007/s10668-023-03638-7

Knapp M, Wong G (2020) Economics and mental health: the current scenario. World Psych 19(1):3–14. https://doi.org/10.1002/wps.20692

Kõlves K, Shaw-Williams D, Krishnamoorthy S, Bayliss L, Hawgood J, Reifels L (2025) From rising temperature to eco-emotions: exploring the impact of climate change on suicidality. Lancet Reg Health West Pacific 55:101491. https://doi.org/10.1016/j.lanwpc.2025.101491

Kunnathepeedikayil S (2020) The long-term mental health consequences of floods and landslides in Thrissur District, Kerala, India. Harvard University

Lai BS, Kelley ML, Harrison KM, Thompson JE, Self-Brown S (2015) Posttraumatic stress, anxiety, and depression symptoms among children after Hurricane Katrina: a latent profile analysis. J Child Fam Stud 24(5):1262–1270. https://doi.org/10.1007/s10826-014-9934-3

Langarizadeh M, Tabatabaei MS, Tavakol K, Naghipour M, Rostami A, Moghbeli F (2017) Tele-mental health care, an effective alternative to conventional mental care: a systematic review. Acta Inform Med AIM J Soc Med Inform Bosnia Herzegovina Casopis Drustva Za Medicinsku Informatiku BiH 25(4):240–246. https://doi.org/10.5455/aim.2017.25.240-246

Lohrey S, Chua M, Gros C, Faucet J, Lee JKW (2021) Perceptions of heat-health impacts and the effects of knowledge and preventive actions by outdoor workers in Hanoi, Vietnam. Sci Total Environ 794:148260. https://doi.org/10.1016/j.scitotenv.2021.148260

Lombe P, Carvalho E, Rosa-Santos P (2024) Drought dynamics in sub-Saharan Africa: impacts and adaptation strategies. Sustainability 16(22):9902. https://doi.org/10.3390/su16229902

Loucks DP (2021) Impacts of climate change on economies, ecosystems, energy, environments, and human equity: a systems perspective. The impacts of climate change. Elsevier, Amsterdam, pp 19–50

Mahbod M, Parnian A (2024) Sustainable development in rural areas: climate change impacts on water, agriculture, energy, and psychological health consequences. In: Integration of core sustainable development goals in rural areas: current practices of water, energy, food, climate change, and ecosystems. Springer, Cham, pp 67–80. https://doi.org/10.1007/978-3-031-60149-1_3

Mampane MR (2020) The effect of a community-based social support programme on the resilience of children from vulnerable families. Early Child Dev Care 190(10):1638–1653. https://doi.org/10.1080/03004430.2019.1670653

Miller AB, Davis SH, Mulder LA, Winer JP, Issa OM, Cardeli E, Ellis BH (2024) Leveraging community-based mental health services to reduce inequities for children and families living in United States who have experienced migration-related trauma. Psychol Trauma Theory Res Pract Policy 16(Suppl 2):S426–S434. https://doi.org/10.1037/tra0001392

Mishra H, Michael G (2024) Unraveling the psychological toll: understanding eco-anxiety as a consequence of climate change. Impact of climate change on social and mental well-being. Academic Press, New York, pp 299–316

Negi S, Singh R, Singh AK (2025) Natural disaster: prevalence of post-traumatic stress and psychological outcome of survivors in Joshimath, Uttarakhand, India. In: Managing the health risks of climate change. IGI Global Scientific Publishing, pp 107–134. https://doi.org/10.4018/979-8-3693-4183-4.ch005

Noor T, Nazeer I, Attique Z, Shahzad M, Baqi A (2021) Global temperature variations since pre industrial era. Int J Innov Sci Technol 3:67–74

Nye C, Winter M, Lobley M (2023) Farmers supporting farmers: livestock auctions as spaces to reconstruct occupational community and counter mental health issues. J Agromedicine 28(3):401–414. https://doi.org/10.1080/1059924X.2023.2176959

Ogbuagu OO, Mbata AO, Balogun OD, Oladapo O, Ojo OO, Muonde MU (2024) Expanding access to mental health in low-resource settings: strategies for policy, supply chain, and implementation. IRE Journals 7(9):407–412

O'Keefe VM, Cwik MF, Haroz EE, Barlow A (2021) Increasing culturally responsive care and mental health equity with indigenous community mental health workers. Psychol Serv 18(1):84–92. https://doi.org/10.1037/ser0000358

Ore ML, Foli KJ (2020) Reintegration for post-treatment cancer survivors: a concept analysis. J Holist Nurs 38(3):300–317. https://doi.org/10.1177/0898010119883760

Orimoloye IR (2022) Agricultural drought and its potential impacts: enabling decision-support for food security in vulnerable regions. Front Sustain Food Syst 6:838824. https://doi.org/10.3389/fsufs.2022.838824

Palinkas LA, O'Donnell ML, Lau W, Wong M (2020) Strategies for delivering mental health services in response to global climate change: a narrative review. Int J Environ Res Public Health 17(22):8562. https://doi.org/10.3390/ijerph17228562

Patrick R, Snell T, Gunasiri H, Garad R, Meadows G, Enticott J (2023) Prevalence and determinants of mental health related to climate change in Australia. Aust N Z J Psych 57(5):710–724. https://doi.org/10.1177/00048674221107872

Perez-Murcia LE (2021) 'Physically sheltered but existentially homeless': losing home in the aftermath of conflict and displacement. Migrat Stud 9(3):1361–1379. https://doi.org/10.1093/migration/mnaa020

Pihkala P (2024) Ecological sorrow: types of grief and loss in ecological grief. Sustainability 16(2):849. https://doi.org/10.3390/su16020849

Rafiey H, Momtaz YA, Alipour F, Khankeh H, Ahmadi S, Sabzi-Khoshnami M, Haron SA (2016) Are older people more vulnerable to long-term impacts of disasters? Clin Intervent Aging 1791:22122. https://doi.org/10.2147/CIA.S122122

Rehling JT (2022) Conceptualising eco-anxiety using an existential framework. South Afr J Psychol 52(4):472–485. https://doi.org/10.1177/00812463221130898

Reiner V, Pathirana NL, Sun YY, Lenzen M, Malik A (2024) Wish you were here? The economic impact of the tourism shutdown from Australia's 2019–2020 'black summer' bushfires. Econ Disast Clim Change 8(1):107–127. https://doi.org/10.1007/s41885-024-00142-8

Ritchie EC, Ehret M, Peake NW, Richter KE (2019) When a disaster disrupts access to psychiatric medications: advance planning can help minimize the risks of withdrawal, other consequences. Curr Psych 18(5):22

Rony MKK, Alamgir HM (2023) High temperatures on mental health: recognizing the association and the need for proactive strategies—a perspective. Health Sci Rep 6(12):e1729. https://doi.org/10.1002/hsr2.1729

Rubin IL, Henry Falk MD (2019) Natural disasters and vulnerable populations: a commentary. Int J Child Health Hum Develop 12(4):303–318

Sabri B, Granger DA (2018) Gender-based violence and trauma in marginalized populations of women: role of biological embedding and toxic stress. Health Care Women Int 39(9):1038–1055. https://doi.org/10.1080/07399332.2018.1491046

Salarieh B, Ugwu IA, Salman AM (2023) Impact of changes in sea surface temperature due to climate change on hurricane wind and storm surge hazards across US Atlantic and Gulf coast regions. SN Appl Sci 5(8):205. https://doi.org/10.1007/s42452-023-05423-7

Setyawan I, Sajidah SA, Sabila A, Wijayanti AP, Dewi KS (2023) A mapping study of psychological preparedness in dealing with extreme temperature changes and natural disasters. In: Proceedings of international conference on psychological studies (ICPsyche), vol 4, pp. 423–459. https://doi.org/10.58959/icpsyche.v4i1.57

Shaban A, Kourtit K, Nijkamp P, Das B (2024) Socio-ecological shocks, weak community support systems, and tragic responses of farmers: a modeling study on India. Reg Sci Policy Pract 16(9):100030. https://doi.org/10.1016/j.rspp.2024.100030

Shareeff N, Pramono RK, Saliha A (2024) Climate change and mental health: exploring the psychological impacts of environmental distress in indonesia. Sci Psych 5(4):605–618

Sharma M, Sharma V, Peeples D (2025) Telepsychiatry, access, and equity: accelerating mental health care for rural and low-income youth. Front Public Health 13:1698682. https://doi.org/10.3389/fpubh.2025.1698682

Shibesh BF, Nagabhatla N (2025) Addressing climate resilience in the African region: prioritizing mental health and psychosocial well-being in disaster preparedness and response planning for mainstream communities and migrants. Climate 13(7):139. https://doi.org/10.3390/cli13070139

Sitorus HP, Silitonga M (2025) The Role of cortisol in the stress response. Int J Ecophysiol 7(1):48–58

Smid GE, Drogendijk AN, Knipscheer J, Boelen PA, Kleber RJ (2018) Loss of loved ones or home due to a disaster: effects over time on distress in immigrant ethnic minorities. Transcult Psych 55(5):648–668. https://doi.org/10.1177/1363461518784355

Smith GS, Anjum E, Francis C, Deanes L, Acey C (2022) Climate change, environmental disasters, and health inequities: the underlying role of structural inequalities. Curr Environ Health Rep 9(1):80–89. https://doi.org/10.1007/s40572-022-00336-w

Smith K, Fearnley CJ, Dixon D, Bird DK, Kelman I (2023) Environmental hazards: assessing risk and reducing disaster. Routledge, New York

Stanley SK, Heffernan T, Macleod E, Lane J, Walker I, Evans O et al (2024) Solastalgia following the Australian summer of bushfires: qualitative and quantitative insights about environmental distress and recovery. J Environ Psychol 95:102273. https://doi.org/10.1016/j.jenvp.2024.102273

Stark L, Robinson MV, Seff I, Gillespie A, Colarelli J, Landis D (2022) The effectiveness of women and girls safe spaces: a systematic review of evidence to address violence against women and girls in humanitarian contexts. Trauma Violence Abuse 23(4):1249–1261. https://doi.org/10.1177/1524838021991306

Sugiura T, Sugiura Y (2016) Relationships between refraining from catastrophic thinking, repetitive negative thinking, and psychological distress. Psychol Rep 119(2):374–394. https://doi.org/10.1177/0033294116663511

Sun L, Deng Y, Qi W (2018) Two impact pathways from religious belief to public disaster response: findings from a literature review. Int J Disast Risk Reduct 27:588–595. https://doi.org/10.1016/j.ijdrr.2017.10.004

Suryana C, Budiandru B, Naibaho KET, Setianti Y, Purwosusanto H (2024) The impact of fake news on public opinion during crisis situations. J Acad Sci 1(4):395–407

Tay JL, Tay YF, Klainin-Yobas P (2018) Effectiveness of information and communication technologies interventions to increase mental health literacy: a systematic review. Early Interv Psych 12(6):1024–1037. https://doi.org/10.1111/eip.12695

Tschakert P, Bourgault du Coudray C, Horwitz P (2024) Walking journeys into everyday climatic-affective atmospheres: the emotional labour of balancing grief and hope. Environ Plan Nat Space 7(1):189–211. https://doi.org/10.1177/25148486231173202

Ullah W, Dong H, Shah AA, Xu C, Alotaibi BA (2025) Unveiling the multi-dimensional vulnerabilities of flood-affected communities in Khyber Pakhtunkhwa, Pakistan. Water 17(2):198. https://doi.org/10.3390/w17020198

Valentine JA, Delgado LF, Haderxhanaj LT, Hogben M (2022) Improving sexual health in US rural communities: reducing the impact of stigma. AIDS Behav 26:90–99

van Eck CW, van den Heijkant L, van der Meer TG (2024) When i say climate change, what do you see? News values and psychological distance in climate visuals of Dutch news media. Commun Clim Hope 30:1325

White BP, Breakey S, Brown MJ, Smith JR, Tarbet A, Nicholas PK, Ros AMV (2023) Mental health impacts of climate change among vulnerable populations globally: an integrative review. Ann Glob Health 89(1):66. https://doi.org/10.5334/aogh.4105

Zangani C, Ostinelli EG, Smith KA, Hong JS, Macdonald O, Reen G et al (2022) Impact of the COVID-19 pandemic on the global delivery of mental health services and telemental health: systematic review. JMIR Mental Health 9(8):e38600. https://doi.org/10.2196/38600

Zimmer-Gembeck MJ, Skinner EA (2016) The development of coping: implications for psychopathology and resilience. Develop Psychopathol 69:1–61. https://doi.org/10.1002/9781119125556.devpsy410

Chapter 7
Drought and Human Health: A Global Scoping Review of Mortality, Morbidity, and Policy Directions

Ejaz Ahmad Khan, Abubakar Abubakar Bello, Gilbert Tayebwa, and Benson Turyasingura

Abstract Drought is a slow-developing climate hazard that places sustained strain on human health by gradually disrupting water supplies, food systems, livelihoods, and health services. Unlike sudden events such as floods or heatwaves, drought unfolds over long periods, which makes its health effects harder to recognise, track, and address, and contributes to the fragmented nature of existing evidence on drought-related mortality and morbidity. This chapter draws on a global scoping review to map and synthesise the available literature on the health impacts of drought. Using the PRISMA Extension for Scoping Reviews (PRISMA-ScR) as a guiding framework, a broad search of peer-reviewed articles, government publications, and grey literature published between 1995 and 2025 was conducted across multiple databases. From an initial 2052 records identified, 1420 were screened after deduplication, resulting in 21 studies that met the inclusion criteria for synthesis. The studies identified reflect a wide range of methodological approaches, highlighting the interdisciplinary character of research at the intersection of drought and health. Collectively, the evidence shows that drought is linked to increased mortality and morbidity through a combination of direct and indirect pathways, including infectious diseases, malnutrition, respiratory and cardiovascular conditions, mental health challenges, and poor maternal and child health outcomes. The health burden is not evenly shared, with

E. A. Khan (✉)
Department of Public Health, Health Services Academy, Islamabad, Pakistan
e-mail: ejaz@hsa.edu.pk

A. A. Bello
Department of Geography, Nigerian Defence Academy, Kaduna, Nigeria
e-mail: bello.abubakara@nda.edu.ng

G. Tayebwa
The International Crane Foundation (Uganda Office), Kampala, Uganda
e-mail: gilbertt@savingcranes.org

B. Turyasingura
Department of Environment and Natural Resources, Faculty of Agriculture and Environmental Sciences, Kabale University, Kabale, Uganda
e-mail: bturyasingura@kab.ac.ug; info@kab.ac.ug

P. Ghosh et al. (eds.), *Climate Crisis and Changing Population Dynamics*, Springer Climate, https://doi.org/10.1007/978-3-032-24533-5_7

young children, older adults, women, and socioeconomically disadvantaged groups consistently facing greater risks. Clear geographical differences also emerge, with the most severe impacts reported in Sub-Saharan Africa and parts of Asia, where drought often coincides with poverty, food insecurity, and limited health system capacity, while studies from higher-income regions more commonly report effects on non-communicable and respiratory diseases. Overall, this review identifies important gaps in the current evidence, particularly in relation to long-term health effects and underrepresented regions, and emphasises the need to better integrate health into drought early warning systems, climate adaptation efforts, and public health planning as climate change intensifies.

Keywords Drought · Morbidity · Mortality · Public health · Climate extremes · Scoping review · Policy

Introduction

One of the most widespread and complicated climate-related risks affecting human populations worldwide is drought. In contrast to sudden occurrences like heat waves or floods, drought develops gradually over months or years, resulting in cumulative effects on the environment, society, and health that are challenging to identify, quantify, and regulate. Its protracted nature impairs access to healthcare, food systems, livelihoods, and water supply, which puts a persistent strain on population health and raises morbidity and mortality in a variety of contexts.

Water scarcity during droughts frequently forces populations to rely on hazardous or contaminated water sources, increasing the risk of waterborne diseases including cholera and dysentery. Simultaneously, decreased agricultural output and livestock losses jeopardise food security, resulting in hunger, weakened immune function, and increased vulnerability to infectious illnesses. Dry circumstances can raise respiratory health concerns because dust storms and airborne particles transport microbiological pathogens and chemicals that damage air quality. Furthermore, drought is frequently associated with high temperatures, which exacerbates heat stress and contributes to dehydration, heat-related diseases, and cardiovascular strain (Khan et al. 2021). These direct health consequences are exacerbated by indirect effects such as population displacement, overcrowding, inadequate sanitation, and overburdened healthcare systems functioning under economic constraints and insufficient resources (Lawrance et al. 2022).

Recent studies show that the frequency, intensity, and spatial extent of droughts have risen in many parts of the world in recent decades, owing primarily to climate change and variability. This escalation has been linked to increased risks of food and water instability, vector- and waterborne infections, respiratory and cardiovascular problems, poor mental health outcomes, and premature death (Mani et al. 2024; Salvador et al. 2023). Over 1.6 billion individuals were impacted by drought between 2000 and 2023, according to global monitoring statistics, highlighting the extent of exposure and the potential scope of health effects (Mani et al. 2024; Xiao et al. 2024) Both direct physiological channels and indirect socioeconomic processes, such as interruption of livelihood, financial stress, and decreased access to critical health services, are manifestations of these consequences (Chen et al. 2013).

There is also evidence of significant regional differences in the health consequences of drought. Moderate to severe drought exposure has been linked to higher respiratory mortality in some parts of North America, with significant geographical variations in risk patterns (Singam et al. 2023). Drought is a major public health concern throughout Asia, especially in South and Southeast Asia, where dry spells have been associated with increased vulnerability and respiratory mortality in older adults and people with pre-existing cardiovascular conditions (Goh et al. 2024; Mojtahedzadeh et al. 2021). Iranian studies also show that the effects of drought go beyond short-term water scarcity to include economic hardship, agricultural disruption, and reduced access to healthcare, all of which lead to negative health outcomes.

Due to the combination of high climate sensitivity, pervasive poverty, food insecurity, and inadequate healthcare infrastructure, Sub-Saharan Africa bears a disproportionate burden of drought-related morbidity and mortality. Several drought-prone nations, such as Somalia, Sudan, Afghanistan, and Yemen, continue to have exceptionally high rates of under-five mortality (Bellizzi et al. 2020). Tens of millions of people live in the Horn of Africa, which is particularly vulnerable due to frequent drought cycles that cause acute malnutrition rates in children that beyond internationally recognised emergency criteria (Tedla et al. 2025). Sub-Saharan Africa is positioned as a worldwide epicentre of climate-related health vulnerability due to these patterns, which show how structural disparities, insufficient adaptation capacity, and limited health systems compound the health costs of drought (Banerjee et al. 2024).

Compared to other climatic extremes like heatwaves and floods, drought has received relatively less attention in the health-climate literature, despite the significant and increasing health burden associated with it. Existing research is frequently dispersed, concentrating on certain health outcomes, demographics, or geographical areas, and the body of data is nonetheless inconsistent and methodologically varied overall. Differential vulnerability based on age, gender, socioeconomic level, and underlying medical issues also makes it more difficult to synthesise findings and restricts the application of evidence in practice and policy. Because of this, there are still significant gaps in our understanding of the entire range of health effects associated with drought and how they vary among populations and geographical areas.

Scoping reviews offer a strong framework for tackling these difficulties by methodically mapping existing knowledge, identifying research gaps, and synthesising findings from various study types and contexts (Munn et al. 2018). As a result, this chapter provides a detailed scoping analysis of the effects of drought on mortality and morbidity, with a special emphasis on spatial disparities, vulnerable populations, and significant health consequences. This chapter intends to improve understanding of the complex pathways linking drought and health by integrating and critically reviewing the global literature. It will also inform future research priorities, public health planning, and climate-resilient policy creation.

Materials and Methods

Study Design and Search Approach

This study used a scoping review design to map and synthesise current evidence on the effects of drought on mortality and morbidity. A comprehensive literature search was conducted from 1995 to 2025, allowing the review to include both foundational studies from the 1990s and more recent research on contemporary climate change dynamics, breakthroughs in health surveillance, and enhanced analytical methodologies. The chosen time span ensured a fair coverage of early conceptual work and contemporary empirical evidence.

The search procedure adhered to established scoping review methodological guidelines, with carefully designed keyword combinations and Boolean operators used to maximise retrieval of relevant papers while retaining search specificity. Medical Subject Headings (MeSH) and free-text keywords were used to provide comprehensive coverage across multiple indexing systems and publication formats (Salvador et al. 2023). To keep the evidence base current, database searches were updated monthly between January 2025 and August 2025, allowing for the incorporation of freshly published studies.

Data Sources and Databases

Several electronic databases were examined in order to obtain thorough coverage across disciplines and geographical contexts. These included the Web of Science Core Collection for interdisciplinary research, Scopus for environmental and medical sciences, PubMed/MEDLINE for clinical and public health literature, ScienceDirect for epidemiological and environmental health studies, and Google Scholar for grey literature and new research. Additional specialised databases were added to improve regional and discipline representation, including EMBASE for European biomedical literature, CINAHL for nursing and allied health views, and the Global Health Database for international public health research. Regional databases, such as African Index Medicus and LILACS, were also examined to ensure that low- and middle-income studies were adequately represented.

Prior to screening, 2052 records from 1995 to 2025 were detected across all databases. The search focused on studies that reported on drought-related mortality and morbidity outcomes at the global, regional, and national levels, as well as climate-health impact assessments and evaluations of health-related adaptation and interventions.

Search Terms and Keyword Strategy

The search approach made significant use of keywords and Boolean operators to capture the multifaceted health effects of drought. The primary search phrases focused on drought exposure and health outcomes, including terms relating to mortality, morbidity, and specific illness categories.

The primary search terms included: ("drought" AND "mortality") AND ("cardiovascular death" OR "respiratory mortality") AND ("morbidity" AND "malnutrition") AND ("infectious disease" AND "drought exposure") AND ("mental health" AND "drought impact") AND ("child mortality" AND "water scarcity") AND ("maternal health" AND "drought stress") AND ("hospital admission" AND "drought period") AND ("heat-related illness" AND "dry season") AND ("dehydration" AND "mortality risk") AND ("food insecurity" AND "health outcomes") AND ("climate change" AND "health burden") AND ("water shortage" AND "disease outbreak") AND ("agricultural drought" AND "public health") AND ("meteorological drought" AND "health crisis") AND ("hydrological drought" AND "mortality rate") AND ("socioeconomic drought" AND "health disparity") AND ("drought severity" AND "health impact") AND ("drought duration" AND "morbidity pattern") AND ("drought frequency" AND "health system") AND ("vulnerable population" AND "drought health") AND ("elderly" AND "drought mortality") AND ("children" AND "drought morbidity") AND ("women" AND "drought health risk") AND ("rural population" AND "drought impact") AND ("urban drought" AND "health effect") AND ("Sub-Saharan Africa" AND "drought mortality") AND ("Asia" AND "drought health burden") AND ("developing countries" AND "drought health") AND ("health intervention" AND "drought response") AND ("adaptation strategy" AND "health resilience") AND ("early warning" AND "health preparedness") AND ("healthcare access" AND "drought emergency") AND ("health system strengthening" AND "climate resilience") AND ("epidemiological surveillance" AND "drought health") AND ("health policy" AND "drought management") AND ("One Health" AND "drought impact") AND ("planetary health" AND "drought consequences") AND ("environmental health" AND "drought exposure") AND ("global health" AND "drought challenge") AND ("health equity" AND "drought vulnerability").

Inclusion and Exclusion Criteria

Drought-related mortality and morbidity studies published between 1995 and 2025 were evaluated. Peer-reviewed journal papers, government reports, and relevant grey literature were all deemed acceptable sources. The studies contained quantitative or qualitative data on the health impacts of meteorological, agricultural, hydrological, or socioeconomic drought, and they covered people of all ages. The health outcomes of interest were death rates, morbidity patterns, illness prevalence, and hospitalisation or use of health services during droughts. Only studies published in English were analysed. Research from both industrialised and developing nations was considered to ensure wide geographic coverage, with an emphasis on Africa, Asia, and other climate-vulnerable regions that corresponded with the review's objectives.

Studies were rejected if they focused solely on agricultural output, economic implications, or drought predictions and meteorological modelling without explicit health outcome measurements. Articles focusing solely on animal or veterinary health, conference abstracts without full-text availability, and duplicate publications were also removed. Furthermore, research that presented exclusively theoretical or conceptual frameworks rather than actual data were excluded. Research that lacked adequate methodological information to allow meaningful interpretation of findings, or that used older analytical methodologies that did not meet current epidemiological standards, was eliminated. Studies on healthcare system readiness or adaption measures that did not report linked health outcomes were also excluded.

Study Selection Process and Methodological Flow

The research selection strategy followed PRISMA (Preferred Reporting Items for Systematic Reviews and Meta-Analyses) guidelines to ensure transparency and reproducibility in literature discovery and screening processes. The systematic selection process is separated into three stages: identification, screening, and inclusion. The identification step included intensive database searching using predetermined search algorithms, which resulted in the first extraction of all possibly relevant documents from various databases and additional sources. Duplicate removal techniques used both automatic software tools and manual inspection to ensure accurate deduplication across many database sources.

The screening step was a two-stage evaluation procedure, with two reviewers screening titles and abstracts independently using specific inclusion and exclusion criteria, followed by a full-text review of potentially suitable publications. When no consensus was reached during screening, disagreements were addressed through discussion and consultation with a third independent reviewer. The inclusion process involved a final selection of publications that satisfied all requirements, followed by data extraction and quality evaluation procedures using standardised forms created specifically for this review. The overall methodological workflow of the scoping review, including database selection, screening, data extraction, and synthesis, is shown in Fig. 7.1.

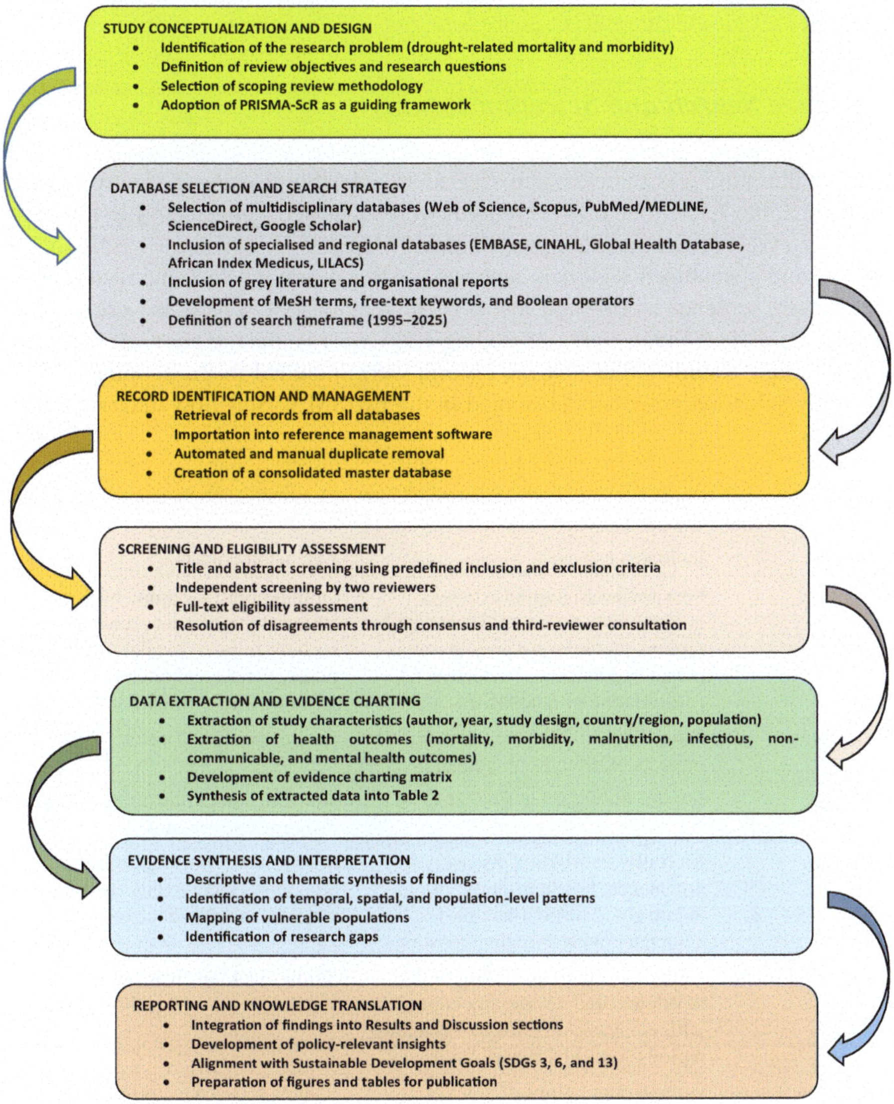

Fig. 7.1 Methodological flowchart

Results

Literature Search and Selection Results

Relevant literature was searched across academic databases using the keywords and criteria in Table 7.1. A total of 2052 records were identified from electronic academic databases (Web of Science, Scopus, PubMed/MEDLINE, Science Direct, Google Scholar, and specialised regional databases). After removing 632 duplicates, 1420 records were screened by title and abstract. The remaining 154 records were selected for full-text review. Finally, after excluding 1266 records that did not meet the inclusion criteria, 21 studies that met the criteria were included in the scoping review. The study selection process is presented in the PRISMA- ScR flow diagram below (Fig. 7.2).

Table 7.1 Inclusion and exclusion criteria

Category	Inclusion criteria	Exclusion criteria
Study type	Peer-reviewed journal articles, government and intergovernmental reports, and relevant grey literature presenting empirical evidence (quantitative or qualitative)	Opinion pieces, commentaries, editorials, conference abstracts without full-text availability, and purely theoretical or conceptual papers without empirical data
Time frame	From 1995 to 2025 (captures recent impacts of drought on human health)	Studies published before 1995
Language	Studies published in English	Non-English publications (due to resource constraints)
Outcomes	Mortality, morbidity, disease prevalence, hospitalisation, or other documented health outcomes associated with drought exposure	Studies not reporting health outcomes or not explicitly linking outcomes to drought exposure
Geographic scope	Global coverage, including both developed and developing countries, with emphasis on climate-vulnerable regions (e.g. Africa and Asia)	Studies lacking clear geographic context
Population	Human populations of any age group (children, adults, elderly)	Studies focused exclusively on animal or veterinary health
Exposure	Meteorological, agricultural, hydrological, or socioeconomic drought	Studies examining drought forecasting, climate modelling, or agricultural impacts without health-related exposure or outcomes
Methodological quality	Studies providing sufficient methodological detail to allow meaningful interpretation of findings	Studies with inadequate methodological reporting or outdated analytical approaches

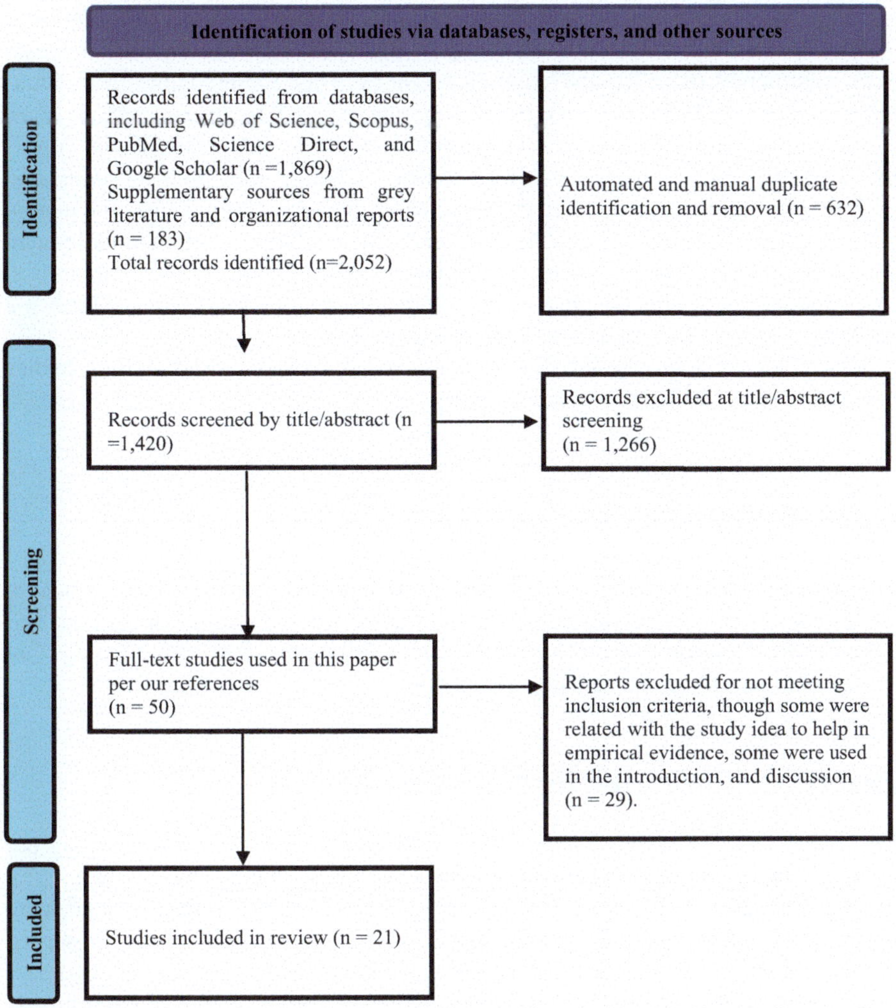

Fig. 7.2 PRISMA-ScR flow diagram of the study selection process

Overview of Included Studies

The selected 21 studies included for the scoping review were published between 2001 and 2025, capturing both historical and contemporary perspectives on drought-health linkages. There was a gradual increase in the number of studies from 2001, reaching the peak in 2007 ($n = 3$) after which it began to decline to ($n = 1$) study per year. A significant increase was observed from 2017 to 2025 with ($n = 2$) studies for the selected years. Methodologically, the studies were dominated by time-series analyses ($n = 7$), narrative and systematic reviews ($n = 5$), ecological studies (n

= 1), household surveys (n = 2), meta-analyses (n = 1), and historical/archival analyses (n = 2), among others. Geographically, studies covered global scope (n = 1), continental (n = 2), and regional scope (n = 1). Others came from countries representing Africa (n = 4), North America (n = 6), South America (n = 1), Europe (n = 1), and Asia (n = 5). The selected studies found the significant contribution of drought to various health outcomes, including increased mortality and morbidity from infectious and non-communicable diseases, malnutrition, and mental health impacts. The most affected populations include infants, children, middle-aged adults, the elderly, women, refugees, displaced persons, indigenous communities, rural residents, and other socially or geographically marginalised groups. Table 7.2 summarises the characteristics of the selected studies.

Source Authors' synthesis based on a scoping review of peer-reviewed studies retrieved from identified database searches

Synthesis of Key Findings

Drought has a complicated web of direct and indirect effects on human health, resulting in considerable mortality and morbidity across varied populations and geographical locations, according to the research synthesised. The findings are presented in important theme categories below.

Temporal Trend in Drought-Health Related Studies

The study of the 21 selected papers over a 30-year period reveals a changing but growing temporal pattern in research on the effects of drought on health (Fig. 7.3). Publications in this field began to appear gradually after 2001, reflecting a growing interest in the effects of drought on health. However, a sharp decline was observed in the trend after 2002 extending until 2006, when the studies witness a significant growth reaching the peak in 2007 with three published studies. The peak was followed by a dip in the trend from 2008 to 2013 (interrupted by only one study in 2011), when no relevant research was identified within this interval. A growing trend continues from 2014 extending through to 2025 with intermittent gap in publication observed in 2016, 2018, and 2019 during which no eligible studies were identified. Overall, the temporal distribution shows an increase in the number of studies published in recent years, with notable variations and periods of limited research activity across the study timeframe.

Study Design and Methodological Approach

The selected studies employed a wide range of study design and methodological approaches as presented in Fig. 7.4. Most studies adopted time-series analyses to

Table 7.2 Key characteristics and findings from selected studies

S. No.	Author, Year [Ref]	Study design	Country/region	Population	Health outcome	Key Findings
1	Charnley et al. (2021)	Narrative review	Africa	Refugees, people in informal settlements, malnourished and displaced groups	Morbidity (cholera outbreak)	Drought conditions often increase the risk of cholera by worsened access to safe water, sanitation, and food security. The study highlights how vulnerable populations face the risks and show clear link between drought and infectious disease outbreaks
2	Gwon et al. (2023)	Two-staged time-series with drought indices	USA	Older adults above 65 years, white and black communities, and non-metro residents	Respiratory mortality	Moderate-to-severe drought increased respiratory mortality risk by up to 15.9% in the upper Midwest and 9% in northern Rokies and plains with variation in regions and group
3	Abadi et al. (2022)	Time-series analysis	Nebraska, USA	Middle-aged white population aged between 25 and 64 living in metro counties and white women between 45 and 54 years in non-metro counties are mostly affected	Mortality	The study found no strong overall association between drought and mortality across the population. However, middle-aged whites, particularly women in rural areas were more vulnerable. There was stronger relationship between longer-term droughts and mortality compared to shorter-term droughts

(continued)

Table 7.2 (continued)

S. No.	Author, Year [Ref]	Study design	Country/region	Population	Health outcome	Key Findings
4	Acuna-Soto et al. (2002)	Historical/ epidemiologic analysis with tree-ring data	Mexico	Indigenous populations	Cocoliztli outbreaks leading to over ten million and two million deaths in 1545 and 1576, respectively	Severe drought coincided with deadly cocolztli epidemics, likely haemorrhagic fevers linked to rodents, were worsened by prolonged drought conditions leading to one of the deadliest mortality in America
5	Alam et al. (2022)	Time-series analysis	Northern Bangladesh	Residents of Saidpur and Rajshahi	Mortality from respiratory diseases and suicide-related deaths	Drought conditions often increase the risk of cholera by worsened access to safe water, sanitation, and food security. The study highlights how vulnerable populations face the risks and show clear link between drought and infectious disease outbreaks
6	Berman et al. (2017)	Time-series analysis	Western USA	Elderly people especially in counties with rare drought	Mortality and hospital admissions from respiratory and cardiovascular diseases	Severe drought raised mortality by 1.6% particularly in areas not used to drought while respiratory admissions dropped by 2% during the period, but the risks rose where drought was less common
7	De Waal et al. (2006)	Household survey comparing drought vs. non-drought areas	Ethiopia	Children under five years and rural household	Child mortality	Child mortality was high in drought-affected areas, linked more to chronic vulnerability than immediate drought

(continued)

Table 7.2 (continued)

S. No.	Author, Year [Ref]	Study design	Country/region	Population	Health outcome	Key Findings
8	Renzaho (2007)	Cluster household survey	Mozambique	Children under five years and pregnant women	Mortality, malnutrition, and lost pregnancy	Mortality of children under five years was high (18.9%); malnutrition prevalent; 7.7% pregnancies were lost due to drought conditions
9	Wu and Li (2007)	Ecological study	China	Rural communities exposed to water scarcity and contaminated surface water	Increased oesophageal cancer mortality	Drought-related reductions in river flow led to poor water quality and increased exposure to carcinogens. Significant association found between drought periods and elevated oesophageal cancer mortality
10	Yusa et al. (2015)	Narrative review	Canada	Rural and indigenous communities, elderly, individuals with pre-existing conditions, and low-income populations	Health effects such as respiratory illnesses, malnutrition, and mental health	Drought is rising public health concern in Canada, which can disrupt food, water, and air, increasing risks for vulnerable population
11	Delbiso et al. (2017)	Meta-analysis	Ethiopia	Children under 5 years in crisis-affected areas	Child mortality	Drought was not directly linked to higher deaths, but malnutrition increased the risk of child deaths
12	Salvador et al. (2021)	Time-series analysis	Lisbon, Portugal	Elderly people above 60 years, especially men	Natural, circulatory, and respiratory deaths	Drought was significantly linked to higher daily deaths. Elderly people, especially men are mostly affected. Circulatory and respiratory deaths rose notably during drought period

(continued)

Table 7.2 (continued)

S. No.	Author, Year [Ref]	Study design	Country/region	Population	Health outcome	Key Findings
13	Lynch et al. (2020)	Time-series analysis	USA	Adults aged between 25 and 64 years, white populations, and southern and western regions.	Mortality	Drought was not strongly linked to mortality across the country. However, some groups like middle-aged adults, white populations, and people in the south and west were more affected, indicating higher death risks during severe drought years
14	Wang et al. (2025)	Cross-sectional analysis	Africa	Infants, especially those in rural areas or tropical regions	Infant mortality	Severe long-term drought increased infant mortality risk by 4%, with new-borns most at risk. Short drought did not have much effect
15	Sena et al. (2014)	Narrative review	Brazil	Women, rural, and semiarid residents, and indigenous communities	Mental health, malnutrition, and waterborne illnesses	Drought leads to under nutrition, mental distress (particular anxiety in women), illness, and forced migration
16	Wakabi, (2006)	News report	Kenya	Rural agro-pastoralists, children, and elderly	Mortality and malnutrition risk	A severe drought led to the death of at least 40 people in rural areas
17	Algur et al. (2021	Systematic review	India	Women and children	Malnutrition and anaemia	Drought increases malnutrition and anaemia, and exposes gendered vulnerabilities
18	Below et al. (2007), Xin et al. (2025)	Global disaster database review	Global	Rural communities, poor households, children, and elderly	Malnutrition and mortality	Over the last century, droughts have been a silent but deadly force, driving food shortages, malnutrition, and widespread of deaths

(continued)

Table 7.2 (continued)

S. No.	Author, Year [Ref]	Study design	Country/region	Population	Health outcome	Key Findings
19	Xin et al. (2025)	Historical spatiotemporal analysis using archival records	China	Broad populations, especially in Northern China	Infectious disease outbreaks and famine	The study reveals the link between the droughts of 1585–1590 deadly disease outbreak. The disease spread from north to south with stronger disease spikes in the northern region. Famine acted as the bridge between drought and diseases
20	Zarocostas (2011)	News report/ situation analysis	Africa (Somalia, Ethiopia, and Kenya)	Children, displaced people, pastoralists, and rural poor	Malnutrition and increased child mortality	The drought left over ten million people facing hunger. Failed rain destroyed crops and livestock, leading to displacement of people, where diseases such as measles and cholera outbreak spread, affecting mostly children
21	Assefa et al. (2001)	Household mortality and nutrition survey	Afghanistan	Entire population, especially children under five years	High child and adult mortality and widespread malnutrition, and infectious diseases	A severe drought during the conflict struck Kohistan, leaving families without food or clean water. Death rates skyrocketed, with nearly six out of every 10,000 with children dying each day. Diarrhoea, measles, and respiratory infections all claimed lives. In addition to hunger, the drought set off a fatal cycle of disease and malnutrition

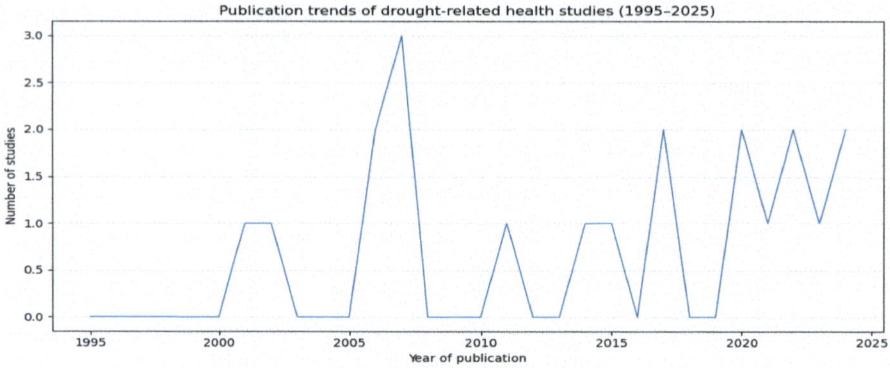

Fig. 7.3 Publication trends of drought-related health studies (1995–2025)

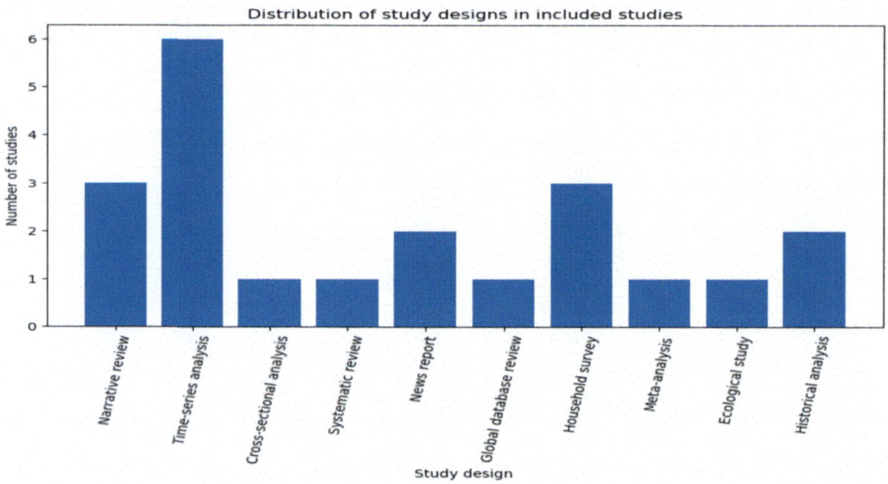

Fig. 7.4 Distribution of study designs in included studies

show the long-term impacts of drought on health with six published studies. A significant studies used narrative review and survey-based methodologies with three studies each. Historical analysis and news reports analysis were represented by two studies each. The remaining five studies used different study design including meta-analysis, ecological study, cross-sectional analysis, database review, and systematic analysis.

Geographical Distribution of Studies

The included studies demonstrate variation in geographic coverage and contextual focus. Only one study adopted a global scope with the remaining focusing on specific countries and regions. Most of the researches concentrated in the African continent with seven studies conducted across Ethiopia, Kenya, Mozambique, and Somalia. North America was the next most represented region, with six studies distributed in the continent, including four from USA and one each from Mexico and Canada.

Asia contributed with five studies, with China accounting for two, while India, Bangladesh, and Afghanistan had one each. Reseasrch coverage in South America and Europe was limited with one study identified each from Brazil and Portugal, respectively. There were no publications identified from Australia and Antarctica on the impacts of drought on human health.

Population Vulnerability to Droughts-Related Health Impacts

Drought was linked to substantial health effects in the examined literature, impacting a diverse population based on age, gender, race, socioeconomic position, and geographic setting. Children under the age of five are especially vulnerable to drought, which increases the risk of mortality from severe malnutrition and infectious illnesses, according to multiple studies. Similarly, older persons are at a higher risk of death, notably from cardiovascular and respiratory disorders exacerbated by heat stress and poor air quality during prolonged drought periods.

Spatial and socioeconomic factors also impacted the distribution of health consequences. Research undertaken in largely rural areas repeatedly found that food shortages, disruption of livelihood activities, and poor access to healthcare services all contributed to compounded hazards. Gender-differentiated results were seen in these settings. Several studies found that women in drought-affected rural areas bear a disproportionate cost, including greater rates of hunger and anaemia, as well as increased reporting of mental health problems. The examined research also highlighted groups living in low-income and conflict-affected areas as having particularly severe health impacts. Drought-related health consequences were exacerbated in areas with unstable infrastructure, limited healthcare systems, and greater governance issues, according to studies from Somalia, Afghanistan, and Mozambique.

Drought-Related Mortality Outcomes Reported in the Literature

The examined studies show a relationship between drought and higher mortality; however, the magnitude of the impact and the primary reasons vary by geography, population category, and drought features. Findings from industrialised countries, such as USA, suggest a more nuanced relationship. For example, a study conducted in Nebraska, USA, found no apparent relationship between drought and total mortality in the general population. Certain groups, however, found to be more vulnerable to the health consequences of drought, most notably middle-aged white populations, with higher effects identified among women residing in rural regions, particularly during prolonged droughts. In contrast, studies that focused on specific causes of death had more consistent results, with respiratory fatalities demonstrating particularly strong relationships. According to data from USA, moderate-to-severe drought conditions led to increases in respiratory mortality of up to 15.9% in some areas, with older persons facing a significantly larger risk. A similar pattern was found in southern Europe, where drought in Portugal was strongly related with an increase in daily fatalities from circulatory and respiratory causes, particularly among elderly males.

Drought mortality was highest among children in poor and middle-income nations. Several studies found notably high mortality rates among children under five in drought-affected countries such as Afghanistan, Pakistan, and Somalia, with rates surpassing 50 fatalities per 1000 live births. Evidence from Sub-Saharan Africa yielded comparable results. Studies conducted in Ethiopia and Mozambique found that drought exposure was associated with extraordinarily high child death rates, with one study finding a mortality rate of 18.9% in Mozambique. During droughts, these mortality were frequently associated to poor nutritional condition and increased susceptibility to infectious illnesses. More recently, a major cross-sectional study throughout Africa assessed this risk, finding that severe, long-term drought was related with a 4% increase in infant mortality, with neonates being the most vulnerable. Historical assessments included in the chapter show that drought can lead to catastrophic death occurrences under certain social and environmental settings. For example, one research linked lengthy megadroughts to haemorrhagic fever outbreaks known as cocoliztli, which killed millions of people in sixteenth-century Mexico. In a similar line, research has shown that the droughts of 1585–1590 in Ming Dynasty China preceded major famine and large-scale infectious disease epidemics, eventually contributing to significant population loss throughout the region.

Morbidity and Other Health Impacts Associated with Drought

According to the reviewed research, drought has a wide-ranging influence on human health, impacting both communicable and non-communicable illnesses, as well as mental well-being. The health repercussions are complicated and multifaceted, with variable levels of severity depending on population and geographic region.

1. Infectious Illness Outbreaks: Several studies have found that infectious illness is more likely to spread during droughts, owing to reduced water availability and sanitation. Studies have found a definite link between drought and an increase in cholera infections in African communities. Similar trends developed in the Horn of Africa, where researchers saw epidemics of cholera and measles moving across drought-stricken regions. Afghanistan provided an even more concerning scenario what researchers referred to as a "deadly cycle" in which drought-related hunger made individuals more vulnerable to diarrhoeal illnesses and respiratory infections, eventually leading to increased mortality rates.
2. Malnutrition and Food Insecurity: A common issue throughout the research was how drought undermines agricultural systems, resulting in malnutrition and food insecurity. Research undertaken in Sub-Saharan Africa and several parts of Asia revealed worrisome increases in acute malnutrition among youngsters living in drought-prone areas. Evidence from India revealed a particularly worrying

picture, demonstrating that women and children endure the largest nutritional burdens as a result of drought. These vulnerable populations had much higher rates of anaemia and overall undernutrition than other sectors of the population.

3. Non-communicable Diseases and Mental Health: Aside from infectious illness, drought has been related to non-communicable diseases and psychosocial effects. In rural China, researchers discovered an ecological link between drought-related water shortages, poor water quality, and an increase in oesophageal cancer mortality. Mental health consequences appeared as a consistent finding across many areas. In Brazil, drought caused severe worry and psychological anguish, particularly among women, and was linked to forced migration. Similarly, increased mental health problems were reported in Canada, particularly among indigenous and rural groups affected by the drought.

Discussion

Summary of Research and Key Findings

This scoping review investigated the global evidence on drought-related mortality and morbidity across a three-decade period (1995–2025), synthesising data from 21 chosen studies from an original pool of over 9000 records. The major goal was to understand the complicated processes that relate drought exposure to health outcomes and quantify the burden in various geographical and socioeconomic circumstances. The evaluation process indicated that drought is not only an agricultural hazard, but a significant public health issue that has affected nearly 1.6 billion people in the previous two decades. The central finding of this synthesis is that drought has a multi-layered impact on health, ranging from immediate water scarcity and malnutrition to complex infectious disease transmission and chronic health effects, with low-income regions bearing a disproportionate burden of mortality, particularly in Sub-Saharan Africa.

Key Findings

Geographical Disparities in Mortality Burden

One of the review's key conclusions is the geographical discrepancies in mortality outcomes, which are primarily dominated by the Global South. Sub-Saharan Africa appears to carry the biggest mortality burden, accounting for the vast majority of worldwide drought-related fatalities, with 23,374 documented deaths (Berman et al. 2017). In contrast, research from high-income countries, such as USA and Portugal, quantify death in terms of increasing ratios for certain non-communicable causes rather than absolute starvation numbers. For example, African populations have under-five death rates that surpass 50 per 1000 live births owing to systemic breakdown, populations in USA.

During a severe drought, the upper Midwest has a 15.9% higher risk of respiratory death (Lombe et al. 2024). This disparity suggests that, while physiological stress from heat and dust affects high-income populations, the profound magnitude of mortality in developing regions is caused by a failure of infrastructure and food systems to buffer the shock, a conclusion supported by the catastrophic excess deaths reported in Somalia (Asmall et al. 2021).

Infectious Disease and Malnutrition Pathways

The synthesis reveals that the synergistic link between hunger and infectious illness is still the leading cause of morbidity in drought-prone developing countries (Charnley et al. 2021). Consistent with earlier research (Lee et al. 2020; Mushwani et al. 2025), our analysis discovered that drought exacerbates water shortages, causing communities to rely on hazardous water resources, resulting in epidemics of waterborne illnesses such as cholera and other diarrheal disorders. One of the most important findings from the included research is the lethal cycle observed in Afghanistan and Ethiopia (Castaño 2025; Dimitrova 2021), in which drought-induced malnutrition considerably raises the case-fatality rate of normally treatable infectious illnesses such as measles and respiratory infections. Historical analyses included in this review support this mechanism by demonstrating that mega droughts have historically triggered large epidemics, such as *cocoliztli* outbreaks in Mexico and pestilence during the Ming Dynasty in China (Xin et al. 2025, Mani et al. 2024), by acting as a force multiplier for pathogen transmission and host vulnerability.

The *Hidden* Burden: Non-communicable Diseases and Mental Health

Aside from the traditional focus on famine and infection, this study emphasises the growing and often overlooked burden of non-communicable diseases (NCDs) and mental health disorders. Methodologies using time-series analysis in USA and Asia successfully identified significant associations between drought severity and cardiovascular or respiratory failure, particularly among the elderly (Chang et al. 2022; Mani et al. 2024). The process here varies from the nutritional pathway, and they are likely caused by poor air quality (dust, fire, smoke) and psychological stress. Notably, ecological research from China connected drought-induced water concentration to enhanced oesophageal cancer mortality, suggesting that decreased river flow may concentrate carcinogens (Mwewa and Mbale 2024). Furthermore, research from Brazil and Canada indicates a large mental health cost, particularly anxiety and distress associated with job loss and forced migration (Brunnet et al. 2018; Shultz et al. 2019). These data show that drought has a far larger health impact than current death numbers suggest.

Methodological Comparisons and Vulnerable Populations

The review included studies with a variety of methodologies, revealing that time-series analyses produced the most reliable quantitative data for NCD risks in developed countries, whereas household surveys were critical for capturing the multi-faceted crisis of malnutrition and child mortality in data-poor regions such as Africa. The comparison of these studies regularly reveals vulnerable populations, including children under the age of five, the elderly, and women. Women have specific hazards for anaemia and mental health, which are typically compounded by gender roles in water collection and home administration. Similarly, the elderly are more sensitive to environmental stresses linked to droughts, such as heat and air pollution. This sensitivity stratification is consistent with the larger climate-health research, but it underlines that susceptibility varies by context, physiological for the elderly in the West, and socioeconomic for children in the Global South.

Policy Recommendations

This review's conclusions have critical policy implications, requiring a transition from reactive crisis management to proactive health system resilience. First, policymakers must incorporate health indicators into national drought early warning systems; existing systems are mostly focused on agriculture, but predictive models could prepare the health sector for hunger and disease epidemics before death occurs. Second, given the identification of specific vulnerable populations, social protection programs such as conditional cash transfers and nutritional support must be repositioned to target women and children in drought-prone regions as soon as the drought is detected. Third, in order to ensure that water security measures prioritise human health outcomes, the One Health approach must be operationalised to close the gap between water management, agriculture, and public health. Lastly, in order to preserve supply chains and functionality during extreme weather events, health systems in high-risk areas need to be climate-proofed.

Limitations and Strengths

Limitations: There are a number of restrictions on this study. Our capacity to provide a single worldwide risk estimate was limited by the heterogeneity of the included research, which ranged from historical archival study to contemporary time-series epidemiology and hindered the feasibility of a statistical meta-analysis. The development of precise threshold-based warning systems is further hampered by the noteworthy lack of research demonstrating a precise dose-response link between drought intensity (e.g. length, severity indices) and certain health effects. Additionally, the review was restricted to English-language publications, which would have left out regional data from drought-prone areas where English is not the primary language.

Strengths: In contrast, a key asset of this review is its extensive chronological and geographical coverage, which includes three decades of data from several continents. The use of several study methods allows for a more comprehensive knowledge of drought consequences, including both acute mortality from humanitarian crises and the chronic illness burden of affluent countries. This study presents a strong, evidence-based framework for connecting scientific discoveries to global development objectives by clearly relating these health outcomes to Sustainable Development Goals (SDG 3, 6, and 13).

Conclusion

By combining the disparate data on drought-related mortality and morbidity into a coherent global narrative, this scoping review marks a substantial development in the area of climate-health research. We have shown that drought is a noisy public health emergency that functions as a hazard multiplier, aggravating pre-existing vulnerabilities across infectious, nutritional, and chronic disease pathways, rather than just a silent agricultural distress signal. The findings supports the urgent need for a paradigm change: we must see drought as a crucial factor in determining human survival and well-being rather than only considering it as a problem with water resources.

This discovery has broader implications that go all the way to the architecture of global health security. The line between environmental and health emergencies will become more hazy as climate change increases the frequency and severity of drought occurrences. Establishing standardised health-centric drought measures and thoroughly assessing the efficacy of adaption treatments must be the main goals of future research. We must now create health systems that are as robust as the people they serve in order to prevent the next drought from unavoidably triggering the next health emergency. The time for observation is over.

Acknowledgements We heartily thank all of the researchers whose work served as the foundation for this chapter. This synthesis was made possible by their dedication and thorough study on the impacts of drought on health. We also like to thank the groups and institutions that maintain scientific databases and repositories, which provided us with access to a wealth of relevant information. Finally, we appreciate the wider scientific community involved in health and climate research, whose ongoing efforts continue to inspire evidence-based public health methods and climate-resilient legislation.

Potential Conflicts of Interest On behalf of all authors, the corresponding author states that there is no conflict of interest.

Informed Consent Informed consent was obtained from all individual participants included in the study.

Authors Contribution All authors have contributed equally to this work.

Data Availability Statement The datasets during and/or analysed during the current study available from the corresponding author on reasonable request.

AI Disclosure Statement AI-assisted tools were used in a limited manner to support language editing and improve clarity and readability. The authors are fully responsible for all the content.

Funding The research was not supported by any funding.

References

Abadi AM, Gwon Y, Gribble MO, Berman JD, Bilotta R, Hobbins M, Bell JE (2022) Drought and all-cause mortality in Nebraska from 1980 to 2014. Sci Total Environ 840:156660

Acuña-Soto R, Stahle DW, Cleaveland MK, Therrell MD (2002) Megadrought and megadeath in 16th century Mexico. Emerg Infect Dis 8(4):360–362

Alam I, Otani S, Nagata A, Khan MS, Masumoto T, Amano H, Kurozawa Y (2022) Short- and long-term effects of drought on selected causes of mortality in northern Bangladesh. Int J Environ Res Public Health 19(6):3425

Algur KD, Patel SK, Chauhan S (2021) Impact of drought on the health and livelihoods of women and children in India: a systematic review. Child Youth Serv Rev 122:105909

Asmall T. The adverse health effects associated with drought in Africa: working towards developing a vulnerability index (2020)

Asmall T, Abrams A, Röösli M, Cisse G, Carden K, Dalvie MA (2021) The adverse health effects associated with drought in Africa. Sci Total Environ 793:148500

Assefa F, Jabarkhil MZ, Salama P, Spiegel P (2001) Malnutrition and mortality in Kohistan District, Afghanistan. JAMA 286(21):2723–2728

Banerjee A, Ariz D, Turyasingura B, Pathak S, Sajjad W, Yadav N, Kirsten KL (2024) Long-term climate change and anthropogenic activities together with regional water resources and agricultural productivity in Uganda using Google earth engine. Phys Chem Earth:103545

Bellizzi S, Lane C, Elhakim M, Nabeth P (2020) Health consequences of drought in the WHO eastern Mediterranean region: hotspot areas and needed actions. Environ Health 19(1):114

Below R, Grover-Kopec E, Dilley M (2007) Documenting drought-related disasters: a global reassessment. J Environ Dev 16(3):328–344

Berman JD, Ebisu K, Peng RD, Dominici F, Bell ML (2017) Drought and the risk of hospital admissions and mortality in older adults in western USA from 2000 to 2013. Lancet Planet Health 1(1):e17–e25

Brunnet AE, Bolaséll LT, Weber JLA, Kristensen CH (2018) Prevalence and factors associated with PTSD, anxiety and depression symptoms in Haitian migrants in southern Brazil. Int J Soc Psychiatry 64:17–25

Castaño VR (2025) Religion over life and death: the Cocoliztli. In: Bodies, remedies, policies from early modern chronicles of the indies to COVID-19 narratives, pp 69–82

Chang AY, Tan AX, Nadeau KC, Odden MC (2022) Aging hearts in a hotter, more turbulent world: impacts of climate change on cardiovascular health of older adults. Curr Cardiol Rep 24:749–760

Charnley GEC, Kelman I, Murray KA (2021) Drought-related cholera outbreaks in Africa and the implications for climate change: a narrative review. Pathog Glob Health 116(1):3–12. https://doi.org/10.1080/20477724.2021.1981716

Chen NX, Kiattisunthorn K, O'Neill KD, Chen X, Moorthi RN, Gattone VH, Allen MR, Moe SM (2013) Decreased microRNA is involved in vascular remodeling abnormalities in chronic kidney disease. PLoS One 8(5):e64558

De Waal A, Taffesse AS, Carruth L (2006) Child survival during the 2002–2003 drought in Ethiopia. Glob Public Health 1(2):125–132

Delbiso TD, Altare C, Rodriguez-Llanes JM, Doocy S, Guha-Sapir D (2017) Drought and child mortality: a meta-analysis of small-scale surveys from Ethiopia. Sci Rep 7:2212
Dimitrova A (2021) Seasonal droughts and the risk of childhood undernutrition in Ethiopia. World Dev 141:105417
Goh RSJ, Chong B, Jayabaskaran J, Jauhari SM, Chan SP, Kueh MTW, Shankar K, Li H, Chin YH, Kong G (2024) The burden of cardiovascular disease in Asia from 2025 to 2050: a forecast analysis. Lancet Reg Health W Pac 49
Gwon Y, Ji Y, Bell JE, Abadi AM, Berman JD, Rau A, Leeper RD, Rennie J (2023) Association between drought exposure and respiratory-related mortality in the United States from 2000 to 2018. Int J Environ Res Public Health 20(12):6076
Khan MZK, Rahman A, Rahman MA, Renzaho AMN (2021) Impact of droughts on child mortality: a case study in southern African countries. Nat Hazards 108(2):2211–2224
Lawrance EL, Thompson R, Newberry Le Vay J, Page L, Jennings N (2022) The impact of climate change on mental health and emotional wellbeing: a narrative review of current evidence and its implications. Int Rev Psychiatry 34:443–498
Lee J, Perera D, Glickman T, Taing L (2020) Water-related disasters and their health impacts: a global review. Prog Disaster Sci 8:100123
Lombe P, Carvalho E, Rosa-Santos P (2024) Drought dynamics in sub-Saharan Africa: impacts and adaptation strategies. Sustainability 16:9902
Lynch KM, Lyles RH, Waller LA, Abadi AM, Bell JE, Gribble MO (2020) Drought severity and all-cause mortality rates among adults in the United States, 1968–2014. Environ Health 19:52
Mani ZA, Khorram-Manesh A, Goniewicz K (2024) Global health emergencies of extreme drought events: historical impacts and future preparedness. Atmos 15(9):1137
Mojtahedzadeh N, Rohwer E, Neumann FA, Nienhaus A, Augustin M, Zyriax B-C, Harth V, Mache S (2021) Health behaviour of German outpatient caregivers in relation to their working conditions: a qualitative study. Int J Environ Res Public Health 18(11):5942
Munn Z, Peters MDJ, Stern C, Tufanaru C, McArthur A, Aromataris E (2018) Systematic review or scoping review? Guidance for authors when choosing between a systematic or scoping review approach. BMC Med Res Methodol 18(1):143. https://doi.org/10.1186/s12874-018-0611-x
Mushwani H, Arabzai A, Ahmad MM, Ullah H (2025) Assessing the socio-economic consequences of drought in Central Afghanistan: a case study of Bamyan Province. GeoJournal 90:1–13
Mwewa T, Mbale J (2024) Exploring cancer risks amidst climate-induced drought and using artificial intelligence to enhance cancer risk prevention strategies. Zambia ICT J 8:54–63
Renzaho AMN (2007) Mortality rates, prevalence of malnutrition, and prevalence of lost pregnancies among the drought-ravaged population of Tete Province, Mozambique. Prehosp Disaster Med 22(1):26–34
Salvador C, Nieto R, Linares C, Díaz J, Gimeno L (2021) Effects on daily mortality of droughts in Galicia (NW Spain) from 1983 to 2016. Sci Total Environ 748:142332
Salvador C, Nieto R, Vicente-Serrano SM, García-Herrera R, Gimeno L, Vicedo-Cabrera AM (2023) Public health implications of drought in a climate change context: a critical review. Annu Rev Public Health 44(1):213–232
Sena A, Barcellos C, Freitas C, Corvalan C (2014) Managing the health impacts of drought in Brazil. Int J Environ Res Public Health 11(10):10737–10751
Shultz JM, Rechkemmer A, Rai A, McManus KT (2019) Public health and mental health implications of environmentally induced forced migration. Disaster Med Public Health Prep 13:116–122
Singam ERA, Durkin KA, La Merrill MA, Furlow JD, Wang J-C, Smith MT (2023) The vitamin D receptor as a potential target for the toxic effects of per- and polyfluoroalkyl substances: an in silico study. Environ Res 217:114832
Tedla MG, Cho Y, Rasmy M (2025) Exploring the dynamics of flood and drought hazards and adaptation strategies in the horn of Africa. Water Int 50(3–4):225–234
Wakabi W (2006) Worst drought in a decade leaves Kenya crippled. Lancet 368(9533):1371

Wang P, Rogne T, Warren JL, Asare EO, Akum RA, Toure NE, Ross JS, Chen K (2025) Long-term drought and risk of infant mortality in Africa: a cross-sectional study. PLoS Med 22(1):e1004516
Wu X, Li Y (2007) Association between esophageal cancer and drought in China using geographic information systems. Environ Int 33(5):603–608
Xiao F, Liu Q, Qin Y, Huang D, Liao Y (2024) Agricultural drought research knowledge graph reasoning using VOSviewer. Heliyon 10(6)
Xin J, Sun Z, He J, Hang X, Zhao Y, Zhang S, Hao Y (2025) Extreme drought events (1585–1590) and spatiotemporal patterns of infectious disease outbreaks in Ming dynasty China. GeoHealth 9(6):e2025GH001386
Yusa A, Berry P, Cheng JJ, Ogden N, Bonsal B, Stewart R, Waldick R (2015) Climate change, drought and human health in Canada. Int J Environ Res Public Health 12(7):8359–8412
Zarocostas J (2011) Famine and disease threaten millions in drought-hit horn of Africa. BMJ 343:d4696

Chapter 8
The Mental Health Crisis Among Farmers: Exploring the Effects of Climate Disasters and Coping Strategies in Vulnerable Agricultural Communities in Odisha

Urmi Pattanayak

Abstract Climatic stressors are increasingly shaping livelihood stability and psychological well-being in agrarian regions, yet limited research integrates climate risks, mental health outcomes and coping behavior within a single analytical framework. This study examines how anxiety, depression, and climate-hazard exposure intersect with socioeconomic vulnerabilities and adaptation strategies among smallholder farmers in Mayurbhanj district, Odisha. Primary data from 200 farmers were analyzed using decision trees, boxplots, and canonical discriminant analysis to identify interaction patterns across mental health indicators, landholding characteristics and coping behaviors. Depression showed the most severe functional impairment, with over half of affected farmers reporting complete loss of work capacity, while anxiety displayed more variable effects. Climate risk was concentrated among farmers owning less than 1.5 ha, who faced higher exposure to floods, droughts and cyclones. Coping behaviors formed three clusters: psychological (faith-based practices, kin support), economic (borrowing during financial strain), and socio-demographic (use of government schemes shaped by education and caste). Faith-based coping emerged as the most distinct discriminant category, contributing to a 64.8% classification accuracy. Findings indicate the need for integrated responses that link mental health support with climate adaptation measures. Priority actions include irrigation expansion and awareness programs in Tiring, forest-based livelihood options in Bangiriposi and watershed and credit initiatives in Khunta, alongside district-level strategies promoting climate-resilient agriculture and accessible psychosocial care.

Keywords Coping strategies · Climate disasters · Farmers · Mental health · Agriculture

U. Pattanayak (✉)
ICMR-Regional Medical Research Centre, Bhubaneswar, Odisha, India

P. Ghosh et al. (eds.), *Climate Crisis and Changing Population Dynamics*, Springer Climate, https://doi.org/10.1007/978-3-032-24533-5_8

Introduction

Agriculture continues to be the backbone of India's rural economy, employing more than half of the country's population and providing a key source of income for millions of small and marginal farmers. Despite its importance, farming has become a more dangerous vocation, especially in climate-vulnerable areas. Odisha, an eastern coastal state blessed with lush terrain and abundant natural resources, exemplifies this contradiction vividly (Nayak and Mishra 2019). Although largely agrarian, the state routinely faces a range of climate-induced calamities, including cyclones, floods, droughts, and unpredictable rainfall (Pattnaik et al. 2020). These incidents have escalated over time, creating major economic shocks and raising concern for rural communities. While the pecuniary costs of such calamities are regularly acknowledged in policy debate, their psychological impacts remain generally disregarded. Extreme weather phenomena such as cyclones e.g., Cyclone Phailin in 2013 and Cyclone Fani in 2019, floods, droughts, and unseasonal rainfall regularly hit Odisha (Berry et al. 2018).

The increasing evidence of mental health problems among Indian farmers highlights the significance of this issue. Climate fluctuation has been linked to high levels of anxiety, sadness, hopelessness, trauma, and professional stress, which frequently leads to an increased risk of suicide. Every year, hundreds of farmers and agricultural laborers commit suicide, according to the National Crime Records Bureau (NCRB 2022). Suicide is the most obvious and sad sign of a mental health issue, yet it is just the beginning.

It hides a deeper, sometimes unseen terrain of emotional suffering, which manifests as persistent anxiety, sadness, sleeplessness, feelings of hopelessness, trauma, and depression (Patel et al. 2018). The psychological load is especially heavy in places like Odisha, where tiny farmers face structural restrictions such as poverty, restricted landholdings, loan reliance, and little social support. Mental health care in rural regions are nonetheless inadequate, stigmatized, and difficult to get, aggravating the vulnerability of afflicted communities. Existing research emphasizes the complicated interaction between environmental stresses and farmer well-being. Changes in the climate and extreme weather events have been associated in studies to mental discomfort, reduced adaptability, and long-term psychological damage.

According to research conducted in India and throughout the world, frequent exposure to catastrophes erodes people's feeling of control, exacerbates financial instability, and increases sensitivity to mental health problems. Gender-based inequities, caste marginalization, and poor institutional assistance intensify these issues. Nevertheless, a large portion of the current research concentrates on either financial losses or suicide consequences, providing little understanding of the wider range of mental health issues and coping strategies among regular farmers. Climate change is no more a distant, abstract danger; it is a daily reality that is directly harming the mental health of individuals, especially in rural economies like Odisha. In addition to disrupting the ecosystem, extreme weather events can also trigger psychological stress (Charlson et al. 2021).

Repeated exposure to disasters reduces people's feeling of control, security, and hope, all of which are important components of psychological resilience (Jena et al. 2021; Kermode et al. 2009). Poverty, illiteracy, caste-based prejudice, and gender inequality all contribute to mourning among Odisha's rural communities. A survey of the literature identifies a number of significant gaps. In the domain of smallholder agriculture, there is theoretically little integration between psychological models and climate risk frameworks. In terms of information, few studies have looked into how farmers in disaster-prone areas like Odisha perceive, experience or express emotional distress. The many psychological effects of climatic fluctuation are not well documented empirically, particularly for vulnerable populations including women farmers, tribal tribes, and landless farmers. Existing research frequently relies on national-level data or suicide statistics, which lack qualitative depth and contextual subtlety. In terms of population, the mental health issues of ordinary farmers, not only those in distress, have received little attention.

To fill these gaps, the current study is directed by the following research questions:

- How do farmers in Odisha's climate-vulnerable areas evaluate the psychological effects of recurring climate-related disasters?
- How do socioeconomic and societal influence their psychological well-being experiences?
- What institutional procedures, coping mechanisms and how well do they work to alleviate mental distress?

The goal of this study is to get a thorough knowledge of the mental health difficulties that farmers in Odisha experience within the wider picture of climate vulnerability. By recording farmers' lived experiences, the undertaking hopes to add to the expanding knowledge on weather-linked mental health in poor countries, as well as give evidence-based insights for implementing mental illness-inclusive adaptation to climate change and rural development strategies. In the end, our study emphasizes the necessity of a people-centered strategy that acknowledges both the obvious economic effects of climate change and the unseen psychological repercussions that influence farmers' resilience and general well-being.

Study Area

Mayurbhanj district in northern Odisha (10,418 km^2) (Fig. 8.1) is predominantly rural and agrarian, with over 92% of its 2.5 million people mainly tribal communities dependent on rainfall-based farming and forest resources. Small and marginal farmers cultivate paddy, pulses, and oilseeds on red and lateritic soils with limited irrigation, making agriculture highly sensitive to climatic variability.

The district is recognized as climate-vulnerable, regularly experiencing delayed monsoons, prolonged dry spells, floods, and crop failures. These environmental stresses intersect with existing socioeconomic constraints such as poverty, traditional cultivation practices, and limited access to health and institutional support. As a result,

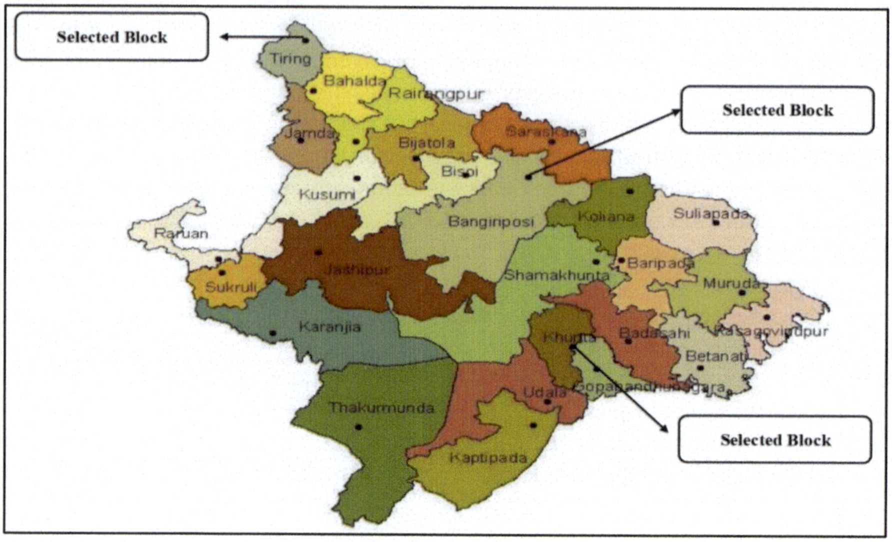

Fig. 8.1 Study area

farmers frequently face heightened livelihood insecurity, indebtedness, and psychological distress. This combination of high climate exposure, livelihood dependence on natural resources, and restricted mental health services makes Mayurbhanj a suitable and significant study area for examining how climate-related stresses contribute to mental health difficulties among farmers in rural Odisha.

Material and Methods

Sample Design and Procedure

Mayurbhanj district comprises 26 blocks. To ensure representativeness and capture spatial variation in climate impacts, on the basis of 10 percent were selected using simple random sampling, resulting in three blocks: Tiring, Bangiriposi, and Khunta. Within these blocks, 200 farmers were surveyed using purposive sampling, focusing specifically on households that had experienced at least one major climate-related event such as cyclones, floods, or droughts during the last ten years. This approach ensured that the sample included respondents who were directly exposed to climate hazards and therefore relevant to the study's objective of assessing mental health impacts.

Data Collection

Primary data were collected through structured questionnaires and in-person interviews, covering socioeconomic characteristics, type and frequency of climate disasters experienced, changes in agricultural productivity, mental health symptoms (anxiety and depression), and coping strategies. Secondary data from Odisha Agriculture Statistics, Annual Report of Natural Calamities, Government of Odisha and District profile were used to contextualize climate exposure.

Outcome and Explanatory Variables

The variables used in the analysis are categorized into socioeconomic characteristics, climate exposure and impacts and mental health outcomes. Table 8.1 summarizes the definition, measurement, and analytical use of each variable.

Analytical Strategies

Three complementary analytical methods were used to examine how climate hazards affect farmers' mental health in Fig. 8.2, such as.

Boxplot: Used to analyze the temporal distribution of compound natural hazards in Mayurbhanj district between 2013 and 2025 and how the size of landholdings (in hectares) varies with the frequency of exposure to climate-related events such as droughts, cyclones, and floods.

Decision Tree Model: Used to predict the likelihood of depression and anxiety based on climate impacts (e.g., severity of productivity loss) and household characteristics. This model helps visualize pathways leading to mental health risks.

Canonical Discriminant Analysis (CDA): Applied to identify the socioeconomic and coping variables that most effectively differentiate farmers based on their level of climate exposure. CDA helps determine which factors contribute most strongly to vulnerability.

Table 8.1 Definition of variables used in the study

Parameters	Define variables	Measurements (units)	Model used
Socioeconomic profile			
Age	Age of the respondents	Continuous (years)	CDA
Gender	Gender of the respondents	Categorical (1 = male, 2 = female)	CDA
Caste	Caste of the respondents	Categorical (1 = general, 2 = OBC, 3 = SC, 4 = ST)	CDA
Education	Education of the respondents	Categorical (0 = illiterate, 1 = primary, 2 = secondary, 3 = higher secondary, 4 = graduate, 5 = post graduate)	CDA
Landholding size	Landholding size of the respondents	Categorical (1 = marginal <1 ha, 2 = small 1 to 2 ha, 3 = medium 2 to 4 ha, 4 = large >4 ha)	Boxplot
Annual income	Annual income of the respondents	Categorical (1 = <50 thousand, 2 = 50 thousand to 1 lakh, 3 = 1 lakhs to 2 lakhs, 4 = 2 lakhs to 5 lakhs, 5= > 5 lakhs)	CDA
Land size	Land in hectares of the respondents	Continuous	CDA
Climate exposure and impact			
Climate disasters	Type of climate hazards	Categorical (1 = drought frequency, 2 = cyclone frequency, 3 = flood frequency)	Boxplot
Impact of productivity	Loss of productivity due to natural hazards	Categorical (1 = mild, 2 = moderate, 3 = severe/complete loss)	Decision tree
Mental health risk			
Depression	Due to natural hazards affect the mental health of the farmers	Binary (1 = severe, 2 = moderate)	Decision tree
Anxiety	Due to natural hazards affect the mental health of the farmers	Binary (1 = severe, 2 = moderate)	Decision tree
Coping strategies	How to manage, adapt, or survive during and after natural disasters or climate-related stresses (like drought, flood, or cyclone)	Categorical (1 = faith, 2 = Govt. scheme, 3 = kin support, 4 = borrow money, 5 = reduce expenses, 6 = others)	CDA

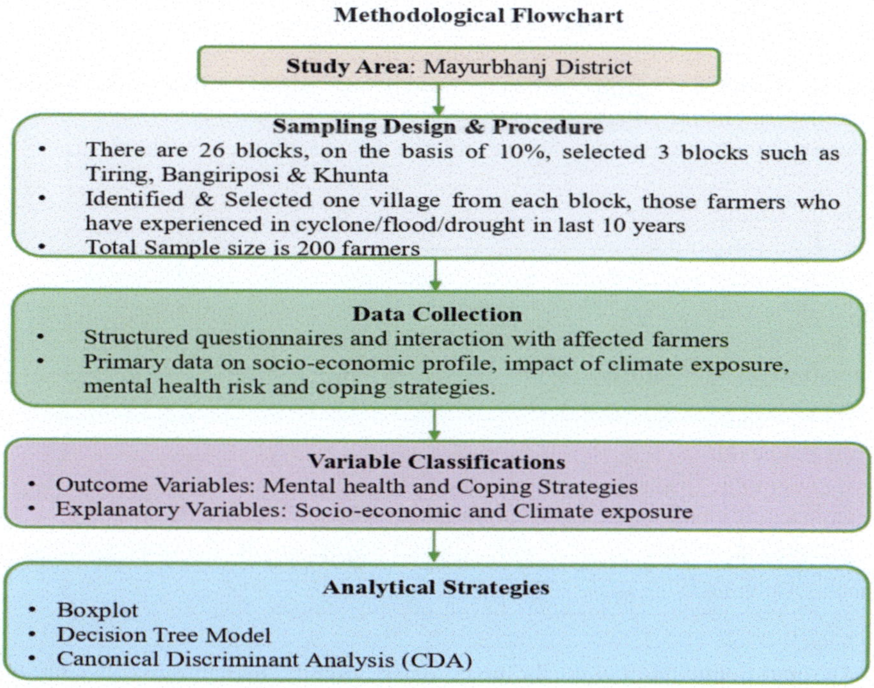

Fig. 8.2 Methodological flowchart

Results

Box Plot

Natural Hazards in Mayurbhanj District

The temporal distribution of compound natural hazards in the Mayurbhanj area from 2013 to 2025 is shown in Fig. 8.3. Floods combined with droughts are concentrated in the most recent years (2022–2024), reflecting increasing monsoon variability and the tendency for the region to experience both excessive rainfall and prolonged dry spells in close succession. Cyclones combined with floods are predominantly associated with earlier years (2013–2019), corresponding to major Bay of Bengal cyclones such as Phailin 2013 and Titli 2018 that generated widespread inland flooding. Cyclones, floods, and droughts together span the widest temporal range (2015–2024), indicating the district's structural exposure to multiple interacting hazards over the last decade. The clustering of recent flood-drought events suggests that climate change is amplifying hydro-climatic extremes, leading to more frequent compound hazards. Overall, the results demonstrate a shift from cyclone-driven flood impacts to climate-driven

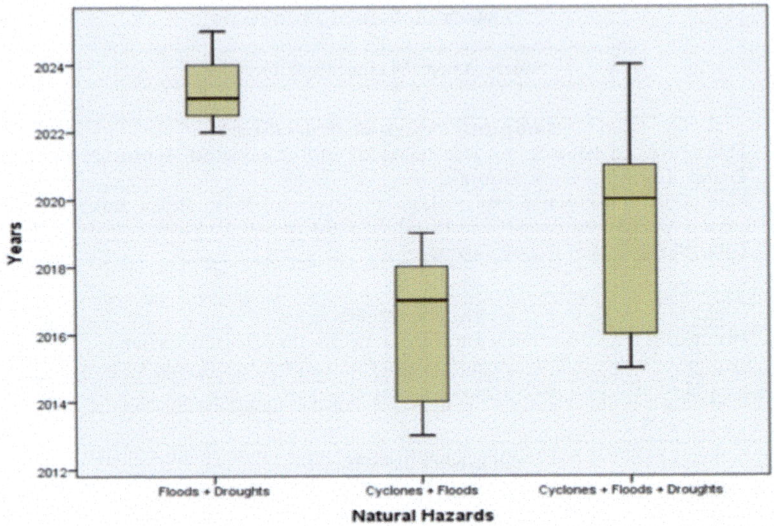

Fig. 8.3 Natural hazards in Mayurbhanj District (2013–2025). *Source* Annual report of natural calamities, Government of Odisha

flood-drought combinations in the early 2020s, highlighting the evolving hazard profile of the district.

Climate Exposure and its Impact on Landholding Size

The extent of landholdings (in hectares) in relation to the frequency of exposure to climate-related phenomena including droughts, cyclones, and floods is examined in Fig. 8.4, a collection of boxplots. The majority of people exposed to these occurrences are small-scale landowners, as seen by the average landholding size remaining below 1.5 ha throughout all three climatic stressors. There is some variance in land size, with a few outliers holding more land, especially among those impacted by cyclones and floods.

.

Decision Tree Model

Effect of Anxiety on Productivity

Figure 8.5 shows a decision tree, analyzing the effect of anxiety severity on productivity. Among the 200 individuals surveyed, 43% reported complete productivity loss, 31.5% had moderate impact, and 25.5% experienced only mild impact. Individuals

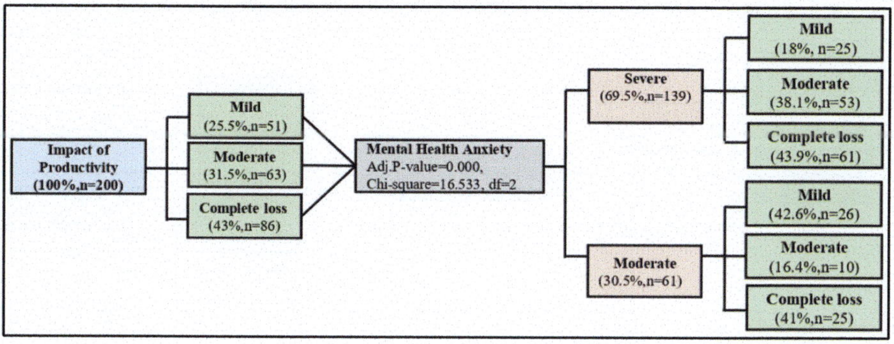

Fig. 8.4 Climate exposure and its impact on landholding size. *Source* Field survey

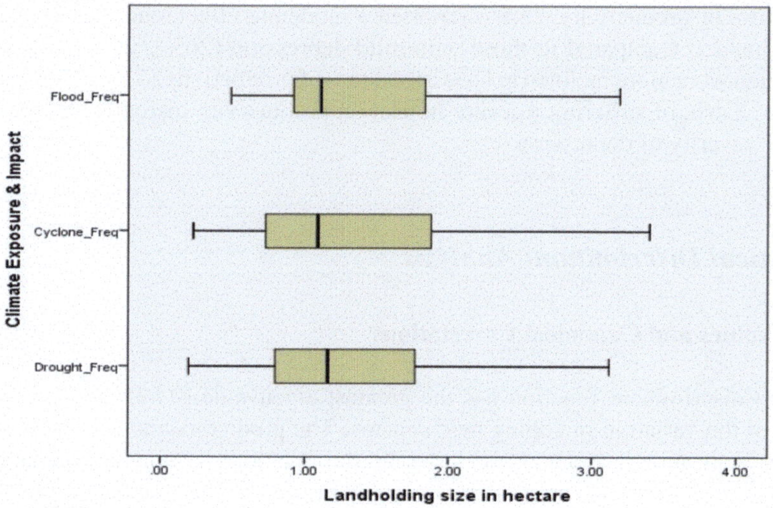

Fig. 8.5 Effect of anxiety on productivity. *Source* Field survey

with severe anxiety had slightly higher rates of complete productivity loss (43.9%) compared to those with moderate anxiety (41%). Interestingly, those with moderate anxiety were more likely to report mild productivity loss (42.6%). While anxiety does impact productivity, the trend is less sharply differentiated between severity levels compared to depression.

Effect of Depression on Productivity

Figure 8.6 depicts a decision tree study that looks at the connection between the degree of depression and how it affects productivity. Of 200 people, 52% reported

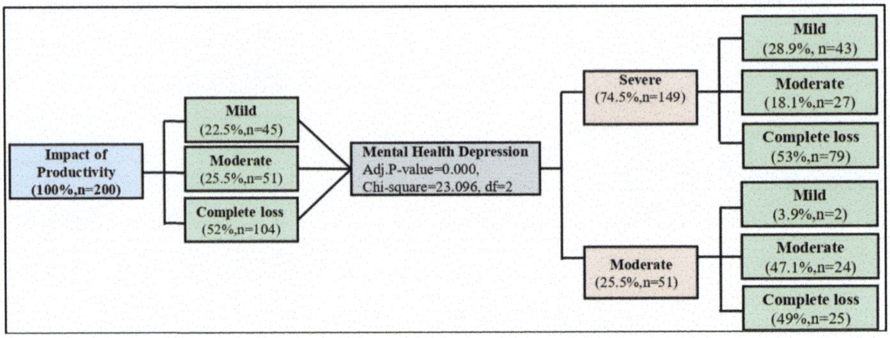

Fig. 8.6 Effect of depression on productivity. *Source* Field survey

a total loss of productivity, 25.5% reported a moderate effect, and 22.5% reported a light impact. Compared to those with mild depression (26%), those with severe depression were more inclined to have a total loss of productivity (53%). This implies that the chance of suffering a severe impact on productivity increases dramatically with the severity of depression.

Canonical Discriminant Analysis

Eigen Values and Canonical Correlations

The first discriminant function has the greatest eigenvalue (0.89), which explains 62.5% of the variation in coping mechanisms. The predictors and the discriminant scores are strongly correlated, as shown by the matching canonical correlation of 0.69. A significant percentage (25.4%) is also explained by the second function, which has a canonical correlation of 0.54. Together, the first two functions account for nearly 88% of the variance, suggesting that the main dimensions of coping are well captured by these functions.

Subsequent functions (3–5) explain only marginal variance (cumulatively ~12%), with weaker canonical correlations (<0.40), indicating that their contribution to distinguishing between coping strategies is limited. Thus, interpretation should focus primarily on the first two canonical functions, which effectively discriminate between farmers relying on coping based strategies versus psychological and economic adjustments (Function 1) and coping based strategies versus socio and demographic adjustments (Function 2). A cursory glance presented in Table 8.2.

Table 8.2 Eigen values and canonical correlations

Function	Eigen value	Variance (%)	Cumulative (%)	Canonical correlation
1	0.89	62.5	62.5	0.69
2	0.42	25.4	87.9	0.54
3	0.15	7.2	95.1	0.36
4	0.08	3.1	98.2	0.27
5	0.04	1.8	100	0.20

Source Field survey

Table 8.3 Wilks lambda

Test of functions	Wilk's lambda	Chi square	df	Significant
1 through 5	0.38	210.5	60	0.000
2 through 5	0.61	132.4	48	0.000
3 through 5	0.79	66.2	36	0.002
4 through 5	0.92	28.4	20	0.099
5	0.97	10.8	8	0.213

Source Field survey

Wilks Lambda

As evident from Table 8.3, the multivariate tests using Wilks' Lambda indicate that the first three discriminant functions are statistically significant, while the last two are not. The test for Functions 1 through 5 yielded Wilks' Lambda = 0.38, $\chi^2(60) = 210.5$, $p < 0.001$, confirming that the discriminant model as a whole is highly significant. The second test (2 through 5) continued to be significant ($\Lambda = 0.61$, $\chi^2(48) = 132.4, p < 0.001$), indicating that the second function has more explanatory power than the first. The third test (3 through 5) was also significant ($\Lambda = 0.79$, $\chi^2(36) = 66.2, p = 0.002$), but the strength of discrimination decreased. Functions 4 and 5, on the other hand, did not contribute significantly to the total discrimination ($p = 0.099$ and 0.213, respectively). When combined, these findings demonstrate that Functions 1 and 2 are the main causes, accounting for about 88% of the variation in coping mechanisms.

Standardized Canonical Discriminant Function Coefficients

According to Table 8.4, the most positive contributions are annual income (0.47), anxiety score (0.52), and depression score (0.58). The size of the landholding (0.35) also makes a significant contribution. This implies that Function 1 mainly differentiates coping techniques according to financial resources and mental health load.

Table 8.4 Standardized canonical discriminant function coefficients

Variable	Function 1	Function 2
Socio and demographic parameters		
Age	0.21	0.18
Gender	−0.06	0.09
Caste	0.12	0.28
Education	0.19	0.41
Farming role	−0.15	0.08
Psychological and economic parameters		
Annual income	0.47	0.23
Landholding size	0.35	−0.14
Depression score	0.58	0.11
Anxiety score	0.52	−0.09

Source Field survey

Farmers who experience greater levels of anxiety and sadness are more likely to use unique coping strategies.

The best indicators are caste (0.28) and education (0.41). This suggests that Function 2 divides groups based on educational achievement, indicating variations in the availability of official coping mechanisms (such as government programs) as opposed to dependence on family or conventional techniques. Age, gender, and farming position are other factors that contribute comparatively little to both functions.

Structure Matrix

The variables that best characterize the canonical discriminant functions in Table 8.5 are more clearly shown by the structural matrix. The scores for depression (0.72), anxiety (0.68), annual income (0.61), and landholding size (0.55) had the strongest associations. This demonstrates that the burden of mental illness and financial ability is the main factors reflected in Function 1. Farmers with better economic foundations and greater psychological anguish have different coping mechanisms mostly defined by caste (0.44) and education (0.58), with a small contribution from farming (0.48). This suggests that Function 2 divides groups according to socio-demographic characteristics, with caste and education determining whether coping mechanisms are more informal (family/traditional) or formal (government/modern). Age (0.17, 0.15) and gender (−0.08, 0.11) show weak correlations, suggesting they play only a minimal role in discriminating among coping strategies. Together, these findings show that psychological and economic factors dominate Function 1 while socio-demographic attributes define Function 2 reinforcing the multidimensional nature of coping among farmers.

Table 8.5 Structure matrix (pooled within groups correlations between variables and canonical functions)

Variable	Function 1	Function 2
Socio and demographic parameters		
Age	0.17	0.15
Gender	−0.08	0.11
Caste	0.19	0.44
Education	0.22	0.58
Farming role	0.48	0.30
Psychological and economic parameters		
Annual income	0.61	0.24
Landholding size	0.55	−0.16
Depression score	0.72	0.18
Anxiety score	0.68	0.09

Source Field survey

Functions at Group Centroids

Table 8.6 analyze that the highest positive score on Function 1 (0.82), indicating strong alignment with psychological distress and resilience-based coping. Negative loading on Function 2 (−0.35) found faith is less tied to socio-demographic supports. Negative on Function 1 (−0.41) but strongly positive on Function 2 (0.62). This indicates that reliance on government programs is institutional in nature, linked more to education and caste than to psychological or economic pressure. Moderately positive on Function 1 (0.35) but negative on Function 2 (−0.28). Furthermore, kin/family support is more associated with psychological coping than formal institutions. Strong negative score on Function 1 (−0.65) with weakly positive Function 2 (0.12) reflects coping rooted in economic compulsion, separated from both faith and institutional supports. Negative on Function 1 (−0.38) but moderately positive on Function 2 (0.44). Thus, expense-cutting is a structured, institutional-style economic adjustment rather than an informal or emotional strategy. Near zero on both functions (0.11, 0.08), suggesting that these may reflect a variety of unique coping strategies and do not substantially correspond with either dimension.

Coping Strategies Adopted by Farmers

The findings of a canonical discriminant analysis are shown in Fig. 8.7, which shows the numerous coping mechanisms people use to deal with different difficulties. Reliance on religion, government programs, family assistance, borrowing money, cutting costs, and other techniques are some of the tactics. Canonical Function 1 is positively correlated with faith and family support, but it is adversely correlated with borrowing money, which may be a reaction motivated by financial hardship. Government programs stand out along Canonical Function 2 because they are a

Table 8.6 Functions at group centroids

Coping strategy	Function 1	Function 2
Faith-based coping	0.82	−0.35
Government scheme	−0.41	0.62
Kin support	0.35	−0.28
Borrow money	−0.65	0.12
Reduce expenses	−0.38	0.44
Other	0.11	0.08

Source Field survey

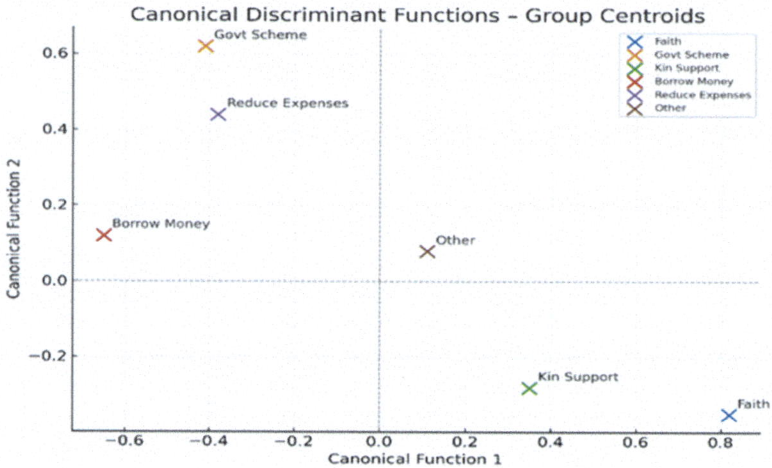

Fig. 8.7 Coping strategies adopted by farmers. *Source* Field survey

unique kind of assistance. Other tactics including cutting costs, are more neutral. These results show that coping mechanisms varies and may be a reflection of social support, personal views, or financial circumstances.

Classification Results (Hit Ratio)

Faith and kin support clustered positively on Function 1, indicating their role in psychologically mediated coping, whereas reliance on government schemes aligned strongly with Function 2, emphasizing its institutional and socio-structural basis, as given in Table 8.7. In contrast, borrowing money had a negative loading on Function 1, indicating that coping was motivated more by financial hardship than by resilience. Compared to the proportionate chance criteria, the total classification accuracy of 64.8% is significantly greater. This demonstrates the strong predictive validity of

Table 8.7 Classification results (hit ratio)

Coping strategy	Faith	Govt. scheme	Kin support	Borrow money	Reduce expenses	Other	% Correct
Faith-based coping	62	5	9	6	4	2	72.1
Government scheme	8	51	6	10	7	3	61.4
Kin support	11	6	49	8	4	2	63.6
Borrow money	5	9	7	48	8	3	60.0
Reduce expenses	4	6	5	9	53	3	65.4
Other	3	4	2	4	5	36	61.0
Overall (%)							**64.8**

Source Field survey

the discriminant functions obtained from the model. Faith has the best classification accuracy (72.1%), indicating that the model most clearly distinguishes responders who use faith-based coping. Both Kin Support (63.6%) and Reduce Expenses (65.4%) have strong predictive power, suggesting that these tactics have stable socioeconomic and psychological characteristics. The Government Scheme (61.4%) and Borrow Money (60.0%) are appropriately categorized at a moderate level, indicating similarities with other methods (e.g., government dependence associated with expenditure reduction, borrowing related with kin assistance). Due to their weak centroid loadings, which reflect their diverse and unique behaviors, other methods (61.0%) are the least distinguishable.

Together, these results demonstrate the many vulnerabilities smallholder farmers confront. Productivity is significantly hampered by mental health issues, particularly depression, and tiny landholdings are more vulnerable to climate shocks. Psychological distress, financial resources, and sociodemographic positioning all strongly influence coping strategies, indicating that interventions should focus on psychosocial support and fair access to institutional resources in addition to material resilience (income, landholding security).

Discussion

This study examines the ways in which coping mechanisms, landholding size, mental health stressors, and compound climatic hazards interact to influence susceptibility among smallholder farmers in the Mayurbhanj area. The study findings, which were obtained using box plots, decision tree models and canonical discriminant analysis, demonstrate a recent trend toward flood-drought compound events, the disproportionate exposure of marginal farmers and the significant influence of psychological distress, particularly depression, on coping behavior and productivity. The results, however, show that farmer vulnerability is multifaceted, involving socio-structural, psychological, economic, and climatic elements. The hazard study shows a shift from flood impacts caused by cyclones from 2013 to 2019 to flood-drought combinations in the early 2020s.

This development is consistent with larger evidence of escalating compound hydro-climatic hazards under climate change in South Asia and reflects increasing monsoon unpredictability (IPCC 2022; Mishra and Singh 2019). Farmers may increasingly confront consecutive and overlapping pressures rather than isolated disasters, making recovery and adaptation more difficult, as seen by the increasing clustering of flood-drought episodes. Smallholder farmers are disproportionately exposed to climate change, according to an analysis of landholding size, with average landholdings continually falling below 1.5 ha across all hazard types. Small land size increases susceptibility to recurring shocks by limiting savings, diversity, and recovery capability. This result is in line with previous research from eastern India that indicates structural economic restrictions cause marginal farmers to incur a disproportionate amount of climate-related losses (Birthal et al. 2014; Panda 2020).

The results of decision trees show that productivity outcomes are significantly impacted by mental health. Although anxiety is linked to decreased productivity, there is no variation in intensity. Depression, on the other hand, has a steep gradient, with more than half of those who are extremely depressed reporting total loss of productivity. This confirms earlier findings showing, especially in rural settings with ongoing stress, depression has more enduring and crippling impacts on occupational functioning than anxiety (Clayton et al. 2017; Evans-Lacko and Knapp 2016). The effects of anxiety seem to be more varied and may be mitigated by coping mechanisms and personal resilience (Gilmour 2014).

Canonical discriminant analysis further indicates that coping techniques are structured along two dominating aspects. Depression, anxiety, income, and landholding size are the primary drivers of the first function, suggesting that farmers' responses to stress are influenced by both psychological load and economic capability. In rural populations that are vulnerable to climate change, similar connections have been shown between mental health distress, financial insecurity, and adaptive behavior (Berry et al. 2018; Hayes et al. 2019).

Caste and education have a major role in defining the second discriminant function, highlighting the significance of socio-structural positioning in gaining access to formal coping strategies. This feature is closely linked to reliance on government programs, indicating that institutional support is mediated more by social capital and educational achievement than by urgent psychological or economic need. This is consistent with research showing that access to climate adaptation initiatives in India is influenced by socioeconomic stratification (Das and Mishra 2021; Rao et al. 2020).

When compared to other approaches, decision tree models are helpful for identifying high-risk groups because they accurately represent the direct and nonlinear impacts of mental health severity on productivity. However, because of its greater explanatory power and classification accuracy, canonical discriminant analysis offers a better understanding of how many factors interact to distinguish coping techniques. When combined, these methods provide complementary viewpoints on adaptability and vulnerability.

The study concludes that the interplay of compound climatic risks, few land resources, psychological anguish, and socio-structural inequality contributes to the vulnerability of smallholder farmers. Therefore, coordinated interventions that incorporate mental health care, fair access to institutional resources, and climate-resilient livelihood assistance are necessary to strengthen resilience. Effective climate adaptation among smallholder farmers requires addressing both material restrictions and emotional stress (IPCC 2022; WHO 2023).

The results provide a number of insights that are pertinent to policy. First, rather than considering catastrophes as discrete occurrences, climate adaptation programs should specifically account for compound risks like flood-drought sequences. Second, in order to lessen structural vulnerability, smallholder farmers must get focused assistance through livelihood diversification, crop insurance, and land security. Third, the necessity to incorporate mental health services into agricultural extension, disaster relief, and rural development projects is highlighted by the substantial correlation between depression and decreased productivity. Early mental health screening and community-based psychosocial intervention might significantly improve adaptive ability. Lastly, differences in access to government programs underscore the necessity of inclusive policy design that tackles caste and educational obstacles, guaranteeing that institutional support reaches the most disadvantaged farmers. The effectiveness of policies may be greatly increased by incorporating social fairness, mental health and climatic resilience.

This study has a number of advantages. It uses a mixed analytical approach that combines multivariate, predictive, and descriptive techniques to provide a comprehensive and multifaceted knowledge of coping and vulnerability. In the context of Indian agriculture, the combination of mental health variables with climatic and socioeconomic determinants is relatively new and extends current research on climate sensitivity. Furthermore, the application of canonical discriminant analysis offers powerful predictive and explanatory insights on coping behavior.

It is important to recognize some restrictions, though. Because the study is based on cross-sectional survey data, temporal generalization and causal inference are limited. Recall or reporting bias may affect self-reported productivity and mental health metrics. The results may not be immediately applicable to other areas with various socioecological characteristics since they are context-specific to the Mayurbhanj district. Longitudinal designs, bigger samples, and comparative geographical analyses might be used in the future studies to confirm and expand on these findings.

Conclusion

This study demonstrates the multifaceted nature of climate vulnerability among smallholder farmers, which is influenced by the interplay of compound climatic hazards, small landholdings, stress related to mental health, and socio-structural disparities. The study goes beyond conventional hazard- or income-based approaches and offers a more comprehensive picture of farmer risk by combining climate exposure with mental health and coping mechanisms. Climate adaptation is both a material and psychosocial process, as evidenced by the emergence of depression as a major factor lowering production and the dependence of coping strategies on psychological state, financial ability, and social standing.

The results have definite policy ramifications. Fair access to institutional resources and mental health care should be part of climate-resilient agriculture. Block-specific initiatives are crucial, such as watershed management, credit access, and primary healthcare integration in Khunta, forest-based livelihoods and tribal-language outreach in Bangiriposi, and irrigation and mental health awareness in Tiring. In order to protect farmers from both environmental and psychological shocks, Mayurbhanj's resilience has to be strengthened via integrated activities that combine climate adaptation, mental health treatment, and socioeconomic assistance.

Acknowledgement The author expresses sincere gratitude to the farming communities, district officials and block officials of Tiring, Bangiriposi and Khunta for their support in providing crucial information and insights into the effects of climatic conditions on farmers' mental health within the agricultural context of the study area.

Potential Conflicts of Interest On behalf of author, the corresponding author states that there is no conflict of interest.

Informed Consent Informed consent was obtained from all individual participants included in the study.

Authors Contribution The study's conception and design, data collection and analysis, and manuscript writing were all done by the author alone.

Data Availability Statement The datasets during and/or analyzed during the current study available from the corresponding author on reasonable request.

AI declaration The author declares that no artificial intelligence (AI) tools were used in the preparation of this chapter.

Funding The research was not supported by any funding.

References

Berry HL, Waite TD, Dear KBG, Capon AG, Murray V (2018) The case for systems thinking about climate change and mental health. Nat Clim Chang 8(4):282–290. https://doi.org/10.1038/s41558-018-0102-4

Birthal PS, Roy D, Negi DS (2014) Assessing the impact of climate change on farm-level profitability in India. J Agric Econ 65(1):1–22. https://doi.org/10.1111/1477-9552.12011

Charlson F, van Ommeren M, Flaxman A, Cornett J, Whiteford H, Saxena S (2021) New WHO estimates on depression and anxiety disorders in the context of climate change and disasters. Lancet Planet Heath 5(7):e440–e448. https://doi.org/10.1016/S2542-5196(21)00084-7

Clayton S, Manning CM, Krygsman K, Speiser M (2017) Mental health and our changing climate: impacts, implications, and guidance. American Psychological Association & eco America

Das S, Mishra A (2021) Social inequality and access to climate adaptation programmes in rural India. World Dev 141:105394. https://doi.org/10.1016/j.worlddev.2021.105394

Evans-Lacko S, Knapp M (2016) Global patterns of workplace productivity for people with depression: absenteeism and presenteeism costs across eight diverse countries. Soc Psychiatry Psychiatr Epidemiol 51(11):1525–1537. https://doi.org/10.1007/s00127-016-1278-4

Gilmour H (2014) Positive mental health and mental illness. Health Rep 25(9):3–9

Hayes K, Blashki G, Wiseman J, Burke S, Reifels L (2019) Climate change and mental health: risks, impacts and priority actions. Int J Ment Heal Syst 13(1):28. https://doi.org/10.1186/s13033-019-0256-2

IPCC (2022) Climate change 2022: impacts, adaptation and vulnerability (contribution of working group II to the sixth assessment report). Cambridge University Press. https://doi.org/10.1017/9781009325844

Jena S, Pradhan B, Mohanty M (2021) Psychosocial impacts of climate-induced disasters in coastal Odisha, India. Int J Disaster Risk Reduction 58:102207. https://doi.org/10.1016/j.ijdrr.2021.102207

Kermode M, Bowen K, Arole S, Joag K, Jorm AF (2009) Community beliefs about causes and risks for mental disorders: a mental health literacy study in a rural area of Maharashtra, India. Int J Soc Psychiatry 55(4):339–350. https://doi.org/10.1177/0020764008095103

Mishra V, Singh D (2019) Hydroclimatic extremes in India: increasing frequency and spatial heterogeneity. Nat Clim Chang 9(9):701–706. https://doi.org/10.1038/s41558-019-0545-6

National Crime Records Bureau (2022) Accidental deaths & suicides in India 2021. Ministry of Home Affairs, Government of India

Nayak SR, Mishra A (2019) Climate change, extreme events and livelihood vulnerability: evidence from coastal Odisha. Indian J Hum Dev 13(2):165–181. https://doi.org/10.1177/0973703019864567

Panda A (2020) Vulnerability to climate variability and drought among smallholder farmers in eastern India. Clim Dev 12(6):547–561. https://doi.org/10.1080/17565529.2019.1651291

Patel V, Saxena S, Lund C, Thornicroft G, Baingana F, Bolton P, Unützer J (2018) The lancet commission on global mental health and sustainable development. Lancet 392(10157):1553–1598. https://doi.org/10.1016/S0140-6736(18)31612-X

Pattnaik S, Sahoo B, Dash PK (2020) Changing rainfall patterns and climate extremes in Odisha, India. Theor Appl Climatol 140:801–814. https://doi.org/10.1007/s00704-020-03122-4

Rao N, Lawson ET, Raditloaneng WN, Solomon D, Angula M (2020) Gendered vulnerabilities to climate change: insights from the semi-arid regions of Africa and South Asia. Clim Dev 12(6):508–523. https://doi.org/10.1080/17565529.2019.1709696

World Health Organization (2023) Mental health and climate change: policy brief. World Health Organization

Chapter 9
Heatwave Vulnerability in Asian Contexts: A Case Study of India's Climatic Extremes and Mortality Correlations

Monika Makwana, Unmesh Patnaik, Kamal Kumar Murari, and Bibhore Kumar Deo

Abstract Heatwaves have emerged as one of the leading causes of climate-related mortality, in India. Despite growing global attention to heat-related health risks, empirical evidences remain limited in the context of Asia that integrates long-term disaster attributed mortality with all-cause excess mortality to assess the impact of heatwave. This chapter addresses this gap by examining long-term mortality pattern associated with heatwaves and compares their severity with other major hazards. We combined 47 years of attributed disaster mortality (1969–2014) with all-cause mortality (2009–2014) to assess temporal trends, identify heatwave-prone states, examine state-level associations between attributed and excess mortality. Descriptive time-series and five-year grouped analyses show a pronounced increase in heatwave mortality in the most recent period, exceeding mortality attributed to other hazards such as floods, cyclones, clod waves, and landslides. The correlation coefficients provide further evidence of heatwaves dominant role. States with high heatwave mortality, like Andhra Pradesh (correlation coefficient of 0.60), exhibit a strong link between heatwaves and excess deaths, while other factors like cold waves and floods have a lesser impact. Similarly, Bihar and Haryana demonstrate significant associations with both heatwaves and cold waves, highlighting the vulnerability of populations to both extremes. By situating heatwaves alongside other hazards and linking mortality outcomes with adaptive capacity and resilience strategies, this chapter contributes empirical evidence to understand how climate change is actively transforming population dynamics in the Global South. These findings underscore

M. Makwana (✉)
School of Public Policy, Indian Institute of Technology, Delhi, India
e-mail: makwana92@sopp.iitd.ac.in

U. Patnaik · K. K. Murari
Centre for Climate Change and Sustainability Studies, School of Habitat Studies, Tata Institute of Social Sciences, Mumbai, India
e-mail: unmesh.patnaik@tiss.ac.in

B. K. Deo
United Nations Development Programme (UNDP), New Delhi, India

P. Ghosh et al. (eds.), *Climate Crisis and Changing Population Dynamics*, Springer Climate, https://doi.org/10.1007/978-3-032-24533-5_9

the urgency of implementing tailored mitigation and adaptation strategies to build resilience across India, including but not limited to public health awareness, early warning systems, and heat action plans.

Keywords Heatwaves · Mortality · Trend · Vulnerability · Awareness · Resilience · Excess deaths

Introduction

India has been witnessing heatwaves for the past few decades, the issue is with the rising severity and frequency and spatial spread, which is now, grappling millions of people across multiple states. As per the IPCC SREX report, '*exposure and vulnerability are key determinants of disaster risk and of impacts when risk is realized*'. In the case of heatwaves. The risk is inherently multilayered and intertwined deeply with structural vulnerability, manifesting through adverse impacts on human health and elevated mortality rate. (IPCC 2012). It is difficult to capture the impacts of heatwaves due to its diffused and cumulative nature, unlike sudden onset disasters the ability to capture correct statistics is also challenging

Asia comprises more than 60% global population that is vulnerable to heat extremes due to factors like rapid urbanization, high population density, and innate structural vulnerability which limits the adaptive capacity. Within Asia, India stands out as climate hotspot where rising temperatures intersects with poverty, groundwater depletion, unequal access to health, and lacking basic social protection (IPCC 2021; IPCC 2012). A growing body of literature has examined the relationship between temperature extremes and mortality, employing various models and excess mortality estimation and epidemiological approaches. Various associations are established between heat exposure and short-term mortality displacement (Hajat et al. 2005; Azhar et al. 2014; Zhu et al. 2014). However, peoples' acclimatization capacity plays a crucial role but does not shield populations from the risks related to extreme heat nor does it define their levels of sensitivity or adaptive capacity to climate shifts, thereby giving rise to a multilayered vulnerability (Dholakia et al. 2015; Zhu et al. 2014). These dynamics are further highlighted by Hajat et al., 2005, they used Poisson time-series models to show that heat acts as a final trigger for vulnerable populations, from his comparative study between London and Delhi. Similarly, Azhar (2014) finds significant excess mortality during the extreme heat spell in Ahmedabad, highlighting the strong impact of prolonged extreme heat, even in hot regions. Particularly, in the case of Indian states, extreme events are well illustrated during 2015 in Andhra Pradesh and Telangana, where reports noted a heat-related mortality going beyond 2300. During this event, the daytime temperature rose by more than 7 °C above the seasonal baseline in coastal Andhra Pradesh, far exceeding the physiological threshold of human comfort (Brink 2015). These events shed light on the unrealized death toll and long-term health impacts, these slow onset climate extremes have, that goes far beyond the sudden onset disasters.

Despite these advances, several research gaps can be found in the literature, giving a foundation to this study. Firstly, an empirical gap remains in integrating disaster attributed mortality with excess deaths in the context of India, extending beyond case studies and anecdotal evidences. Secondly, methodological gap that explores the true burden of heatwaves, when compared with other natural hazards. Thirdly, a comparative assessment is limited in literature, very few studies compare mortality across various hazards, systematically positioning heatwaves. And finally, a policy and sustainability gap, with limited studies triangulating scientifical evidences to policy. Addressing these gaps are essential for understanding heatwaves not merely as episodic disasters, but as structural driver of demographic vulnerability under climate change.

This study addresses the questions of how have heat-related mortality patterns evolved in India over long term relative to other major climatic hazards? Which are the most vulnerable states to heatwaves and to what extent are heatwave attributed deaths associated with excess deaths at the state level to understand the demographic impacts of extreme heat.

This chapter directly contributes to the theme of Climate Crisis and Changing Population Dynamics: Emerging Issues and Sustainability by empirically demonstrating how intensifying heat extremes are reshaping patterns of mortality and vulnerability in one of the world's most climate-exposed regions. By integrating long-term mortality data with climate variables and triangulating to the need of transformational adaptation and resilience, the study positions heatwave mortality as an emerging demographic issue with profound implications for the sustainable development goals and strategies of public health across India.

Data

The severity of any event can be measured by the damage it caused to the society. In the present study, the severity is identified in the sense of mortality due to the adverse effects of extreme heat. As we know, the mortality due to heatwaves is recorded less significantly; hence, the use of attributed mortality can produce undermining results. Therefore, to minimize the error in reporting the deaths, both excess mortality and attributed mortality rates are used. And by correlating them we can come across to a less biased number. For the analysis, secondary data is taken from the National Crime Record (NCR) Bureau and Civil Registration Service reports (CRS). NCR data is released by the government of India, every year, which consists of data on accidental death due to natural disasters/extreme events. NCR data deals with attributed mortality due to disasters like flood, cyclone, heatwave, cold wave, landslide, etc. This data is available for 47 years starting from the year 1969 till the year 2014 and was used for analysing the attributed mortality. The CRS data is also released by the government of India, providing data regarding total mortality rate, however, available for the much shorter period (2009–2014). Moreover, there are certain advantages and disadvantages of using this data which is discussed in the following points:

1. Since the effect of heatwaves on human health is not directly observable, mortality due to heatwave could manifest in the form of severe fever, heart attack, and severe dehydration.
2. In view of this, the main disadvantage is that it is attributed mortality which might be under-reported during a heatwave.
3. Another issue with NCR data is inconsistency in the category head under which mortality was reported. These categories have changed over the years during 1969–2004. In recent years, more categories were added under which accidental mortality is recorded.
4. The CRS data is all-cause mortality data that is reported for a period. A major advantage hence is that it is carefully examined. Excess mortality can be derived from this dataset which may point towards the occurrence of natural events that might have caused excess mortality. Assessment of the mortality burden of a heatwave is not straightforward because severe complications triggered by extreme heat, such as respiratory and cardiovascular, are often not diagnosed particularly as heatwave-induced mortality. Many heat-related deaths are therefore not implied as heat induced but rather as underlying respiratory or chronic conditions. The method to assess the mortality impact of a heatwave is to calculate excess mortality during heatwave period, as the sum of deaths exceeding a baseline of expected deaths in the absence of extreme heat activity.

For positioning heatwaves, secondary data was used from the National Crime Record Bureau consisting a time series of 47 years of data, from 1967 to 2014. This data is published by the Government of India consisting of state-wise categorization from the year 1997–2014. The recorded data is based on the mortality of population due to natural disasters or extreme events, like earthquake, landslide, flood, cyclone, heatwave, cold wave, etc. Data from all Indian States and Union Territories (UTs) from the sample frame, in this case, is taken. The analysis is also further downscaled to study the nature of extremes and disaster in selected states and cities.

Methodology

The rationale for this analysis is to study the impact of heatwaves on the population on a temporal scale. In doing so the study attempts firstly, to analyse the frequency and severity of these events on a quantitative scale. Secondly, an attempt is also made to examine statistically what has been the trend in impacts of various other disasters in India and the contribution of heatwaves in these. The approach for analysing these consists of simple descriptive statistics, cross-tabulations, and tests of means for analysing meteorological and weather data (Wang et al. 2018; Ray et al. 2021). Figure 9.1 summarizes the methodological flow of this study:

Analysis of temporal trends for positioning heatwaves.

Most vulnerable states are selected based on the available mortality dataset. Figure 9.2 shows the time series of mortality per million of the population. The

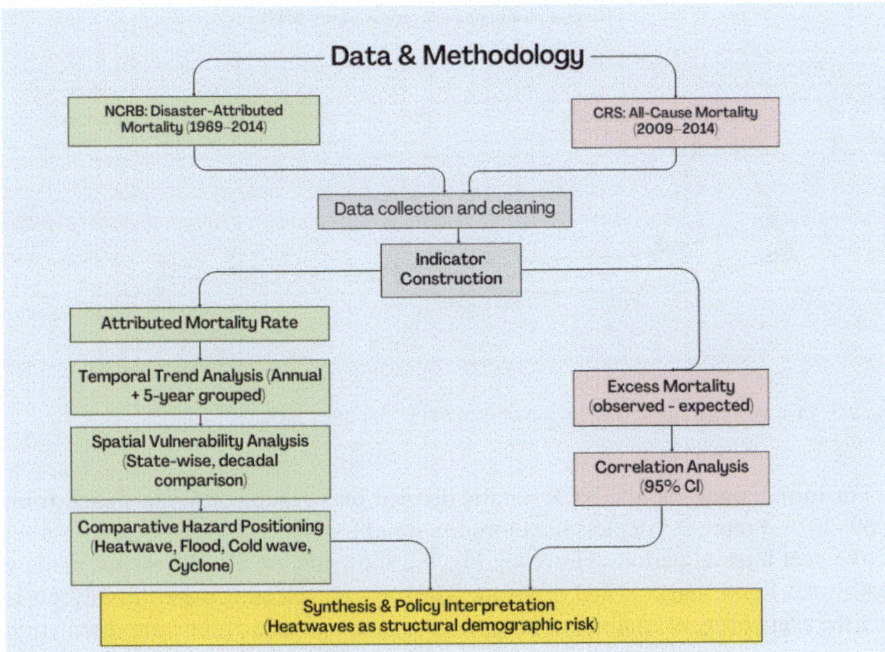

Fig. 9.1 Representation of data and methodological flow chart

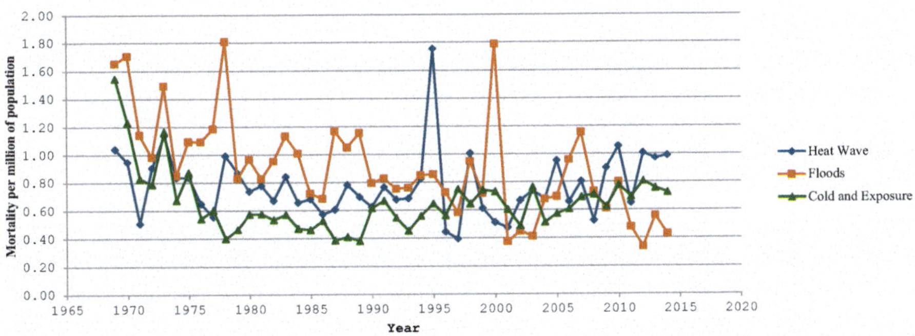

Fig. 9.2 Time-series plot representing the behaviour of a heatwave, cold wave, and flood

figure depicts the year-wise mortality for three main climate-based disasters, namely heatwave, cold wave, and flood. The rationale for selecting these three disasters is their availability of data and also to minimize the effect of point-based mortality.[1]

[1] It is very high in case of disasters like cyclones and landslides, and appendix represents the time-series pattern of mortality per million of population with the inclusion of the cyclone and landslides.

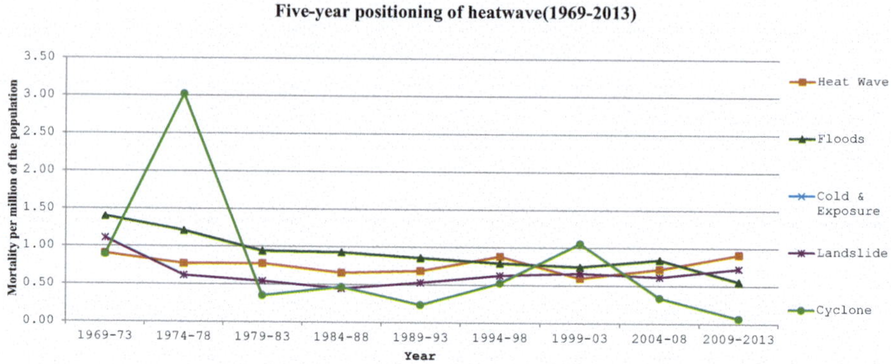

Fig. 9.3 Five years grouped, positioning of heatwave with other hazards

For further analysis, the NCR data is divided into groups of 5 years each from 1969–2013. Figure 9.3 depicts the mortality trend for five major disasters over a set of five-year interval periods. However, Fig. 9.2 shows the trends of 3 events, namely heat stroke, flood, and cold and exposure. The basis for selecting these three events is first, the availability of continuous long-term, and second, it is a proper representation of extremes. Although for disasters like cyclone and landslide data is also available but because of their nature of sudden onset, mortality is point-based, which could lead to biases in the proper representation of extremes. Hence, they are not included in the analysis for Fig. 9.2. However, they have been included for analysis shown in Fig. 9.3. From the state-wise data, a trend line is calculated for disasters reflecting the fatality of each event based on the rate of mortality. From these two figures, it can be inferred that there is evidence of mortality increasing due to temperature variability specifically over the last decade.

Analysing the severity of heatwave in states during the past one decade.

From Fig. 9.2, it is clearly observed that the severity of heatwave-related deaths is increasing on a temporal scale. Therefore, to concentrate on the recent phenomena, the second part of analysis deals with representing a state-wise mortality scenario for heatwaves only. Figures 9.3 and 9.4 depict the distribution of state-wise mortality rate in India from 2004 to 2013. In both the figures, the x-axis depicts the states while the y-axis shows the mortality per million of the population. (For calculating mortality per million of population, the number of deaths in a particular year and a particular state is divided into the population in million the state comprises in that particular year.)

Similarly, Figs. 9.5 and 9.6 correspond to the depiction of heatwave-induced mortality rates for select states depending on the severity of these events. These states are Andhra Pradesh, Bihar, Delhi, Gujarat, Haryana, Madhya Pradesh, Maharashtra, Orissa, Punjab, Rajasthan, Tripura, Uttar Pradesh, Uttarakhand, and West Bengal. Further, to examine the overall decadal trends in mortality trends for major extreme events vis a vis the heatwaves the cumulative mortality per million of the

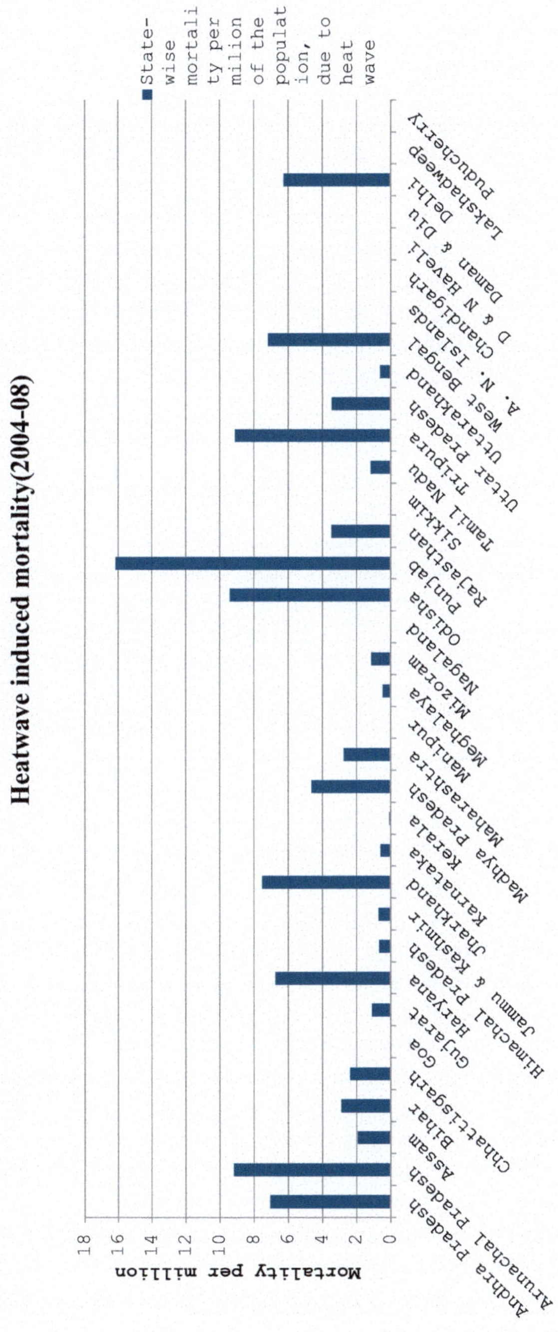

Fig. 9.4 Representing the heatwave-induced mortality in states of India, between the time period of 2004–2008

population is calculated for three time periods 2000–2004, 2005–2009, and 2010–2014, respectively, over major heat-wave-prone states in India. The results of this analysis are shown in Fig. 9.6.

Correlating the attributed and excess mortality.

The concluding part of this study analyses the trends in the correlation between the attributed and excess mortality. If the correlation coefficient (at 95% confidence limit) is more than 0.5 than the mortality during that time period is hypothesized to be highly correlated because of the particular event. In this analysis, the correlation not only for heatwave but also for other disasters is calculated as it is helpful to know why the degree of association is weak and what are the factors affecting the linear relationships. Correlation coefficients between the attributed and excess mortality are presented in Table 9.1. Since Table 9.1 presents the analysis for few states, Appendix presents the profile of all states of India (* represent significance).

Results

The results of the time-series charts shown in Figs. 9.2 and 9.3 reveal that the severity of heatwave in the recent decades has surged upwards, which is even higher than a cold wave, floods, landslide, and cyclones. If Fig. 9.2 has been observed closely then, from 2009–2014, a significant rise in mortality due to heatwave can be observed. Moreover, a five-year grouped trend is also providing a similar outcome with the combination of another disaster. Proving the evidence that out of all the major disasters (considered in this study) heat extreme has a significant impact on the mortality of the vulnerable population.

Further, in-depth analysis for each of the Indian states unveils the prevalence of heatwave in states like Andhra Pradesh, Bihar, Delhi, Gujarat, Haryana, Madhya Pradesh, Maharashtra, Orissa, Punjab, Rajasthan, Tripura, Uttar Pradesh, Uttarakhand, and West Bengal, during the past decade. In these same states, when the initial and later five-year mortality trends are observed, it is clearly projecting an increase in mortality every passing year with increased climate variability.

Table 9.1 provides valuable insights into the dominant role of heatwaves in driving excess mortality in India. Most of the states which were showing high mortality due to the heatwave in Table 9.1 have their correlation coefficient significant and highly correlated, i.e. the excess mortality in that particular states in those particular years is highly related or resultant from the extreme impacts of heat. For instance, the correlation coefficient is 0.60 Andhra Pradesh, which signifies a strong relationship between the crude mortality and heatwaves-induced mortality. At the same time, the mortality is less affected by other events like cold and exposure, floods, and landslide. Similar to Andhra Pradesh, Bihar exhibits a strong correlation with heatwaves (0.56) and floods (0.30). Haryana and Punjab show statistically significant correlations with both heatwaves and cold waves, highlighting the potential impact of both extremes on mortality. Madhya Pradesh has a significant correlation with heatwaves

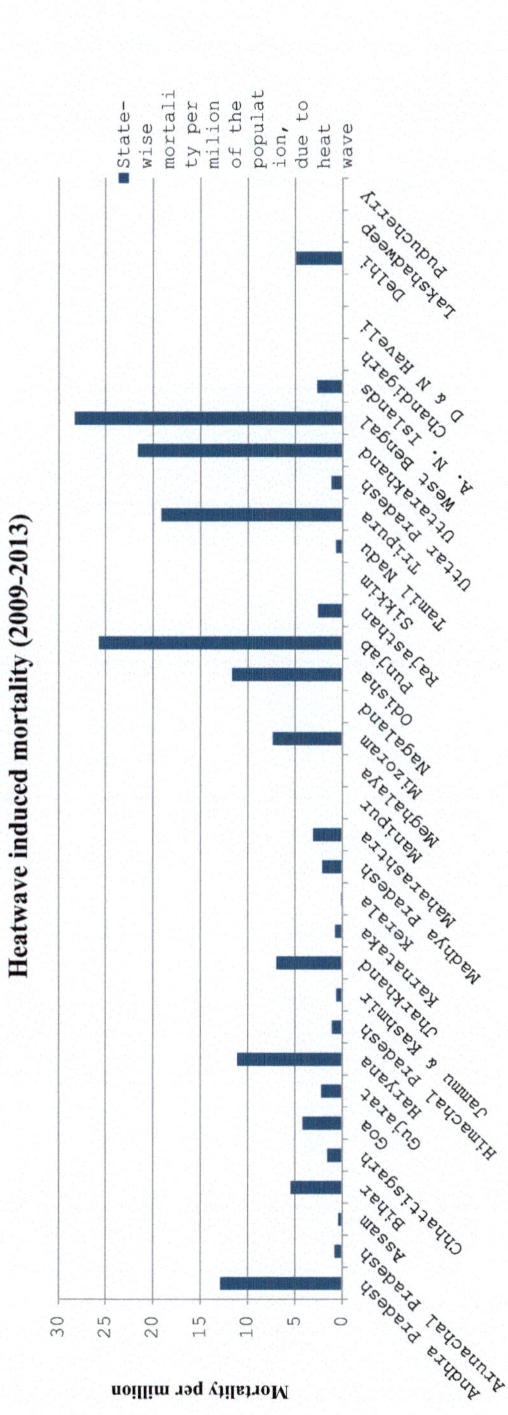

Fig. 9.5 Representing the heatwave-induced mortality in states of India, between the time period of 2009–2013

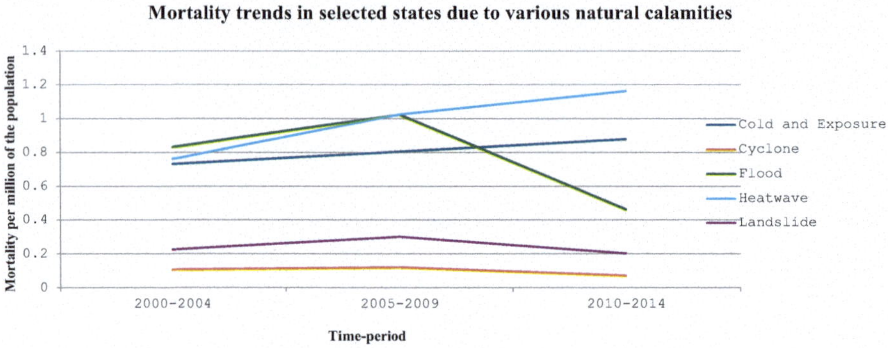

Fig. 9.6 Trends of various disasters in heatwave exposed states

Table 9.1 Correlation between attributed and excess mortality, in heatwave exposed states

States	Heatwave		Cold and exposure		Flood		Landslide	
	Cor. coeff.	Sig. at 95%	Cor. coeff.	Sig. at 95%	Cor. coeff.	Sig. at 95%	Cor. coeff.	Sig. at 95%
Andhra Pradesh	0.60	0.07	0.17	0.64	0.07	0.85	0.32	0.37
Bihar	0.56	0.09	0.05	0.89	0.04	0.93	0.30	0.43
Chhattisgarh	0.32	0.36	0.16	0.68	0.31	0.42	0.50	0.18
Gujarat	0.42	0.23	0.50	0.17	0.39	0.30	0.82*	0.01
Haryana	0.65*	0.04	0.84*	0.00	0.45	0.23	0.19	0.63
Madhya Pradesh	0.33	0.36	0.32	0.40	0.70*	0.03	0.16	0.68
Odisha	0.39	0.26	0.17	0.66	0.74*	0.02	0.70*	0.04
Punjab	0.64*	0.04	0.86*	0.00	0.67*	0.05	0.69*	0.00
Rajasthan	0.13	0.72	0.33	0.39	0.30	0.44	0.13	0.75
Tamil Nadu	0.50	0.14	0.34	0.37	0.51	0.16	0.19	0.62
Uttar Pradesh	0.35	0.33	0.31	0.42	0.07	0.85	0.14	0.71
West Bengal	0.33	0.35	0.18	0.64	0.14	0.72	0.35	0.36
Delhi	0.41	0.25	0.05	0.89	0.26	0.50	0.26	0.85

(0.33) and floods (0.70). While the correlation with heatwaves is moderate (0.39), in case of Odisha which shows a more significant association with floods and landslides (0.74 and 0.70, respectively). In contrast to other states, Rajasthan exhibits a weak correlation with heatwaves (0.13) and stronger associations with cold waves (0.33). Delhi shows a moderate correlation with heatwaves (0.41) and a significant association with cold waves (0.84). By understanding the specific vulnerabilities of each state and implementing tailored mitigation strategies, India can build resilience against this growing threat and protect its population from the deadly consequences of heatwaves.

Discussion

With rising atmospheric temperature, heatwaves have become a norm in the global context. Heatwaves have surpassed natural disasters and extremes such as cold waves, floods, cyclones, and landslides in terms of mortality across India. The amalgamation of extreme heat and vulnerability created a recipe for societal suffering and mortality, thereby posing a public health concern that requires strategic long-term planning and intervention. The results of this study provide a multi-year grouped assessment indicating an upward trajectory of heatwaves mortality, highlighting its dominant effects particularly on vulnerable populations such as children, elderly, immune-compromised and those with cardiovascular and related disabilities (Amirkhani et al. 2022; Climate Impact Lab 2020). This rise is not merely a statistical anomaly; it reflects a complex interplay of climatic factors and societal vulnerabilities. Studies suggest that rising global temperatures, driven by anthropogenic greenhouse gas emissions, are leading to more frequent and intense heatwaves in India (Arulalan et al. 2023; Ravindra et al. 2024). Additionally, factors like rapid urbanization, deforestation, and water scarcity exacerbate the heat island effect and amplify the human stress response to thermal extremes (Ebi & Meehl 2007; UNISDR 2015). The geographical spread of this heatwave threat is equally concerning. In-depth analysis for each Indian state reveals a worrying pattern: over a dozen states, including Andhra Pradesh, Bihar, Delhi, Gujarat, and West Bengal, have witnessed a prevalence of heatwaves in the past decade (Guleria & Gupta, 2018). These states, when further examined through initial and five-year mortality trends, exhibit a consistent and alarming rise in deaths year after year, mirroring the increasing climate variability and highlighting the need for regionalized adaptation strategies. To delve deeper into the severity of this issue, two crucial scientific exercises were conducted. Figure 9.5 showcases a time-series plot comparing the magnitude of heatwaves with other disasters in selected states. The trend is undeniable—heatwave-related mortality has been steadily climbing since 2000, with cold wave mortality also showing a concerning increase (Nogueira et al. 2020). This necessitates a shift in disaster preparedness and response, prioritizing heatwave risk reduction alongside other natural hazards. But numbers alone cannot fully capture the extent of the problem. Statistical analysis, often neglected due to limitations in heatwave-related mortality data, provides valuable insights. Figure 9.6 reveals that most states with high heatwave mortality also exhibit statistically significant correlation coefficients (Guha-Sapir et al. 2019). For instance, Andhra Pradesh, with a correlation coefficient of 0.60, demonstrates a robust link between crude mortality and heatwave-induced deaths, suggesting that other factors like cold waves and floods have a lesser impact in this region (Ryan et al. 2015). This highlights the need for geographically-specific mitigation strategies tailored to the dominant hazard in each region.

The climate sensitivity can easily observe with the health outcomes. Therefore, it is very important to include climate variability and its outcomes in the public health policy for the country like India where increasing poverty and population are the weak spot for implementation measures of public health, which further worsened by the increased instability of climate (Bush et al. 2011). Nevertheless, we need a set of objectives to list out the complex health outcomes, risk by risk. After that, we might able to reach a desirable outline for implementing the adaptation and mitigation plans. The World Health Organisation has proposed a methodology to calculate the impact of climate sensitivity, the DPSEEA (Driving Force Pressure State Exposure Effect Action) model, a climate model taking the exposure as an input variable, which is incorporating the influence of social, economic, and cultural factors on how exposure is calculated for different groups of individuals (WHO 2000; Gentry-Shields and Bartram 2014). This, in turn, produces the current health outcomes and future impacts of climate sensitivity on health.

Literature like (Das & Smith 2012) has focused on how heatwave awareness campaigns can significantly reduce heat stroke deaths during heatwaves in a developing country. In the study Bernard and Mcgeehin (2004) suggested an organization of plan for city heatwaves, based on the fundamental features of: working under the supervision of leading agencies and organization, standard alarming system sensitive to weather conditions, awareness programs for elderly and children, response actions focusing on most vulnerable population, information evaluation, and frequently updating the plan. Also, the Antwerp city adaptation plan reflects integrated planning with the lowest level of society, bottom-up response with community-based decision-making that has provided a local weather forecast and early warning system for heat-related stress. Therefore, especially in case of adaptation to climate change, it is context, area, and culture-based approach. Planned adaptation is usually focused on,post- disaster relief and preparing for such potential events. While these community-based adaptations are often autonomous and can lead to long-term resilience, they still benefit from supportive system and monitoring. Autonomous adaptation works beyond the intersectionality of gender, caste, or religion, being inclusive as it considers the diverse risk faced by all groups regardless of their identity. While planned adaptation if not properly managed, can risk benefiting some, and often forgetting who is left behind. Does that mean planned adaptation is not good for communities? No, they work as well as autonomous adaptation, if monitored and guided by subsidiary principles, while implementing and planning. Then, the question arises, How can autonomous adaptation lead to tranformative outcomes? Autonomous adaptation has more probability to transform the way of living, also because it is adopted within and not imposed by the system. It's important to know that plans inclusive of vulnerable groups often turn out to be the most effective in addressing disaster risk reduction and achieving adaptation goals (Dale 2022; Kang et al., 2020). The increasing focus on nature-based adaptation solutions offers a crucial link between restoring ecosystems and mitigating the impact of heatwaves on populations. Also, the role of communities in tackling climate extreme is crucial and empowering. These steps taken by communities reflect proactive adaptation and resilience. Building on the previous example, Antwerp's communities uses their private weather station to forecast heat waves up to a week in advance, allowing them to implement tailored measures that reduce risk. (Dale 2022). The escalating heatwave threat demands immediate and comprehensive action on various fronts. Here are some key scientific and policy priorities:

1. Strengthening and Building Early Warning Systems and Heat Action Plans: Robust and human centric systems tailored to specific regions are crucial for timely preparedness and response, taking in consideration the subsidiary principle is a crucial step. Heat action plans should include public awareness campaigns, heatwave shelters, and access to medical resources, with a focus on vulnerable populations such as the elderly, children, and those with immune-compromised system and chronic health conditions (WHO 2021; NDMA 2021). These also involve the aspect of finance, to build necessary infrastructure. But even in absence of state support, proactive communities can build their private EWS to build adaptive capacity.
2. Promoting Climate-Resilient Infrastructure and Urban Planning: Embedding nature-based solutions into infrastructure planning by building heat-resistant structures with green roofs and reflective surfaces, expanding green spaces, and adopting sustainable urban design practices like permeable pavements and urban forests can significantly reduce the impact of heatwaves (WHO 2004). Implementing nature-based solutions (NBS) within urban planning can enhance ecological resilience while providing co-benefits such as improved air quality, biodiversity conservation, and enhanced public well-being (Dale 2022). Furthermore, integrating the adaptation gradient in infrastructure can ensure long-term sustainability. This approach includes eight key measures: (1) ecological enhancement, (2) infrastructure stability, (3) adaptation over time, (4) reduced exposure to hazards, (5) cost-effectiveness, (6) greenhouse gas (GHG) reduction, (7) participatory decision-making, and (8) social benefits. Policies must be designed to avoid maladaptation, ensuring that climate resilience efforts do not lead to gentrification and the displacement of marginalized communities. This requires equitable access to green infrastructure, rent control mechanisms, and participatory urban planning to integrate community voices in decision-making (Dale 2022).
3. Investing in Heatwave Research and Development: Further research into heat-related illnesses, vulnerability mapping, and effective mitigation is the need for the hour.
4. Investing in Innovation and Capacity Building for Heatwave Resilience, along with integrated planning: Advancing research on heat-related illnesses, vulnerability mapping, and mitigation strategies is crucial for informed policy interventions. However, beyond research, fostering innovation in capacity-building practices is essential to enhance community resilience. This includes developing localized training programs for urban planners, health workers, and emergency responders, promoting technology-driven solutions such as AI-based heat forecasting, and strengthening institutional frameworks for climate governance.

Encouraging participatory knowledge-sharing platforms can empower communities to implement adaptive strategies effectively, ensuring long-term preparedness against extreme heat events.

Conclusion

Climate crisis is exhibiting in various forms of extremes and disasters, affecting the population dynamics in profound ways. It is well established that heatwaves have severe health impacts, not only exacerbating morbidity and mortality among vulnerable populations but also affecting otherwise healthy individuals or reducing longevity. The effectiveness of adaptation efforts largely depends on public awareness, preparedness, and the integration of localized strategies, both incremental and transformational strategies can benefit from these factors. While numerous studies have explored temperature variability and its consequences globally, research in India has primarily focused on the broader impacts and future projections of heatwaves. By situating heatwave mortality alongside other hazards and integrating disaster attributed deaths, this chapter advances the understanding of how climate change is actively reshaping the population dynamics.

A decade-long analysis reveals a concerning trend: mortality due to heatwaves has steadily increased, with certain states facing disproportionately higher exposure. This underscores the need for targeted, context-specific adaptation strategies that prioritize both nature-based solutions (NbS) and community-driven approaches. Implementing NbS—such as urban greening, water-sensitive design, and climate-responsive architecture—can enhance resilience by reducing heat stress and improving ecological stability. However, for these solutions to be effective, they must be co-designed with local communities, incorporating traditional knowledge and participatory decision-making.

In a country as diverse as India, where socio-economic and environmental conditions vary widely, a bottom-up approach to adaptation is essential. Strengthening community-led adaptation initiatives, promoting decentralized governance, and integrating indigenous climate knowledge into urban and rural planning can help mitigate heatwave risks more equitably. Addressing location-specific vulnerabilities through inclusive and sustainable strategies will be key to ensuring long-term resilience against rising temperatures. As heat stress intensify under growing climate change, the Indian experience offers a deep insight for other climate vulnerable regions: safeguarding population from extreme heat is not just disaster management strategy but a central pillar of sustainable development and a sign of demographic resilience in a warming world.

Appendix

States	Heatwave		Cold and Exposure		Flood		Landslide	
	Cor. coeff.	Sig. at 95%	Cor. coeff.	Sig. at 95%	Cor. coeff.	Sig. at 95%	Cor. coeff.	Sig. at 95%
Andhra Pradesh	0.60	0.07	0.17	0.64	0.07	0.85	0.32	0.37
Arunachal Pradesh	0.05	0.89	0.34	0.33	0.11	0.76	0.15	0.69
Assam	0.83	0.00	0.37	0.33	0.27	0.49	0.37	0.32
Bihar	0.56	0.09	0.05	0.89	0.04	0.93	0.30	0.43
Chhattisgarh	0.32	0.36	0.16	0.68	0.31	0.42	0.50	0.18
Goa	0.19	0.60	0.33	0.39	0.29	0.45	0.45	0.23
Gujarat	0.42	0.23	0.50	0.17	0.39	0.30	0.82	0.01
Haryana	0.65	0.04	0.84	0.00	0.45	0.23	0.19	0.63
Himachal Pradesh	0.05	0.89	0.28	0.46	0.43	0.25	0.07	0.86
Jammu & Kashmir	0.32	0.32	0.19	0.63	0.13	0.73	0.02	0.96
Jharkhand	0.14	0.70	0.19	0.62	0.33	0.39	0.02	0.96
Karnataka	0.14	0.71	0.02	0.96	0.39	0.30	0.35	0.36
Kerala	0.23	0.52	0.15	0.71	0.09	0.82	0.20	0.61
Madhya Pradesh	0.33	0.36	0.32	0.40	0.70	0.03	0.16	0.68
Maharashtra	0.27	0.45	0.68	0.04	0.68	0.04	0.68	0.42
Manipur	0.18	0.62	0.18	0.65	0.25	0.52	0.30	0.43
Meghalaya	0.29	0.42	0.42	0.26	0.27	0.47	0.41	0.27
Mizoram	0.63	0.05	0.27	0.44	0.60	0.09	0.67	0.05
Nagaland	0.17	0.63	0.47	0.20	0.47	0.20	0.11	0.78
Odisha	0.39	0.26	0.17	0.66	0.74	0.02	0.70	0.04
Punjab	0.64	0.04	0.86	0.00	0.67	0.05	0.69	0.00
Rajasthan	0.13	0.72	0.33	0.39	0.30	0.44	0.13	0.75
Sikkim	0.29	0.42	0.18	0.65	0.50	0.17	0.65	0.06
Tamil Nadu	0.50	0.14	0.34	0.37	0.51	0.16	0.19	0.62
Tripura	0.09	0.80	0.25	0.51	0.57	0.11	0.21	0.59
Uttar Pradesh	0.35	0.33	0.31	0.42	0.07	0.85	0.14	0.71
Uttarakhand	0.22	0.53	0.45	0.23	0.47	0.20	0.03	0.94
West Bengal	0.33	0.35	0.18	0.64	0.14	0.72	0.35	0.36
A. N. Islands	0.37	0.30	0.17	0.65	0.17	0.65	0.82	0.01

(continued)

(continued)

States	Heatwave		Cold and Exposure		Flood		Landslide	
	Cor. coeff.	Sig. at 95%	Cor. coeff.	Sig. at 95%	Cor. coeff.	Sig. at 95%	Cor. coeff.	Sig. at 95%
Chandigarh	0.18	0.62	0.29	0.45	0.33	0.39	0.33	0.39
D & N Haveli	0.48	0.16	0.07	0.86	0.80	0.01	0.07	0.86
Daman & Diu	0.42	0.23	0.09	0.82	0.89	0.82	0.89	0.82
Delhi	0.41	0.25	0.05	0.89	0.26	0.50	0.26	0.85
Lakshadweep	0.09	0.82	0.63	0.07	0.63	0.07	0.63	0.07
Puducherry	0.02	0.97	0.06	0.88	0.06	0.88	0.06	0.88

Declaration Statement Competing Interest: The authors declare that they have no competing interests.

Availability of Data and Materials: The datasets generated and/or analysed during the current study are available from the corresponding author upon reasonable request.

No artificial intelligence (AI) tools were used for data collection, data analysis, interpretation of results, or generation of original scientific arguments in this chapter. Limited use of an AI-based language tool (ChatGPT, OpenAI; accessed in 2025 via https://chat.openai.com) was made solely for grammatical editing, sentence restructuring, and improvement of language clarity under the direct supervision of the authors. All conceptualization, methodological design, analysis, interpretation, and conclusions presented in this chapter are entirely the original work of the authors. The authors take full responsibility for the content of this manuscript.

References

Amirkhani M, Ghaemimood S, von Schreeb J, El-Khatib Z, Yaya S (2022) Extreme weather events and death based on temperature and CO2 emission–A global retrospective study in 77 low-, middle-and high-income countries from 1999 to 2018. Prev Med Rep 28:101846

Arulalan T, Achutarao K, Sagar A (2023) Climate science to inform adaptation policy: heat waves over India in the 1.5°C and 2°C warmer worlds. Clim Change 176(5):64

Azhar GS, Mavalankar D, Nori-Sarma A, Rajiva A, Dutta P, Jaiswal A, Sheffield P, Knowlton K, Hess JJ (2014) Heat-related mortality in India: excess all-cause mortality associated with the 2010 Ahmedabad heat wave. PloS one 9(3):e91831

Bernard SM, McGeehin MA (2004) Municipal heat wave response plans. Am J Public Health 94(9):1520–1522

Brink GW (2015) India heat wave deaths: what heat does to the human body. Natl Geogr. http://news.nationalgeographic.com/2015/05/150528-india-heat-wave-deaths-human-body-health-science

Bush KF, Luber G, Kotha SR, Dhaliwal RS, Kapil V, Pascual M et al (2011) Impacts of climate change on public health in India: future research directions. Environ Health Perspect 119(6):765

Climate Impact Lab (2020) Global death rate from rising temperatures projected to surpass the current death rate of all infectious diseases combine
Dale L (2022) Climate Change Adaptation: An Earth Institute Sustainability Primer. Columbia University Press
Das S, Smith SC (2012) Awareness as an adaptation strategy for reducing mortality from heat waves: evidence from a disaster risk management program in India. Clim Change Econ 3(02):1250010
Dholakia HH, Mishra V, Garg A (2015) Predicted increases in heat related mortality under climate change in urban India. Indian Institute of Management, Ahmedabad, India
Ebi KL, Meehl GA (2007) The heat is on: climate change and heatwaves in the Midwest. In: Regional impacts of climate change: four case studies in the United States, pp 1–14
Gentry-Shields J, Bartram J (2014) Human health and the water environment: Using the DPSEEA framework to identify the driving forces of disease. Sci Total Environ 468:306–314
Guha-Sapir D, Below R, Philippe H (2019) Université catholique de Louvain. The emergency events: 2019. the emergency events. https://uclouvain.be/en/index.html. [Google Scholar]
Guleria S, Gupta AK (2018) Heat wave in India documentation of state of Telangana and Odisha (2016). National Institute of Disaster Management, New Delhi, p 124
Hajat S, Armstrong BG, Gouveia N, Wilkinson P (2005) Mortality displacement of heat-related deaths: a comparison of Delhi, Sao Paulo, and London. Epidemiology 16(5):613–620
IPCC (2012) Summary for policymakers. In: Managing the risks of extreme events and disasters to advance climate change adaptation Field CB, Barros V, Stocker TF, Qin D, Dokken DJ, Ebi KL, Mastrandrea MD, Mach KJ, Plattner G-K, Allen SK, Tignor M, Midgley PM (eds). A special report of working groups I and II of the intergovernmental panel on climate change. Cambridge University Press, Cambridge, UK, and New York, NY, USA, pp 3–21
IPCC (2021) Climate Change 2021: the physical science basis. Masson-Delmotte V, Zhai P, Pirani A, Connors SL, Péan C, Berger S, Caud N, Chen Y, Goldfarb L, Gomis MI, Huang M, Leitzell K, Lonnoy E, Matthews JBR, Maycock TK, Waterfield T, Yelekçi O, Yu R, Zhou B (eds). Contribution of working group I to the sixth assessment report of the intergovernmental panel on climate change. Cambridge University Press, Cambridge, United Kingdom and New York, NY, USA, In press. https://doi.org/10.1017/9781009157896
Kang C, Park C, Lee W, Pehlivan N, Choi M, Jang J, Kim H (2020) Heatwave-related mortality risk and the risk-based definition of heat wave in South Korea: a nationwide time-series study for 2011–2017. Int J Environ Res Public Health 17(16):5720
NDMA. (2021). National Disaster Management Authority. Golechha, M., Mavalankar, D., & Bhan, S. C. (2021). India: Heat Wave and Action Plan Implementation in Indian Cities. Urban Clim Sci Plan Healthy Cities, 285–308
Nogueira LM, Yabroff KR, Bernstein A (2020) Climate Change and Cancer. CA Cancer J Clin 70(4):239–244. https://doi.org/10.3322/caac.21610
Ravindra K, Bhardwaj S, Ram C, Goyal A, Singh V, Venkataraman C, Bhan SC, Sokhi RS, Mor S (2024) Temperature projections and heatwave attribution scenarios over India: a systematic review. Heliyon 10(4)
Ray K, Giri RK, Ray SS, Dimri AP, Rajeevan M (2021) An assessment of long-term changes in mortalities due to extreme weather events in India: A study of 50 years' data, 1970–2019. Weather Clim Extrem 32:100315
Ryan B, Franklin RC, Burkle FM, Aitken P, Smith E, Watt K, Leggat P (2015) Identifying and describing the impact of cyclone, storm and flood related disasters on treatment management, care and exacerbations of non-communicable diseases and the implications for public health. PLoSCurr 7(DISASTERS). https://doi.org/10.1371/currents.dis.62e9286d152de04799644dcca47d9288
The United Nations Office for Disaster Risk Reduction (UNISDR) (2015) The human cost of weather-related disasters 1995–2015. Centre for Research on the Epidemiology of Disasters (CRED). https://www.unisdr.org/files/46796_cop21weatherdisastersreport2015.pdf

Wang S, Zeng J, Huang Y, Shi C, Zhan P (2018) The effects of urbanization on CO2 emissions in the Pearl River Delta: A comprehensive assessment and panel data analysis. Appl Energy 228:1693–1706. https://doi.org/10.1016/j.apenergy.2018.06.155

WHO (2021) World Health Organization. Heat and health. https://www.who.int/news-room/fact-sheets/detail/climate-change-heat-and-health

World Health Organization (2000) Environmental health indicators: development of a methodology for the WHO European Region: interim report, 6 November 2000 (No. EUR/00/5026344). WHO Regional Office for Europe, Copenhagen

World Health Organization (WHO) (2021) Climate change and health. In: Issues in environmental science and technology Vol. 2021-Janua(50). 10.1039/9781839160431-00353

World Health Organization (WHO)—Europe 2004 Heat waves: risks and responses

Zhu Q, Liu T, Lin H, Xiao J, Luo Y, Zeng W et al (2014) The spatial distribution of health vulnerability to heat waves in Guangdong province. China. Glob, Health Action

Chapter 10
Beyond the Rubble: Unveiling the Mental Health Crisis in the Wake of Climate Disasters and Building Psychosocial Resilience for Equitable Recovery

Daisy Gohain, Tiny Tanushree Gohain, and David Paul

Abstract As climate-related disasters increase in frequency and severity, such as floods, wildfires, heatwaves, and cyclones, their serious psychological effects require immediate attention from researchers and policymakers. This study looks closely at the impacts of mental health related to recent environmental disasters, including the wildfires in Maui in 2023 and the 2022 floods in Pakistan. Both these events caused widespread trauma, post-traumatic stress disorder (PTSD), and anxiety among affected people. In India, the 2023 Assam floods severely disrupted mental well-being in rural areas, especially among socially and economically disadvantaged communities. The current study highlights the psychological burden faced by vulnerable groups, particularly women, children, and indigenous populations, who have very limited access to mental health support. It investigates the significant gaps in how climate science, mental health resources, and disaster response work together. The study strongly supports adding community-focused, trauma-informed mental health practices to broader climate resilience plans. It views mental health as an essential part that demands recovery after disasters and promotes strategies that prioritize fairness and well-being. The chapter recommends adding preparedness of mental health to disaster risk plans, improving climate education, and creating recovery systems that support emotional resilience and physical healing.

Keywords Climate change · Mental health · Trauma · Stress · Recovery · Disasters · Resilience

D. Gohain (✉)
School of Social Sciences and Languages, Vellore Institute of Technology, Chennai, India
e-mail: daisy.gohain@vit.ac.in

T. T. Gohain
Department of Management, Brainware University, Barasat, West Bengal, India

D. Paul
Department of English, Kristu Jayanti (Deemed to be University), Bengaluru, Karnataka, India

P. Ghosh et al. (eds.), *Climate Crisis and Changing Population Dynamics*, Springer Climate, https://doi.org/10.1007/978-3-032-24533-5_10

Introduction

Climate change is increasingly recognized as a pressing challenge globally, not only for its environmental and economic repercussions but also for its deep psychological and social impacts (Clayton et al. 2017). Extreme weather events such as the sporadic rise of floods, wildfires, landslides, cyclones, and prolonged heatwaves have increased in frequency and severity over the past decade, affecting millions of people worldwide. In addition to the immediate physical damage and economic loss, these events can have long-lasting effects on mental health, such as post-traumatic stress disorder (PTSD), depression, anxiety, and other stress-related disorders (Palinkas and Wong 2020; Lowe et al. 2025).

The mental health impact of catastrophic events caused by climate change is not evenly spread out. Those who are vulnerable, such as women, children, the elderly, indigenous communities, and economically marginalized groups, are disproportionately impacted due to preexisting social inequalities, restricted access to health care, and inadequate coping mechanisms (Ansari 2024; American Psychological Association [APA] 2023). The 2023 Assam floods in India, for example, hit rural areas more severely than urban areas, especially those that relied on farming for their income. This demonstrates how environmental risk and social and economic vulnerability can come together (Cosh 2024). The wildfires in Maui, USA, in 2023 and the floods in Pakistan in 2022 also showed that this is a global problem, with survivors suffering from higher rates of trauma, anxiety, and depression (Walinski 2023).

Even though people are becoming more aware of how climate disasters affect mental health, mental health issues are often not given enough attention in disaster risk reduction (DRR) policies and climate adaptation strategies. Climate resilience initiatives concentrate on infrastructural readiness, early warning systems, and post-disaster assistance; however, the incorporation of mental health support is still limited and inconsistent (Clayton 2023). This inattention is particularly troubling because untreated trauma can have long-term effects on people's social, economic, and cognitive health, which can make it harder for communities to recover and adapt.

This research seeks to bridge this crucial gap by exploring recent climate-related disasters' mental health impacts, focusing on vulnerable populations. By scrutinizing case studies of varied geographical settings, such as the 2023 Maui wildfires, the 2022 Pakistan floods, and the 2023 Assam floods, the research explores the psychological effects of catastrophic events, recognizes systematic gaps in mental health response, and assesses the impact of community-based, trauma-informed interventions.

The study in this chapter has three main goals: first, to look into how disasters related to climate affect mental health in different cultural and social settings; second, to find the groups that are most likely to suffer from mental health problems; and third, to give practical advice on how to include mental health support in larger climate resilience plans. The research employs an interdisciplinary framework, incorporating trauma theory, ecopsychology, and intersectionality, to articulate mental health as an essential element of disaster planning and recovery.

By placing mental health at the center of climate adaptation, this study adds to a more integrated concept of resilience. It highlights the importance of incorporating psychological well-being into emergency planning, response, and recovery processes and hence into equitable and sustainable climate resilience initiatives. In the end, treating mental health alongside physical recovery and infrastructural development is key to building resilient and adaptive communities in the context of a changing climate.

This study's originality is its comparative, cross-regional examination of climate-related disasters and their effects on mental health in the Global North and South, highlighting underrepresented perspectives that are frequently disregarded in disaster discourse. The study intends to investigate psychological effects like PTSD and anxiety, identify systemic gaps in disaster response and mental health care, and promote community-based, trauma-informed mental health integration within frameworks for disaster preparedness and climate resilience by fusing climate science with trauma studies.

Literature Review

Climate Disasters and Psychological Impacts

Climate disasters are becoming better understood as prime drivers of psychological distress, and empirical evidence has shown a considerable link between high-magnitude environmental events and outcomes of mental health (Clayton 2023; Lowe et al. 2025). Flooding, bushfires, landslides, cyclones, and heatwaves not only destabilize livelihoods and physical infrastructure but also lead to acute and chronic psychological disorders such as post-traumatic stress disorder (PTSD), anxiety, depression, and eco-anxiety (Ansari 2024). Recent data from the 2023 Maui wildfires indicated that survivors had high rates of PTSD symptoms, with prominent emotional distress remaining months following displacement and loss of property (Walinski 2023). Likewise, research after the 2022 Pakistan floods indicated broad psychological morbidity, including anxiety, depression, and sleep issues, most especially in groups that had extended displacement and loss of livelihood (Cosh 2024).

These mental health issues are made worse by the occurrence, intensity, and unpredictability of climatic disasters. For instance, extended heatwaves not only strain physical health but also correlate with enhanced psychological stress levels, as demonstrated by study correlating extreme temperatures to increased hospitalizations for mental health crises (Palinkas and Wong 2020). These results imply that climate change is a public mental health emergency as well as an environmental one, calling for integrated approaches that address psychological and physical vulnerabilities.

Vulnerable Populations

The psychological effects of climate catastrophes are not equally spread; some groups bear a disproportionate amount of the burden. The young and adolescents are especially vulnerable because of their developmental vulnerability and restricted capacities for coping (APA 2023). Children in disaster-hit areas in Assam reported greater anxiety, depression, and post-traumatic stress symptomatology after the 2023 Assam floods, where age-specific interventions are most necessary (Ansari 2024).

Women are also exposed to increased psychological risk during disasters. Socially constructed gender roles, caregiving responsibilities, and lack of resource access enhance stress, anxiety, and trauma after environmental disasters (Clayton et al. 2017). Indigenous communities are subject to even greater risk factors, such as cultural disconnection, damage to traditional ways of life, and loss of ancestral territories, causing collective trauma and profound ecological sorrow. Economically vulnerable populations, particularly those who are reliant upon climate-sensitive industries like agriculture and fisheries, are also disproportionally impacted because they suffer the double burden of economic loss along with restricted availability of mental health interventions (Islam and Winkel 2017).

Socioeconomic Determinants and Disparities

Socioeconomic status, healthcare access, education, and social capital are prominent predictors of mental health outcomes in the aftermath of climate disasters. Research conducted in Pakistan and India indicates that those from low-income families tend to have increased levels of PTSD, depression, and anxiety after a disaster because they lack insurance, economic security, and social support (Lowe et al. 2025; Cosh 2024). Moreover, rural areas tend to have poor access to mental health facilities, further exacerbating the long-term psychological cost (Ansari 2024). The findings indicate the nexus of social inequality and environmental exposure, leading to the conclusion that disaster response systems need to tackle structural inequalities to adequately counteract mental health outcomes.

Climate Change, Eco-Anxiety, and Long-Term Impacts

Aside from direct post-disaster trauma, climate change causes anticipatory anxiety, or eco-anxiety, which is constant concern regarding ecological deterioration and impending climate occurrences (Clayton 2023). Eco-anxiety affects especially young generations, who are increasingly cognizant of long-term implications of climate change on their existence and enterprise. A 2024 study revealed that teenagers who were exposed to climate education programs reported increased eco-anxiety but also exhibited greater participation in pro-environmental behavior, which suggests that psychological distress can drive adaptive action if supported appropriately (Walinski 2023).

Long-term psychological effects encompass chronic depression, chronic stress, and social network disruptions. For example, displaced persons usually experience prolonged uncertainty over housing, work, and social reintegration, which sharpen trauma and decrease community resilience (Palinkas and Wong 2020). The evidence underscores the importance of incorporating mental health interventions not just in the acute phase of disaster but also in recovery and adaptation processes.

Integration of Mental Health into Disaster Planning

In the face of increasing evidence of climate disaster's mental health impact, inclusion in disaster risk reduction (DRR) planning is still marginal. Research has pointed to policy gaps, where most DRR plans focus on physical infrastructure, relief services, and early warning systems with little consideration for psychological preparedness (Clayton et al. 2017; Ansari 2024). Recent efforts in Australia, the US, and India have sought to integrate trauma-informed care into disaster response, such as psychological first aid, community-based counseling, and mental health training for responders. Early findings indicate these interventions enhance resilience, decrease PTSD symptoms, and increase overall community recovery (Lowe et al. 2025).

Literature Gaps

The current chapter on climate disasters and mental health, though growing, contains a number of key gaps that inhibit thorough understanding and policy response. To begin with, there is a significant lack of longitudinal studies that monitor psychological effects like trauma, depression, and post-traumatic stress disorder over longer periods. Most recent studies use cross-sectional data or short-term measurements, which are valuable snapshots but do not capture the long-term and cumulative mental health loads that disasters tend to put, especially in low-resource environments where recovery is lengthy and support networks are weak.

Second, marginalized groups are far too underrepresented in both policy discussions and scholarly research. Women, children, and indigenous peoples disproportionately experience the psychosocial impacts of climate crises owing to prior socioeconomic marginalization, caregiving roles, and dislocation from culture. Their everyday experiences, however, are often missing from mainstream scholarship. This under-representation not only constrains the research to be more inclusive but also threatens to reinforce existing inequalities by generating responses to disasters that fail to appropriately meet the needs of those who are most vulnerable.

Third, mental health measures are still inadequately integrated into disaster preparedness, response, and rehabilitation. Although physical infrastructure reconstruction and economic restoration are prioritized, psychosocial recovery is relegated to secondary status. The lack of systematic, trauma-informed practice in disaster management protocols represents an abiding gap between climate science, mental health practice, and policy translation. Closing this gap will depend on interdisciplinarity and acknowledgment of mental health as an integral element of resilience-building.

Collectively, these gaps highlight the urgency of developing research and policy frameworks that foreground marginalized voices, extend temporal scope, and institutionalize mental health within disaster governance. Addressing them is essential for building equitable, holistic responses to the escalating psychological toll of climate change. This study seeks to address these gaps by combining case study analysis, qualitative insights, and interdisciplinary theory to generate actionable recommendations for policy and practice.

Finally, by illustrating how climate-related disasters alter population resilience, vulnerability, and well-being, this study closely relates to the book's topic, "Climate Crisis and Changing Population Dynamics—Emerging Issues and Sustainable Solutions." By disproportionately harming women, children, indigenous groups, and low-income communities, climate stressors including floods, heatwaves, and relocation change demographic stability and exacerbate socioeconomic and health disparities.

The logicality and its relationship with the theme of the book can be succinctly summarized in the following points:

1. Climate catastrophes diminish population stability and production while harming mental health.
2. Demographic disparity rises as a result of the greater suffering experienced by women, children, indigenous people, and impoverished communities.
3. Population patterns are altered by climate stress, which encourages migration and displacement.
4. Youth eco-anxiety affects their future involvement in society and the environment.
5. The psychological effects of climate change are exacerbated by poverty and limited access to healthcare.
6. Communities that are emotionally resilient are necessary for sustainable solutions, not simply physical rehabilitation.
7. Long-term population resilience is strengthened when mental health is incorporated into catastrophe strategy.

Methodology

This study uses an interdisciplinary, qualitative research framework that combines comparative case study analysis with a thorough evaluation of the literature. In order to identify prevailing patterns and theoretical gaps, peer-reviewed research on climate change, disaster psychology, trauma, stress, resilience, and recovery was critically analyzed. To capture lived psychological and socio-ecological repercussions, factual case studies of climate-related disasters were analyzed. An integrated understanding of how environmental crises combine with structural inequalities to produce distinct mental health vulnerabilities and adaptive capabilities is made possible by the analytical framework, which is informed by trauma theory, ecopsychology, and intersectionality theory.

Theoretical Framework

Understanding the mental health consequences of climate-related disasters necessitates an interdisciplinary theoretical lens that takes into account the complexities of human–environment interactions, socio-cultural vulnerabilities, and trauma experiences. This study primarily uses trauma theory, ecopsychology, and intersectionality theory to provide a coherent framework for studying psychological outcomes, resilience, and policy implications.

Trauma Theory

Trauma theory provides the underlying framework to explain the psychological impact of disastrous incidents. According to it, exposure to severe stressors like floods, fires, or cyclones can impair emotional regulation, cognitive processing, and social functioning. In the case of climate catastrophes, trauma is usually added to displacement, loss of income, and protracted uncertainty, giving rise to PTSD, anxiety, and depression (Lowe et al. 2025). With the use of trauma theory, this research frames the mental health effects of disasters as dimensional, including both acute and residual psychological trauma, and identifies that recovery requires trauma-informed responses.

Ecopsychology

Ecopsychology stresses the interrelatedness of human well-being and ecological systems. In contrast to mainstream psychology, which tends to unlink individual experiences, ecopsychology locates mental health within environmental and social contexts, underscoring the psychological impact of environmental degradation and loss (Clayton 2023). The theory presents the term solastalgia—a type of distress generated by environmental change in one's home environment—which is highly applicable to communities undergoing climate-induced displacement. Ecopsychology therefore offers a framework for comprehension of how climate catastrophes can bring about individual and collective bereavement and impact community cohesion, cultural identity, and adaptive capacities.

Intersectionality Theory

Intersectionality theory provides a framework for exploring how intersecting social identities—gender, age, caste, ethnicity, and socioeconomic status—impact exposure to climate disasters and mental health. For instance, low-income rural women might experience increased psychological distress owing to caregiving roles, restricted mental health services, and financial vulnerability (Clayton et al. 2017; Ansari 2024). In a similar vein, indigenous groups face compounded stress due to environmental deterioration, displacement of ancestral lands, and institutional marginalization. Intersectionality provides room for an understanding of differential vulnerability that ensures that mental health intervention is not just trauma-informed but socially just.

Integration of Theories

Integrating trauma theory, ecopsychology, and intersectionality allows for a comprehensive understanding of mental health concerns related to climate. Trauma theory helps identify psychological symptoms and distress pathways. Ecopsychology situates these symptoms within environmental and social systems, highlighting the importance of ecological loss and community upheaval. Intersectionality ensures that analyses and responses account for structural disparities, showing which groups are most vulnerable and why. Together, these ideas create a solid conceptual framework for examining case studies, creating interventions, and making policy recommendations.

Application to the Present Study

This theoretical foundation drives the research methodology for analyzing the 2023 Maui wildfires, the 2022 Pakistan floods, and the 2023 Assam floods. It guides the choice of research variables such as psychological symptoms, coping mechanisms, social support systems, and systemic barriers to mental health services. In integrating these viewpoints, the research focuses on community-based, trauma-aware, and equity-driven methods to call for policies that respond to both the environmental and psychosocial aspects of climate catastrophes.

Figure 10.1 highlights the theoretical framework—trauma theory, ecopsychology, and intersectionality theory to understand environmental trauma (Fig. 10.2).

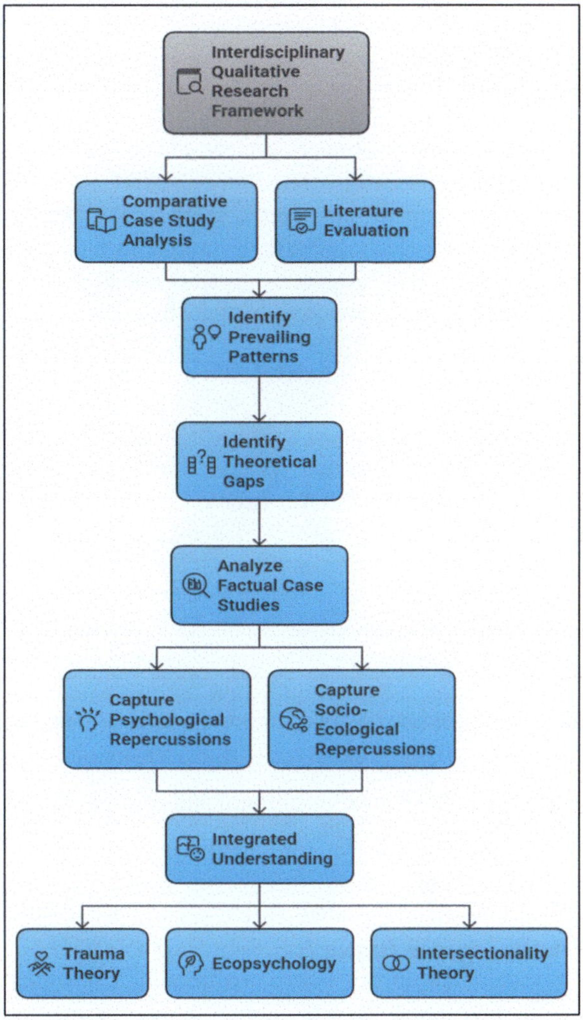

Fig. 10.1 Flow of methods employed for analyzing this study

Case Studies: Climate Disasters and Mental Health—The 2023 Maui Wildfires (Hawai'i, USA)

Background and Scale of the Disaster

In August 2023, wildfires swept across Maui, Hawai'i, devastating the historic town of Lahaina and claiming over 100 lives, with thousands displaced (US Fire Administration 2024). The fires, fueled by hurricane-driven winds and exacerbated by prolonged drought conditions linked to climate change, represented one of the deadliest wildfires in modern U.S. history. Beyond the physical destruction, the event

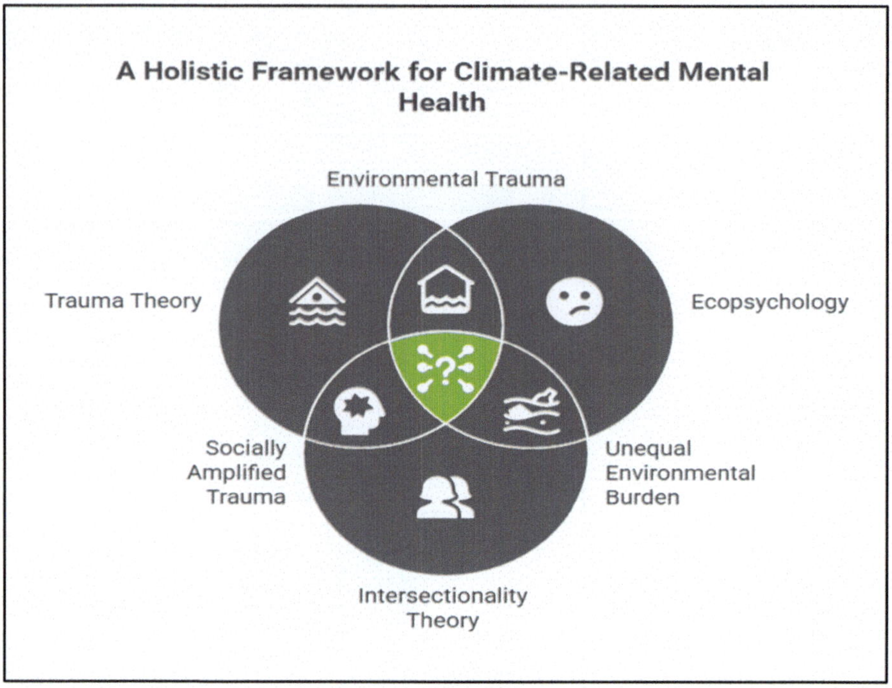

Fig. 10.2 Theoretical framework to understand environmental trauma

deeply unsettled Hawai'i's cultural and spiritual fabric, as ancestral lands and sacred sites were reduced to ashes.

Psychological Impacts

Survivors showed acute stress reactions, sleeplessness, survivor's guilt, and symptoms of post-traumatic stress disorder (PTSD) (Miller et al. 2024). Children were especially susceptible, and school disruptions and destruction of safe havens heightened anxiety. Indigenous Hawaiian populations, which were already dealing with histories of colonial displacement, suffered a cumulative sense of grief and identity loss, echoing the phenomenon of solastalgia.

Interventions and Gaps

The U.S. federal and state governments deployed swift emergency interventions, such as temporary housing and counselling services. Yet, numerous survivors identified stigma, cost, and cultural incompatibility of Western psychological models as obstacles to receiving ongoing mental health care (Kawano and Harada 2024). Indigenous healing rituals, like ho'oponopono (a shared reconciliation ceremony), were critical sources of emotional strength but under integrated into official disaster recovery processes.

Policy Implications

The Maui wildfires highlight the importance of culturally grounded, trauma-informed catastrophe responses. Policy frameworks should formalize collaborations between healthcare practitioners and indigenous healers, secure funding for long-term mental health recovery, and strengthen climate adaption strategies to prevent future crises from intensifying.

The 2022 Pakistan Floods

Background and Scale of the Disaster

Between June and October 2022, record monsoon rainfall caused floods that inundated almost a third of Pakistan's landmass. More than 33 million people were impacted, with eight million having been displaced and estimated economic damages over USD 30 billion (UNICEF 2023). Entire rural communities lost crops, livestock, and houses, resulting in widespread food insecurity and long-term disruption to livelihoods.

Psychological Impacts

The floods ushered in a mental health emergency, with documented cases of PTSD, depression, and increased risk of suicide among survivors (Shafiq et al. 2023). Women, especially in rural conservative settings, bore unequal burdens—taking on caregiving responsibilities in congested relief camps while suffering gender-based violence and limited access to health care. Children presented with symptoms of trauma such as nightmares, regression in behavior, and decreased cognitive function resulting from extended school closures (Khalid et al. 2023).

Interventions and Gaps

Although humanitarian agencies offered psychosocial interventions, Pakistan's weak mental health infrastructure of fewer than 500 psychiatrists for a population of 220 million was found to be insufficient (WHO 2023). Community-led responses, including peer support groups and religious coping strategies, brought transient relief but were not formally coordinated. Lack of mental health literacy also facilitated the perpetuation of stigma, which deterred people from using professional services.

Policy Implications

The Pakistan floods have highlighted the importance of incorporating mental health into national disaster management policy. Scalable, low-cost treatments, such as educating frontline health workers in psychological first aid, could help close key service gaps. Furthermore, investments in mental health literacy initiatives and gender-sensitive recovery practices are critical to long-term resilience.

The 2023 Assam Floods (India)—Background and Scale of the Disaster

Heavy incessant rainfall worsened the floods in western Assam, impacting 1.2 lakh people across 20 districts. About 780 villages were submerged and many acres of agriculture cropland was damaged. Barama (Baksa) and Nalbari district were the worst hits. Four relief camps, 14 shelters, and 17 centres was operated, while infrastructure across several districts remained severely affected.

Policy Implications

The Assam floods highlight the necessity of integrating mental health services into India's disaster risk reduction strategy. It involves reinforcing community health systems, equipping Accredited Social Health Activists (ASHAs) with basic psychosocial care skills, and developing localized, culturally appropriate frameworks for the recovery from trauma. Climate adaptation policy needs to include the psychosocial aspects of recurring disasters, especially among socially and economically marginal populations (Table 10.1).

Table 10.1 The consolidated scale of disaster at Maui, Pakistan, and Assam

Disaster and year	Scale of disaster	Key psychological impacts	Interventions and gaps	Policy implications
Maui wildfires (2023, USA)	100+ deaths, thousands displaced; Lahaina devastated	PTSD, survivor's guilt, insomnia, solastalgia (esp. indigenous communities)	Shelters, counseling; cultural misalignment with Western models; traditional healing underused	Integrate indigenous healing, fund long-term recovery, strengthen adaptation
Pakistan floods (2022)	33 M affected, 8 M displaced; USD 30B loss	PTSD, depression, suicide risk; women & children most vulnerable	Limited psychiatrists (<500 nationwide); peer/religious coping; stigma	Train health workers in psychological first aid; gender-sensitive recovery; mental health literacy
Assam floods (2023, India)	4.2 M displaced; recurring seasonal floods	Anxiety in women, trauma in children, cultural loss in indigenous groups	Relief camps with little/no mental health support; small NGO art therapy efforts	Embed mental health in disaster risk reduction; train ASHAs; localized trauma frameworks

Discussion

The Maui wildfires (2023), Pakistan floods (2022), and Assam floods (2023) case studies all contribute to shedding light on how climate disasters are not just ecological and economic crises but also deeply psychological events. Each disaster, contextually distinct as it is, unfolds to show how the climate crisis breaches the intimate domain of human existence, transforming memory, identity, and collective welfare. Based on ecocritical theory, trauma studies, and climate psychology, this discussion places the results within wider debates concerning climate-related mental health and resilience.

Intersecting Vulnerabilities and Unequal Burdens

In all three instances, vulnerable groups most notably women, children, and indigenous people were found to be disproportionately impacted. For Pakistan, women performed caregiving responsibilities in congested camps under conditions of limited access to mental health care (Khalid et al. 2023). For Assam, women and children suffered most from repeated displacements, reminiscent of the extant literature on the feminization of environmental exposure (Rao et al. 2024). Indigenous Hawaiians, in turn, experienced double trauma: the direct loss of homes and loved ones, and the loss of ancestral lands that form the foundation of cultural identity (Kawano and Harada 2024). These observations reiterate that climate-related mental health issues cannot be separated from structural class, caste, gender, and colonial history inequalities.

Cultural Coping and the Limits of Biomedical Models

The case studies also uncover the resilience of culturally embedded coping mechanisms. Ho′oponopono, the traditional Hawaiian practice, Islamic faith-based resilience in Pakistan, and community art therapy in Assam all offered survivors meaning frameworks, belonging, and recovery. These practices are nevertheless frequently neglected or marginalized in Western biomedical models of trauma. Theoretical ecopsychology and transcultural psychiatry interventions serve as a reminder that recovery needs to be understood in terms of a dialogue between cultural memory and mental health care, and not as the imposition of templated therapeutic practices (Watts et al. 2022).

Trauma, Solastalgia, and Ecological Loss

Another potent trend is the occurrence of solastalgia, which refers to the distress of environmental degradation in one's home environment (Albrecht 2020). In both Maui and Assam, survivors mourned not just physical loss of houses but also the vanishing of holy spaces, ancestral points, and farming livelihoods. Trauma here becomes psychological and ecological as well, dissolving the lines of human suffering and environmental collapse. This challenges the conventional psychiatric models that confine trauma to a singular pathology; on the contrary, climate-related trauma has to be understood as collective, intergenerational, and site-specific.

Policy Gaps and Pathways Forward

Notwithstanding some efforts, persistent gaps exist in mental health infrastructures at the system level. Pakistan's scarce psychiatric human resources indicate how mental health is continued to be neglected in disaster management systems (WHO 2023). Assam's relief camps, while responding to immediate survival requirements, disregarded psychosocial well-being altogether, illustrating a pervasive silencing of mental health in South Asian disaster governance (Baruah and Das 2024). Even in the United States, where resources are relatively higher, survivors of the Maui wildfires faced obstacles to ongoing and culturally appropriate care (Miller et al. 2024).

These lacunae highlight the need for policy convergence—where climate adaptation, disaster response, and mental health care are not siloed domains. Immediate actions include training community health workers in psychological first aid, making investments in mental health literacy, and institutionalizing culturally nuanced care. At a structural level, governments need to integrate psychosocial resilience into

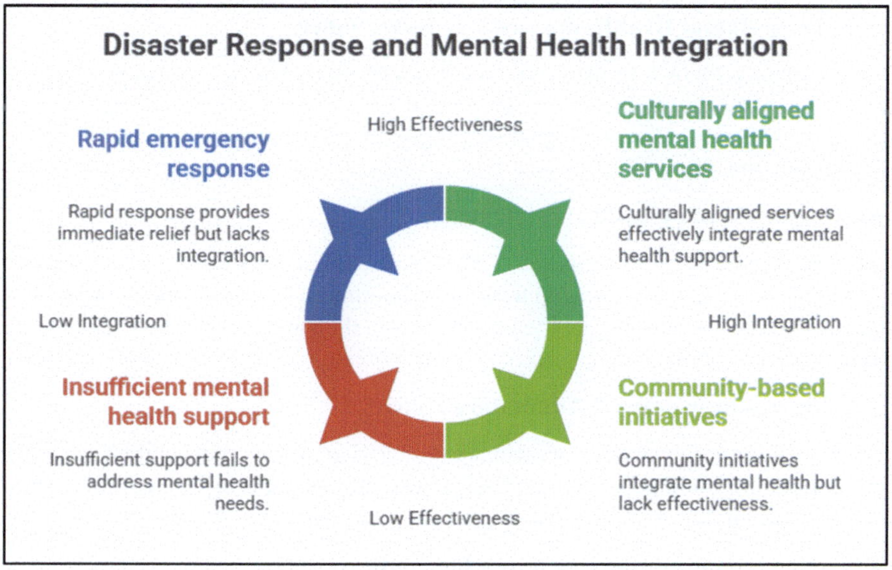

Fig. 10.3 Disaster response and mental health integration relationship

climate action policies with a recognition of mental health as both a social justice and climate adaptation priority (Fig. 10.3).

Toward a Humanized Climate Resilience Framework

The evidence points toward a need for climate-associated trauma to be addressed by a humanized, multidisciplinary framework that encompasses environmental humanities, psychology, and policy studies. Theoretically, this corresponds to an ecocritical-psychosocial model where the environment is not the background, but a co-actor in determining mental health trajectories. Practically, it demands intersectional climate policy that acknowledges the multiplicative vulnerabilities of various groups, while maintaining cultural modes of healing.

In that respect, these three case studies call for a paradigm shift: from considering mental health in disasters as an afterthought, to viewing it as the key to climate resilience. Such a shift is essential, not only for healing current communities, but also for equipping societies for the increasing psychological costs of a heating world.

Policy Recommendations

Policy-wise, some imperative actions come into view:

Integrating Mental Health into Disaster Governance

Disaster management systems need to incorporate mental health care at all stages—prevention, response, and recovery. This entails cross-sectoral coordination between ministries of health, climate adaptation programs, and community-based organizations (Islam and Winkel 2022).

Investing in Mental Health Infrastructure

Low- and middle-income countries (LMICs) are still under-resourced significantly in psychiatric services. Scaling up training of community health workers in psychological first aid and building resilience can bridge the gaps (WHO 2023; Patel et al. 2018).

Fostering Equity-Centered Adaptation Policies

Climate-related trauma is disproportionately shared among women, children, indigenous populations, and poor communities. Intersectional approaches in policies are a must to maintain fair recovery and special support (Rao et al. 2024).

Supporting Community-Led Coping Mechanisms

Governments and NGOs should amplify culturally embedded practices—such as hoʻoponopono in Hawaiʻi, faith-based resilience in Pakistan, or storytelling traditions in Assam—as integral forms of healing. This ensures interventions are not only effective but also culturally legitimate (Watts et al. 2022).

Embedding Mental Health in Climate Action Frameworks

Global climate policy platforms like the UNFCCC should officially recognize psychosocial well-being as a cornerstone of climate adaptation. Integrating climate finance with mental health programming will enhance resilience both locally and globally (Clayton and Karazsia 2020).

Moving the Science Forward on Climate-linked Trauma

Although research on ecological grief and solastalgia is growing, longitudinal, cross-cultural studies are still needed that follow the intergenerational impact of climate-related trauma. Interdisciplinary academic partnerships among environmental humanities, psychiatry, and policy studies are necessary to enrich this question (Berry et al. 2018; Doherty and Clayton 2011).

Conclusion

Mental anguish is linked to land loss, interrupted livelihoods, and impaired social cohesiveness in areas of South Asia hit by flooding and Indigenous lands in Hawaii devastated by fire. According to Cianconi et al. (2020), these patterns support the idea that climate-related discomfort is not only a clinical problem but also a structurally generated and collectively experienced one. Therefore, building resilience requires more than just infrastructure; it also requires trauma-informed recovery, community-based mental health assistance, and culturally grounded healing practices. In order to ensure social fairness and a sustainable recovery, it is crucial to incorporate psychological care into climate adaption measures (Clayton 2023).

This work theoretically validates the need for an ecocritical-psychosocial approach that emphasizes the interdependence of natural systems and human happiness. Biomedical therapies are still vital, but they should be supplemented with culturally sensitive, community-based approaches that recognize the deep connection between people and their surroundings. Global mental health initiatives should include collective memory work, indigenous healing techniques, and culturally adaptable therapies (Fernando 2022; Kawano and Harada 2024).

To reduce climate change without addressing its psychological impact risks creating adaptation systems that are physically resilient but emotionally fragile. By incorporating mental health into disaster planning and climate policy, countries may improve not just infrastructure resilience, but also psychological, cultural, and humane resilience.

Acknowledgements We, the authors, gratefully acknowledge the support, guidance, and encouragement of our mentors, colleagues, and family in the completion of this chapter.

Authors Contribution All authors have contributed equally to this work.

AI Declaration The author(s) acknowledge using ChatGPT, a generative artificial intelligence tool, throughout the creation of this chapter for language editing and content structure. The author(s) accept full accountability for the final content and affirm that no AI technology was utilized to create original study data or results.

References

Albrecht G (2020) Earth emotions: new words for a new world. Cornell University Press

American Psychological Association (2023) Mental health and our changing climate: children and youth. https://www.apa.org/pubs/reports/climate-change-mental-health-children-2023

Ansari E (2024) Mental health amid climate crisis: a narrative review. *PMC*. https://pmc.ncbi.nlm.nih.gov/articles/PMC11708969/

Baruah P, Das M (2024) Recurrent floods and rural mental health in Assam: an ethnographic perspective. Indian J Soc Psychiatry 40(1):45–54. https://doi.org/10.4103/ijsp.ijsp_83_23

Berry HL, Waite TD, Dear KBG, Capon AG, Murray V (2018) The case for systems thinking about climate change and mental health. Nat Clim Chang 8(4):282–290. https://doi.org/10.1038/s41558-018-0102-4

Cianconi P, Betrò S, Janiri L (2020) The impact of climate change on mental health: a systematic descriptive review. Front Psych 11:74. https://doi.org/10.3389/fpsyt.2020.00074

Clayton S (2023) The effects of climate change on mental health. Annu Rev Clin Psychol. https://doi.org/10.1146/annurev-clinpsy-081423-025932

Clayton S, Karazsia BT (2020) Development and validation of a measure of climate change anxiety. J Environ Psychol 69:101434. https://doi.org/10.1016/j.jenvp.2020.101434

Clayton S, Manning CM, Krygsman K, Speiser M (2017) Mental health and our changing climate: impacts, implications, and guidance. American Psychological Association

Cosh SM (2024) The relationship between climate change and mental health. PMC https://pmc.ncbi.nlm.nih.gov/articles/PMC11577747/

Doherty TJ, Clayton S (2011) The psychological impacts of global climate change. Am Psychol 66(4):265–276. https://doi.org/10.1037/a0023141

Fernando GA (2022) Cultural dimensions of climate trauma: expanding therapeutic interventions. Transcult Psychiatry 59(5):693–708. https://doi.org/10.1177/13634615221075922

Islam R, Winkel J (2022) Integrating mental health into disaster risk reduction: evidence from South Asia. Int J Disaster Risk Sci 13(2):213–225. https://doi.org/10.1007/s13753-022-00405-8

Kawano H, Harada Y (2024) Indigenous healing and post-disaster resilience in Hawai'i: a cultural psychology perspective. Transcult Psychiatry 61(2):189–207. https://doi.org/10.1177/13634615231234567

Khalid A, Shafiq F, Malik R (2023) Mental health consequences of the 2022 floods in Pakistan: a scoping review. Int J Disaster Risk Reduct 91:103639. https://doi.org/10.1016/j.ijdrr.2023.103639

Lowe SR et al (2025) Climate disasters and mental health: directions for future research. Lancet Public Health. https://doi.org/10.1016/S2468-2667(25)00097-0

Miller T, Ka'ai H, Johnson D (2024) Psychological impacts of the Maui wildfires: early evidence from affected communities. J Trauma Stress 37(1):11–25. https://doi.org/10.1002/jts.23044

Palinkas LA, Wong M (2020) Global climate change and mental health. Curr Opin Psychol 32:12–16. https://doi.org/10.1016/j.copsyc.2019.06.023

Patel V, Saxena S, Lund C, Thornicroft G, Baingana F, Bolton P et al (2018) The lancet commission on global mental health and sustainable development. Lancet 392(10157):1553–1598. https://doi.org/10.1016/S0140-6736(18)31612-X

Rao N, Lawson ET, Bhadra C (2024) Gendered dimensions of climate vulnerability: insights from South Asia. World Dev 176:106367. https://doi.org/10.1016/j.worlddev.2024.106367

Shafiq F, Khalid A, Malik R (2023) Post-disaster trauma and resilience among flood survivors in Pakistan. Asian J Psychiatr 87:103754. https://doi.org/10.1016/j.ajp.2023.103754

UNICEF (2023) Pakistan floods 2022: humanitarian situation report. United Nations International Children's Emergency Fund

US Fire Administration (2024) After-action report: 2023 Maui wildfires. US Fire Administration

Walinski A (2023) The effects of climate change on mental health. PMC. https://pmc.ncbi.nlm.nih.gov/articles/PMC10154789/

Watts A, Wilson S, Singer J (2022) Transcultural psychiatry and ecological grief: expanding frameworks of climate-linked trauma. Cult Med Psychiatry 46(3):412–431. https://doi.org/10.1007/s11013-021-09757-3

WHO (2023) Mental health country profile: Pakistan. World Health Organization

Part III
Climatic Extremes, Migration Patterns, Policy Implications, and Future Directions

Chapter 11
Climatic Disaster-Driven Economic Losses and Migration Dynamics: Evidence from Indian Households

Jadab Munda, Amiya Saha, and Bittu Mandal

Abstract A climatic hazard is any natural or human-caused situation, process, or event in the environment that has the potential to endanger human life, health, property, livelihoods, or ecosystems. These hazards arise from interactions between natural systems and human activities and can cause disasters when communities are exposed and vulnerable. There have been several previous studies whose findings reveal that, due to the environmental hazard and climatic disaster, people are eager to practice temporary and permanent migration, rural-to-urban migration, and movement from hazard-affected areas to non-hazard areas. According to the Centre for Research on the Epidemiology of Disasters, globally 2,03,89,962 people are affected by environmental hazards. For instance, from 2000 to 2023, more than 110 million people in India were severely affected over the last two decades. The study data were based on the India Human Development Survey-II (IHDS-II), conducted in 2011–12. This study examines climate disaster-induced economic loss and migration using binary logistic regression. The results indicate significant financial losses and out-migration in states such as Madhya Pradesh (21.85%), Bihar (14.42%), Uttar Pradesh (10.50%), and Rajasthan (8.27%). Marginalized groups, including scheduled castes (24%) and tribes (18%), experienced the greatest migration burden. Households that experienced economic loss were more likely to migrate, influenced by factors such as income source (41.25%) and household size (47.30%). The findings underscore disaster-related economic loss as a primary driver of involuntary migration in India.

J. Munda (✉)
Department of Migration and Urban Studies, International Institute for Population Sciences, Mumbai, India
e-mail: jadabmunda1996@gmail.com

Department of Geography, Magrahat College, University of Calcutta, Kolkata, India

A. Saha
Department of Public Health and Mortality Studies, International Institute for Population Sciences (IIPS), Govandi Station Road, Deonar, Mumbai, Maharashtra, India
e-mail: amiyasaha4444@gmail.com

B. Mandal
School of Humanities and Social Sciences, Indian Institute of Technology Indore, Indore, India
e-mail: mbittu545@gmail.com

P. Ghosh et al. (eds.), *Climate Crisis and Changing Population Dynamics*, Springer Climate, https://doi.org/10.1007/978-3-032-24533-5_11

Focusing on our study findings, the government should strengthen the integration of disaster risk reduction, climate adaptation, and the social protection framework.

Keywords Migration · Natural hazard · Climate · Economic impact · Disaster · India

Introduction

Climatic hazards and disasters are natural phenomena that disrupt the economic system, significantly impacting assets, production factors, and employment (Hallegatte and Przyluski 2010; Pörtner et al. 2022). According to the Centre for Research on the Epidemiology of Disasters, at the worldwide level, 2,03,89,962 people are affected by climatic disasters (EMDAT 2021). However, developing and least-developed countries around the world are more susceptible to these natural events. For instance, from 2000 to 2023, more than 110 million people in India were severely affected over the last two decades. Among them, 1.36 million people were rendered homeless, 1.98 million were injured, and 88,060 lost their lives (Datar et al. 2013; Tulloch et al. 2016; van Aalst 2006). On the other hand, according to the Internal Development Monitoring Centre (IDMC) reports in 2023, 2,507,000 people were internally displaced due to climatic disasters in India (IDMC 2023). Climatic disasters have a substantial impact on individuals, communities, and even entire country, driving migration patterns such as rural-to-urban migration and movement from hazard-affected areas to non-hazard areas (Ishtiaque and Nazem 2017; Maharjan et al. 2020a, b), and most of the time, these incidents are commonly related to the developing and least-developed countries (Marchiori et al. 2012a, b; Pradhan and Narayanan 2022a, b). The association between environmental hazards and migration has been studied globally, leading to some interesting findings; research on climate disaster migration reveals several key themes. Studies show that asset losses due to climatic disasters often lead to permanent migration, while livestock and crop failures typically result in temporary migration (Afifi et al. 2014a, b; Bardsley and Hugo 2010a, b; Gray and Mueller 2012a, b). Natural disasters, such as floods and crop failures, are strongly linked to intra-state and inter-state migration (Dallmann and Millock 2017; Maharjan et al. 2020a, b; Singh 2019; Singh and Basu 2020).

There has been some important work done by the various scholars around the worlds, and these studies collectively highlight that climatic disasters significantly influence migration decisions, with droughts, floods, and other disasters prompting either temporary or permanent relocation. However, a study was done by Kartiki on Climate Change and Migration: A Case Study from Rural Bangladesh. The study findings show deterministic narratives that present climate change as a direct and uniform cause of migration, rather than emphasizing migration as a contextual, socially rooted, and economically mediated event (Kartiki and Kartiki 2011). The authors, like Marchiori, Maystadt, and Schumacher, investigated how the weather anomalies influence migration patterns in sub-Saharan Africa using a macro-empirical approach. The study findings show that positive temperature anomalies, particularly rises in average temperatures, have a substantial impact on migration rates, particularly from rural and agricultural areas (Marchiori et al. 2012b). The VanWey examines the role of land ownership in shaping the temporary and circular migration pattern in Nang Rong district, a rural agrarian region of Thailand. The findings of the study show land ownership has a nonlinear and indirect effect on migration. The individuals from landless households are more likely to be engaged in temporary migration due to the limited employment opportunities in their native areas (Vanwey 2003a). On the other hand, Xiong et al. examine the "determinants of migration decision-making for rural households: a case study in Chongqing, China." The study findings suggest that economic factors like income levels, employment opportunities, and expected wage differentials are the most significant drivers of migration. However, the study also highlighted that climatic disasters indirectly affect migration by undermine agriculture productivity and livelihood security in China (Xiong et al. 2020a). At national levels, various research scholars have done some significant work on climatic disasters like droughts, floods, and cyclones that threaten livelihoods by increasing poverty, limiting access to resources, reducing the ability to adapt, and decreasing agricultural productivity in the affected areas (Jennath and Paul 2022, 2024; Rahman et al. 2024a, b; Singh and Basu 2020). Due to this, the affected population seeks alternative adaptation strategies for individuals and households, which forces people to migrate in search of diversified income sources, better living standards, social stability, and alternative employment opportunities (Cattaneo et al. 2019; Dallmann and Millock 2017; Deshingkar 2012; Maharjan et al. 2020a, b; Rahman et al. 2024a, b). This migration occurred in the form of short or long term, rural to urban, inter-districts, and inter-states in India (Sarkar et al. 2022; Tiwari et al. 2015; Tripathi 2017). While existing research offers valuable insights into the impact of climatic disasters on migration patterns globally, there is a gap in the literature regarding the specific phenomenon of natural disasters in the notion of economic loss-induced migration in the context of India. Despite the substantial economic repercussions of climatic disasters in the country, including asset loss, damage to production factors, and loss of employment opportunities, there is a scarcity of studies that systematically analyze the relationship between these economic losses and subsequent migrant's movements within India. This gap in the literature calls for urgent attention, as understanding the drivers and patterns of economic loss-induced migration is crucial for informing effective policy interventions and disaster management strategies. Therefore, the objective of this study is to address this gap by examining the nexus between disaster-induced economic losses and migration in India, providing insights that can contribute to more targeted and resilient disaster response and recovery efforts.

Materials and Methods

Data

The study data were based on the India Human Development Survey-II (IHDS-II), 2011–12, a nationally representative survey of households conducted jointly by the National Council of Applied Economic Research (NCAER), New Delhi, and the University of Maryland. The survey covers India's states and union territories except for Andaman & Nicobar and Lakshadweep. The survey is conducted among 42,152 sample households, spread across 33 states and union territories, 384 districts, 1420 villages, and 1042 urban blocks in 276 towns and cities. Both the rural and urban samples are collected through stratified random sampling. These data are re-interviews of 83% of households interviewed for IHDS-I in 2004–05, and an additional 2134 households are included in IHDS-II (Desai and Vanneman 2018).

Study Sample

This study includes households that migrated due to large expenditure/loss in environmental hazards (drought, floods, fire, and crop failure) in the last six to seven years. The total sample in this study is 42,152 households. Among the total sample, 757 households were regarded as misreporting migration status, 41,395 households with any migration status, 67 households with miss reporting in expenditure/loss due to the environmental hazard from the total sample, and 42,085 households with expenditure/loss due to environmental hazard. The households with miss reporting in any covariate is 194 households, and for the final study, the sample includes 41,0134 households with complete information on migration status and amount of expenditure/loss due to environmental hazard. The complete sample selection criteria for this study are shown in Fig. 11.1.

Variables Description

Outcome Variable

The outcome variable in this study is migration status, which was based on the question: "Have you or any member of your household left to find seasonal/short-term work during the last five years and returned to live here?" The response was categorized as yes and no (Mandal et al. 2023).

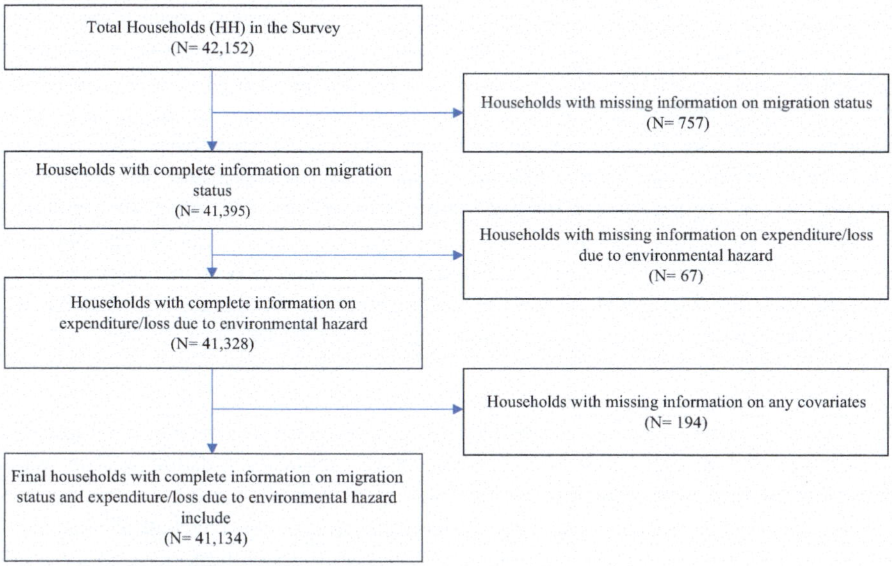

Fig. 11.1 Selection of the study sample. *Source* Authors

Key Explanatory Variable

The key explanatory variables in this study are formed from one main question: "Is there any incidence of large amount of expenditure/loss in your household during last six or seven years since we visited you?" Based on these questions, two variables were included for the study: i. Drought, Flood, Fire, etc. (No = 0, Yes = 1) and ii. Crop failure (No = 0, Yes = 1). From these above variables, we constructed a new key explanatory variable name "climatic disasters," which had responses that households with no amount of expenditure/loss due to the climatic disasters were coded as "0" and households with an amount of expenditure/loss due to the climatic disasters were coded as "1." Many previous studies also used drought, floods, fire, and crop failure to measure the explanation of climatic disasters, which is similar to our study (Afifi et al. 2014a, b; Bardsley and Hugo 2010a, b; Gray and Mueller 2012a, b; Kartiki and Kartiki 2011; Pradhan and Narayanan 2022a, b).

Other Covariates

Based on the previous extensive literature reviews, this study also includes different types of independent covariates. These independent covariates were categorized into the three following sections.

Demographics Status

The demographic status was categorized into six sub-groups such as households' size was categorized as "≤ 5 person lives in a house and > 5 persons live in a house." Highest households' adults' education was categorized as no schooling, primary, secondary, and secondary and above (Hoang et al. 2023; Saha et al. 2022). The caste category was categorized as scheduled caste "SC," scheduled tribes "ST," others below caste "OBC," and others (Munda et al. 2024; Samanta and Munda 2023). Religion was categorized as Hindu, Muslim, and Others. The place of residence was categorized as rural and urban (Saha, Muhammad et al. 2023a, b). The place of residence was measured by "Urban residence from census 2011"; the response was categorized as rural or urban (Saha et al. 2024). Regions were categorized as North, Central, East, West, Northeast, and South (Samanta et al. 2023).

Economic Status

The economic status was categorized into three sub-groups: Main sources of income was categorized as cultivation, wage labor, small business, and others. The main sources of income were measured by "What is the principal source of income for the household?"; the response was categorized as cultivation, allied agriculture, agricultural wage labor, non-agricultural wage labor, artisan/independent, petty shop/small business, organized trade/business, salaried employment, profession NEC, pension/rent/dividend, and others (specify). Land holding was categorized as yes and no (Chandrasekhar and Sahoo 2019). The land holding was measured by "Does your household own or cultivate any agricultural land?"; the response was categorized as yes and no. Wealth was categorized as poor and non-poor. The wealth was measured by expenditure/capita < poverty line (Panagariya and Mukim 2014).

Governmental Schemes Status

The status of civil rights and governmental schemes was categorized into three sub-groups: such as ration cards, MGNREGA job card, and income from government sources. The ration card was measured by "Does the household have a ration card?"; the response was categorized as yes or no. MGNREGA job card was categorized as yes or no. The MGNREGA job card was measured by "How many MGNREGA (NREGA) job cards does your household have?"; the response was categorized as "0 = no" and "1–5 = yes" (Ahuja et al. 2011; Nandy et al. 2021). Income from government sources was categorized as yes or no. The income from government sources was measured by "Income from any govt. sources, such as Draught/Flood compensation, Insurance payout or Other schemes?" the response was categorized as "0 rupees = no" and "70–180,000 rupees = yes."

Statistical Analysis

The statistical analysis was applied to explain the study results, such as the sociodemographics characteristics of the study population (migrants and non-migrants), which were examined by descriptive statistics. The spatial analysis was examined through ArcMap 10.7. The significance level of expenditure/loss due to the natural disaster across sociodemographic and economic status of migrants and non-migrant households was examined through chi-square test. Finally, the determinants of migration due to the expenditure/loss during environmental hazards are examined by binary logistic regression. The equation of the logistic regression is as follows:

The regression coefficient in this example was β1, β2.... $\beta_{k,}$ and they showed the relative effect of explanatory variables and sociodemographic factors on the dependent variable, with the coefficients changing depending on the context of the studies. The unadjusted and adjusted regression models were developed to determine the association between explanatory, covariates, and outcome variables. The result was presented in the form of odds ratio (OR) and with a 95% confidence interval (CI) (Mandal and Pradhan 2024; Mandal et al. 2023a, b).

$$\text{Logit}(y) = \ln(p/(1-p)) = \alpha + \beta_1 X_1 + \beta_2 X_2 + \beta_3 X_3 ... \beta_k X_k \tag{11.1}$$

All the statistical analyses were computed using StataSE-18 (StataCorp 2023).

Results and Discussion

To contribute to the understanding of the underlying mechanisms by which severe climatic events can impact human migration, this study investigated the relationship between economic losses caused by climatic disasters and migration in search of livelihood. Within the context of developing countries, a disaster-induced economic loss impacts a household's capital by disrupting the functioning of the local economy and destroying productive assets and employment (Chen and Lee 2022; Loebach 2016). The subsequent migration in search of a livelihood is the only feasible ex-post household adaptation response (Black et al. 2013). This process is complicated, however, as disaster-induced economic loss has the potential to catalyze the motivation to migrate (McLeman and Hunter 2010). The destruction of household assets and the disruption of incomes vectored by a disaster may make migration imperative for the household (Mbaye 2023). The outcomes of our study are consistent with previous empirical studies assessing the relationship between economic loss caused by climatic disasters and migrations, which typically found a positive relationship between the occurrence of such events and migrations (Ishtiaque and Nazem 2017; Maharjan et al. 2020a, b; Trinh et al. 2021). The sociodemographic profile of migrant and non-migrant households has been shown in Table 11.1. Almost 30 percent of the households consist of more than five members (non-migrant 29.02% vs. migrant 39.57%). 25.96% among the migrant households had no formal education, which is more than the non-migrant households. The distribution of major source of income from wage labor was almost 55% among the migrant household which was higher than the non-migrant household (33%). Almost 55% of the migrant households had MGNREGA job card, and 91% of the migrant households were from the rural areas, respectively.

Table 11.1 Sociodemographic profile of the study population

Background variables	Non-migrants		Migrants		Total	
	Sample (n)	Percent	Sample (n)	Percent	Sample (n)	Percent
Households size						
≤ 5	27,014	70.98	1,857	60.43	28,871	70.19
> 5	11,047	29.02	1,216	39.57	12,263	29.81
Level of education						
No formal education	7,196	18.91	798	25.96	7,994	19.43
Primary	4,977	13.08	693	22.54	5,669	13.78
Secondary	13,573	35.66	1,093	35.56	14,666	35.65
Higher	12,315	32.36	490	15.94	12,805	31.13
Caste						
SC	8,172	21.47	929	30.23	9,101	22.12
ST	2,998	7.88	397	12.93	3,395	8.25
OBC	16,079	42.24	1,279	41.63	17,358	42.20
Others	10,812	28.41	467	15.21	11,280	27.42
Religion						
Hindu	31,463	82.66	2,602	84.66	34,065	82.81
Muslim	4,302	11.30	377	12.27	4,679	11.37
Others	2,296	6.03	94	3.07	2,391	5.81
Main sources of income						
Cultivation	9,401	24.70	902	29.36	10,303	25.05
Wage labor	12,668	33.28	1,687	54.90	14,355	34.90
Small business	4,903	12.88	203	6.59	5,106	12.41
Others	11,090	29.14	281	9.16	11,371	27.64
Land holding						
No	21,483	56.44	1,328	43.21	22,811	55.45
Yes	16,578	43.56	1,745	56.79	18,323	44.55
Wealth						
Poor	6,388	16.78	783	25.47	7,171	17.43
Non-Poor	31,673	83.22	2,290	74.53	33,963	82.57
Ration card						
No	5,142	13.51	455	14.80	5,597	13.61
Yes	32,919	86.49	2,618	85.20	35,537	86.39
MGNREGA job card						
No	26,706	70.17	1,384	45.02	28,090	68.29
Yes	11,355	29.83	1,689	54.98	13,044	31.71

(continued)

Table 11.1 (continued)

Background variables	Non-migrants		Migrants		Total	
	Sample (n)	Percent	Sample (n)	Percent	Sample (n)	Percent
Income from govt. sources						
No	37,070	97.40	2,971	96.67	40,041	97.34
Yes	991	2.60	102	3.33	1,093	2.66
Place of residence						
Rural	25,272	66.40	2,807	91.35	28,080	68.26
Urban	12,789	33.60	266	8.65	13,054	31.74
Regions						
North	5,488	14.42	271	8.83	5,759	14.00
Central	7,838	20.59	967	31.47	8,805	21.40
East	8,196	21.53	1,018	33.14	9,215	22.40
West	5,581	14.66	245	7.98	5,826	14.16
North East	1,465	3.85	69	2.24	1,534	3.73
South	9,493	24.94	502	16.33	9,995	24.30
Total	38,061		3,073		41,134	

Source India Human Development Survey-II (IHDS-II), 2011–12

Table 11.2 presents the prevalence of loss due to climatic disasters (e.g., drought, flood, fire, crop failure) across sociodemographic variables of migrant and non-migrant households. Households with more than five members have seen more migration (47.30%) of any member than the non-migrants' households (36.22%) experiencing loss due to climatic disasters. On top of that, the loss of livelihood assets and prospects affected poor households by diminishing their ability to migrate (Mallick et al. 2023; Mallick and Schanze 2020). Our research also suggests that household size played a significant role in driving migration; this finding was consistent with the results presented by Xu et al. (2015). It is recognized that households with limited resources for labor face substantial challenges in pursuing diverse livelihood activities and accumulating capital (Wang et al. 2019). Consequently, even after facing climatic disasters, rural households with smaller members tend to prioritize staying in local areas and working in their local neighborhood (Hoang et al. 2023). On the other hand, when there were enough people to do farm work or other jobs in the local area, or when there were not many opportunities available locally, extra family members chose to migrate to cities or other places. They do this to help their family financially by diversifying economic activities and to rebuild the assets damaged by climatic disasters. Conversely, when the household labor force satisfied the need for agriculture requirement or local economic activities or when local livelihood opportunities were not available, most of the surplus household members choose to migrate to work in either cities or towns to maximize family benefits and reduce the risk posed by the loss due to climatic disasters (Hoang et al. 2023; Wang et al. 2019; Xiong et al. 2020b). Significant difference was also observed among scheduled caste and scheduled tribe households. Almost 24% of scheduled caste and 18% of scheduled tribe had experienced movement of any member which was greater than the non-migrants' households. This is due to SC/ST communities often residing in economically vulnerable areas with limited access to resources and infrastructure. Thus, belonging to marginalized sections exacerbates post-disaster migration due to their pre-existing lack of basic infrastructure and employment opportunities. These communities often struggle with inadequate access to essential services, pushing them to migrate in search of basic living conditions and livelihoods, which were reported by other empirical researchers as well (Hallegatte et al. 2020). Households where wage labor was the primary source of income experienced a higher rate of member's mobility (41.25%) than the non-migrant households (25.10%). It was also observed that the non-migrants' households had others source of income (14.50%) which was greater than the migrants' households (3.71%). According to a study by (Mallick et al. 2023), short-distance migrants tend to be either heavily dependent on agriculture or have low levels of capital ownership and income. Disaster have a significant, positive impact and cause financial harm to people whose dependence on agricultural income is average to high (Dallmann and Millock 2017; Joarder and Miller 2013). It was not surprising that we had evidence of very few people migrating out because of financial losses from climatic disasters when they had small businesses or other sources of income. Our finding was in line with the above-mentioned and several other studies, indicating that agricultural households were particularly vulnerable to loss due to climatic disasters and thus migrated more (De Haen and Hemrich 2007; Maharjan et al. 2020a, b; Xu et al. 2015). Those who had an MGNREGA job card and also experienced the loss due to climatic disasters had more migration from their households (migrant households 63.20% vs. non-migrant households 46.04%) than their counterparts. This pattern suggests that MGNREGA was implemented in rural India to give livelihood security and reduce distress migration. But the protective capacity of this scheme was substantially weakened when households were exposed to environmental shocks (Bharadwaj et al. 2022). Regional differences were also observed in Table 11.2, where people from central (38.81%) and east (26.51%) areas had more migration from households experiencing loss due to climatic disasters. In the previous studies, findings suggest that regional disparities in agro-vulnerability, livelihood dependence, and development infrastructure strongly influence migration during environmental shocks. Whereas in the central and eastern areas, many parts are often characterized by rainfed agriculture, recurrent drought, land degradation, and limited non-farm employment opportunities, which drives households' livelihood insecurity when climatic disasters occur (Deshingkar and Akter 2009; Rajan and Bhagat 2017).

Table 11.3 depicts the unadjusted and adjusted odds of migration by key explanatory variable (i.e., loss due to climatic disasters) and other covariates. Five models have been used to perform the association between loss due to climatic disasters and the migration of any member from any households. Model 1 indicates the unadjusted effect of loss due to climatic disasters on migration. Model 2 includes sociodemographic adjusted effect of loss on migration. Model 3 shows adjusted economic and land holding effect of loss due to climatic disasters on migration. Model 4 includes

Table 11.2 Prevalence of expenditure/loss due to the climatic disasters across sociodemographic variables of migrants and non-migrants' households

Covariates	Non-migrants HH		Migrants HH		Total HH	
	Expenditure/ loss (%)	p-value	Expenditure/ loss (%)	p-value	Expenditure/ loss (%)	p-value
Households size						
≤ 5	63.78	*≤ 0.001*	52.70	*≤ 0.001*	62.46	*≤ 0.001*
> 5	36.22		47.30		37.54	
Level of education						
No formal education	18.80	*≤ 0.001*	26.51	*0.651*	19.72	≤ 0.001
Primary	15.28		20.25		15.88	
Secondary	38.35		36.48		38.13	
Secondary and above	27.57		16.76		26.27	
Caste						
SC	16.34	*≤ 0.001*	23.97	*≤ 0.001*	17.25	*≤ 0.001*
ST	11.53		17.71		12.27	
OBC	43.69		44.54		43.79	
Others	28.45		13.79		26.69	
Religion						
Hindu	86.94	*≤ 0.001*	89.71	*≤ 0.001*	87.27	*≤ 0.001*
Muslim	8.81		8.59		8.78	
Others	4.25		1.70		3.95	
Main sources of income						
Cultivation	54.27	*≤ 0.001*	51.43	*≤ 0.001*	53.93	*≤ 0.001*
Wage labor	25.10		41.25		27.03	
Small business	6.13		3.61		5.83	
Others	14.50		3.71		13.21	
Land holding						
No	15.33	*≤ 0.001*	13.68	*≤ 0.001*	15.13	*≤ 0.001*
Yes	84.67		86.32		84.87	
Wealth						
Poor	17.45	*≤ 0.001*	26.94	*0.567*	18.58	*≤ 0.001*
Non-poor	82.55		73.06		81.42	
Ration card						
No	11.62	*≤ 0.001*	11.77	*≤ 0.001*	11.64	*≤ 0.001*
Yes	88.38		88.23		88.36	
MGNREGA job card						
No	53.96	*≤ 0.001*	36.80	≤ 0.001	51.91	*≤ 0.001*

(continued)

Table 11.2 (continued)

Covariates	Non-migrants HH		Migrants HH		Total HH	
	Expenditure/ loss (%)	p-value	Expenditure/ loss (%)	p-value	Expenditure/ loss (%)	p-value
Yes	46.04		63.20		48.09	
Income from govt. sources						
No	90.05	*≤ 0.001*	93.00	≤ 0.001	90.40	*≤ 0.001*
Yes	9.95		7.00		9.60	
Place of residence						
Rural	92.00	*≤ 0.001*	97.14	≤ 0.001	92.61	*≤ 0.001*
Urban	8.00		2.86		7.39	
Regions						
North	17.62	*≤ 0.001*	10.60	≤ 0.001	16.78	*≤ 0.001*
Central	25.75		38.81		27.31	
East	20.63		26.51		21.34	
West	11.06		6.79		10.55	
Northeast	6.08		2.55		5.66	
South	18.86		14.74		18.37	

Source India Human Development Survey-II (IHDS-II), 2011–12

adjusted effect of government schemes of loss on migration, and model 5 shows adjusted effect of all the covariates of loss on migration.

The result from model 1 showed that the unadjusted prevalence of migration was estimated to be 2.46 times (95%CI: 2.25, 2.67) experiencing loss due to climatic disasters. After controlling sociodemographic factors such as household size, level of education, caste, religion, and place of residence and region, the odds of migration were estimated to be 1.84 times (95%CI: 2.25, 2.67) (model 2) for those experiencing loss due to climatic disasters. Following the studies conducted by Drzewiecki et al. (2020) and Joseph et al. (2021), we also found that low levels of education often compel households to migrate following disaster-led economic losses. This can be explained by the fact that having a limited education makes it harder to find higher-paying employment, making it more challenging to recover from economic losses from a disaster. Additionally, those with less education might need more abilities to manage jobs that are resilient to disasters. They often relied on jobs that were affected the most by disasters, like wage labor or cultivation. Consequently, individuals with limited educational backgrounds found migration to be a compelling option when faced with the economic losses caused by disasters. Moreover, climatic disasters predominantly damage the livelihood of individuals residing in impoverished circumstances. Extensive research findings consistently demonstrated that these disasters directly impact the employment and income levels of poor populations (Berlemann and Steinhardt 2017; Drabo and Mbaye 2015). Results from the model 3 showed the odds of migration were estimated to be 2.08 times (95%CI:

Table 11.3 Unadjusted and adjusted odds of migration by key explanatory variables and other covariates

Key explanatory variable	Model 1		Model 2		Model 3		Model 4		Model 5	
	OR	95% CI	OR	95% CI	OR	95% CI	OR	95% CI	OR	95% CI
Expenditure/loss										
No®										
Yes	2.46***	[2.26–2.67]	1.84***	[1.63–2.08]	2.08***	[1.82–2.37]	2.10***	[1.86–2.38]	1.90***	[1.66–2.17]
Households size										
> 5®										
≤ 5			0.68***	[0.60–0.77]					0.67***	[0.59–0.76]
Level of education										
No Schooling®										
Primary			1.39***	[1.17–1.64]					1.39***	[1.18–1.64]
Secondary			0.99	[0.85–1.16]					1.08	[0.92–1.27]
Secondary and above			0.66***	[0.55–0.78]					0.85	[0.71–1.02]
Caste										
Others®										
SC			2.30***	[1.92–2.77]					1.76***	[1.44–2.14]
ST			2.46***	[2.00–3.04]					2.01***	[1.62–2.49]
OBC			1.56***	[1.34–1.82]					1.46***	[1.25–1.70]
Religion										
Others®										
Hindu			1.76**	[1.23–2.54]					1.60*	[1.12–2.29]
Muslim			2.57***	[1.74–3.81]					2.19***	[1.48–3.25]

(continued)

Table 11.3 (continued)

Key explanatory variable	Model 1		Model 2		Model 3		Model 4		Model 5	
	OR	95% CI	OR	95% CI	OR	95% CI	OR	95% CI	OR	95% CI
Main sources of income										
Others®										
Cultivation					2.30***	[1.78–2.97]			1.58***	[1.21–2.06]
Wage labor					4.57***	[3.57–5.85]			2.56***	[1.95–3.36]
Small business					1.63**	[1.21–2.21]			1.24	[0.90–1.70]
Land holding										
No®										
Yes					1.37***	[1.18–1.59]			0.93	[0.79–1.10]
Wealth										
Poor®										
Non-Poor					0.76***	[0.67–0.87]			1.11	[0.97–1.27]
Ration card										
No®										
Yes							0.76***	[0.65–0.89]	0.89	[0.75–1.05]
MGNREGA job card										
No®										
Yes							2.58***	[2.29–2.90]	1.48***	[1.30–1.69]
Income from govt. sources										
No®										
Yes							0.82	[0.63–1.07]	0.82	[0.62–1.09]

(continued)

Table 11.3 (continued)

Key explanatory variable	Model 1		Model 2		Model 3		Model 4		Model 5	
	OR	95% CI	OR	95% CI	OR	95% CI	OR	95% CI	OR	95% CI
Place of residence										
Rural®										
Urban			0.30***	[0.26–0.35]					0.39***	[0.33–0.48]
Regions										
North®										
Central			1.93***	[1.62–2.30]					1.88***	[1.56–2.26]
East			1.91***	[1.62–2.26]					1.83***	[1.55–2.17]
West			1.07	[0.85–1.34]					1.08	[0.86–1.36]
North East			0.77	[0.51–1.15]					0.73	[0.48–1.11]
South			1.19	[0.98–1.43]					1.03	[0.85–1.24]

Note **CI**: Confidence Interval; ***p** ≤ 0.05; ** $p \leq 0.005$; ≤ 0.001; **SC**: Scheduled Caste; **ST**: Scheduled Tribe; **OBC**: Other Backward Class. *Source* India Human Development Survey-II (IHDS-II), 2011–12

1.82, 2.37) while adjusting to economic factors such as source of income, land holding, and wealth. In addition to that, previous studies have demonstrated a plausible negative relationship between land ownership and migration in the context of economic losses brought on by climatic disasters (McLeman and Hunter 2010; Vanwey 2003b). In accordance with the above-mentioned studies, we also found that people who suffer economic losses as a result of climatic disasters, particularly those without secure land tenure, were more likely to migrate. This connection resulted from landless populations' increased susceptibility to the economic shocks brought on by disasters, which forces them to look for safer and more reliable employment elsewhere. At the same time, model 4 has been adjusted to government schemes such as having a ration and MGNREGA job card and income from any government sources, and the results showed that the odds of migration were 2.10 times (95%CI: 1.86, 2.38). The study found that having a ration card or a government job acts as a resistance for the household to migrate. Although the results were insignificant, the possession of a ration card can act as a potential deterrent against migration in the aftermath of post-disaster economic loss. This can be explained in many ways; the most important reason behind this phenomenon might be that a ration card ensures access to subsidized food and essential commodities, providing a safety net during times of economic distress (Ishtiaque and Nazem 2017). This reduces the immediate need for individuals to migrate in search of sustenance. On the other hand, government jobs often provide stability and reliable income, which becomes critical when local economies and income opportunities are disrupted by hazards. Alternatively, the government jobs allow for more savings, which enables persons to utilize that in time of economic loss, which in turn prevents them from migrating elsewhere (Ahuja et al. 2011). Model 5 has been adjusted to all the covariates, and the results showed that the odds of migration among those who experienced loss due to climatic disasters were 1.90 times (95%CI:1.66, 2.17).

Figure 11.2 presents the state-wise differences in migration of those who experienced loss due to climatic disasters in India. A positive relation has been observed between migration and major incidence among the states. Bihar, among all the states, experienced the highest percentage of loss (14.46%) due to climatic disasters, followed by Uttar Pradesh (11.12%), Andhra Pradesh (9.42%), and Madhya Pradesh (7.72%). At the same time, migration from those states was also high. Among all the states, Madhya Pradesh had the highest (21.85%) migration due to loss followed by Bihar (14.42%), Uttar Pradesh (10.50%), and Rajasthan (8.27%). Also, it has been observed that states like Haryana, Maharashtra, Punjab, and Tamil Nadu were less experienced both in terms of loss and migration for those particular reasons. These states are largely experiencing floods, riverbank erosion, waterlogging, drought, heat waves, agrarian distress, coastal flooding, and cyclones, which resulted in distress migration (Black et al. 2011; Rajan and Bhagat 2017; Singh et al. 2017).

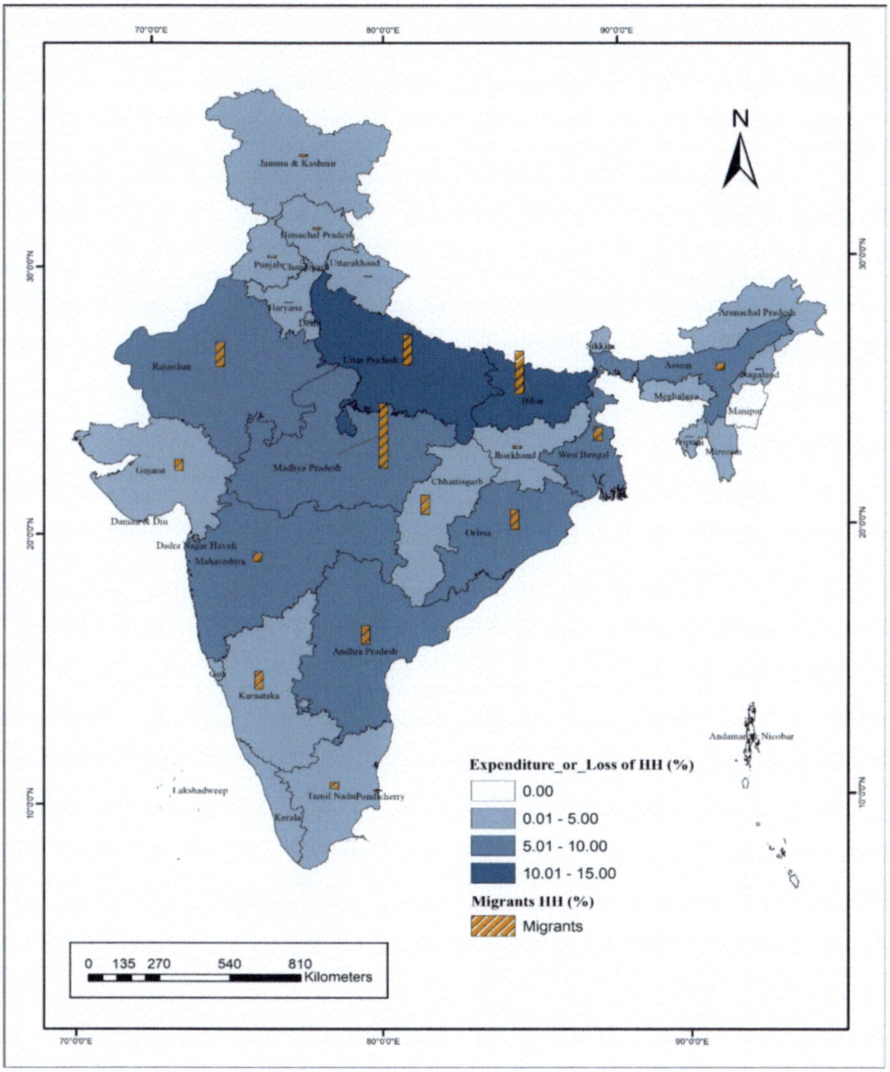

Fig. 11.2 Choropleth map shows the state-wise loss or expenditure due to environmental hazards, and bars show the percentage share of migrants from the states. *Source* Authors

Limitation of the Study

While our study provides valuable insights into the relationship between disaster-driven economic losses and migration dynamics in India, it is not free from limitations. Several factors could influence the findings and interpretations, which are important to consider. The analysis is based on cross-sectional data from the India

Human Development Survey-II. This limits the ability to establish causality between economic losses due to climatic disasters and subsequent migration. Longitudinal data would be more effective in capturing the temporal relationship and the long-term impact of these losses on migration decisions. The data rely on respondents' recall of economic losses and migration experiences over the past six to seven years. This could introduce recall bias, particularly regarding the accuracy of the timing and magnitude of losses, as well as the reasons for migration. The study uses a binary variable to capture economic losses due to climatic disasters, which might oversimplify the varying degrees of losses experienced by households. A more nuanced measure that captures the magnitude and type of loss could provide deeper insights into how different levels of economic loss influence migration decisions.

Conclusion

In conclusion, the study contributes to the field of climate change and human migration. This study reveals that climatic disasters play a significant role in shaping households' economic vulnerability and migration decisions in India. The findings of the study show that households experiencing disaster-related economic losses are more likely to migrate as a livelihood coping strategy. Furthermore, this study reveals that various sociodemographic, economic, land holding, and government schemes significantly determine the disaster-related economic losses that drive migration in India. Focusing on our study findings, the government and other national and world-level NGOs should strengthen the integration of disaster risk reduction, climate adaptation, and the social protection framework at both the national and international levels. Reducing forced migration can be achieved by strengthening rural livelihoods through timely pay, increased guaranteed employment days, skill diversity, and localized non-farm employment alternatives. Additionally, measures that safeguard migrant households while addressing the underlying causes of disaster-induced displacement can be developed by acknowledging migration as an adaptive reaction rather than a failure of development.

AI Disclosure Statement

The authors used 17 percent artificial intelligence tools to improve grammatical correction, sentence restructuring, and formatting consistency in the final manuscript. All the substantive introduction, materials and methods, interpretation of results and discussion, limitations, and conclusion were conducted by the authors, who take full responsibility for the accuracy and originality of the work.

Data Availability Statement

The IHDS data from wave II are available for download from the ICPSR website: https://www.icpsr.umich.edu/web/DSDR/studies/36151.

References

Afifi T, Liwenga E, Kwezi L (2014a) Rainfall-induced crop failure, food insecurity and 5529. https://doi.org/10.1080/17565529.2013.826128

Afifi T, Liwenga E, Kwezi L (2014b) Rainfall-induced crop failure, food insecurity and out-migration in Same-Kilimanjaro Tanzania. Climate Develop 6(1):53–60. https://doi.org/10.1080/17565529.2013.826128

Ahuja UR, Tyagi D, Chauhan S, Chaudhary KR (2011) Impact of MGNREGA on rural employment and migration: a study in agriculturally-backward and agriculturally-advanced districts of Haryana. Agric Econ Res Rev 24:495–502. https://doi.org/10.22004/AG.ECON.119403

Bardsley DK, Hugo GJ (2010a) Migration and climate change: examining thresholds of change to guide effective adaptation decision-making. Popul Environ 32(2):238–262. https://doi.org/10.1007/S11111-010-0126-9/FIGURES/1

Bardsley DK, Hugo GJ (2010b) Migration and climate change: examining thresholds of change to guide effective adaptation decision-making. Popul Environ 32(2):238–262. https://doi.org/10.1007/s11111-010-0126-9

Berlemann M, Steinhardt MF (2017) Climate change, natural disasters, and migration—A survey of the empirical evidence. Cesifo Economic Studies 63(4):353–385. https://doi.org/10.1093/CESIFO/IFX019

Bharadwaj R, Raj N, Karthikeyan N, Shanker R, Topno J, Kaur D (2022) Climate change social protection and informal job market reform for tackling the climate migration nexus produced by IIED's climate change group. www.iied.org

Black R, Arnell NW, Adger WN, Thomas D, Geddes A (2013) Migration, immobility and displacement outcomes following extreme events. Environ Sci Policy 27:S32–S43. https://doi.org/10.1016/J.ENVSCI.2012.09.001

Black R, Bennett SRG, Thomas SM, Beddington JR (2011) Migration as adaptation. Nature 478(7370):447–449

Cattaneo C, Beine M, Fröhlich CJ, Kniveton D, Martinez-Zarzoso I, Mastrorillo M, Millock K, Piguet E, Millock FK (2019) Human migration in the era of climate change "human migration in the era of climate change

Chandrasekhar S, Sahoo S (2019) Land-holding, participation in agriculture and short-term migration in rural India. Asian Popul Stud 15(2):209–227. https://doi.org/10.1080/17441730.2019.1600324

Chen THY, Lee B (2022) Income-based inequality in post-disaster migration is lower in high resilience areas: evidence from U.S. internal migration. Environ Res Lett 17(3):034043. https://doi.org/10.1088/1748-9326/AC5692

Dallmann I, Millock K (2017) Climate variability and inter-state migration in India. Cesifo Economic Studies 63(4):560–594. https://doi.org/10.1093/cesifo/ifx014

Datar A, Liu J, Linnemayr S, Stecher C (2013) The impact of natural disasters on child health and investments in rural India. Soc Sci Med 76(1):83–91. https://doi.org/10.1016/J.SOCSCIMED.2012.10.008

De Haen H, Hemrich G (2007) The economics of natural disasters: implications and challenges for food security. Agric Econ 37(S1):31–45. https://doi.org/10.1111/J.1574-0862.2007.00233.X

Desai S, Vanneman R (2018) India human development survey-II (IHDS-II), 2011–12. Inter-university consortium for political and social research [distributor]. https://doi.org/10.3886/ICPSR36151.v6
Deshingkar P (2012) Environmental risk, resilience and migration: implications for natural resource management and agriculture. Environ Res Lett 7(1). https://doi.org/10.1088/1748-9326/7/1/015603
Deshingkar P, Akter S (2009) Migration and human development in India
Drabo A, Mbaye LM (2015) Natural disasters, migration and education: an empirical analysis in developing countries. Environ Dev Econ 20(6):767–796. https://doi.org/10.1017/S1355770X14000606
Drzewiecki DM, Wavering HM, Milbrath GR, Freeman VL, Lin JY (2020) The association between educational attainment and resilience to natural hazard-induced disasters in the West Indies: St. Kitts & Nevis. Int J Disaster Risk Reduct 47:101637. https://doi.org/10.1016/J.IJDRR.2020.101637
EMDAT (2021) Missing data on economic losses variables from EM-DAT. CRED Crunch, Issue No. 63
Gray CL, Mueller V (2012a) Natural disasters and population mobility in Bangladesh. Proc Natl Acad Sci USA 109(16):6000–6005. https://doi.org/10.1073/PNAS.1115944109/SUPPL_FILE/SAPP.PDF
Gray CL, Mueller V (2012b) Natural disasters and population mobility in Bangladesh. Proc Natl Acad Sci USA 109(16):6000–6005. https://doi.org/10.1073/pnas.1115944109
Hallegatte S, Przyluski V (2010) The economics of natural disasters. Cesifo Forum 11(2):14–24. https://doi.org/10.4337/9781785365980
Hallegatte S, Vogt-Schilb A, Rozenberg J, Bangalore M, Beaudet C (2020) From poverty to disaster and back: a review of the literature. Econ Disasters Clim Chang 4(1):223–247. https://doi.org/10.1007/S41885-020-00060-5
Hoang TX, Nguyen HT, Tran TQ (2023) Heterogeneous effects of natural disasters on migration and household well-being in rural Vietnam: a panel data analysis. Environ Develop Sustain, 1–21. https://doi.org/10.1007/S10668-023-03751-7/TABLES/8
IDMC (2023) Global report on internal displacement 2023 IDMC global report on internal displacement 2023 IDMC II III
Ishtiaque A, Nazem NI (2017) Household-level disaster-induced losses and rural–urban migration: Experience from world's one of the most disaster-affected countries. Nat Hazards 86(1):315–326. https://doi.org/10.1007/s11069-016-2690-5
Jennath A, Paul S (2022) Migration as a response to climate disasters in coastal areas: insights from Kerala, India. Ocean. 2022-Chennai, 1–10
Jennath A, Paul S (2024) Trajectories of environmental migrants: understanding migration destinations of the coastal population in Kerala India. Hum Ecol 52(1):99–113
Joarder MAM, Miller PW (2013) Factors affecting whether environmental migration is temporary or permanent: Evidence from Bangladesh. Glob Environ Chang 23(6):1511–1524. https://doi.org/10.1016/J.GLOENVCHA.2013.07.026
Joseph J, Irshad SM, Alex AM (2021) Disaster recovery and structural inequalities: a case study of community assertion for justice. Int J Disast Risk Reduct 66:102555. https://doi.org/10.1016/J.IJDRR.2021.102555
Kartiki K, Kartiki K (2011) Climate change and migration : a case study from rural Bangladesh climate change and migration : a case study from rural Bangladesh, 37–41. https://doi.org/10.1080/13552074.2011.554017
Loebach P (2016) Household migration as a livelihood adaptation in response to a natural disaster: Nicaragua and Hurricane Mitch. Popul Environ 38(2):185–206. https://doi.org/10.1007/S11111-016-0256-9/sS/5
Maharjan A, de Campos RS, Singh C, Das S, Srinivas A, Bhuiyan MRA, Ishaq S, Umar MA, Dilshad T, Shrestha K, Bhadwal S, Ghosh T, Suckall N, Vincent K (2020a) Migration and household

adaptation in climate-sensitive hotspots in South Asia. Curr Clim Chang Reports 6(1):1–16. https://doi.org/10.1007/s40641-020-00153-z
Maharjan A, de Campos RS, Singh C, Das S, Srinivas A, Bhuiyan MRA, Ishaq S, Umar MA, Dilshad T, Shrestha K, Bhadwal S, Ghosh T, Suckall N, Vincent K (2020b) Migration and Household adaptation in climate-sensitive hotspots in South Asia. Current Climate Change Reports 6(1):1–16. https://doi.org/10.1007/S40641-020-00153-Z/FIGURES/4
Mallick B, Best K, Carrico A, Ghosh T, Priodarshini R, Sultana Z, Samanta G (2023) How do migration decisions and drivers differ against extreme environmental events? Environ Hazards 22(5):475–497. https://doi.org/10.1080/17477891.2023.2195152
Mallick B, Schanze J (2020) Trapped or voluntary? Non-migration despite climate risks. Sustainability 12(11):4718. https://doi.org/10.3390/SU12114718
Mandal B, Pradhan KC (2024) A comparative study of health outcomes between elderly Migrant and non-migrant population in India: exploring health disparities through propensity score matching. Ssm–populat Health 25:101619. https://doi.org/10.1016/J.SSMPH.2024.101619
Mandal B, Pradhan KC, Mohanty P, Muhammad T (2023) Migration status, physical limitations and associated self-rated health: a study of older Indian adults. BMC Geriatr 23(1):1–16. https://doi.org/10.1186/s12877-023-04002-0
Marchiori L, Maystadt J-F, Schumacher I (2012a) The impact of weather anomalies on migration in sub-Saharan Africa. J. Environ. Econ. Manage. 63(3):355–374
Marchiori L, Maystadt J-F, Schumacher I (2012b) The impact of weather anomalies on migration in sub-Saharan Africa. J Environ Econ Manag 63(3):355–374
Mbaye LMOA (2023) Climate change, natural disasters, and migration. IZA World of Labor. https://doi.org/10.15185/IZAWOL.346
McLeman RA, Hunter LM (2010) Migration in the context of vulnerability and adaptation to climate change: insights from analogues. Wiley Interdisciplinary Reviews: Climate Change 1(3):450–461. https://doi.org/10.1002/WCC.51
Munda J, Adhikary M, Saha A (2024) Untouchability-induced migration in India: exploring the trajectory of scheduled caste migrants. https://doi.org/10.1177/2455328X241259822
Nandy A, Tiwari C, Kundu S (2021) India's rural employment guarantee scheme—How does it influence seasonal rural out-migration decisions? J Policy Model 43(6):1181–1203. https://doi.org/10.1016/J.JPOLMOD.2021.09.001
Panagariya A, Mukim M (2014) A comprehensive analysis of poverty in India. Asian Dev Rev 31(1):1–52. https://doi.org/10.1162/ADEV_A_00021/ASSET/IMAGES/LARGE/ADEV.2014.00021-FIGURE05.JPEG
Pörtner H-O, Roberts DC, Tignor MMB, Poloczanska E, Rama B (2022) Working group II contribution to the sixth assessment report of the intergovernmental panel on climate change. https://report.ipcc.ch/ar6/wg2/IPCC_AR6_WGII_FullReport.pdf
Pradhan KC, Narayanan K (2022a) Does climatic risk induce labour migration? Evidence from semi-arid tropics region of India. J Public Aff 22(1):e2323. https://doi.org/10.1002/PA.2323
Pradhan KC, Narayanan K (2022b) Does climatic risk induce labour migration? Evidence from semi-arid tropics region of India. J Public Aff 22(1). https://doi.org/10.1002/pa.2323
Rahman A, Singh C, Srinivas A (2024a) Migration along Bengaluru's Rural–Urban continuum: implications for household well-being and climate change adaptation. Indian J Hum Dev, 09737030241244572
Rahman A, Singh C, Srinivas A (2024b) Migration along Bengaluru's Rural–Urban continuum: implications for household well-being and climate change adaptation. Indian J Hum Dev, 09737030241244572
Rajan SI, Bhagat RB (2017) Climate change, vulnerability and migration. Taylor & Francis
Saha A, Mandal B, Muhammad T, Ali W (2024) Decomposing the Rural–Urban differences in depression among multimorbid older patients in India: evidence from a cross-sectional study. BMC Psychiatry 24(1):1–18. https://doi.org/10.1186/S12888-023-05480-7/TABLES/4
Saha A, Mandal B, Muhammad T, Barman P, Ahmed W (2023a) Gender-specific determinants of overweight and obesity among older adults in India: evidence from a cross-sectional survey,

2017–18. BMC Public Health 23(1):1–15. https://doi.org/10.1186/S12889-023-17156-8/TABLES/3
Saha A, Muhammad T, Mandal B, Adhikary M, Barman P (2023b) Socio-demographic and behavioral correlates of excess weight and its health consequences among older adults in India: Evidence from a cross-sectional study, 2017–18. PLoS ONE 18(10):e0291920. https://doi.org/10.1371/JOURNAL.PONE.0291920
Saha A, Rahaman M, Mandal B, Biswas S, Govil D (2022) Rural urban differences in self-rated health among older adults: examining the role of marital status and living arrangements. BMC Public Health 22(1):1–15. https://doi.org/10.1186/S12889-022-14569-9/TABLES/4
Samanta R, Munda J (2023) Husband's migration status and contraceptive behaviors of women: evidence from Middle-Ganga Plain of India. BMC Women's Health 23(1):1–13. https://doi.org/10.1186/S12905-023-02325-Z/TABLES/6
Samanta R, Munda J, Mandal S, Adhikary M (2023) Health-care utilisation among India's middle and older aged migrants: scrutinizing the status and predictors using Andersen's simplified healthcare utilisation framework. Int J Migrat Health Social Care 19(2):142–156. https://doi.org/10.1108/IJMHSC-07-2022-0068/FULL/XML
Sarkar B, Dutta S, Singh PK (2022) Drought and temporary migration in rural India: a comparative study across different socio-economic groups with a cross-sectional nationally representative dataset. PLoS ONE 17(10):e0275449. https://doi.org/10.1371/JOURNAL.PONE.0275449
Singh C (2019) Migration as a driver of changing household structures: implications for local livelihoods and adaptation. Migr. Dev. 8(3):301–319. https://doi.org/10.1080/21632324.2019.1589073
Singh C, Basu R (2020) Moving in and out of vulnerability: interrogating migration as an adaptation strategy along a Rural–Urban continuum in India. Geogr J 186(1):87–102. https://doi.org/10.1111/geoj.12328
Singh C, Deshpande T, Basu R (2017) How do we assess vulnerability to climate change in India? A systematic review of literature. Reg Environ Change 17(2):527–538
StataCorp (2023) Statistical software for data science|Stata
Tiwari PC, Joshi B, Access O (2015) Climate change and rural out-migration in Himalaya, pp 8–25. https://doi.org/10.1515/cass-2015-0002
Trinh TA, Feeny S, Posso A (2021) The impact of natural disasters on migration: findings from Vietnam. J Demograp Eco 87(3):479–510. https://doi.org/10.1017/DEM.2020.14
Tripathi A (2017) Socioeconomic backwardness and vulnerability to climate change: evidence from Uttar Pradesh state in India. J Environ Planning Manage 60(2):328–350. https://doi.org/10.1080/09640568.2016.1157059
Tulloch O, Machingura F, Melamed C (2016) Health, migration and the 2030 agenda for sustainable development. ODI Brief., 1–16.
van Aalst MK (2006) The impacts of climate change on the risk of natural disasters. Disasters 30(1):5–18. https://doi.org/10.1111/j.1467-9523.2006.00303.x
Vanwey LK (2003a) Land ownership as a determinant of temporary migration in Nang Rong, Thailand. Eur J Population 19(2):121–145
Vanwey LK (2003b) Land ownership as a determinant of temporary migration in Nang Rong Thailand. European J Populat 19(2):121–145. https://doi.org/10.1023/A:1023307100719/METRICS
Wang P, Yan J, Hua X, Yang L (2019) Determinants of livelihood choice and implications for targeted poverty reduction policies: a case study in the YNL river region, Tibetan Plateau. Ecol Ind 101:1055–1063. https://doi.org/10.1016/J.ECOLIND.2019.02.007
Xiong S, Wu Y, Wu S, Chen F, Yan J (2020a) Determinants of migration decision-making for rural households: a case study in Chongqing China. Nat Hazards 104(2):1623–1639
Xiong S, Wu Y, Wu S, Chen F, Yan J (2020b) Determinants of migration decision-making for rural households: a case study in Chongqing China. Nat Hazards 104(2):1623–1639. https://doi.org/10.1007/S11069-020-04236-W/TABLES/2

XuD, Zhang JF, Xie FT, Liu SQ, Cao MT, Liu EL (2015) Influential factors in employment location selection based on "push-pull" migration theory—A case study in three gorges reservoir area in China. J Mount Sci 12(6):1562–1581. https://doi.org/10.1007/S11629-014-3371-Z/METRICS

Chapter 12
Climate-Induced Disasters and Its Impact on Coastal Migration: An Empirical Study on the Southern Part of Bangladesh

Md. Mohiuddin, Samshad Nowreen, Mallik Akram Hossain, and Mesbah Uddin

Abstract Climate-induced disasters pose a significant and devastating impact on migration across various parts of the Earth. The study focused on climate change-induced disaster-driven migration in the southern coastal areas of Bangladesh. The study collected data from the semi-structured questionnaire survey and secondary sources regarding climate change-induced disasters and their impact on migration. The research has observed the perceptions of coastal communities regarding climate change-induced migration following disaster devastation. However, climate change-induced disasters, including cyclones, floods, erosion, and salinity intrusion, have affected the coastal communities and their livelihoods. About 51% of the coastal people agreed that people migrate due to frequent cyclone occurrences and devastation. Moreover, coastal people face cyclones and associated surges and floods, which impact their livelihoods and income opportunities, such as agricultural work, fishing activities, aquaculture, business, daily labor, seasonal labor, and other daily work. Additionally, most of the coastal population (53%) faces severe flooding and associated disasters, including erosion and salinity. Consequently, the coastal people lost their lives and property, especially settlements, agricultural land, crops, livestock, and fishing farms, due to coastal disasters. Furthermore, the study found that people migrate due to the loss of their livelihood, property, and infrastructure during disasters, and many have no financial backup.Therefore, the global community, researchers, and experts would gather knowledge from the research on the

Md. Mohiuddin (✉) · M. A. Hossain
Department of Geography and Environment, Jagannath University, Dhaka, Bangladesh
e-mail: mahigeo.jnu@gmail.com

S. Nowreen
Department of Tourism and Hospitality Management, University of Dhaka, Dhaka, Bangladesh

M. Uddin
ADHVAN Inc, Shibuya City, Tokyo 150-0043, Japan
e-mail: mesbah.uddin@adhvan.io

P. Ghosh et al. (eds.), *Climate Crisis and Changing Population Dynamics*, Springer Climate, https://doi.org/10.1007/978-3-032-24533-5_12

impact of climate-induced disasters on migration. Government, NGOs, and policy-makers of Bangladesh may obtain information and take the initiative to minimize the effect of disasters on coastal communities by providing shelters, financial support, preparedness, and awareness-raising.

Keywords Climate change · Disasters · Migration · Coastal area · Bangladesh

Introduction

Climate change-induced disasters and associated migration are a current issue worldwide (Celekli et al., 2023). The global climate change-induced migration is increasing rapidly (Milán-García et al, 2021). The South Asian and Southeast Asian countries face adversity and migrate to other destinations (Nolos, 2025). Bangladesh is a land of migration due to climate-induced disasters, especially cyclones, floods, erosion, storm surges, and salinity (Hossain et al., 2022). Climate change is a challenge for the twenty-first century (Tajrin and Hossain, 2017). Developing countries face significant problems, particularly due to their limited resources and large populations (Yousafzai et al., 2022). The climatic components are changing, and the results are evident in the increasing frequencies and intensities of disasters, especially floods, cyclones, erosion, and storm surges (Hossain et al., 2020b). The impact of climate change is severe on the environment, economy, and social and political aspects of coastal and char land communities (Uddin et al., 2022). Climate change affects the lives and livelihoods of people worldwide, including economically advanced countries (Wrathall et al., 2019). The number of migrants is increasing, with climate change being a significant factor (Wang et al., 2020). The other migrants are not bigger than the climate change-induced displaced people (Brzoska and Froshlich, 2016).

Climate change and environmental degradation influence human migration, particularly in the aftermath of ecological and climatic depilation (IOM, 2009). The number of disasters has doubled over the last two decades, from 200 to 400 events, and nine out of ten disasters are climate-related, especially floods, cyclones, and storm surges. According to the Norwegian Refugee Council (NRC), approximately 20 million people migrated due to climate-induced sudden natural disasters (Elverland, 2009). Sudden onset disasters, including floods, cyclones, storm surges, and earthquakes, resulted in approximately 165 million people migrating between 2008 and 2013 (Islam and Shamsuddoha, 2017). Climate change-induced migration is projected to reach 150 million by 2050 (IPCC, 2019).

Climate change affects three ways: firstly, sea level rise destroys low-lying land, particularly affecting soil fertility, clean water, and salinity, as well as causing submersion; secondly, warming and drying on the Earth's surface create trouble for the ecosystem; and thirdly, heavy precipitation leads to riverine floods and flash floods (Islam and Shamsuddoha, 2017). Climate change-induced disasters are occurring due to rapid urbanization, high population density, and increased settlement (MoEF, 2008). The socioeconomic and cultural factors that encourage people to migrate to destinations (Mallick and Vogt, 2014). It is difficult for Bangladesh, as a developing country, to settle migrants originating from climate change-induced disasters (Ahsan et al., 2011).

People affected by disasters caused by climate change, who experience economic losses to their livelihoods, homes, and lives, then flee to other destinations within the country (Islam et al., 2022). Local communities are migrating not to fulfill their basic needs due to the regular occurrence of disasters in the char lands and coastal areas of Bangladesh (UN-Habitat, 2015). Poor and vulnerable individuals are disproportionately affected during disasters, particularly children, women, and older adults (Cutter, 2017). Push–pull factors are responsible for climate change-induced migration in worldwide (Penning-Rowsell et al., 2013).

Bangladesh faces a significant threat from climate change-induced disasters, and it is assumed that if sea level rise exceeds 80 cm, 30 million people will be forced to leave their homes (Barua et al., 2017). Every five years, 50 million people are affected by disasters, and every three years, Bangladesh faces a major cyclone (Dasgupta et al., 2014). The southern part of Bangladesh faces challenges to its income and property due to regular cyclones and associated tidal surges, floods, and salinity (Nowreen and Mohiuddin, 2017).For example, 61% of the country was inundated, and 45 million people became homeless in the 1998 flood (Riaz and Mohiuddin, 2025; Alam et al., 2017). Climate change and migration are interconnected; climate change-induced disasters influence the displacement of coastal communities (Mallick and Vogt, 2014; Kniveton et al., 2008). The migration is a consequence of socioeconomic factors, especially social security, health hazards, income, poverty, livelihood, and unemployment (Hasan and Foliente, 2015; Felsenstein and Lichter, 2014). People migrated internally to nearby areas during Cyclone Aila due to the loss of physical resources, occupation, and financial vulnerability (Islam and Hasan, 2016). Fishing communities in southern Bangladesh faced challenges during Cyclone Aila (Islam and Herbeck, 2013).

Young women and children are facing sexual harassment during the disasters, which instigates the displacement of coastal people (IOM, 2009). Following Cyclone Aila, approximately 100,000 people migrated from the four upazilas Koyra, Paikgacha, Dacope, and Batiaghata to seek employment (Islam and Hasan, 2016). The coastal people faced different types of adversities, and people lost their livelihood and property due to climate change (Islam et al., 2024). The loss of livelihoods and property, primarily agricultural land, fishing farms, livestock, and businesses, resulted in many settlements being abandoned (Mohiuddin and Nowreen, 2023). Various types of disasters affected the coastal area in Bangladesh, including cyclones, floods, storm surges, salinity, and erosion (Tandrila et al., 2020; McNamara et al., 2018; Mohiuddin and Latif, 2013). As a result, coastal residents lost their property and decided to migrate temporarily or permanently as an adaptation strategy (Ghani and Hidayati, 2024; Bardsley and Hugo, 2010). People are displaced mainly due to the lack of private and government services in coastal areas (Paul and Islam, 2015). The study highlighted migration scenarios resulting from climate change-induced disasters in the southern coastal area in Bangladesh (Fig. 12.1).

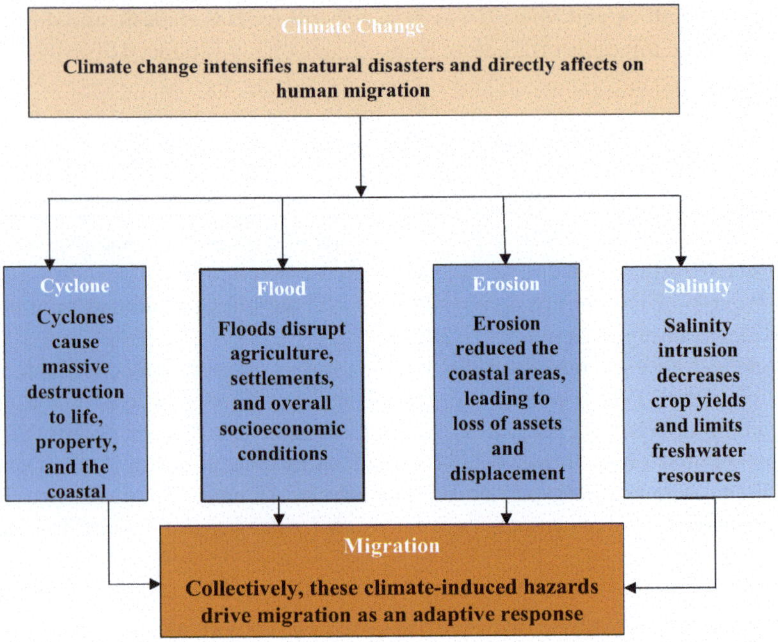

Fig. 12.1 Migration from climate change-induced disasters (Author prepared, 2025)

The objective of the study is to identify the migration scenarios resulting from the climate-induced disasters in the southern coastal area. The goal highlighted the impact of climate-induced disaster risk reduction on the local and international communities (Fuso et al., 2019). Sustainable Development Goals (SDGs) comprise 17 goals, of which the 13th focuses on climate action. Climate action (13th goal of Sustainable Development Goals, SDGs) is connected to the 17 goals of the Sustainable Development Goals (SDGs), and climate-induced disasters affect communities (Filho et al., 2023). The study found that people migrated from the coastal area due to cyclone devastation. The frequent and severe cyclone impacts on livelihood and income. Moreover, people of southern Bangladesh face flooding, erosion, and salinity every year. During cyclones and surges, coastal areas are submerged when the embankments break down, and high tides with heavy rain lead to flooding, erosion, and salinity. The communities lost their agricultural fields, fishing farms, businesses, jobs, and working opportunities due to disastrous situation stemming from climate change. The coastal people are migrating to other areas for shelter, food, and income. The impact of climate-related disasters is unavoidable for the world communities (Ripple et al., 2024). The worldwide impact of climate change caused disasters in the coastal area of Bangladesh as well (Donatti et al., 2024). The study's findings would help to fill the gaps, enabling policymakers and researchers to address these challenges effectively in the local community and world communities as well.

Methodology

Data Sources

The study has used both primary and secondary data. Secondary data have been collected from various journals, books, and online sources in the first step. The secondary data and literature review helped to develop draft questionnaire in the second step of primary data collection. The sampling was determined using tools and techniques prepared in the third step along with questionnaire survey and image analysis. The images of the study area were collected to determine the disasters, obtained from the USGS. After the questionnaire survey, the collected data were analyzed using descriptive statistics. The image analysis helped to triangulate the result and information in the fourth step (Fig. 12.2).

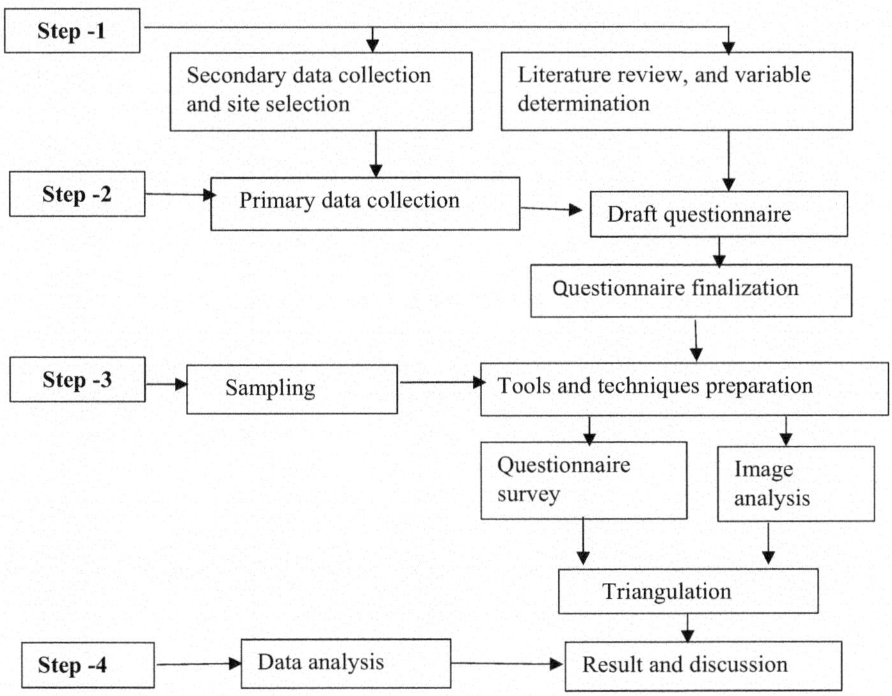

Fig. 12.2 Flowchart of methodology (Author prepared, 2025)

Sample Design and Data Collection

(a) **Sampling**

The sample site has been selected as the three upazilas from the three districts: Assasuni upazila from Satkhira district, Koyra upazila from Khulna district, and Sarankhola upazila from Bagerhat district. The study randomly selected three unions among the three upazilas. The total population of the three unions is 82088, and the population was surveyed proportionately according to each area's population size (Table 12.1).

The study used the Yamane formula $n = N/(1 + N^*e^2)$, where n = sample size, N = total population, and, e = margin of error (Sultana et al., 2024). Now, N = 82,088, e = 0.05 (5%), for 95% confidence level, and n = 398. Approximately 400 questionnaires were collected from the field through purposive sampling (Mohiuddin, 2022).

(b) **Data collection tool and techniques**

Questionnaire and Focus Group Discussion (FGD) have been used for primary data collection. The semi-structured and FGD focused on the climate change-induced disasters' impact in the coastal area as follows.

The FGD was conducted for the development of the semi-structured questionnaire. Three Focus Group Discussions (FGDs) were conducted in the field, with each FGD comprising more than seven (7) persons from different occupational backgrounds. The information from FGDs pertained to migration resulting from climate change-induced disasters, explicitly identifying the disasters responsible for the adversity and migration. The semi-structured questionnaire consisted of three sections: demographic information, climate-induced disasters, and the impact of disasters on migration. In the questionnaire, a five-point Likert scale (strongly agreed = 1, agreed = 2, neutral = 3, disagreed = 4, and strongly disagreed = 5) has been used (Mohiuddin and Hossain, 2023; Kuvan & Akan, 2005). The respondents in the questionnaire were from diving occupational backgrounds and were aged 18 to 60. The questionnaire was surveyed from 15 October 2022 to 30 October 2022.

Table 12.1 Sampling population

Upazila	Union	Population	Number of samples
Koyra	Koyra	33,230	161
Assasuni	Baradal	28,037	136
Sarankhola	Dhansagar	20,821	101
Total population = 82,088			

Source BBS, 2011

Outcome and Explanatory Variables

(I) Variable determination

The variable determination is based on the literature review (Table 12.2). The study found that climate change-induced disasters, especially cyclones, are associated with increasing magnitude, loss, and damage. The coastal people face not only flooding but also cyclones and storm surges, which inundate low-lying areas in the southern Bangladesh (Table 12.2). Erosion damaged the embankment and agricultural land, and many people lost their settlement as a result. Salinity has a long-term effect on crop fields and homestead vegetation originating from cyclones, tidal surges, and flooding (Table 12.2).

Analytical Strategies

(i) Primary data analysis

The primary data was collected through a semi-structured questionnaire. The data are scrutinized and organized for analysis. The event climate change induced major disasters, cyclones, floods, erosion, and salinity impact on migration. Variables related to disasters were measured using tools, primarily questionnaires. The analysis was conducted using SPSS 26.0. The chart and table (descriptive statistics) present disasters-induce migration in coastal areas (Table 12.3).

(ii) Image analysis

The images for preparing the Cyclone Affected Area Map, Storm Surge Risk Level Map, Soil Erosion Hotspots Map, and Salinity Status has been developed through

Table 12.2 Variable determination

Event	Disasters	Variables	Reference
Climate change	Cyclone	Cyclone frequency, cyclone magnitude, cyclone impact, loss of occupation, income, property loss	Mallick and Vogt, 2012; Islam et al.2024
	Flood	Flood inundation, loss of agricultural farm, fishing farm, loss of embankment	Kaczan and Orgill-Meyer, 2020; Raiz and Mohiuddin, 2025; Fei al al., 2025
	Erosion	Loss of embankment, loss of settlement, agricultural land	Rahman et al.2025; Zhang and Zhuang, 2019; Mohiuddin, 2022
	Salinity	Loss of crops, loss of fertility, loss of homestead vegetation	Chen and Mueller, 2018; Rakib et al., 2019

Source Author prepared, 2025

Table 12.3 Analytical strategies

Event	Disasters	Variables	Mesuring scale	Tool and techniques	Analysis
Climate change	Cyclone	Cyclone frequency, cyclone magnitude, cyclone impact, loss of occupation, income, and property loss	Numeric and image	Image analysis, questionnaire survey	Descriptive
	Flood	Flood inundation, loss of agricultural farm, fishing farm, loss of embankment	Numeric and image	Image analysis, questionnaire survey	Descriptive
	Erosion	Loss of embankment, loss of settlement, agricultural land	Numeric and image	Image analysis questionnaire survey	Descriptive
	Salinity	Loss of crops, loss of fertility, loss of income	Numeric and image	Image analysis, questionnaire survey	Descriptive

Source Author prepared, 2025

a systematic literature review from which related data on cyclone affected locations, surge heights, erosion rates, and salinity values together with some qualitative descriptions have been extracted from published studies and reports (Islam et al., 2017). The information has then been geocoded to administrative units or geographic features and standardized into comparable scales. Qualitative descriptors, such as 'high', 'moderate', and 'low', and quantitative values were translated into severity scores using set thresholds (Fang et al., 2014). Studies were reconciled by weighting of recency and quality. The scores were then aggregated and spatialized through interpolation (IDW/Kriging), kernel density estimation (where the data was point/line based), or direct assignment to the polygon, followed by a classification scheme that could be either natural breaks or threshold values into low, moderate, and high classes (Needham et al., 2015). Companion confidence layers were based on evidence from visually interpreted patterns in secondary sources, rather than on primary measurements (Table 12.3).

Site Selection for Study

Bangladesh is a disaster-prone country (Nahar et al., 2025; Mohiuddin et al., 2021; Tondrila et al., 2020; Toufique and Islam, 2014). The coastal area of Bangladesh comprises nineteen districts, covering approximately 2.85 million hectares, and has a length of 200 km, with crisscrossing rivers totaling 1,48,000 km (BBS, 2011). The population is concentrated in low-lying areas and has a high population density in the coastal districts (World Bank, 2012). The southern three coastal districts, including Khulna, Satkhira, and Bagherhat, are most vulnerable to floods, cyclones, storm surges, and salinity (Akhter et al., 2012). Most of the cyclones that have hit this area, especially Sidr, Nargis, Amphan, and Aila, have resulted in cyclone-related disasters, including floods, storm surges, and salinity intrusions in the coastal area (IOM, 2009). The migration is part of the adaptation process in the southern coastal area of Bangladesh (Mallick and Etzold, 2015). Bangladesh is a disaster-prone country, where the southern part, especially the Satkhira, Khulna, and Bagherhat districts, is most vulnerable to climate change-induced disasters (Fig. 12.3). This area has been selected for the study.

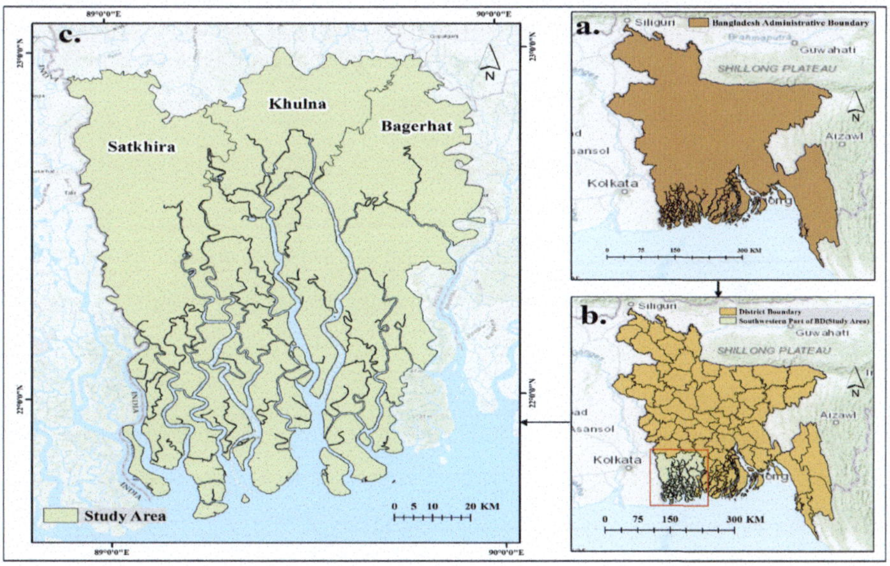

Fig. 12.3 Study area (Author prepared, 2025)

Results

The study highlighted major disasters, particularly cyclones, floods, erosion, and salinity in the southern part of Bangladesh. The study revealed the result as follows.

(i) **Migration by coastal cyclone**

Most (51%) of the people in the coastal area affected by the cyclone strongly agreed that the cyclone was the main reason for their migration, while 47.5% strongly agreed that the cyclone's impact was severe in the area. Climate change impacts people's lives directly or indirectly in several ways, and displacement is one such indirect effect. The cyclone affected coastal people who had migrated to nearby areas due to their financial vulnerability, loss of physical resources, insecurity, unemployment, lack of housing, health problems, and limited access to employment. Most people who study the issue believe that population migration is increasing daily due to climate-induced cyclones in coastal areas (Fig 12.4).

The Fig. 12.5 shows the cyclone-affected areas in Khulna, Bagerhat, and Satkhira districts in southern Bangladesh. The most vulnerable and highly affected areas are located very close to the coast. The affected areas are marked in red on the map, very near the coast; the moderate area is marked yellow in the middle of the map; and the green area is found in the Satkhira district. People living in red and yellow zones are most vulnerable to cyclones, and seasonal and permanent migration is present in coastal areas (Fig. 12.5).

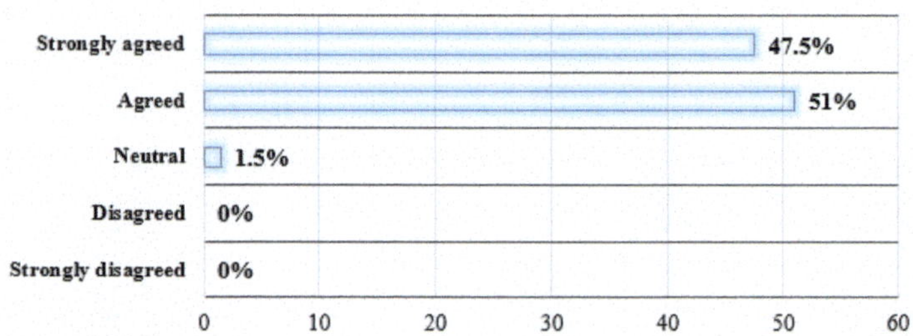

Fig. 12.4 Population migration by cyclone (*Source* Field survey, 2022)

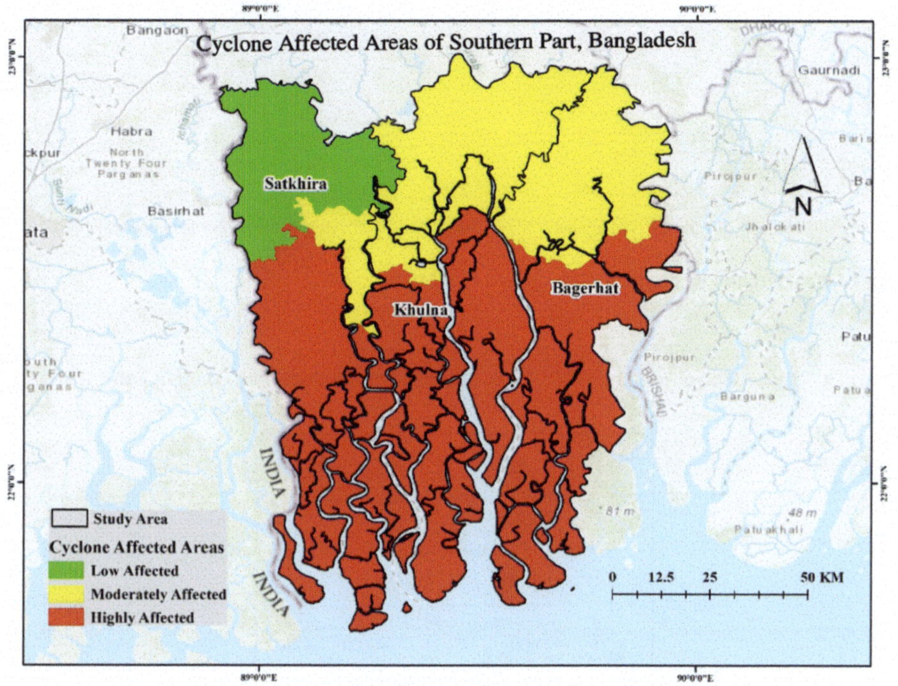

Fig. 12.5 Cyclone-affected area (Author developed, 2025)

(ii) Migration by coastal flood

Figure 12.6 shows that the maximum (53%) of people in the coastal area strongly agreed that the coastal flood extensively damages or causes losses of houses, property, and infrastructure (house, sanitation facilities, drinking water sources, etc.), human life, livestock, and agriculture and livelihoods. 46% of the respondents agreed that the agricultural fields, fishing farms, livestock, and businesses were affected by flood inundation. That is why people migrated from one place to another to avoid the loss. Repeated exposure to such disasters gradually declines the household's assets and property. Therefore, people migrate from this place to escape this problem.

Figure 12.7 shows the flood-affected area in the southern part of Bangladesh, where map-a shows the 1987, map-b shows the 1988, and map-c shows the 1998 flood in the country, along with the inundation area. In 1987, the flooding was a remarkable event. In the southern portion, it was more severe in 1988 and 1998 compared to 1987. People lost their property and occupational opportunities, especially settlements, agricultural fields, and social vegetation.

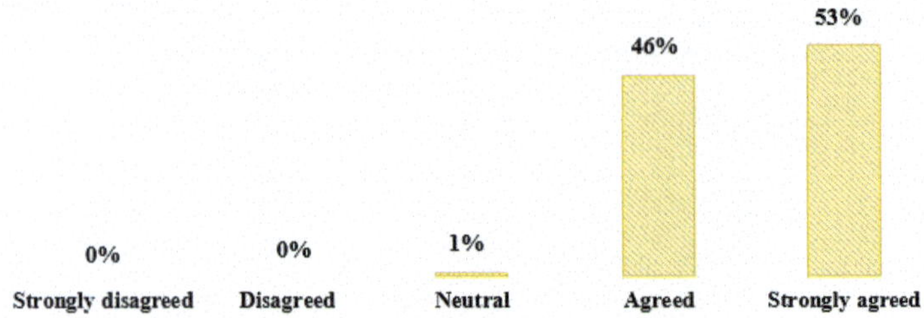

Fig. 12.6 Migration by coastal flood (*Source* Field survey, 2022)

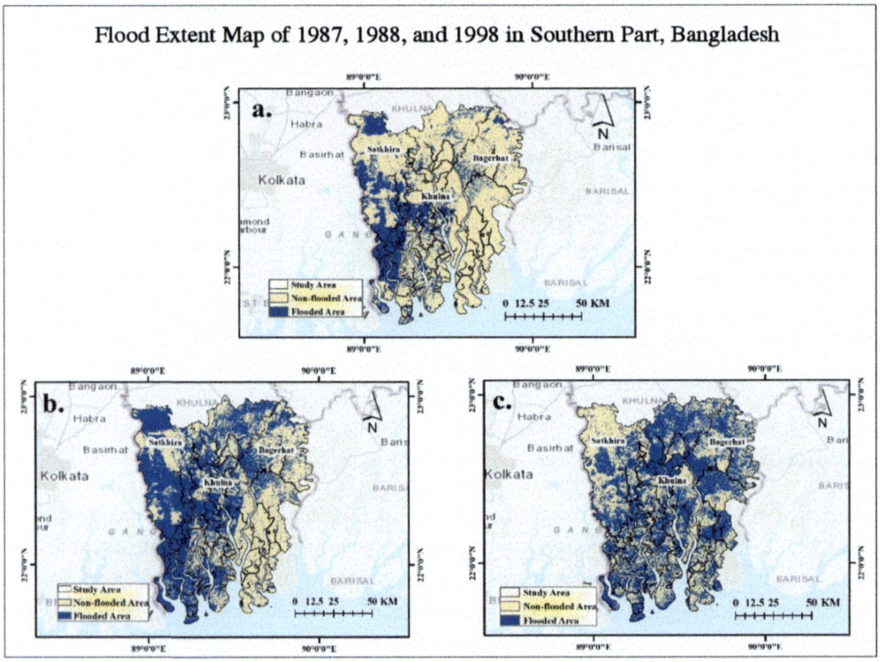

Fig. 12.7 Flood-affected area (Author developed, 2025)

Coastal people lost their income sources during the flood, especially in agriculture, fishing, and business. People are trying to survive from the adversity by migrating to safe place in the area. The maps show the flooding conditions in the area in 1987, 1988, and 1998.

(iii) **Coastal migration by erosion**

Approximately 47% of the respondents agreed that the riverbank erosion caused by the river stream had made them homeless. The southern coastal region of Bangladesh is affected by erosion from river streams during tidal surges, and heavy rain leads to flooding, which is a significant problem related to climate change. River stream

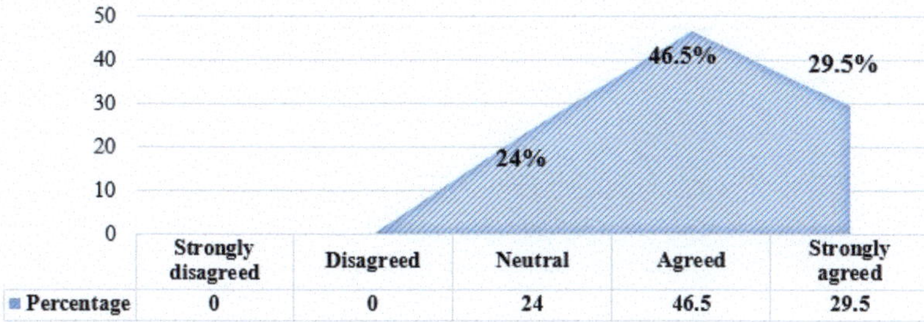

Fig. 12.8 Migration by river bank erosion (*Source* Field survey, 2022)

erosion damages standing crops and seed beds, causing death, injuries, and outbreaks of waterborne diseases in the study area. To avoid this uncertain situation, people are migrating from here to different destinations at a higher rate than in any other region of Bangladesh. 29.5% of the coastal people strongly agreed that they lost their settlement and land, and that the embankment was damaged by erosion in the coastal region. Thus, coastal people migrate to safer places to overcome adversity (Fig. 12.8).

Coastal erosion is increasing due to flooding and storm surges, particularly in southern Bangladesh, including the districts of Satkhira, Khulna, and Bagerhat. The southern Satkhira district, especially the partial areas of Assasuni and Shyamnagar Upazilas eroded during the rainy season; some parts of Koyra Upazila in Khulna district eroded due to tidal surges, and Soronkola Upazila in Bagerhat district eroded during the flooding season as well (Fig. 12.9).

(iv) **Migration by coastal salinity**

The majority of respondents (62.5%) strongly agreed that people are migrating for various climatic-related reasons, including coastal salinity. Salinity primarily affects land and water in coastal areas (Fig. 12.10). As a consequence of climate change, it is gradually extending towards inland water and soil. Coastal salinity intrusion renders agricultural land unsuitable for crop, and fodder production, and freshwater fisheries.

The study reveals that coastal residents are migrating due to a lack of drinking water, as most surface water sources in the coastal area are inundated by saline water. Every year, coastal people face challenges in their agricultural fields due to saline water inundation, which makes the farmland unfertile. Soil salinity is increasing in the coastal districts, particularly in Khulna, Satkhira, and Bagerhat. Coastal people are migrating due to rising salinity in Khulna, Satkhira, and Bagerhat districts. Local people lost their agricultural fields, fishing farms, and social vegetation due to saline water intrusion in the mainland during the monsoon and tidal flood. People lost their income sources and economic activities due to cyclone and surge-induced flooding, and people migrated to avoid the adversity in the coastal area (Fig. 12.11).

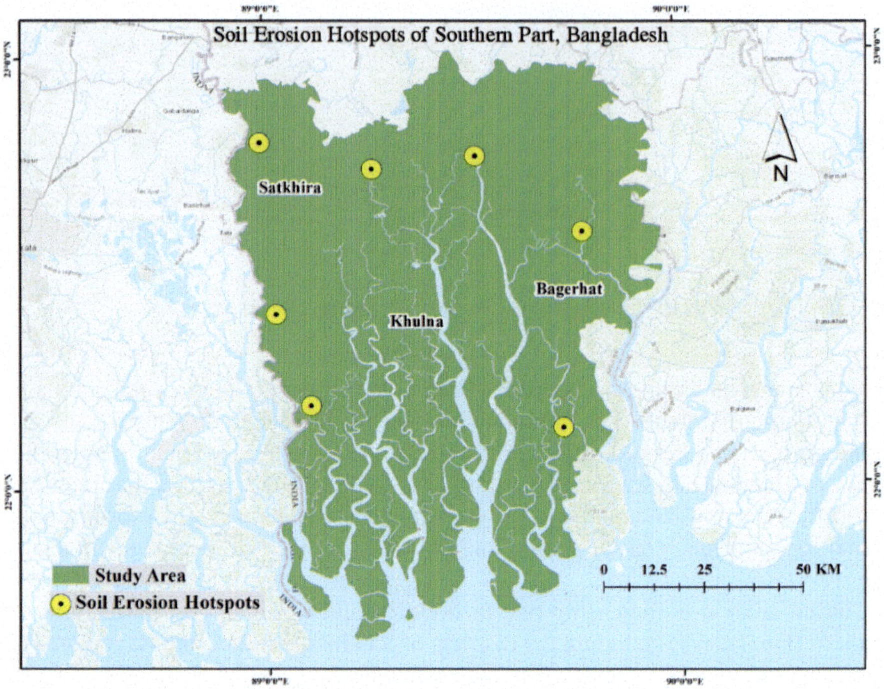

Fig. 12.9 Erosion in the southern part of Bangladesh (Author developed, 2025)

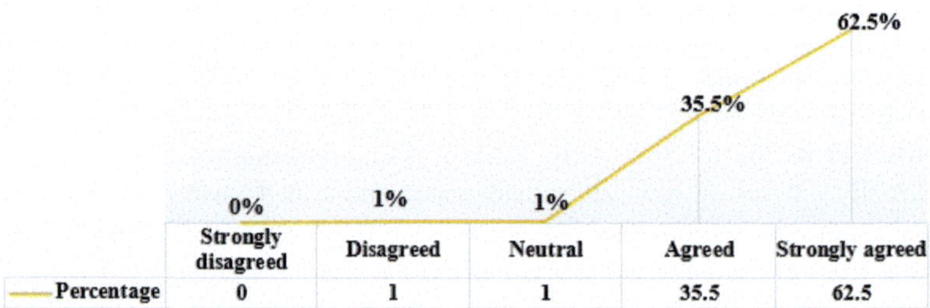

Fig. 12.10 Migration by coastal salinity (*Source* Field survey, 2022)

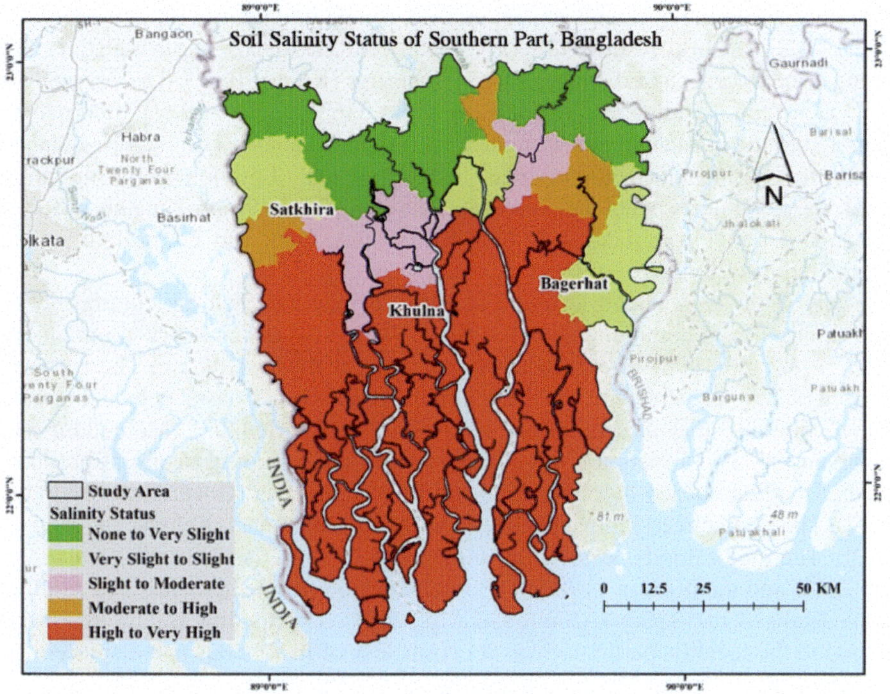

Fig. 12.11 Salinity-affected area (Author developed, 2025)

Discussion

The study focused on the impact of climate change-related disasters on migration in the coastal area of Bangladesh. Gradually, adversity is increasing because of the changing climate, especially cyclones, floods, erosion, and salinity. People are migrating due to cyclone devastation in coastal areas, while flood inundation also occurs. Another reason for migration in coastal areas is erosion that happens during surges and flooding. Coastal residents are migrating to avoid the increased coastal salinity associated with intensifying climate change. Coastal disasters are increasing due to global warming, including cyclones, floods, erosion, tidal surges, and salinity (Nowreen and Mohiuddin, 2017). The loss and damage are increasing because of the intensification of coastal disasters, particularly evident in the loss of houses, agricultural land, and property (Mohiuddin and Latif, 2013). Climate-induced migration is increasing because exposed and working people cannot avoid adversity (Islam et al., 2024).

The majority of people in the coastal area affected by the cyclone lost their settlement, communication, and income, according to the respondents. People are migrating to other places to avoid the loss and damage. The study showed that Khulna and Bagerhat districts are more affected than Satkhira district. People are migrating due to the cyclone's impact on the coastal community (Mallick and Vogt, 2012). Cyclone Gorky resulted in approximately 138,000 deaths due to high tides and strong wind speeds, and damaged many properties (Khalil, 1993). The devastation of Cyclone Sidr resulted in the deaths of about 3363 people, and 6,50,000 people migrated to the coastal area of Bangladesh (EM-DAT, 2016). The cyclone-induced storm surge during Aila in 2009 resulted in approximately 1 million people migrating and 1,750,000 homes being damaged (Higgins et al., 2014). Approximately 400,000 hectares of agricultural land were lost due to inundation, and 2.5 million people migrated during Cyclone Amphan (Kelman et al., 2018).

Many of the respondents strongly agreed that they lost their agricultural fields, fishing farms, and communication by flood. Flood disasters are very common but 1987, 1988 and 1998 were particularly notable for the loss of life. The research showed on map the adversities in the southern coastal area. Coastal people migrated due to flooding that affected their settlements and agriculture (Mohiuddin and Nowreen, 2023). Farmers are migrating to avoid the loss of agricultural products during flooding caused by climate change in coastal areas (Government of India, 2017). The flood directly impacted livelihoods and settlements (Kelman and Ahmed 2020). The major floods occurred in Bangladesh in 1974, 1987, 1988, 1998, 2004, and 2017, and the flood in 1998 left about 25 million people homeless and resulted in the deaths of 1,050 people (Simmond et al., 2021). The flooding condition is detrimental to the locality, particularly as it causes loss of livelihoods and displacement of settlement (Riaz and Mohiuddin, 2025).

The southern coastal people affected by erosion agreed that almost fifty percent of people lost their cropland and settlements to erosion. The research showed that the small islands and coastal areas eroded due to heavy streamflow during the rainy season, surges, and flooding. Local people are migrating due to adversity. The coastal people of Bangladesh, particularly those in the southwestern region, are migrating due to the loss of agricultural land caused by soil salinity (Chen and Mueller, 2018). The char land people faced challenges because it is disconnected from the mainland and is a rapidly developing area prone to erosion and deposition (Karim and Thiel, 2017). Coastal people are migrating due to the loss of settlements, agricultural land, and property caused by erosion (Hasnat et al., 2022). Another group of coastal people who live outside the embankments and char land is affected by salinity, especially as they lose their agricultural land, drinking water sources, and livestock (Islam et al., 2023).

Salinity is a significant challenge for coastal communities in the southern portion of Bangladesh. Many people migrating for their loss especially fertility, cropland, vegetation, drinking water and settlement were found in the study. The coastal people lose their agricultural land, shrimp farms, and livestock, and they face scarcity of drinking water, food, education, and communication (Rahman et al., 2014). The migrants were both temporary and permanent, seeking survival in the area (Mallick et al., 2011). Another group of coastal people who live outside the embankments and char land is affected by salinity, especially as they lose their agricultural land, drinking water sources, and livestock (Islam et al., 2023).

Policy and Recommendation

The study's findings would contribute to knowledge of the local community's coping strategies. Policymakers can obtain guidelines from the research results. The international community can compare to others in terms of the scenerios of the climate change-induced disasters impact resulting in migration in the coastal area Bangladesh. Climate change is a global issue that was highlighted at the UN Climate Change Conference (COP26), held in the United Kingdom, with a focus on achieving zero carbon emissions and limiting global warming to 1.5 °C (COP26, 2021). According to Climate Action Tracker, none of the governments are capable of maintaining the limit (CAT, 2021). Natural hazards increased with changes in climatic components (Ahmed, 2018). The temperature is increasing by 0.20 °C per decade, indicating a gradual rise (Rahman and Lateh, 2017). Bangladesh is considered one of the most climate-vulnerable countries in the world, and its temperature is expected to increase by 1.0 °C in 2030 and 1.4 °C in 2050 (Roth et al., 2021; IPCC, 2019). Bangladesh is one of the five most climate change-vulnerable countries in the world, as the country faces climate change-induced disasters every year, including floods, cyclones, storm surges, erosion, salinity, drought, and landslides (Rakib et al., 2018). Climate change has a profound impact on displaced people in Bangladesh, particularly on the severity and intensity of the disasters (Mohiuddin et al., 2021; Alam et al., 2017).

Strengths and Limitations

The study identified the causes of migration in the southern part of Bangladesh, which will help coastal communities, researchers and policymakers. The limited budget and time were the challenges for the study. The further scope of the study would be to examine the intensity and magnitude of cyclones, floods, erosion, and salinity, and their impact on the local communities and coping strategies.

Conclusion

Climate-induced disasters impact on migration in the coastal area of Bangladesh. Every year, coastal communities experience climate-induced disasters, including floods, cyclones, erosion, storm surges, and salinity, which are often attributed to climate change. The southern part of Bangladesh mainly faces challenges because of losses and damages from climatic disasters. The frequencies and intensity of cyclones are increasing in coastal areas, especially the southern part of Bangladesh. People lost their livelihoods and property due to cyclones, for example, Cyclone Gorky, Sidr, Amphan, and Aila. Migration increases in coastal area because of frequent cyclones. Flood, erosion, and salinity are associated with cyclone disasters in coastal areas. Coastal flooding also affects the loss of agricultural products, jobs, fishing farms, and businesses. The remarkable floods were in 1987, 1988, 1998, and 2004 in Bangladesh. Coastal people migrate to the mainland due to flood. Erosion results from flooding and streams in the coastal region inundate farmland and settlements, prompting migration. Salinity is a challenge in the southern part of the country, driven by cyclones, flooding, and erosion. Saline waters inundate some areas during coastal flooding. Salinity is destroying agricultural fields and water sources. Coastal people migrate to safer places to avoid the salinity. The goal of SDG 13, climate action, aligns with the study's results. The world community will gain practical knowledge from Bangladesh's coastal communities. Global researchers and policymakers can get the knowledge and compare the spatial variation in the impacts of climate-induced disasters on migration. The government and policymakers will receive the information to be cautious about the problems that threaten coastal communities.

Potential Conflicts of Interest

The corresponding author, on behalf of all authors, states that there are no conflicts of interest.

Informed Consent

Informed consent was obtained from all individual participants included in the study.

Data Availability Statement

The datasets used and analyzed during the current study are available from the corresponding author.

AI Discloser

This study did not get support from Artificial Intelligence (AI) and ChatGPT.

Acknowledgements At first, I am grateful to Almighty Allah for giving me the aptitude to complete the research. I would like to thank the authorities of Jagannath University for permitting me to conduct this research work.

I extend my heartfelt thanks to the people of the study area for sharing their information and knowledge regarding the impact of climate change-induced disasters and migration. I sincerely acknowledge Maharab Hossain Moon and the research team members who have assisted cordially and honestly throughout the research process.

Authors' Contribution All authors have contributed equally to this work.

Funding The research was funded by Jagannath University, Bangladesh.

Appendix 1: Questionnaire

Climate Induced Disasters and Its Impact on Coastal Migration: An Empirical Study on the Southern Part of Bangladesh

Sample ID……………… ***/………/ 2021***	***Date: ……..***

Name of interviewee: ……………………………………………………… Age: ………… Sex: ……………

Household Head (H/H) Name: ……………………………………… Age: ………… Sex: ……………

Present Address; ***Past Address;***

Village: ………………………………………… Village: …………………………………………

Union: ………………………………………… Union: ……………………………………………

Thana: ………………………………………… Thana: ……………………………………………

Sample ID……………… ***/………/ 2021***	***Date: ……..***

Name of interviewee: ……………………………………………………… Age: ………… Sex: ……………

Household Head (H/H) Name: ……………………………………… Age: ………… Sex: ……………

Present Address; ***Past Address;***

Village: ………………………………………… Village: …………………………………………

Union: ………………………………………… Union: ……………………………………………

Thana: ………………………………………… Thana: ……………………………………………

District: ……………………………………… District: ……………………………………………

Section A: Demographic Information

Sl no	Household's name	Relation with H/H	Age	Sex	Education	Occupation	Income	Origin
		1.Household head 2.Husband/wife 3. Father/Mother 4. Son/daughter 5.Grandson/granddaughter 6.Brother in law/ Sister in law 5. Brother/ Sister 6. Nephew/nice 7. others		1. Male 2.Female 3.Common	1. Illiterate 2.Primary School 3.Secondary School 4. SSc 5. HSC 6.Graduate 7. other	1. Government service 2. Private/NGO 4. Farmer (Land owner) 5.Farmer (Leaser) 6. Farmer (Day labor) 7. Business 8.Business retails 7. others		*1.origin name* *2. Ancestor*

***1.4** Do you face climate change?*	***Yes***	***No***

If Yes,

a) *How is the nature of climate change?............................*

b) *How do you understand the climate change?....................................*
c) *Do you know the people are migrating from your locality by climate change?...........*
d) *Did you get any help from* ***Government/NGOs***

If yes, How they help you?.............................

Section B: The Scenarios of climate change and disasters at coastal area

1. Do you see any climate change there? Yes or no, if yes give the answer of the following questions;

1. **Changing Nature of Climate Components.**

Qs	Scale/measurement of the parameters (5 = Strongly Agreed, 4 = Agreed, 3 = Neutral (neither Agreed nor Disagreed), 2 = Disagreed and 1 = Strongly disagreed					
	Climate Components Change	Level of measurement related to questions				
		5	4	3	2	1
1	Temperature is gradually being high					
2	Humidity is steadily higher day by day					
3	Wind speed is being higher					
4	Wind movement is changing					

1. Nature of climate change-induced disasters

Qs	Scale/measurement of the parameters (5 = Strongly Agreed, 4 = Agreed, 3 = Neutral (neither Agreed nor Disagreed), 2 = Disagreed and 1 = Strongly disagreed					
	Nature of climate change induced disasters	Level of measurement related to questions				
		5	4	3	2	1
1	Cyclone is increasing from previous time					
2	Cyclone frequency is increasing					
3	Coastal flood frequency is increasing					
4	Erosion is increasing					
5	Erosion intensity is increasing					
6	Coastal salinity rate is increasing					
7	Coastal salinity intrusion is increasing					

Section C: The Scenarios of migration by climate change induced disasters at coastal area

1. Migration by direct impact of climate change induced disasters

Qs	Scale/measurement of the parameters (5 = Strongly Agreed, 4 = Agreed, 3 = Neutral (neither Agreed nor Disagreed), 2 = Disagreed and 1 = Strongly disagreed					
	Migration by direct impact of climate change induced disasters	Level of measurement related to questions				
		5	4	3	2	1
1	Population displacement is increasing by Cyclone					
2	Migration is happening by frequent Cyclone					
3	Coastal Migration is increasing by coastal flood					
4	Migration is alarming issue in coastal area by flood for high rainfall					
5	Migration is increasing by river stream erosion					
6	Migration by coastal salinity is increasing					
7	Migration is increasing for water scarcity by coastal salinity					

2. Migration by loss from the climate change induced disasters

Qs	Scale/measurement of the parameters (5 = Strongly Agreed, 4 = Agreed, 3 = Neutral (neither Agreed nor Disagreed), 2 = Disagreed and 1 = Strongly disagreed					
	Migration by loss from the climate change induced disasters	Level of measurement related to questions				
		5	4	3	2	1
1	Coastal migration is happening by occupation loss					
2	Migration is happening by income loss					
3	Coastal migration is occurring by shifting their traditional occupation					
4	Migration is happened for settlement loss					
5	Migration is caused for taking loan for cropping loss					
6	Migration is occurred by fishing farm loss					
7	Migration is increasing day by day in southern part of coastal area in Bangladesh					

Thank you

References

Ahmed B (2018) Who takes responsibility for the climate refugees? Int J Clim Chang Strateg Manag 10:5–26. https://doi.org/10.1108/IJCCSM-10-2016-0149

Akhter MN, Chakraborty TK, Ghosh GC, Ghosh P, Jahan S (2012) Migration due to climate change from the south-west coastal region of Bangladesh: a case study on shymnagor upazilla, satkhira district. Am J Environ Prot 5(6):145–151

Alam GMM (2017) Livelihood cycle and vulnerability of rural households to climate change and hazards in Bangladesh. Environ Manag 59:777–791. https://doi.org/10.1007/s00267-017-0826-3

Bardsley DK, Hugo GJ (2010) Migration and climate change: examining thresholds of change to guide effective adaptation decision-making. Popul Environ 32(2):238–262. https://doi.org/10.1007/s11111-010-0126-9

Barua P, Rahman SH, Molla MH (2017) Sustainable adaptation for resolving climate displacement issues of south eastern islands in Bangladesh. Int J Clim Change Strateg Manag 9:790–810. https://doi.org/10.1108/IJCCSM-02-20170026

BBS (2011) Bangladesh-population-and-housing-census-2011. Bureau of Statistics (BBS), Bangladesh, p 2011

Brzoska M, Fröhlich C (2016) Climate change, migration and violent conflict: vulnerabilities, pathways and adaptation strategies. Migr. Dev. 5(190):210. https://doi.org/10.1080/21632324.2015.1022973

CAT (2021) Climate action tracker, climate analytics and NewClimate institute. https://climateactiontracker.org/documents/871/CAT_2021-09_Brefing_GlobalUpdate

Çelekli A, Yaygir S, Zariç ÖE (2023) A review of climate change-induced migration. Acta Biol Turc 36(2):3–1

Chen J, Mueller V (2018) Coastal climate change, soil salinity and human migration in Bangladesh. Nat Clim Chang 8:981–985. https://doi.org/10.1038/s41558-018-0313-8

COP26 (2021) UN climate change conference, United Nations framework convention on climate. https://ukcop26.org

Cutter SL (2017) The forgotten casualties redux: women, children, and disaster risk. Glob Environ Chang 42:117–121. https://doi.org/10.1016/j.gloenvcha.2016.12.010

Dasgupta S, Huq M, Khan ZH, Ahmed MMZ, Mukherjee N, Khan MF, Pandey K (2014) Cyclones in a changing climate: the case of Bangladesh. Climate Dev 6(2):96–110. https://doi.org/10.1080/17565529.2013.868335

Donatti CI, Nicholas K, Fedele G, Delforge D, Speybroeck N, Moraga P, Zvoleff A (2024) Global hotspots of climate-related disasters. Int J Disaster Risk Reduct 108:104488. https://doi.org/10.1016/j.ijdrr.2024.104488

Elverland S (2009) 20 million climate displaced in 2008. Norwegian Refugee Council, 8 June. Available at: www.nrc.no/did=9407544

EM-DAT(2016) Percentage of people affected by disaster in bangladesh from 1981 to 2016. International Database for Disasters, Louvain, Université Catholique de Louvain

Environmental justice foundation (2017) Country master, United Nations department of economic and social affairs: population division, New York, NY 10017 USA

Fang J, Sun S, Shi P, Wang J (2014) Assessment and mapping of potential storm surge impacts on global population and economy. Int J Disaster Risk Sci 5(4):323–331. https://doi.org/10.1007/s13753-014-0035-0

Fei K, Du H, Lei CC, Gao L (2025) Assessing the impact of migration activities on flood exposure and evaluating sustainable development areas based on hydrodynamic simulations and gridded migration data. J Hydrol 133425. https://doi.org/10.1016/j.jhydrol.2025.133425

Felsenstein D, Lichter M (2014) Social and economic vulnerability of coastal communities to sea-level rise and extreme flooding. Nat Haz 71(1):463–491. https://doi.org/10.1007/s11069-013-0929-y

Filho WL, Wall T, Salvia AL, Dinis MAP, Mifsud M (2023) The central role of climate action in achieving the United Nations' sustainable development goals. Sci Rep 13(1):20582. https://doi.org/10.1038/s41598-023-47746-w

Fuso Nerini F, Sovacool B, Hughes N, Cozzi L, Cosgrave E, Howells M, Milligan B (2019) Connecting climate action with other sustainable development goals. Nat Sustain 2(8):674–680. https://doi.org/10.1038/s41893-019-0334-y

Ghani MW, Hidayati I (2024) Adapting to sea level rise: migration decisions and disaster mitigation strategies in small-scale fishing communities. In: IOP conference series: earth and environmental science, vol 1314(1), p. 012013. IOP Publishing

Government of India (2017) The economic survey 2016–2017. https://www.thehinducentre.com/multimedia/archive/03193/E

Habitat UN (2015) Informal settlement. UN Habitat, New York

Hasan S, Foliente G (2015) Modeling infrastructure system interdependencies and socioeco nomic impacts of failure in extreme events: emerging R&D challenges. Nat Hazards 78(3):2143–2168. https://doi.org/10.1007/s11069-015-1814-7

Hasnat MA, Chowdhury MA, Abdullah-Al-Mamun MM (2022) Perception of people on climate-induced migration issues in coastal areas of Bangladesh. Migration and Development 11(1):142–162. https://doi.org/10.1080/21632324.2020.1742504

Higgins SA, Overeem I, Steckler MS, Syvitski JPM, Seeber L, Akhter SH (2014) InSAR measurements of compaction and subsidence in the Ganges-Brahmaputra delta Bangladesh. J Geophys Res Earth Surf 119:1768–1781. https://doi.org/10.1002/2014JF003117

Hossain B, Sohel MS, Ryakitimbo CM (2020b) Climate change induced extreme flood disaster in Bangladesh: implications on people's livelihoods in the Char Village and their coping mechanisms. https://doi.org/10.1016/j.pdisas.2020.100079

Hossain B, Shi G, Ajiang C, Sarker MNI, Sohel MS, Sun Z, Yang Q (2022) Climate change induced human displacement in Bangladesh: Implications on the livelihood of displaced riverine island dwellers and their adaptation strategies. Front Psychol 13:964648. https://doi.org/10.3389/fpsyg.2022.964648

IOM (international organization for migration) (2009) Migration, environment and climate change: assessing the evidence. International Organization for Migration, Geneva

IOM (2009) Migration, environment and climate change: assessing the evidence. International Organization for Migration (IOM), Geneva, Switzerland

IPCC (2019) Global warming of 1.5 °C. Available online at: https://www.ipcc.ch/. Accessed 05 Jan 2021

Islam MR, Hasan M (2016) Climate induced human displacement: a case study of cyclone Aila in the Southwest coastal region of Bangladesh. Nat Hazards 81(2):1051–1071. https://doi.org/10.1007/s11069-015-2119-6

Islam MM, Herbeck J (2013) Migration and trans local livelihoods of coastal small-scale fishers in Bangladesh. J Develop Stud 49(6):832–845. https://doi.org/10.1080/00220388.2013.766719

Islam MR, Hossain D (2014) Island char resources mobilization (ICRM): changes of livelihoods of vulnerable people in Bangladesh. Soc Indic Res 117(3):1033–1054. https://doi.org/10.1007/s11205-013-0375-y

Islam MR, Shamsuddoha M (2017) Socioeconomic consequences of climate induced human displacement and migration in Bangladesh. Int Sociol 32(3):277–298. https://doi.org/10.1177/0268580917693173

Islam MdA, Hasan MdA, Farukh MA (2017) Application of GIS in general soil mapping of Bangladesh. J Geogr Inf Syst 09(05):604–621. https://doi.org/10.4236/jgis.2017.95038

Islam ARMT, Nabila IA, Hasanuzzaman M, Rahman MB, Elbeltagi A, Mallick J et al (2022) Variability of climate-induced rice yields in northwest Bangladesh using multiple statistical modeling. Theor Appl Climatol 147(1263):1276. https://doi.org/10.1007/s00704-021-03909-1

Islam MY, Mohiuddin M, Tanvir Hossain K, Salauddin M, Farin S (2024) Trend of heat waves in Dhaka Metropolitan City and its impact on livelihood and health of exposed people. Arab J Geosci 17(8):232. https://doi.org/10.1007/s12517-024-12027-x

Islam R, Ahmed R, Dey B, Haque MS, Aktar S, Bhuiyan MS, Saha N (2023) Salinity hazard drives the alteration of occupation, land use and ecosystem service in the coastal areas: evidence from the southwestern coastal region of Bangladesh. Heliyon 9(8). https://doi.org/10.1016/j.heliyon.2023.e18512

Kaczan DJ, Orgill-Meyer J (2020) The impact of climate change on migration: a synthesis of recent empirical insights. Clim Change 158(3):281–300. https://doi.org/10.1007/s10584-019-02560-0

Karim MR, Thiel A (2017) Role of community based local institution for climate change adaptation in the Teesta riverine area of Bangladesh. Clim Risk Manag 17:92–103. https://doi.org/10.1016/j.crm.2017.06.002

Kelman I, Ahmed B, Esraz-Ul-Zannat M, Saroar MM, Fordham M, Shamsudduha M (2018) Warning systems as social processes for Bangladesh cyclones. Disaster Prev Manag 27:370–379. https://doi.org/10.1108/DPM-12-2017-0318

Khalil GM (1993) The catastrophic cyclone of april 1991: its impact on the economy of Bangladesh. Nat Hazards 8:263–281. https://doi.org/10.1007/BF00690911

Kniveton D, Schmidt-Verkerk K, Christopher S, Richard B (2008) Climate change and migration: improving methodologies to estimate flows. Geneva: Int Organ Migr. https://doi.org/10.18356/6233a4b6-en

Kuvan Y, Akan P (2005) Residents' attitudes toward general and forest-related impacts of tourism: the case of Belek. Antalya Tourism Manag 26(5):691–706

Mallick B, Vogt J (2012) Cyclone, coastal society and migration: empirical evidence from Bangladesh. Int Dev Plan Rev 34(3):217–240. https://doi.org/10.3828/idpr.2012.1

Mallick B, Vogt J (2014) Population displacement after cyclone and its consequences: empirical evidence from coastal Bangladesh. Nat Hazards 73(2):191–212. https://doi.org/10.1007/s11069-013-0803-y

Mallick B, Rahaman KR, Vogt J (2011) Coastal livelihood and physical infrastructure in Bangladesh after cyclone Aila. Mitig Adapt Strat Glob Change 16(6):629–648. https://doi.org/10.1007/s11027-011-9285-y

Mallick B, Ahmed B, Vogt J (2017) Living with the risks of cyclone disasters in the south-western coastal region of Bangladesh. Environments 4(1):13. https://doi.org/10.3390/environments4010013

Mallick B, Etzold B (2015) Environment, migration and adaptation. Evidence and politics of climate change in Bangladesh, 1–14

McNamara KE, Bronen R, Fernando N, Klepp S (2018) The complex decision-making of climate-induced relocation: adaptation and loss and damage. Climate Policy 18(1):111–117. https://doi.org/10.1080/14693062.2016.1248886

Milán-García J, Caparrós-Martínez JL, Rueda-López N, de Pablo Valenciano J (2021) Climate change-induced migration: a bibliometric review. Glob Health 17(1):74

Mohiduddin M, Nowreen S, Uddin MF, Hossain MA (2021) A geographical analysis of thunderstorm in Southern Bangladesh, NGJI. An Int Peer-Rev J NGSI-BHU 67(3):308–321. https://doi.org/10.48008/ngji.1779

Mohiuddin M (2022) Impact of riverbank erosion on livelihood of char land people in Jamalpur District of Bangladesh. Jagannath Univ J Life Earth Sci 7(2):95–108

Mohiuddin M, Latif MB (2013) Housing condition of coastal area in bangladesh: a case study of kutubdia. Cox's Bazzar, J Environ Sci Nat Resour 6(1):15–20. https://doi.org/10.3329/jesnr.v6i1.22034

Mohiuddin M, Hossain MA (2023) Determining the impact of Rohingya forced migration on the natural environment of tourism destination in Bangladesh. GeoJournal Tour Geosites 50(4):1506–1516. https://doi.org/10.30892/gtg.50430-1148

Mohiuddin M, Nowreen S (2023) Climate change-induced livelihood vulnerabilities in south western coastal region of Bangladesh. Jagannath Univ J Life Earth Sci 9(2):163–178. https://doi.org/10.3329/jnujles.v9i2.72922

Mohiuddin M, Mesbah UddinMM, NowreenS, Nahian A (2021) Tackling climate change effects and sustainable tourism products and services development through innovation in bangladesh, technology application in the tourism and hospitality industry of Bangladesh, Azizul Hassan(eds.), Springer Nature Singapore Pte Ltd. Singapore, pp 284–301. https://doi.org/10.1007/978-981-16-2434-6

Nahar N, Mohiuddin M, Nowreen S, Hossain MA (2025) Influential factors of landslide occurrence in the south-eastern part of Bangladesh. IDRiM J 15(2). https://doi.org/10.5595/001c.143100

Needham HF, Keim BD, Sathiaraj D (2015) A review of tropical cyclone-generated storm surges: global data sources, observations, and impacts. Rev Geophys 53(2):545–591. https://doi.org/10.1002/2014RG000477

Nolos R (2025) Drivers and patterns of disaster-induced migration in Southeast Asia. Asian Pac Migr J 34(3):547–559

Nowreen S, Mohiuddin M (2017) An assessment on application of indigenous knowledge as disaster risk reduction strategies in Kutubdia. Cox's Bazar, Bangladesh, Dhaka Univ J Sci 65(2):125–132. https://doi.org/10.3329/dujs.v65i2.54520

Paul S, Islam MR (2015) Ultra-poor char people's rights to development and accessibility to public services: a case of Bangladesh. Habitat Int 48:113–121. https://doi.org/10.1016/j.habitatint.2015.03.018

Penning-Rowsell EC, Sultana P, Thompson PM (2013) The 'last resort'? population movement in response to climate-related hazards in Bangladesh. Environ Sci Policy 27(Suppl. 1):S44–S59. https://doi.org/10.1016/j.envsci.2012.03.009

Rahman MA, Hossain ML, Foysal MA, Asma Akter AA (2014) Impact of climate change on food security in south-west coastal region of Bangladesh. ISSN: 2231–0606

Rahman MN, Sharifuzzaman SASM, Sarker MNI, Lo K(2025) Socioeconomic impacts and migration dynamics of riverbank erosion. NPJ Clim Action 4(1):102. https://doi.org/10.1038/s44168-025-00306-6

Rahman MR, Lateh H (2017) Climate change in Bangladesh: a spatio temporal analysis and simulation of recent temperature andrainfall data using GIS and time series analysis model. Theor Appl Climatol 128:27–41. https://doi.org/10.1007/s00704-015-1688-3

Rakib MA, Sasaki J, Matsuda H, Fukunaga M (2019) Severe salinity contamination in drinking water and associated human health hazards increase migration risk in the southwestern coastal part of Bangladesh. J Environ Manage 240:238–248. https://doi.org/10.1016/j.jenvman.2019.03.101

Riaz R, Mohiuddin M (2025) Application of GIS-based multi-criteria decision analysis of hydrogeomorphological factors for flash flood susceptibility mapping in Bangladesh. Water Cycle 6:13–27. https://doi.org/10.1016/j.watcyc.2024.09.003

Ripple WJ, Wolf C, Gregg JW, Rockström J, Mann ME, Oreskes N, Crowther TW (2024) The 2024 state of the climate report: perilous times on planet Earth. Bioscience 74(12):812–824. https://doi.org/10.1093/biosci/biae087

Roth N, Jaramillo F, Wang-Erlandsson L, Zamora D, Palomino-Ángel S, Cousins SAO (2021) A call for consistency with the term's 'wetter' and 'drier' in climate change studies. Environ Evidence 10:1–7. https://doi.org/10.1186/s13750021-00224-0

Sammonds P, Shamsudduha M, Ahmed B (2021) Climate change driven disaster risks in Bangladesh and its journey towards resilience. J British Acad 9(s8):55–77. https://doi.org/10.5871/jba%2F009s8.055

ESRI storymap (2021) Bhola cyclone in 1970. https://www.arcgis.com/apps/MapJournal/index.html

Sultana N, Mohiuddin M, Shanta MA (2024) Assessing the impact of Covid-19 on workers of the informal sector: evidence from urban slums in dhaka north city corporation, Bangladesh. Jagannath Univ J Life Earth Sci 10(2):125–142. https://doi.org/10.3329/jnujles.v10i2.85248

Tajrin MS, Hossain B (2017) The socio-economic impact due to cyclone aila in the coastal zone of Bangladesh. Int J Law Hum Soc Sci 1:60–67. ISSN:2521–0793

Tandrila TA, Mohiuddin M, Hossain MA, Rahman MS, Samad A (2020) Causes of landslides in Chittagong hill tracks Bangladesh, NGJI. An Int Peer-Rev J NGSI-BHU 66(1):42–50

Toufique KA, Islam A (2014) Assessing risks from climate variability and change for disaster-prone zones in Bangladesh. Int J Disaster Risk Reduct 10:236–249. https://doi.org/10.1016/j.ijdrr.2014.08.008

Uddin MJ, Wahiduzzaman M, Islam ARMT, Eibek KU, Nasrin ZM (2022) Impacts of climate modes on temperature extremes over Bangladesh using statistical methods. Meteorol Atmos Phys 134:1–18. https://doi.org/10.1007/s00703022-00868-8

Wang N, Shi G, Zhou X (2020) To move or not to move: How farmers now living in flood storage areas of China decide whether to move out or to stay put. J Flood Risk Manag 13:1–12. https://doi.org/10.1111/jfr3.12609

Wrathall DJ, Mueller V, Clark PU, Bell A (2019) Meeting the looming policy challenge of sea-level change and human migration. Nat Clim Change 9:898–901. https://doi.org/10.1038/s41558-019-0640-4

Yousafzai MT, Nawaz M, Han H, Ariza-montes A, Molina-sánchez H, Vega-muñoz (2022) Assessing socioeconomic risks of climate change on tenant Farmers in Pakistan. Front Psychol 13:870555. https://doi.org/10.3389/fpsyg.2022.870555

Zhang H, Zhuang L (2019) The impact of soil erosion on internal migration in China. PLoS ONE 14(4):e0215124. https://doi.org/10.1371/journal.pone.0215124

Chapter 13
People's Perception of Flood as a Geo-Environmental Driver of Residential Displacement: A Field-Based Study on Katwa Municipality, West Bengal, in India

Tanmoy Basu and Biraj Kanti Mondal

Abstract Floods continue to be one of the most disruptive dangers in riverine and deltaic areas, where communities and livelihoods are intricately linked to hydrological processes, and with climate variability, their effects increasingly result in repeated residential displacement. Notwithstanding comprehensive embankment building, the enduring efficacy of structural flood control measures in mitigating household-level displacement, especially in semi-urban environments, remains inadequately comprehended. This study investigates flood-induced residential displacement to Katwa Municipality from the nearby areas and resultant land-use changes, and household vulnerability. A comprehensive methodological framework integrating multi-temporal geospatial analysis (Landsat and IRS imagery 2001, 2012, and 2023) with perception-based household questionnaires was implemented. Twenty-two indicators were assessed for 60 displaced households utilizing a five-point Likert scale and the analytical hierarchy process (AHP), facilitating the calculation of exposure, sensitivity, and adaptive capacity indices via weighted aggregation and consistency validation. Results indicate significant prior to Ajay Bandh establishment susceptibility, with 98.33% of families undergoing flood-related displacement, primarily over short distances (< 5 km). Over 82% recognized recurrent flooding and extended waterlogging as principal drivers of displacement, while 70–74% reported significant losses in housing and agricultural lands. The AHP results demonstrate significant exposure (0.72), moderate to high sensitivity (0.68), and moderate adaptive capacity (0.54). Despite 78% of households indicating enhanced perceived safety after establishing the Ajay Bandh, the composite vulnerability score remains elevated at 0.86. According to the study, while embankments reduce the perceived severity of floods, practical displacement management calls for comprehensive approaches that

T. Basu (✉)
Department of Geography, Katwa College, Katwa, Purba Bardhaman, West Bengal, India
e-mail: tanmoybasu.2017@gmail.com

B. K. Mondal
Department of Geography, Netaji Subhas Open University, Kolkata, India

P. Ghosh et al. (eds.), *Climate Crisis and Changing Population Dynamics*, Springer Climate, https://doi.org/10.1007/978-3-032-24533-5_13

incorporate structural protection, secure tenure, participatory planning, and targeted livelihood support.

Keywords Flood-induced displacement · Land use land cover · Perception · AHP-vulnerability framework · SWOC · Katwa

Introduction

Displacement due to flooding has become a significant and intricate worldwide geo-environmental issue, especially considering rising climatic variability, fast urbanization, and uneven socio-economic development (IPCC 2022; UNDRR 2021). Floods disproportionately affect low- and middle-income nations with weak institutional resilience and limited capacity for adaptation, accounting for approximately one-third of all disaster-related economic losses and population displacements globally (Warner and Head 2011; Kakinuma et al. 2020; IDMC 2023). Climate change is expected to increase riverine flooding's frequency, intensity, and geographic extent, which will raise the risk of long-term socio-economic instability, livelihood disruption, and involuntary displacement (Ghimire et al. 2015; Castells-Quintana et al. 2022; IPCC 2022). As a result, population dynamics, shifting mobility patterns, settlement patterns, and socio-demographic structures are all significantly influenced by flood-induced displacement in at-risk areas (Black et al. 2011; McLeman 2018). The displacement brought about by floods is one of the most prominent human–environment interactions in South and Southeast Asian deltaic and alluvial river systems (Mirza 2011; Sarker et al. 2019). Empirical studies from Bangladesh, Pakistan, and India show that persistent flooding deteriorates physical landscapes through bank erosion and channel migration, as well as changing governance responses, livelihood systems, and long-term population redistribution dynamics (Roy and Misra 2019; Alam et al. 2020). Although the literature has grown, it has primarily concentrated on large metropolitan regions or rural floodplains, leaving semi-urban towns along the urban–rural border somewhat understudied. Transitional areas play an important role as absorption zones for displaced people, influencing regional demographic reconfiguration, socio-economic integration, and growing patterns of vulnerability (Bezboruah et al. 2021). In accordance with state and national flood management plans, river training initiatives and embankments have long been marketed as crucial structural interventions to lower flood risks in India (NDMA 2008; Sinha 2016). Nonetheless, an increasing amount of data indicates that these interventions frequently reallocate flood risk both spatially and temporally, creating new exposure zones, secondary displacement, and increased demographic pressure in protected areas, instead of removing it (Roy 2012; Mukhopadhyay et al. 2018; Mucherera and Spiegel 2022). This issue is illustrated in the Ajay River Basin in West Bengal. Because of its significant seasonal flow variability, lateral channel movement, and noticeable bank erosion, the non-perennial Ajay River has historically caused frequent flooding and residential displacement in Katwa Municipality

(Roy 2012; Roy and Misra 2019). Since the construction of the Ajay Bandh, the city center has experienced significantly less widespread flooding; however, overbank flooding, localized waterlogging, and monsoon-induced riverbank erosion persist, leading to intermittent displacement, monetary losses, and ongoing insecurity among marginalized riverine populations (District Disaster Management Plan, Purba Bardhaman 2023; Census of India 2011). These dynamics enable localized population redistribution, livelihood diversification, and modifications to settlement stability, like more general climate-induced demographic shifts (de Sherbinin et al. 2012; McLeman and Gemenne 2018).

The literature on flood hazards and displacement is growing both nationally and internationally, but there are still several important study gaps. According to Tacoli (2017) and Bezboruah et al. (2021), semi-urban towns are vital for climate-induced population adaptation, but they are often overlooked in displacement studies that focus on either large urban centers or rural floodplains. Second, research in river basins like the Ajay has mostly concentrated on hydrological and geomorphological processes, with little attention paid to household-level perceptions, post-embankment adaptations, and socio-spatial displacement dynamics, important factors that influence displacement (Adger et al. 2013; Elum and Lawal 2022). Third, despite the importance of risk perception, governance efficacy, and community resilience being highlighted by recent research, there are still few methodologically integrated approaches that combine geospatial analysis with structured, perception-based decision frameworks, especially in the semi-urban context of India (Mallick and Vogt 2014; Sarker et al. 2019).

The Analytical Hierarchy Process (AHP)-based vulnerability assessment offers a conceptual and methodologically valid approach in this respect. AHP closely aligns with the IPCC's vulnerability concept, which defines vulnerability as a function of exposure, sensitivity, and adaptive capacity (IPCC 2007, 2014, 2022). AHP-based methodologies have been widely used in flood risk and displacement studies to operationalize this framework by assigning relative weight to various physical, socio-economic, and institutional indicators based on household perceptions and expert opinion (Vaidya and Kumar 2006; Malczewski 2006). Empirical evidence from South and Southeast Asia shows that AHP effectively captures unique risk characteristics and migratory behavior in a variety of data-constrained scenarios (Cutter et al. 2008; Rahman et al. 2018; Sarker et al. 2019). Because it enables the integration of qualitative perception data with quantitative environmental indicators while ensuring logical consistency through consistency ratio validation, AHP is particularly suitable for semi-urban riverine settings, unlike purely hydrological or single-variable models (Saaty 1980; Bhushan and Rai 2004; Elum and Lawal 2022; Adger 2016). Considering this, the current study provides a novel and thorough assessment of climate-related residential displacement in Katwa Municipality by combining multi-temporal geospatial analysis of land use and land cover change with perception-based household data evaluated using a five-point Likert scale and the AHP framework. The study focuses on 60 flood-displaced households from Begunkola Mouza that were moved to Ward No. 3 and looks at pre- and post-embankment trends across 22 socio-environmental characteristics.

The specific objectives of the study are:

(i) To examine the climatic situation during the years of major flood events and land use and land cover changes in 2001, 2012, and 2023 in Katwa Municipality.
(ii) To assess the socio-economic conditions and vulnerability framework of flood-induced displaced households.
(iii) To analyze community perceptions on the flood-induced displacement situation of the residents.
(iv) To identify major challenges induced by flooding and propose resilient mitigation strategies toward future sustainability and development. The primary focus of the chapter is directly related to a clear connection between household relocation, settlement reconfiguration, and adaptive responses to flood hazards caused by climate change. By offering globally applicable, transferable insights into how climate stress alters population systems in semi-urban environments, the Katwa case adds to global discussions on climate-induced mobility, adaptive governance, and sustainable resettlement strategies in flood-prone areas of South Asia, Africa, and other developing contexts (UN-Habitat 2020; IPCC 2022).

Materials and Methods

Study Area

Katwa (23.65° N, 88.13° E) is a flood-prone municipality located in the Purba Bardhaman district of West Bengal, India, with an average elevation of around 16.077 m above mean sea level. It is particularly vulnerable to recurrent monsoon-induced riverine flooding because of its position between the Ajay River to the west and the Bhagirathi–Hugli River to the east and north (Ghosh 2012; Bhattacharya and Das 2021). The municipality encompasses 19 administrative wards and has a population density of 9,681 individuals per km^2, covering an area of 8.53 km^2 (Census of India 2011). Katwa, historically noteworthy as the location where Sri Chaitanya Mahaprabhu achieved Diksha in 1510 (Chakraborty 2017), has consistently been susceptible to flooding and riverbank erosion (Banerjee et al. 2018; Bardhan 2015). Although Ajay Bandh's (critically, an earthen embankment to protect the confluence area of the lower Ajay basin near Katwa which consists of an approximate 145 km to 152 km total length in West Bengal) building mitigated extensive floods in certain areas of the town in the years of post-2000, it concurrently modified the hydrological system, resulting in backflow, sediment accumulation, drainage impediment, and localized flooding during heavy rainfall events. Moreover, adjacent low-lying wards, including Ward Nos. 3 and 4, exhibit heightened susceptibility to flooding owing to the degradation of embankments and inadequate maintenance (Mukhopadhyay 2009; Dutta 2010; Nandy 2013; Sinha 2016). Ongoing floods and bank erosion along the Ajay River have compelled numerous households in Begunkola Mouza to

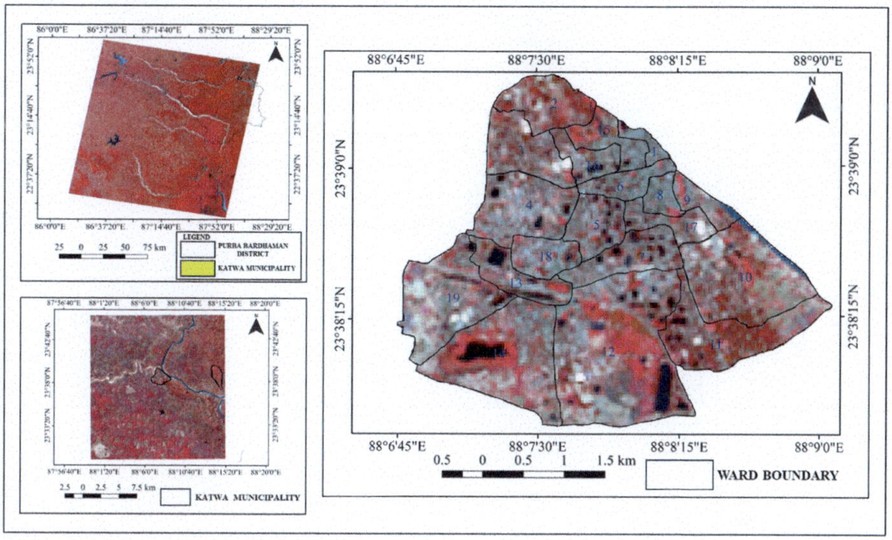

Fig. 13.1 Location map of the study area

relocate either permanently or semi-permanently to Ward Nos. 3 and 4, where land is comparatively inexpensive yet environmentally susceptible (Bardhan 2015; Field Survey, 2023). Recent field study indicates that between 100 and 150 new residences have been constructed in these wards, predominantly by low-income daily-wage families, rendering flood-related displacement a persistent socio-economic challenge rather than a transient relocation (Rahman 2017; Sen 2016; de Sherbinin et al. 2012). Katwa Municipality, particularly Ward Nos. 3 and 4, serve as a pertinent semi-urban case study for analyzing climate-induced flood displacement and evolving population dynamics due to its history of recurrent displacements, risk redistribution from embankments, and constrained adaptive capacity (Tacoli 2017), as illustrated in Figs. 13.1, 13.2, 13.3, 13.4, 13.5, 13.6 and 13.7.

Data Sources

This study investigates the spatial–temporal dynamics of floods and household-level residential displacement in Katwa Municipality through the systematic integration of primary and secondary data sources. Land use land cover (LULC) patterns, flood-prone regions, and elevation attributes were mostly analyzed through secondary databases, whilst primary data offered insights into community perspectives, displacement histories, and socio-economic effects. We acquired multi-temporal satellite imagery from two reputable and commonly utilized platforms: the National Remote Sensing Centre (NRSC) Bhuvan portal and the United States

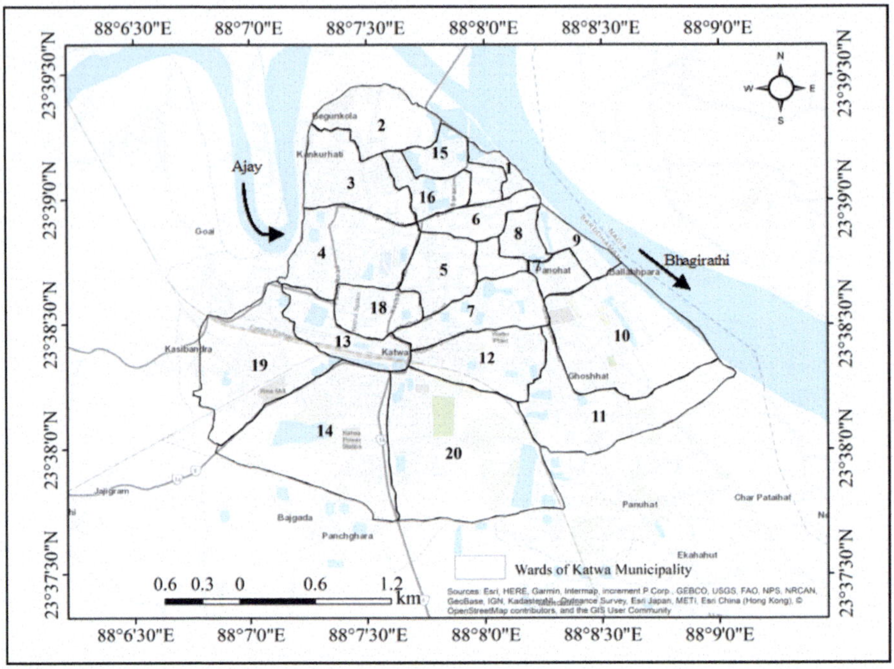

Fig. 13.2 Ward map of Katwa Municipality

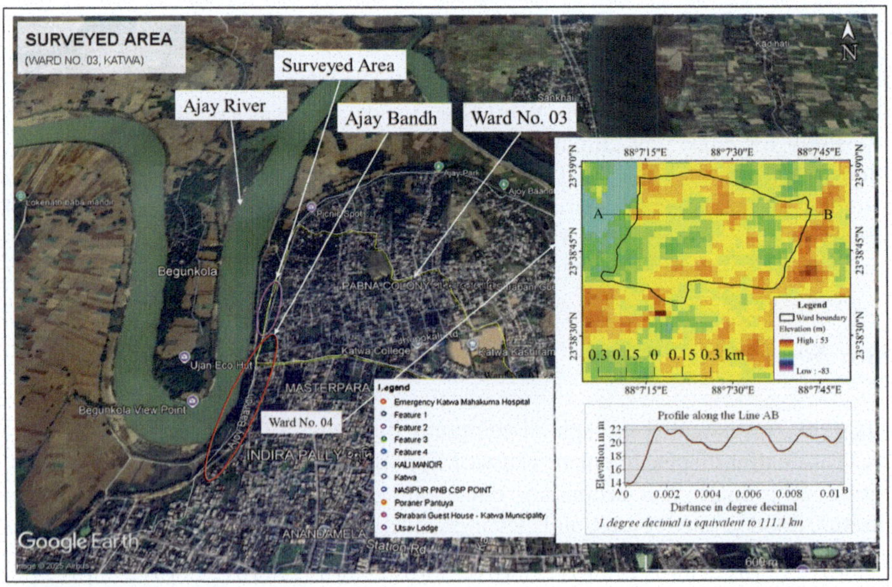

Fig. 13.3 Google Earth image of the survey area

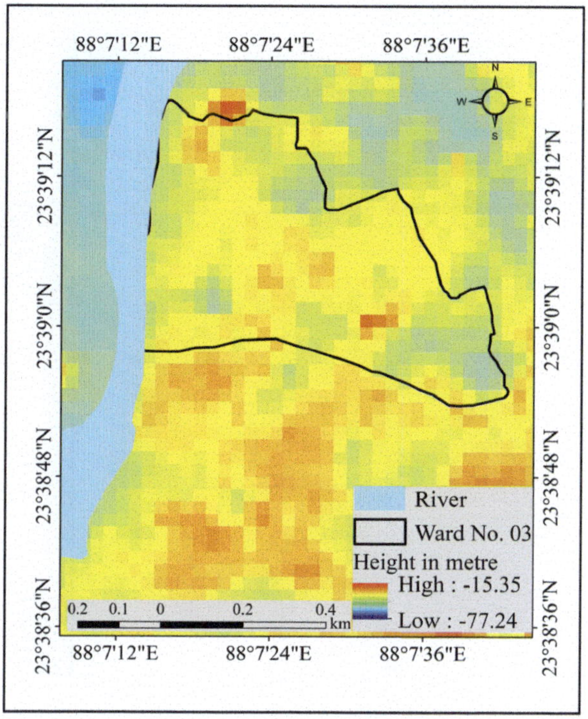

Fig. 13.4 DEM and relief profile of Ward No. 03

Fig. 13.5 Ajay River near Katwa Municipality

Geological Survey (USGS) Earth Explorer. In 2001, historical land use and land cover (LULC) conditions were established utilizing Landsat ETM + imagery with a spatial resolution of 30 m; in 2012, intermediate land cover conditions were illustrated using IRS Resourcesat-1 AWiFS data with a spatial resolution of 56 m; and in 2023, contemporary land-use dynamics were recorded using Landsat OLI/TIRS imagery with a

Fig. 13.6 Ajay Bandh near Ward No. 03

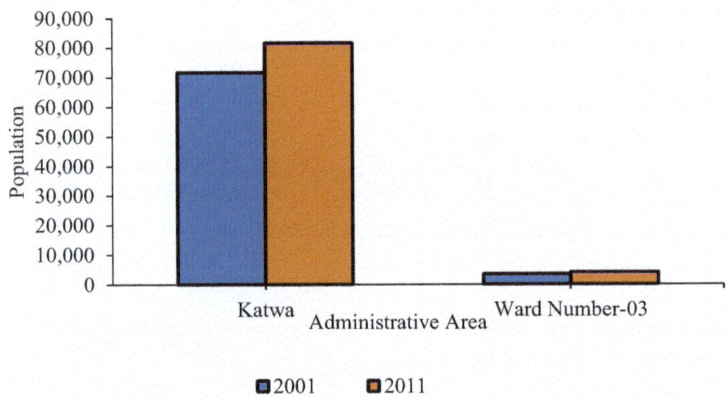

Fig. 13.7 Population of Katwa and Ward Number 03 (2001 and 2011)

spatial resolution of 30 m (USGS 2001; NASA and USGS 2023; NRSC 2012). Digital elevation model (DEM) data with a resolution of 30 m from the Shuttle Radar Topography Mission (SRTM) (USGS 2000) were utilized to generate elevation profiles and pinpoint low-lying regions susceptible to flooding, thereby examining topographical characteristics that affect flood susceptibility (Moore et al. 1991). Weather data were acquired from IMD (2000, 2015, and 2023) to analyze the rainfall and relative

Strongly Disagree	Disagree	Neutral	Agree	Strongly Agree
1	2	3	4	5

Fig. 13.8 5-point rating scale

humidity patterns across specific dates. To contextualize population density, urban pressure, and vulnerability patterns, secondary demographic and settlement-related data were sourced from the 2011 Census of India (Census 2001; 2011). To maintain geographical and temporal uniformity, all secondary datasets underwent geometric rectification, were projected onto a unified coordinate system, and were incorporated into a GIS framework (Coppock and Rhind 1991).

A systematic field survey was conducted to gather primary data in March 2023 in Ward No. 3 of Katwa Municipality, adjacent to the Ajay Bandh. A 5-point Likert scale (Likert 1932) (Fig. 13.8) was employed in a purposive sampling strategy to identify respondents, focusing on households who had undergone residential displacement from the adjacent Begunkola Mouza due to repeated flooding incidents. A survey of sixty homes was deemed sufficient for perception-based analysis in a relatively homogeneous neighborhood impacted by displacement and appropriate for forthcoming Likert-scale and analytic hierarchy process (AHP) analysis (Kothari 2004). To reduce response uncertainty and facilitate participation from respondents with limited literacy, data were gathered via in-person structured interviews employing a pre-tested questionnaire. The questionnaire encompasses flood exposure history, frequency, and duration of relocation, losses of housing and livelihoods, perceived efficacy of flood protection measures, and socio-psychological stress associated with flooding (Patton 1999). Upon encoding, survey results were amalgamated with geographical datasets for analysis. The general methodological framework of the study is represented in Fig. 13.9.

Methods and Techniques

Raster Analysis and Percentage Calculation

Geographic information systems and remote sensing methodologies were employed to generate land use and land cover maps for Katwa Municipality for the years 2001, 2012, and 2023. In 2001, the standard false colour composite (SFCC) technique was employed to visually distinguish agricultural land, vegetation, water bodies, and urban areas based on their spectral reflectance characteristics, utilizing Landsat ETM + data (Lillesand et al. 2015). Comparable categorization methodologies were employed for the Landsat OLI/TIRS (2023) and IRS AWiFS (2012) datasets to facilitate a coherent analysis of temporal land cover alterations. SRTM DEM data were utilized to assess elevation-related flood susceptibility, facilitating the identification

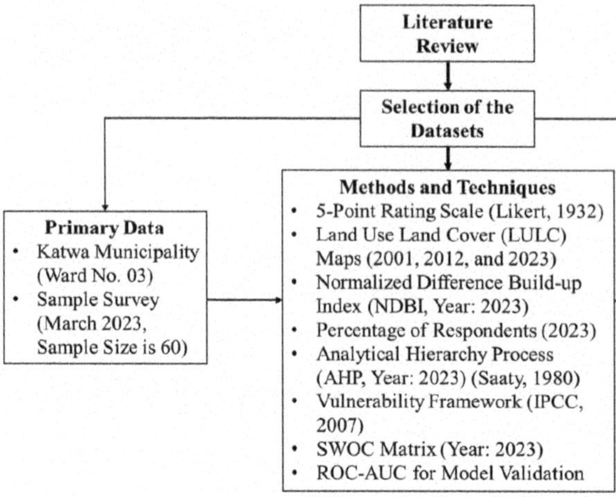

Fig. 13.9 General methodological framework of the present study

of low-lying, inadequately drained regions that are more vulnerable to extended flooding during monsoon events. The land use land cover (LULC) classes were delineated and classified employing the maximum likelihood classification (MLC) method and the semi-automatic classification plugin (SCP) in QGIS (version 3.16.4) (Lillesand et al. 2015; Congedo 2021).

The normalized difference built-up index (NDBI) was calculated using the formula published by Zha et al. (2003) to enhance the identification of developed areas in flood-prone regions.

$$\text{NDBI} = \frac{(\text{SWIR} - \text{NIR})}{(\text{SWIR} + \text{NIR})}, \tag{13.1}$$

where

NIR denotes the near-infrared band, SWIR stands for the short-wave infrared band. While low NDBI values often suggest vegetation or aquatic environments, high positive values indicate urbanized areas. According to Lu and Weng (2007), NDBI works especially well in semi-urban settings because developed areas frequently spectrally overlap with bare or uncultivated terrain.

A five-point Likert scale was used to evaluate community perceptions of flood impacts and displacement, with responses being categorized as strongly disagree, disagree, neutral, agree, and strongly agree (Yang 2023). Statistical analysis was used to calculate the percentage of respondents in each response category.

$$\text{Percentage} = \frac{\text{Number of respondents in a response category}}{\text{Total number of respondents}} \times 100. \tag{13.2}$$

This approach rendered it possible to evaluate social perception patterns clearly.

AHP-Vulnerability Framework and Model Validation

The analytic hierarchy process (AHP) was used as a multi-criteria decision-making (MCDM) tool in accordance with Saaty (1980) to rank the perceived causes of flood-induced household displacement. Using a five-point Likert scale, respondents assessed six criteria connected to displacement; mean scores ($\overline{x}_i$) were calculated to reflect the collective view of each criterion. A reciprocal pairwise comparison matrix was created using these mean values.

$$A = \left[a_{ij}\right], \tag{13.3}$$

where

$a_{ij} = \overline{x}_i/\overline{x}_j$ and $a_{ji} = 1/a_{ij}$.. The priority vector (eigenvector) and the matrix were then normalized by dividing each element by the sum of its columns,

$$w = (w_1, w_2, \ldots, w_n) \tag{13.4}$$

was acquired by calculating the average of the normalized rows, which show the relative importance of each displacement driver. The calculation of the consistency index (CI) was done as

$$\mathrm{CI} = \frac{\lambda_{\max} - n}{n - 1} \tag{13.5}$$

where

n is the number of criteria, and λ-max. is the matrix's major eigenvalue. To calculate the consistency ratio (CR),

$$\mathrm{CR} = \frac{\mathrm{CI}}{\mathrm{RI}} \tag{13.6}$$

where

According to Saaty (1980), RI is the random index that corresponds to n. Respondent judgments were deemed to be logically consistent if the CR value was less than 0.10 (Vaidya and Kumar 2006). An AHP-based paradigm combining exposure, sensitivity, and adaptive capacity was used to evaluate flood vulnerability at the household level. These components comprised twenty-two indications obtained from survey data on a five-point Likert scale. To provide robust component-wise vulnerability indices, pairwise comparison matrices were built, normalized weights were produced using the primary eigenvector, and consistency was checked using CI and CR (CR < 0.10) (Saaty 1980). The functional relationship was used to compute overall vulnerability (V).

$$V = (E + S) - \mathrm{AC}, \tag{13.7}$$

where

V is the vulnerability.
E is the exposure.
S is the sensitivity.
AC is the adaptive capacity.

In this case, vulnerability is decreased by increased adaptive capacity and increased by increased exposure and sensitivity. This concept is consistent with well-known frameworks for vulnerability assessment that highlight the compensating function of adaptive capacity in systems that are vulnerable to hazards (IPCC 2014). The resulting AHP-based vulnerability scores were then utilized for policy analysis and comparative interpretation.

This study evaluated the discriminative ability of the vulnerability scores generated by the AHP using the receiver operating characteristic–area under the curve (ROC–AUC) method, without employing a regression-based prediction model. ROC–AUC assesses a continuous metric's capacity to distinguish between binary outcome categories across various threshold values (Hanley and McNeil 1982; Fawcett 2006). The binary responses from the household survey, based on perception and impact, were directly compared with the normalized AHP composite scores. AUC values ranging from 0.70 to 0.75 indicate adequate model validity; a value of 0.5 signifies a lack of discrimination (Hosmer et al. 2013).

Strengths–Weaknesses–Opportunities–Challenges (SWOC) Matrix

A SWOC analysis was employed to derive policy-relevant insights and to enhance the AHP-based vulnerability assessment. Household surveys and field observations in Katwa Municipality were employed to ascertain strengths, shortcomings, possibilities, and challenges. These were subsequently assessed concerning exposure, sensitivity, and adaptive capacity. Strategic recommendations for managing flood-induced displacement were formulated utilizing a SWOC interaction matrix (S–O, S–C, W–O, W–C) (Hill and Westbrook 1997). Consistent with established methodologies in geo-environmental and disaster displacement research, the integration of multi-temporal remote sensing, GIS-based spatial analysis, structured household surveys, Likert-scale perception evaluation, and AHP-based prioritization with consistency verification offers a scientifically robust and replicable framework for investigating flood-induced residential displacement in a semi-urban riverine context (Kiker et al. 2005; Cutter et al. 2008).

Results

Rainfall, Flooding Pattern, and LULC Changes in Katwa

Katwa Municipality experiences a tropical wet–dry (Aw) climate characterized by significant monsoonal precipitation from June to September. This environment is defined by low elevation (about 16.077 m above mean sea level), which markedly enhances river discharge and flood vulnerability (Figs. 13.10, 13.11, 13.12, 13.13, 13.14 and 13.15). Significant flooding events transpired in 1999, 2000, and 2015, although moderate flooding was observed in 2006–2007, and small incidents were recorded in 2020, as per an examination of historical flood data. This indicates a persistent hydroclimatic stress regime instead of sporadic extremes. A multi-temporal land use and land cover (LULC) analysis reveals significant alterations in the landscape under this climate pressure (Figs. 13.16, 13.17, 13.18, 13.19, 13.20, 13.21, 13.22, 13.23, 13.24, 13.25 and 13.26). The water bodies increased from 2.31% in 2001 to 4.58% in 2012 due to prolonged flooding and waterlogging but subsequently decreased to 2.40% in 2023 because of partial drainage and land conversion. Fallow or barren land reached its zenith in 2012, corresponding with post-flood land abandonment, and subsequently diminished as resettlement and urban expansion escalated, while agricultural land exhibited a continuous decline throughout the study period. Flooding catalyzes environmental transformation and alters settlement patterns, as evidenced by these geographical changes. Approximately 558 individuals, primarily from Begunkola Mouza, have been moved due to repeated flooding; the majority have resettled in Wards 3 and 4, thereby establishing a direct link between hydroclimatic stress and localized population displacement. In the Ganga–Brahmaputra delta and Mekong floodplains, flooding simultaneously alters land use, livelihood sustainability, and settlement stability, with analogous climate-induced land system and mobility responses recorded (Mallick and Vogt 2014; Rahman et al. 2018; IPCC 2021). The book's principal argument, that the climate crisis modifies population dynamics rather than merely inducing temporary environmental disturbances, is substantiated by these findings, which underscore the necessity for sustainable solutions in flood-prone semi-urban systems that integrate flood management with land-use planning, livelihood adaptation, and settlement regulation (UNDRR 2019; IPCC 2022).

Socio-Economic and Household Conditions of the Respondents

Flood-induced displacement in Katwa Municipality disproportionately affects socio-economically deprived communities, as evidenced by household survey data (Tables 13.1 and 13.2; Figs. 13.27, 13.28, 13.29, 13.30, 13.31, 13.32, 13.33, 13.34,

Fig. 13.10 Rainfall–temperature graph (Year: 2000)

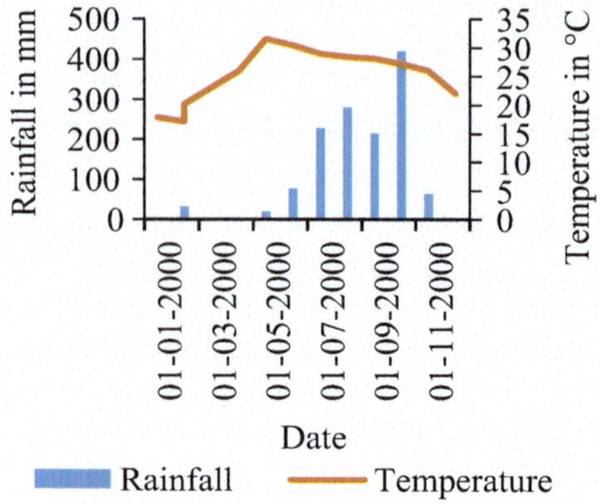

Fig. 13.11 Rainfall–temperature graph (Year: 2015)

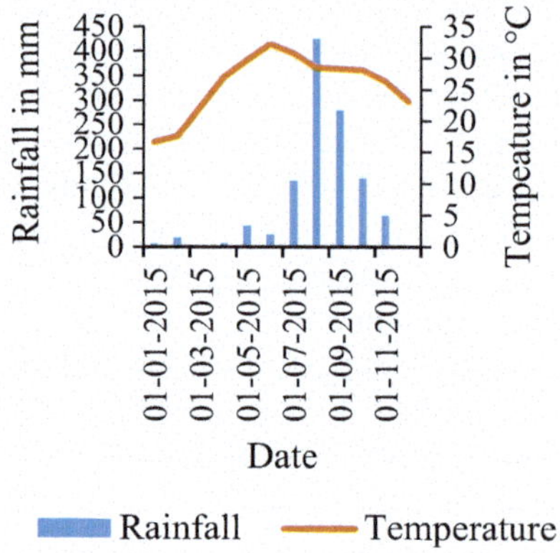

Fig. 13.12 Rainfall–temperature graph (Year: 2020)

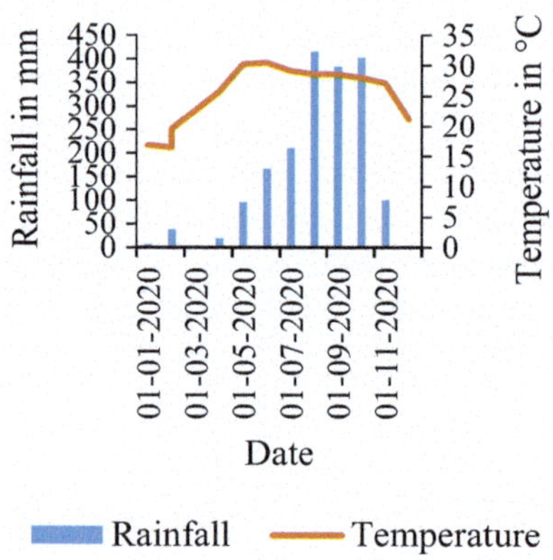

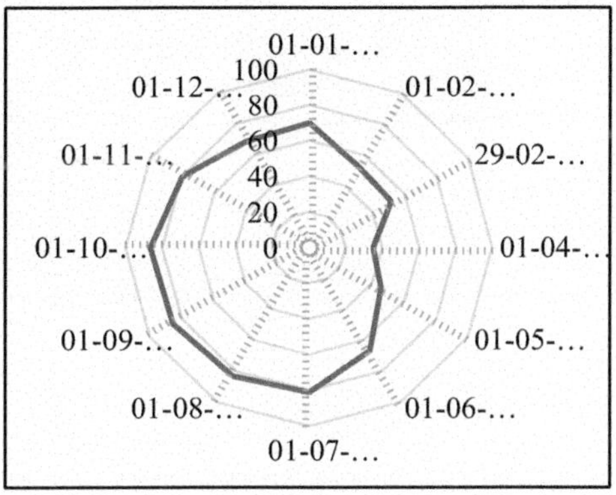

Fig. 13.13 Humidity conditions (Year: 2000)

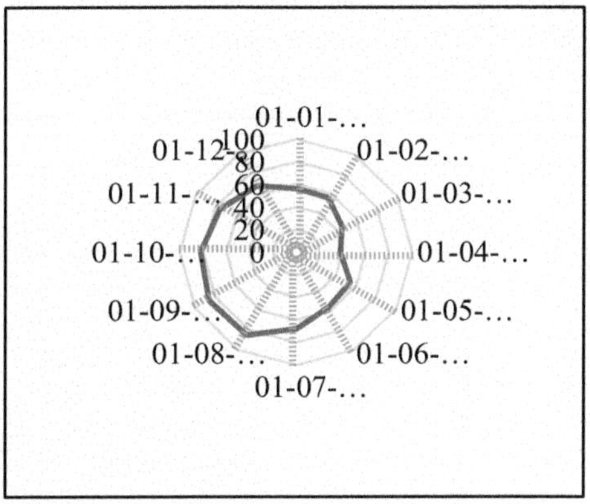

Fig. 13.14 Humidity conditions (Year: 2015)

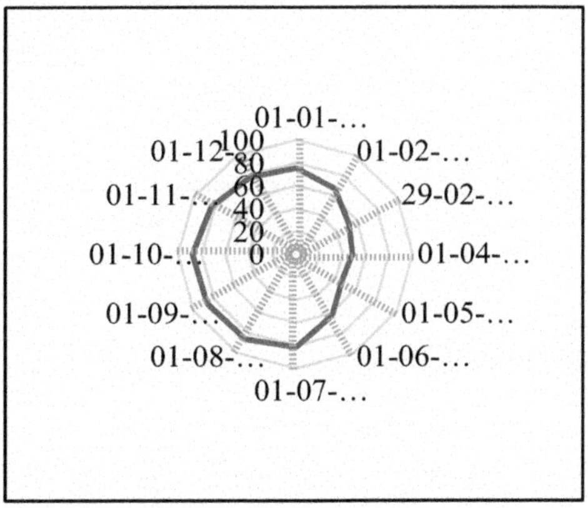

Fig. 13.15 Humidity conditions (Year: 2020)

Fig. 13.16 Water level measurement site near the Ajay River

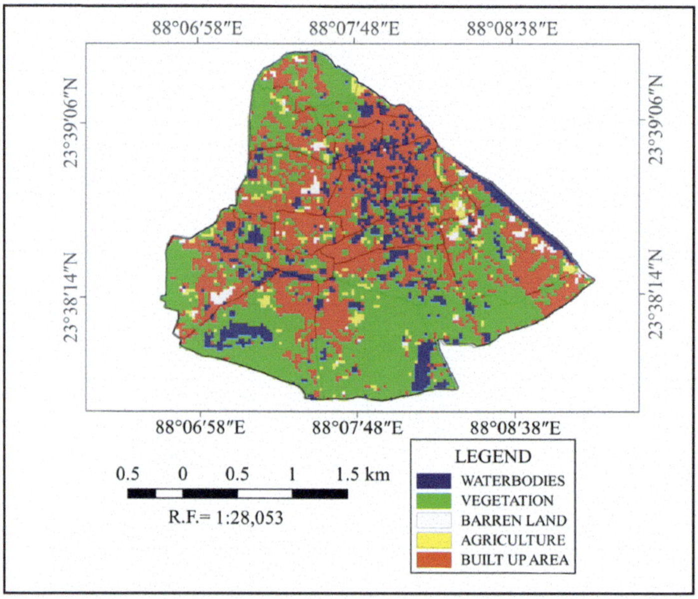

Fig. 13.17 Land use land cover map of Katwa Municipality (2001)

13.35, 13.36, 13.37, 13.38, 13.39, 13.40, 13.41 and 13.42). Households from scheduled castes (50%) and scheduled tribes (10%) constitute a substantial segment of the displaced population, indicating considerable social diversity in sensitivity to climatic risks. Many family structures are nuclear (41.67%) or joint (58.33%), with low monthly incomes, minimal savings, and a significant dependence on main and informal employment serving as indicators of limited economic resilience. Significantly, 63.33% of the households surveyed are classified as below the poverty

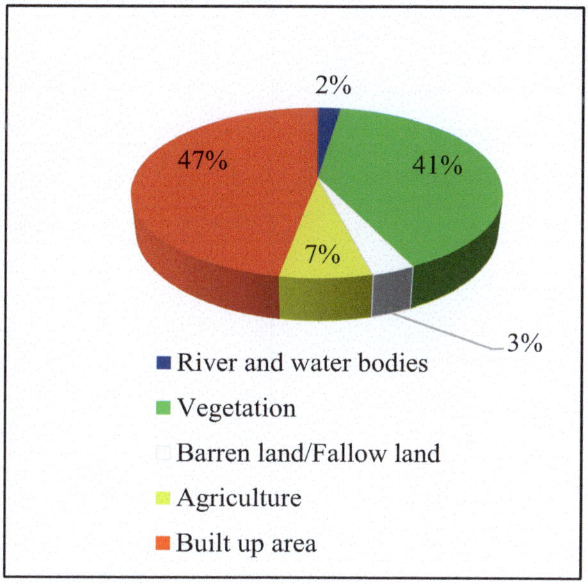

Fig. 13.18 Land use land cover of Katwa Municipality (2001)

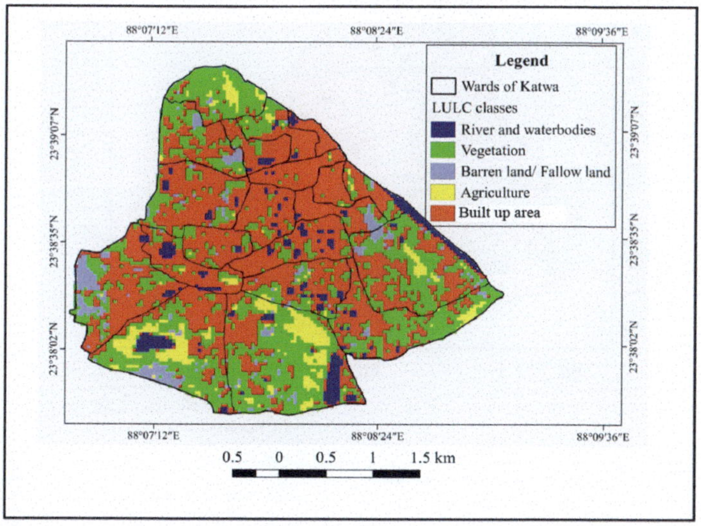

Fig. 13.19 Land use land cover map of Katwa Municipality (2012)

line (BPL), underscoring the relationship between environmental vulnerability and economic adversity. Despite the prevalence of Internet connectivity (88.33%), mobile phones (96.67%), and power (100%), substantial infrastructure deficiencies persist, particularly with housing quality, drainage, and sanitation. Alongside drainage systems that are entirely unpaved, just 18.33% of households have access to tap

Fig. 13.20 Land use land cover of Katwa Municipality (2012)

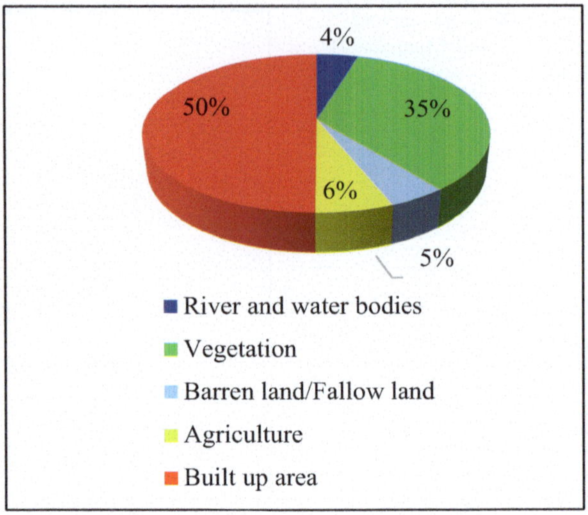

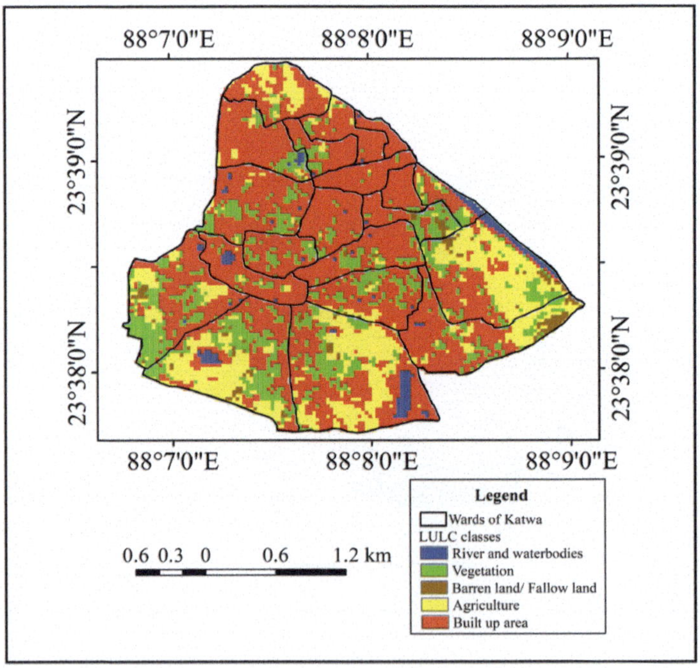

Fig. 13.21 Land use land cover map of Katwa Municipality (2023)

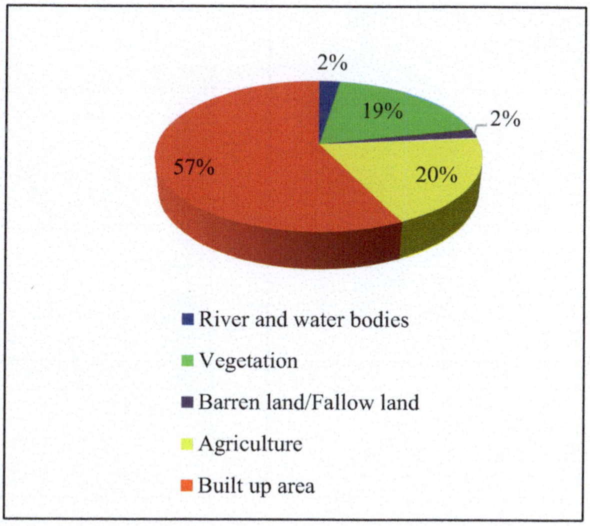

Fig. 13.22 Land use land cover of Katwa Municipality (2023)

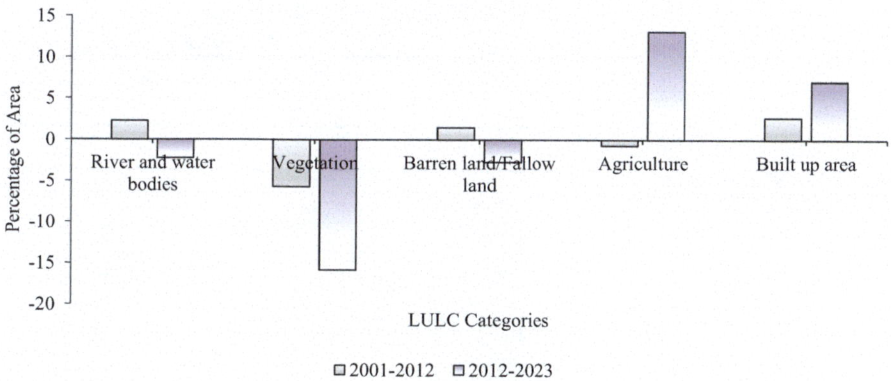

Fig. 13.23 Land use land cover changes of Katwa Municipality (2001–2012 and 2012–2023)

water, exacerbating waterlogging during monsoon floods. 55% of residents are categorized as kutcha or mixed structures, indicating precarious housing conditions that heighten susceptibility to water damage. The socio-economic characteristics indicate that poverty, caste-based marginalization, and infrastructural neglect are significant mediators of displacement in Katwa, rather than hydrological hazards alone. Comparable patterns, wherein marginalized people experience heightened vulnerability and diminished adaptation capability despite proximity to urban amenities, have been documented in flood-prone semi-urban areas of Bangladesh, eastern India, and the Mekong delta (Cutter et al. 2008; Adger 2006; Tacoli 2017). The findings substantiate the primary assertion of the study, which associates socio-economic

Fig. 13.24 SFCC of Ward No. 03 (2023)

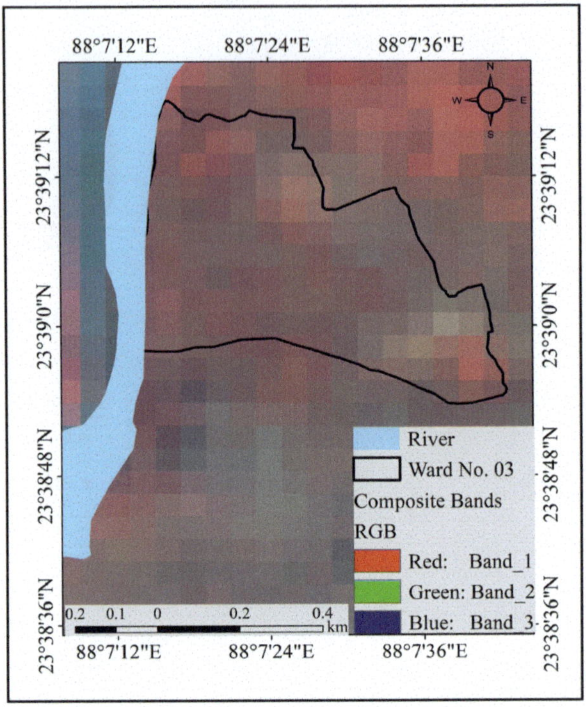

Fig. 13.25 NDBI map of Ward No. 03 (2023)

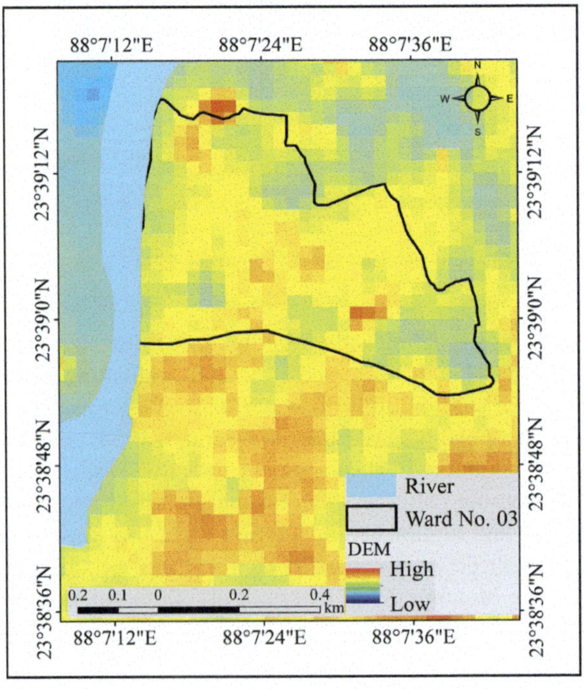

Fig. 13.26 Barren land situated in the survey area

vulnerability with changing population dynamics due to the climate crisis and advocates for sustainable solutions to climate-induced displacement that integrate flood risk mitigation, poverty alleviation, essential service provision, and social inclusion.

Scenarios of Residential Displacement Before and After the Establishment of the Ajay Bandh

The survey results clearly indicate that flood-related displacement in Katwa Municipality is predominantly short-distance, recurrent, and motivated by livelihood factors (Tables 13.3, 13.4 and 13.5). Mobility functions predominantly as a limited adaptive response rather than a permanent migration, as seen by the substantial 98.33% of households reporting displacement due to flooding, with 98.33% relocating within 5 km and merely 1.67% migrating beyond 10 km. Most displacement events occurred during the catastrophic flood disasters of 2000 and 2015, when flood depths often surpassed 5–10 feet, resulting in extensive destruction of residences and productive assets, along with total property loss. 91.69% of households reported farmland damage due to flooding, with over two-thirds experiencing partial or whole land loss, significantly affecting agricultural livelihoods and jeopardizing food security and stable earnings. 51.67% of households indicated an increase in income attributed to enhanced access to non-farm and urban employment opportunities, while 35% experienced income reductions. Additionally, 85% reported heightened household expenditures, especially on food, health care, housing construction, and debt repayment. The livelihood transitions post-resettlement reveal distinct adaptation pathways. These findings indicate that displacement in Katwa serves as an adaptive compromise rather than a sustainable solution, as it reduces immediate flood risk while imposing new economic limitations. In Bangladesh, Assam, and the Mekong Delta, peri-urban flood resettlement areas have demonstrated similar

Table 13.1 General information of the respondents

Category	Percentage of the total number of respondents
1. Total respondents	60 (Total number)
2. Male respondents	53.33
3. Female respondents	46.67
4. Total family members	
< 3	13.33
3–5	76.67
> 5	10.00
5. Age	
< 30	18.33
30–60	66.67
> 60	15.00
6. Religion	
Hindu	100.00
Muslim	0.00
7. Caste	
General	30.00
SC	50.00
ST	10.00
OBC-A	0.00
OBC-B	10.00
8. Education	0.00
Literate	80.00
Illiterate	20.00
9. Occupation	
Farmer	46.67
Labor	23.33
Shopkeeper	1.67
Driver (of any transport mode)	10.00
Businessman	3.33
Serviceman	0.00
Others	15.00
10. Monthly Income	
< 5000	25.00
5000–10,000	56.67
> 10,000	18.33
11. Monthly expenditure	
< 5000	61.67

(continued)

Table 13.1 (continued)

Category	Percentage of the total number of respondents
5000–10,000	31.67
> 10,000	6.67
12. Monthly savings	
< 5000	70.00
5000–10,000	30.00
> 10,000	0.00
13. Monthly debt	
< 5000	83.33
5000–10,000	16.67
> 10,000	0.00
14. Joint family	58.33
15. Nuclear family	41.67
16. BPL	
Yes	63.33
No	36.67

Source Authors' calculation based on field survey (March 2023)

livelihood reorganization and increased expenses. Proximity to metropolitan labor markets improves income potential but also elevates living costs and livelihood instability (Black et al. 2011; Kakinuma et al. 2020; Tacoli 2017). Besides mitigating flood risk, the findings together underscore the necessity for sustainable strategies addressing climate-induced displacement, which include livelihood diversification, income stabilization, and social protection measures. This immediately reinforces the book's central premise, which posits that climate stress modifies population dynamics through complex trade-offs among economic stability, mobility, and safety.

AHP-Based Prioritization of the Vulnerability Framework and Model Validation

The AHP analysis, utilizing 22 perception-derived indicators (Q4–Q25), statistically identifies the primary factors contributing to flood-induced relocation in Katwa Municipality as proximity to rivers, inadequate drainage, and recurrent monsoonal flooding (Figs. 13.42, 13.43, 13.44 and 13.45; Tables 13.6 and 13.7). The internal logical coherence and reliability of respondent evaluations were validated by the pairwise comparison matrices, which adhered to consistency criteria (CR < 0.10). Indicators of substantial structural significance and strong consensus were further delineated by quadrant analysis, establishing a hierarchy of flood vulnerability criteria that

Table 13.2 Household statuses

Category	Percentage of the total number of respondents
1. Household conditions	
Kuccha	16.67
Pucca	45.00
Mixed	38.33
2. Drinking water sources	
Tube well	76.67
Tap water	18.33
Others	5.00
3. Number of rooms	
< 2	16.67
2 to 3	33.33
> 3	33.33
4. Kitchen	
Joint	31.67
Separated	68.33
5. Latrine and bathroom	
Joint	35.00
Separated	65.00
6. Latrine and bathroom	
Pucca	80.00
Mixed	20.00
7. Drainage/sewage system	
Concretized	0.00
Not Concretized	100.00
8. Sources of fuel	
Timbers	56.67
Cow dung	1.66
Coal	0.00
Gas	41.67
Electric heater	0.00
Others	0.00
9. Electricity	
Yes	100.00
No	0.00
10. TV	
Yes	81.67
No	18.33

(continued)

Table 13.2 (continued)

Category	Percentage of the total number of respondents
11. Computer	
Yes	6.67
No	93.33
12. Mobile	
Yes	96.67
No	3.33
13. Internet	
Yes	88.33
No	11.67
12. Motorbike/Scooty	
Yes	21.67
No	78.33

Source Authors' calculation based on field survey (March 2023)

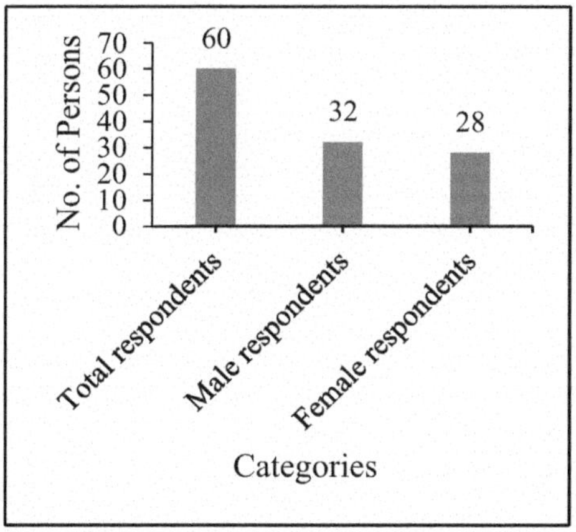

Fig. 13.27 Number of total respondents, male and female respondents

is directed toward decision-making. This statistically robust prioritization illustrates the strong alignment between objective hazard conditions and community perception, underscoring the significance of perception-integrated vulnerability assessment in climate adaptation research (Saaty, 1980; Cutter et al. 2008; UNDRR 2015). Exposure has the most significant influence (0.72), as indicated by component-wise aggregation, which illustrates persistent flood interaction via repeated displacement, prolonged waterlogging, and recurrent flooding, particularly during the pre-Ajay Bandh establishment period. The continued dependence on flood-prone assets results in a high sensitivity level (0.68), attributed to residential loss, land degradation, disruption of livelihoods, and income volatility. This indicates a constrained

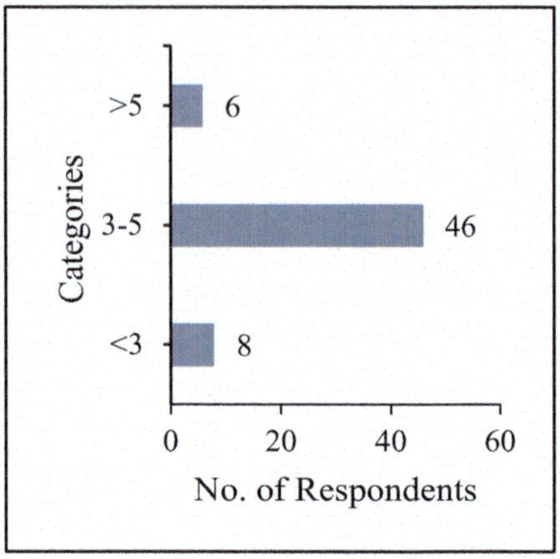

Fig. 13.28 Family members of the respondents

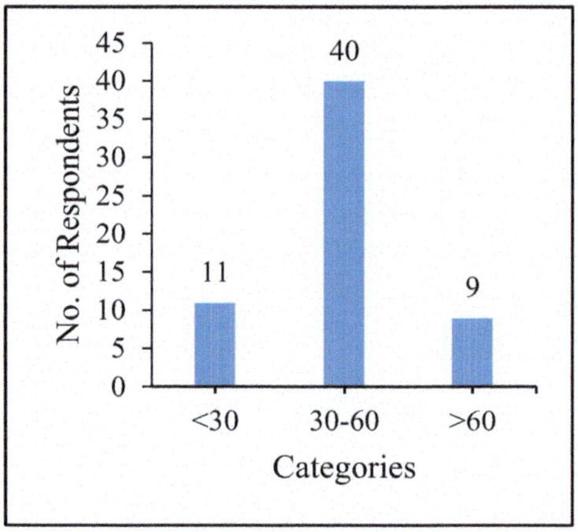

Fig. 13.29 Age of the respondents

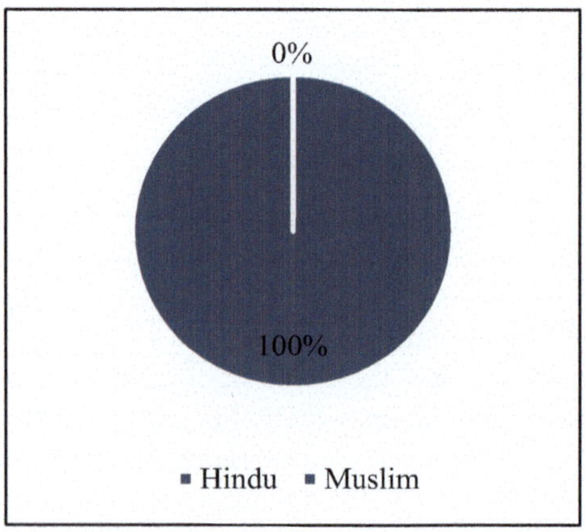

Fig. 13.30 Religions of the respondents

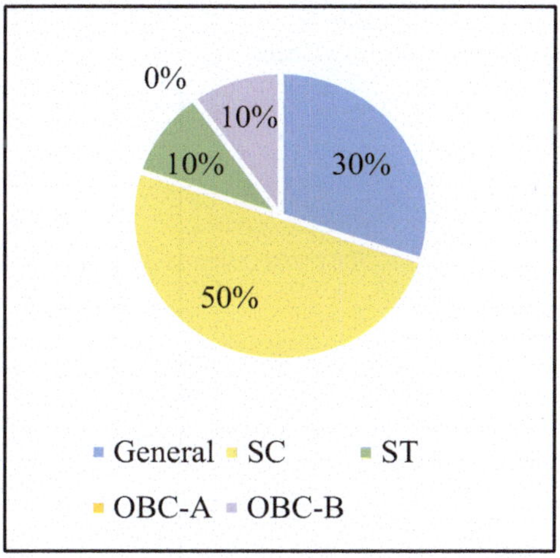

Fig. 13.31 Castes of the respondents

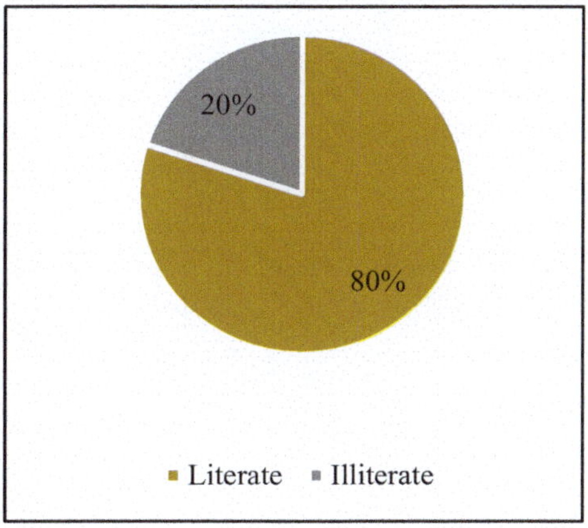

Fig. 13.32 Respondents: literates and illiterates

capacity to endure flood shocks. Post-embankment construction, adaptive capability shows a modest enhancement (0.54), supported by increased perceived safety and a partial implementation of flood-resilient housing techniques. Nonetheless, its alleviating impact is inadequate to offset heightened sensitivity and exposure. The composite vulnerability score remains elevated at 0.86, demonstrating that structural flood mitigation alone cannot sustainably diminish risk without simultaneous institutional and socio-economic interventions. The ROC–AUC analysis (Fig. 13.46) validation results indicated that the AHP-derived index has a satisfactory discriminative capacity in differentiating between relatively vulnerable households and those that were not. The readings varied between 0.70 and 0.75. The intermediate AUC values suggest that vulnerability is influenced by broader behavioral, governance, and livelihood factors that are not entirely captured by structural metrics alone, despite these results affirming the reliability of the selected indicators. To attain

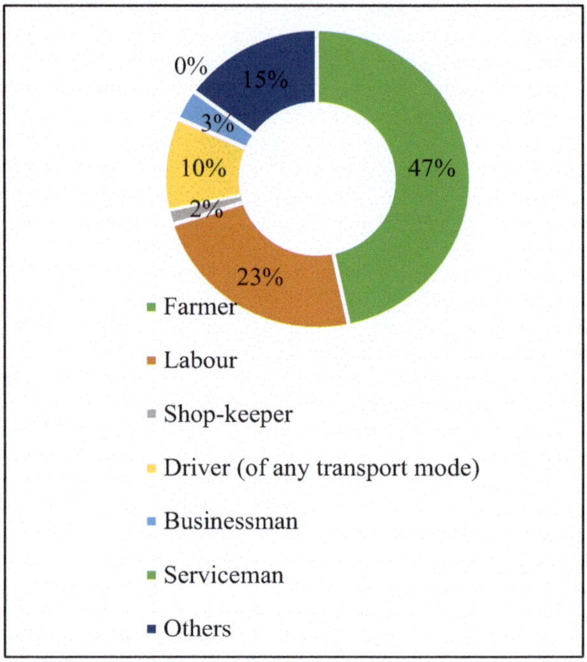

Fig. 13.33 Occupational categories of the respondent

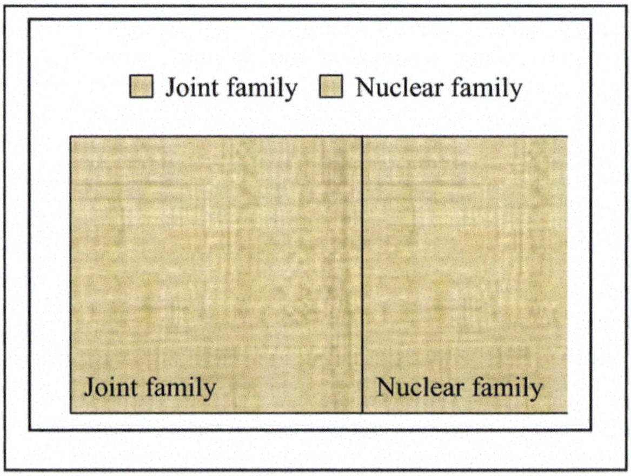

Fig. 13.34 Family status of the respondents

climate-resilient population outcomes, analogous findings from flood-prone regions of Bangladesh, eastern India, and the Mekong delta underscore the imperative of combining structural protection with livelihood diversification, secure tenure, and participatory governance (Sinha et al. 2014; Rahman et al. 2018; IPCC 2022).

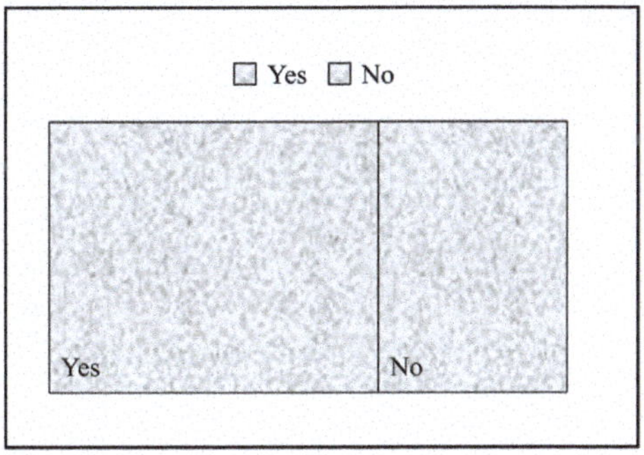

Fig. 13.35 BPL status of the respondents

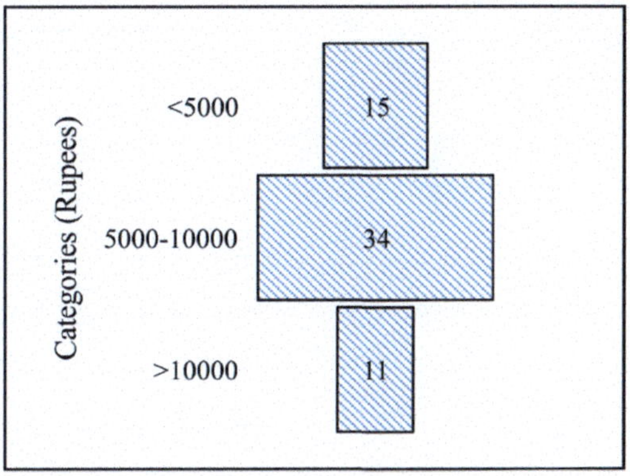

Fig. 13.36 Income status of the respondents

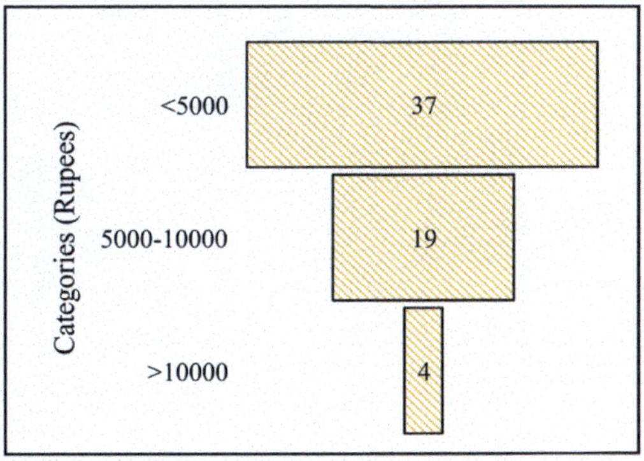

Fig. 13.37 Expenditure status of the respondents

Respondent-Specific Perceptions of Flood-Induced Displacement

The perception-based study in Katwa Municipality (Table 13.8; Figs. 13.47, 13.48, 13.49, 13.50, 13.51, 13.52, 13.53, 13.54, 13.55, 13.56, 13.57, 13.58, 13.59, 13.60,

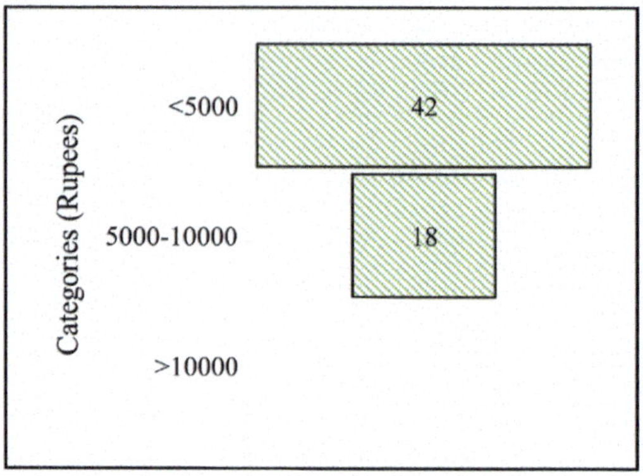

Fig. 13.38 Savings status of the respondents

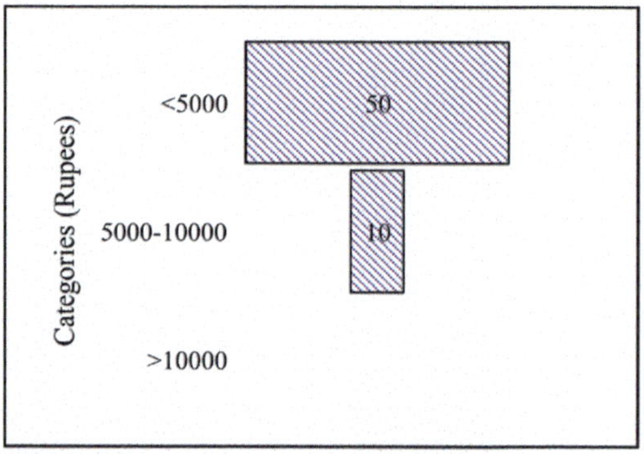

Fig. 13.39 Debt status of the respondents

Fig. 13.40 Mixed-category house in the survey area

Fig. 13.41 Kuccha road in the survey area

Fig. 13.42 Domestic animals in a household in the survey area

13.61, 13.62, 13.63, 13.64 and 13.65) reveals a markedly diverse pattern of post-displacement outcomes among affected households, indicating differing adaptation experiences in response to climate stress. Over 81.67% of respondents reported satisfaction with their current residence, suggesting a perceived reduction in immediate flood risk post-displacement, despite the majority recognizing flood-related displacement, habitat loss, and accompanying psychological distress. Notably, 68.33% of respondents indicated that the Ajay Bandh contributed to mitigating flood intensity, underscoring the perceived efficacy of structural interventions in enhancing physical safety. However, improvements in overall socio-economic well-being remain inconsistent. A minor fraction of households reported an enhanced overall quality of life or financial stability, whereas a substantial number indicated rising expenses for housing, food, and health care. Furthermore, more than 83% of participants emphasized the necessity for enhanced institutional support and aid allocation during and following flood disasters, highlighting persistent deficiencies in governance and

Table 13.3 Information about flood and displacement

Category	Percentage of the total number of respondents
1. Do you migrate from anywhere?	
Yes	98.33
No	1.67
2. Year of migration	
Before 2000	18.33
2000–2015	66.67
After 2015	15.00
3. Causes of migration	
Flood	100.00
Others	0.00
4. Distance from the previous residence	
< 5 km	98.33
5–10 km	0.00
> 10 km	1.67
5. Total number of flooding days (2015–2022)	
< 10	68.33
10 to 20	23.33
> 20	8.33
6. The durability of the flood (days)	
< 5	25.00
5 to 10	35.00
> 10	40.00
7. Large floods during the last 25 years	
< 3	35.00
3 to 5	45.00
> 5	20.00
8. Did your displacement cause by flood conditions?	
Yes	98.33
No	1.67
9. Nature of flood in your locality	
Mild	10.00
Severe	31.67
Extreme	58.33
10.The water level during a flood	
< 5 feet	43.33

(continued)

Table 13.3 (continued)

Category	Percentage of the total number of respondents
5–10 feet	56.67
> 10 feet	0.00
11. Loss of properties during a flood	
Yes	100.00
No	0.00
12. Preventive measures for the flood	
Yes	41.67
No	58.33
13. Any relief/ grant assistance by the Government/NGOs	
Yes	3.33
No	96.67

Source Authors' calculation based on field survey (March 2023)

service provision. This discrepancy in perception regarding economic security and physical safety underscores a significant climate adaptation dilemma: while structural flood mitigation may reduce hazard exposure, it does not invariably lead to enduring livelihood resilience or social welfare. In Bangladesh, eastern India, and Southeast Asia, peri-urban resettlement situations exhibit analogous perception tendencies. Although displacement mitigates immediate risk, it also sustains long-term vulnerability owing to inadequate services, unstable tenure, and restricted livelihood diversification (Wisner et al. 2004; Tacoli 2017; IPCC 2022). The results underscore the necessity of integrating structural measures with social protection, livelihood assistance, and participatory governance to attain sustainable solutions in climate-affected population systems, reinforcing the study's broader assertion that climate-induced displacement serves as a transitional adaptation strategy rather than a permanent resilience outcome.

Figures 13.48, 13.49, 13.50, 13.51, 13.52, 13.53, 13.54, 13.55, 13.56, 13.57, 13.58, 13.59, 13.60, 13.61, 13.62, 13.63, 13.64 and 13.65 is based on the original question numbers 4–25 mentioned in the survey questionnaire.

SWOC Analysis of Post-Displacement Conditions

The SWOC analysis (Tables 13.9 and 13.10; Figs. 13.64, 13.65 and 13.66) encapsulates the post-displacement conditions of residences impacted by flooding and translates empirical findings into strategic insights. The primary strength found is enhanced access to healthcare services, education, and transportation infrastructure in the resettlement wards, which exemplifies the socio-spatial benefits associated with

Table 13.4 Economic conditions before and after displacement

Category	Percentage of the total number of respondents
1. Present landholding size (total)	
< 5 bigha	75.00
5–10 bigha	25.00
> 10 bigha	0.00
2. Was your agricultural land affected by the flood?	
Yes	91.67
No	8.33
3. Amount of land affected by the flood	
< 5 bigha	83.33
5–10 bigha	16.67
> 10 bigha	0.00
4. Were settlements affected by the flood?	
Yes	93.33
No	6.67
5. Number of houses affected by the flood	
< 20	45.00
20–50	33.33
> 50	21.67
6. Were domestic animals affected by the flood?	
Yes	58.33
No	41.67
7. Number of animals affected by the flood	0.00
< 5	53.33
5–10	33.33
> 10	13.33
8. Any land/property that existed in the previous residence	0.00
Yes	58.33
No	41.67
9. Use of those lands	
Agriculture	65.00
Settlements	13.33
Poultry farm	0.00
Cottage industry	0.00
Others	5.00
Vacant	16.67

(continued)

Table 13.4 (continued)

Category	Percentage of the total number of respondents
10. Did you want to sell those lands?	
Yes	93.33
No	6.67
11. Do you daily/occasionally visit to look after your property in the previous residence?	
Yes	68.33
No	31.67

Source Authors' calculation based on field survey (March 2023)

displacement. Ongoing deficiencies, such as the absence of a potable water supply, inadequate drainage systems, and poor road conditions, continue to constrain daily living conditions and exacerbate enduring vulnerability. The development of flood-resistant infrastructure, enhanced relief systems, and diversification of livelihoods beyond flood-prone agriculture, especially via non-farm and urban employment, all offer potential options. Conversely, ongoing challenges that hinder the transformation of advancements in physical safety into enduring socio-economic resilience encompass recurrent flooding, entrenched poverty, escalating living costs, and governance limitations. The coexistence of structural defects and developmental achievements indicates that displacement has altered vulnerability patterns rather than eliminated them.

Discussion and Policy-Relevant Implications

The study has substantial theoretical, empirical, and policy implications for understanding climate change-induced relocation outside the Katwa setting. An expanding corpus of international literature characterizes climate hazards as persistent socio-environmental stressors instead of isolated shocks, as evidenced by the ongoing flood-induced residential displacement in Katwa Municipality, despite substantial structural interventions such as the Ajay Bandh (Adger 2006; Wisner et al. 2004; IPCC 2021). This perspective highlights the interconnection between prolonged climate-society interactions, where repeated exposure progressively undermines livelihoods, settlement stability, and adaptive capacity, and relocation in flood-prone regions. Comparable patterns have been noted in the Indus floodplains of Pakistan, where recurrent short-distance displacements have transformed peri-urban settlement systems without achieving enduring resilience (Ghimire et al. 2015), and in the Ganga–Brahmaputra delta of Bangladesh, where embankments have reduced flood frequency but heightened residual risk and settlement density (Mallick and Vogt 2014; Sarker et al. 2019). The observed parallels indicate that Katwa is not an isolated case, but rather part of a broader global phenomenon in which restricted,

Table 13.5 Information regarding occupation, economy, and living status before and after displacement

Category	Percentage of the total number of respondents
1. Occupation before displacement	
Farmer	65.00
Labor	13.33
Shopkeeper	6.67
Driver (any)	0.00
Business	6.67
Service	0.00
Others	8.33
Unemployed	0.00
2. Occupation after displacement	0.00
Farmer	51.67
Labor	20.00
Shopkeeper	5.00
Driver (any)	8.33
Business	3.33
Service	0.00
Others	11.67
Unemployed	0.00
3. Occupation changed	0.00
Yes	38.33
No	61.67
4. Income	0.00
Increased	51.67
Decreased	35.00
Remain same	13.33
5. Expenditure	
Increased	85.00
Decreased	5.00
Remain same	10.00
6. Expenditure changes for (if any)	
Food	21.67
Cloth	6.67
Education	11.67
Medication	18.33
Housing construction	16.66
Investment	0.00

(continued)

Table 13.5 (continued)

Category	Percentage of the total number of respondents
Debt repayment	15.00
Buying anything	0.00
Agriculture	5.00
Occasion/Entertainment	0.00
Any other	5.00

Source Authors' calculation based on field survey (March 2023)

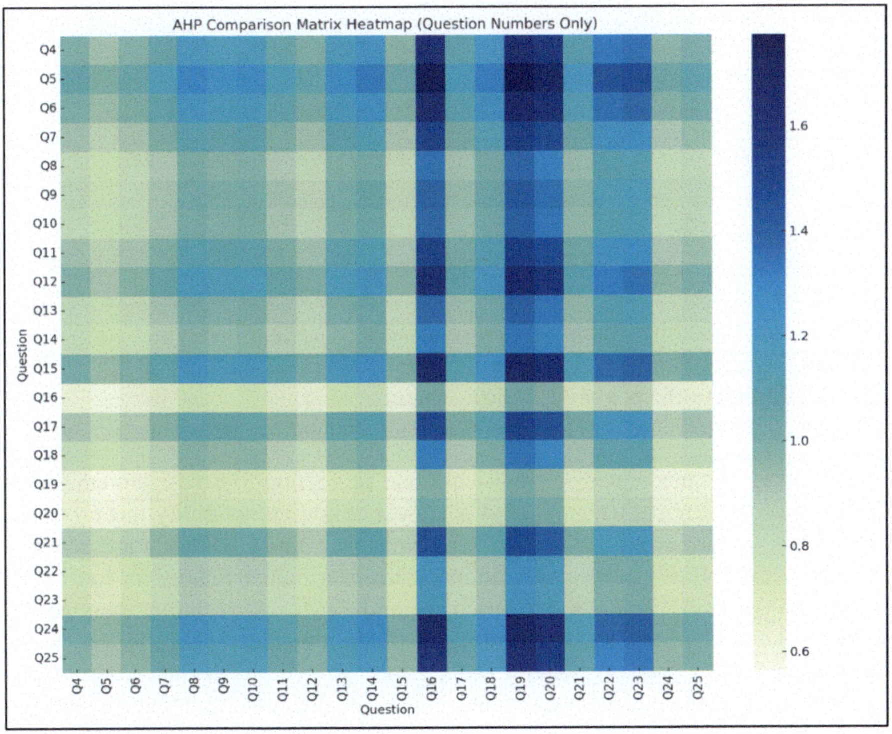

Fig. 13.43 AHP comparison matrix (based on the original question numbers 4–25 of the questionnaire)

recurrent, and socially stratified migration induced by climate change alters population dynamics (Black et al. 2011; McLeman 2018). The vulnerability paradigm grounded in AHP provides a systematic rationale for the ongoing danger of displacement. The elevated sensitivity (0.68) and exposure (0.72) scores indicate that households remain entrenched in flood-prone socio-ecological systems where hydrological hazards intersect with vulnerable livelihoods, unstable housing, and inadequate

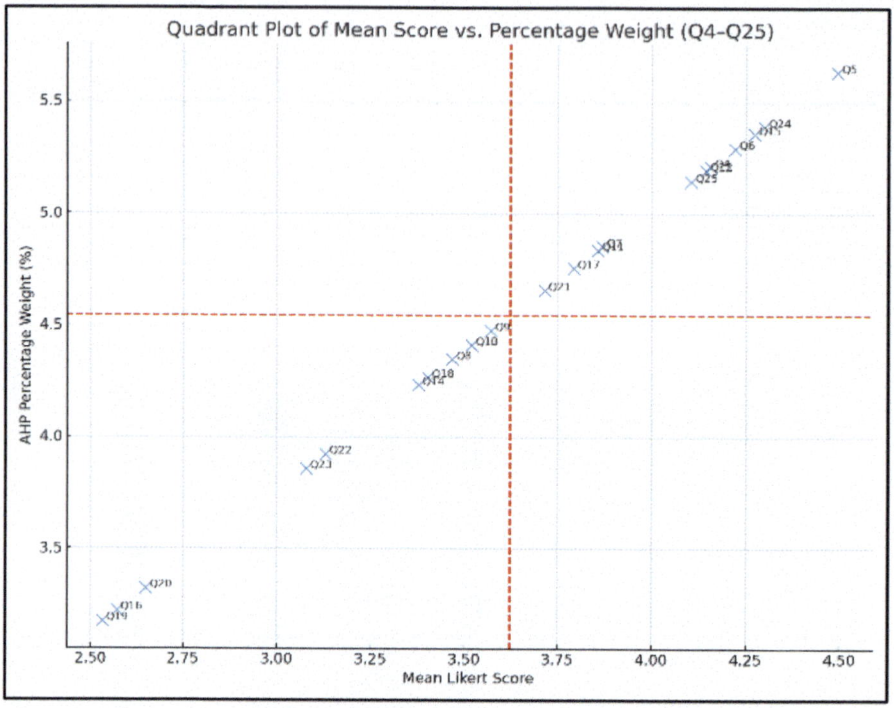

Fig. 13.44 Quadrant plot based on AHP (question number only)

infrastructure. The inadequacy of physical flood protection to isolate human settlements from hydrological hazards amid socio-economic vulnerability has profound implications for flood risk governance. In semi-urban and peri-urban areas where governance capabilities and service delivery are insufficient relative to settlement growth, analogous AHP- and index-based research from eastern India, Bangladesh, and the Mekong Delta indicates a comparable predominance of exposure and sensitivity over adaptive capacity (Cutter et al. 2008; Vaidya and Kumar 2006; Sinha et al. 2014; Rahman et al. 2018). Despite 78% of respondents indicating a sense of increased safety, Katwa's moderate adaptive capacity (0.54) underscores the limitations of infrastructure-centric adaptation. This finding aligns with IPCC evaluations that stress the importance of social inclusion, livelihood diversity, and institutional efficacy, beyond mere structural interventions—as critical determinants of resilience (IPCC 2014; 2022; Folke 2006). The methodological contribution of the work is further substantiated by the ROC–AUC validation results (0.70–0.75). This analysis illustrates that vulnerability is inherently multidimensional and context-dependent, shaped by behavioral responses, risk perception, governance quality, and social capital, hence affirming the commendable discriminative capacity of the AHP-derived vulnerability index. The assertion that mixed-method, perception-integrated

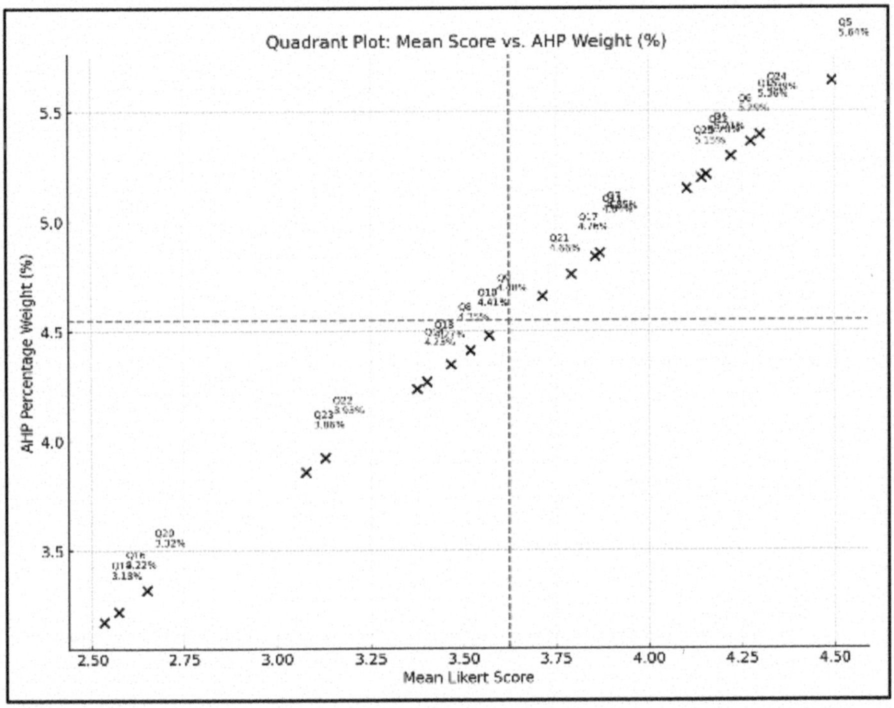

Fig. 13.45 Quadrant plot based on AHP (value only) (based on the original question numbers 4–25 of the questionnaire)

frameworks are essential for elucidating the intricacies of climate-induced displacement is corroborated by analogous validation studies in Bangladesh, Vietnam, and sub-Saharan Africa, which similarly indicate moderate predictive efficacy for composite vulnerability indices (Fawcett, 2006; Kakinuma et al. 2020; Elum and Lawal 2022). The socio-economic consequences of Katwa's relocation exemplify the disparities in climate adaptation. A decline in income for certain households highlights the continuation of inequality, the informalization of livelihoods, and inadequate infrastructure in resettlement areas, whilst a gain in income for others indicates enhanced access to non-farm employment and urban services. This pattern mirrors experiences in the peri-urban resettlement areas of Dhaka, Kolkata, and Hooghly, where displacement exacerbates social marginalization, insecure tenure, and service deficiencies, while simultaneously generating new economic opportunities (Tacoli 2017; Bezboruah et al. 2021). The findings corroborate broader evidence that displacement engenders transient resilience, which is vulnerable to subsequent climate disturbances, functioning as an adaptive compromise rather than a lasting resolution (Wisner et al. 2004; Adger et al. 2013). Ultimately, the integrated AHP–SWOC findings underscore the global relevance of the levee effect by demonstrating how embankments can unintentionally foster prolonged risk accumulation and urban

Table 13.6 AHP scores[a,b]

Original question numbers	Mean score	Weighted score	Percentage (%)
4. Flooding was a common problem in Ward No. 3 & 4 before constructing Ajay Bandh	4.155844	1.14658	5.211726
5. Floodwaters remained stagnant for several days or weeks before constructing Ajay Bandh	4.493506	1.239739	5.635179
6. The floods severely damaged houses, roads, and infrastructure before constructing Ajay Bandh	4.220779	1.164495	5.29316
7. Many residents were forced to temporarily or permanently relocate due to floods before constructing Ajay Bandh	3.87013	1.067752	4.85342
8. The floods negatively impacted agricultural activities and local businesses before constructing Ajay Bandh	3.467532	0.956678	4.348534
9. The local administration took proactive flood control measures before constructing Ajay Bandh	3.571429	0.985342	4.478827
10. People had to adopt flood-resistant construction techniques before constructing Ajay Bandh	3.519481	0.97101	4.413681
11. The frequency of floods has decreased significantly after constructing Ajay Bandh	3.857143	1.064169	4.837134
12. The severity of floods has reduced after constructing Ajay Bandh	4.142857	1.142997	5.19544
13. The risk of flood-related displacement has reduced due to Ajay Bandh construction	3.519481	0.97101	4.413681
14. Economic activities (farming, trade, etc.) have improved due to reduced flood risk	3.376623	0.931596	4.234528
15. Property values in Ward No. 3 & 4 have increased due to reduced flood risks	4.272727	1.178827	5.358306
16. The quality of roads and infrastructure has improved after Ajay Bandh construction	2.571429	0.709446	3.224756
17. The Ajay Bandh has provided long-term flood protection for this area	3.792208	1.046254	4.7557
18. Flooding is no longer a major concern for residents in Ward No. 3 & 4	3.402597	0.938762	4.267101
19. The existing drainage system is sufficient to handle excess rainwater and prevent flooding	2.532468	0.698697	3.175896
20. Houses are now being built with flood-resistant designs	2.649351	0.730945	3.322476
21. Additional flood control measures (e.g., embankment maintenance, better drainage) are necessary	3.714286	1.024756	4.65798
22. The local government is actively involved in maintaining Ajay Bandh	3.12987	0.863518	3.925081
23. The community is better prepared for future flood risks due to awareness programs and policies	3.077922	0.849186	3.859935

(continued)

Table 13.6 (continued)

Original question numbers	Mean score	Weighted score	Percentage (%)
24. Climate change is expected to impact flood patterns in this region in the future	4.298701	1.185993	5.390879
25. I feel safer from floods now compared to the situation before Ajay Bandh construction	4.103896	1.132248	5.14658

Source Authors' calculation based on field survey (March 2023)
[a]Based on the calculation, the Random Index (RI) is 1.80 (Saaty 1980), [b]The manual exclusion process has been applied during questionnaire validation

Table 13.7 AHP-based vulnerability assessment and policy interpretation

Dimension	Indicators considered (from AHP scores)	Normalized index value	Interpretation	Policy implication
Exposure (E)	Flood frequency, waterlogging duration, flood severity, occurrence of displacement, and damage to infrastructure and agriculture	0.72	High exposure is a result of frequent displacement, extended flooding, and frequent flooding before Ajay Bandh establishment	Flood forecasting systems, enhanced drainage, and ongoing embankment repair
Sensitivity (S)	Loss of land and residences, interruption of livelihood, fluctuating income, depreciation of real estate, and ineffective drainage	0.68	Strong reliance on assets vulnerable to flooding and a low buffering capacity are indicative of moderate–high sensitivity	Ensure land tenure, provide compensation, and assist with housing rebuilding
Adaptive capacity (AC)	Flood-resistant dwellings, institutional assistance, safety enhancements, perceived flood decrease following the Bandh, and awareness	0.54	Ajay Bandh developed moderate adaptive capacity, although it is still uneven because to tenure and livelihood restrictions	Diversifying sources of income, developing skills, obtaining credit, and obtaining insurance
Vulnerability (V)	$V = (E + S) - \text{AC}$	0.86	Despite structural flood prevention, the overall risk is still high	Integrated flood control that incorporates structural, socio-economic, and participative methods

Source Authors' calculation based on field survey (March 2023)

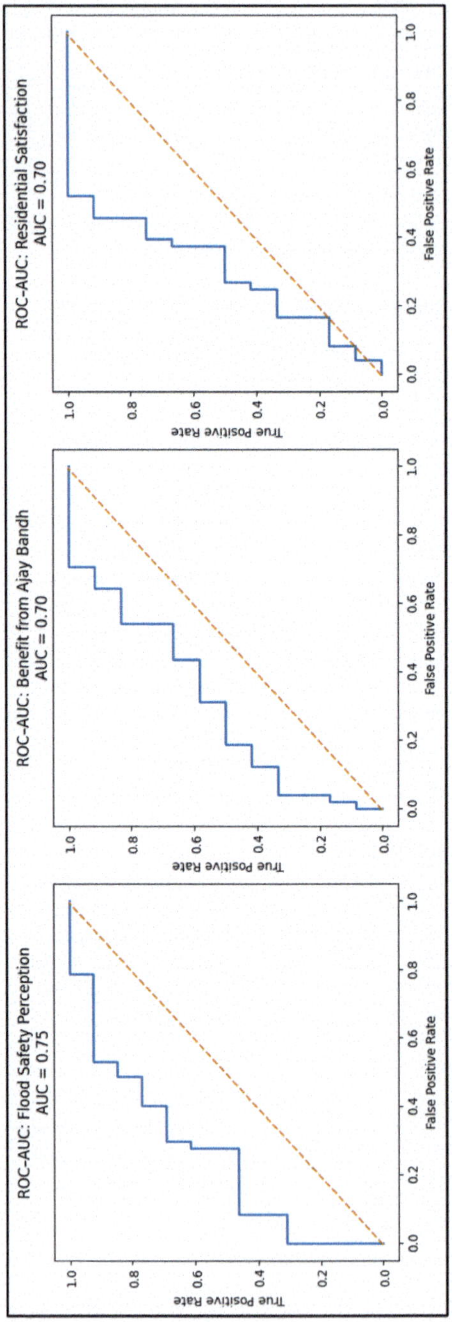

Fig. 13.46 Model validation (ROC-AUC)

Table 13.8 Respondent-specific perceptions

Category	Percentage of the total number of respondents
1. Do you agree that your present condition is changing?	
Yes	58.33
No	25.00
No Response	16.67
Strongly disagree	8.33
Disagree	10.00
Neutral	16.67
Agree	23.33
Strongly agree	41.67
2. Did your family member support the change of residence?	
Yes	88.33
No	0.00
No Response	11.67
Strongly disagree	0.00
Disagree	0.00
Neutral	11.67
Agree	28.33
Strongly agree	60.00
3. Your economic condition is disputed due to displacement	
Yes	53.33
No	21.67
No Response	25.00
Strongly disagree	6.67
Disagree	18.33
Neutral	21.67
Agree	43.33
Strongly agree	10.00
4. The family's education is disputed due to displacement	
Yes	18.33
No	66.67
No Response	15.00
Strongly disagree	33.33
Disagree	33.33
Neutral	15.00

(continued)

Table 13.8 (continued)

Category	Percentage of the total number of respondents
Agree	11.67
Strongly agree	6.67
5. Health condition disputed due to displacement	
Yes	43.33
No	38.33
No Response	18.33
Strongly disagree	10.00
Disagree	28.33
Neutral	18.33
Agree	23.33
Strongly agree	20.00
6. Any flood condition/Amphan maximized the risk during COVID-19	
Yes	70.00
No	16.67
No Response	13.33
Strongly disagree	10.00
Disagree	6.67
Neutral	13.33
Agree	45.00
Strongly agree	25.00
7. Are you satisfied with your present residence?	
Yes	81.67
No	10.00
No Response	8.33
Strongly disagree	6.67
Disagree	3.33
Neutral	8.33
Agree	35.00
Strongly agree	46.67
8. Any flood situation hit destructively at the present location in 2015 and later?	
Yes	70.00
No	15.00
No response	15.00
Strongly disagree	8.33

(continued)

Table 13.8 (continued)

Category	Percentage of the total number of respondents
Disagree	6.67
Neutral	15.00
Agree	41.67
Strongly agree	28.33
9. Did the residents benefit from the establishment of Ajay Bandh?	
Yes	68.33
No	11.67
No response	16.67
Strongly disagree	11.67
Disagree	0.00
Neutral	16.67
Agree	36.67
Strongly agree	35.00
10. Are you psychologically aware of the displacement matter	
Yes	60.00
No	25.00
No response	15.00
Strongly disagree	6.67
Disagree	18.33
Neutral	15.00
Agree	33.33
Strongly agree	26.67
11. Any land/ property that exists till on the previous location was affected by the flood	
Yes	60.00
No	26.67
No response	13.33
Strongly disagree	15.00
Disagree	11.67
Neutral	13.33
Agree	45.00
Strongly agree	15.00
12. Do you think that flooding is a major cause of habitat loss?	
Yes	100.00
No	0.00

(continued)

Table 13.8 (continued)

Category	Percentage of the total number of respondents
No response	0.00
Strongly disagree	0.00
Disagree	0.00
Neutral	0.00
Agree	38.33
Strongly agree	61.67
13. Do you know about any deaths during previous floods (2000 or later)?	
Yes	40.00
No	41.67
No response	18.33
Strongly disagree	25.00
Disagree	16.67
Neutral	18.33
Agree	40.00
Strongly agree	0.00
14. Do you agree that flood and displacement harm the natural and social environment?	
Yes	85.00
No	6.67
No response	8.33
Strongly disagree	6.67
Disagree	0.00
Neutral	8.33
Agree	41.67
Strongly agree	43.33
15. Do you want to return to your previous residence?	
Yes	33.33
No	51.67
No response	15.00
Strongly disagree	18.33
Disagree	33.33
Neutral	15.00
Agree	25.00
Strongly agree	8.33

(continued)

Table 13.8 (continued)

Category	Percentage of the total number of respondents
16. Do you want far better development of relief distribution and the condition of displaced people during and after flood situations?	
Yes	83.33
No	0.00
No response	16.67
Strongly disagree	0.00
Disagree	0.00
Neutral	16.67
Agree	40.00
Strongly agree	43.33
17. Did your economic condition improve after displacement?	
Yes	23.33
No	36.67
No response	40.00
Strongly disagree	0.00
Disagree	23.33
Neutral	40.00
Agree	30.00
Strongly agree	6.67
18. Did your standard of living decrease after displacement?	
Yes	33.33
No	46.67
No response	20.00
Strongly disagree	20.00
Disagree	26.67
Neutral	20.00
Agree	15.00
Strongly agree	18.33

Source Authors' calculation based on field survey (March 2023) (original question numbers 4–25 based on the questionnaire)

Respondent-specific perception (R)								
Positive response			Negative response			Neutral response		
R1	R2	R3	R1	R2	R3	R1	R2	R3
R4	R5	R6	R4	R5	R6	R4	R5	R6
R7	R8	R9	R7	R8	R9	R7	R8	R9
R10	R11	R12	R10	R11	R12	R10	R11	R12
R13	R14	R15	R13	R14	R15	R13	R14	R15
R16	R17	R18	R16	R17	R18	R16	R17	R18
Positive response Classes (Percentage)			Negative response Classes (Percentage)			Neutral response Classes (Percentage)		
<50	Low		<50	Low		<50	Low	
50-80	Moderate		50-80	Moderate		50-80	Moderate	
>80	High		>80	High		>80	High	

Fig. 13.47 Categories of respondent-specific perception

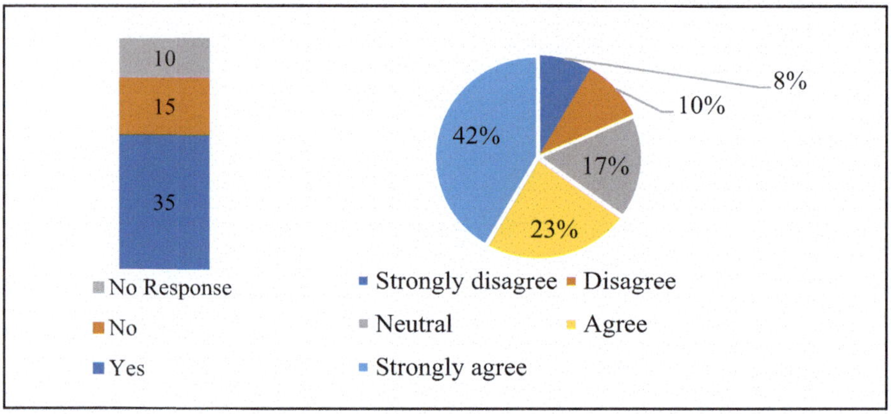

Fig. 13.48 Respondents' perception of Question-1

settlement concentration (UNDRR 2015). Similar patterns observed along the Kosi, Damodar, Brahmaputra, and Mekong rivers underscore the necessity for integrated, human-centered resilience strategies instead of hazard control for sustainable flood management (Mukhopadhyay et al. 2018). These strategies require the integration of structural protection with improved fundamental services, livelihood diversification, participatory planning, and governance reform. The Katwa story illustrates that enduring adaptation necessitates comprehensive socio-institutional transformation rather than solely technological solutions, significantly contributing to global discourse on climate issues and changing population dynamics, and effectively supporting SDG targets. The Katwa case illustrates that flood-induced displacement constitutes a prolonged demographic phenomenon rather than a transient hazard reaction, as evidenced by public opinions and the AHP–SWOC findings. Consequently, sustainable policy solutions should:

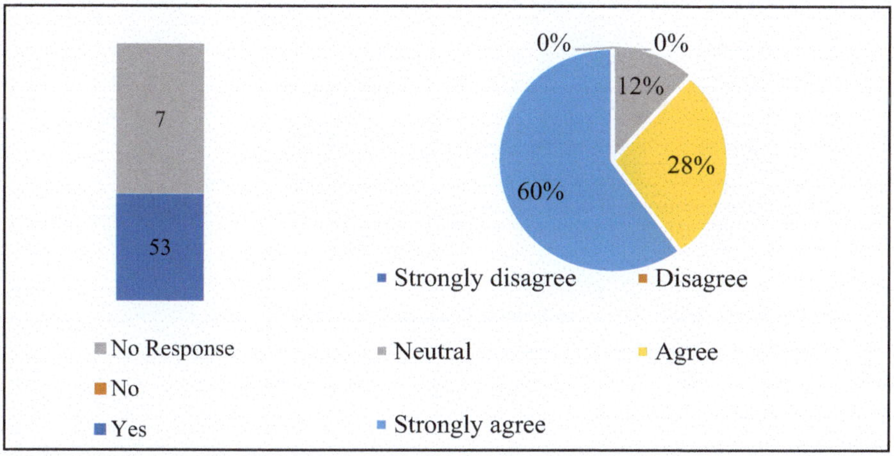

Fig. 13.49 Respondents' perception of Question-2

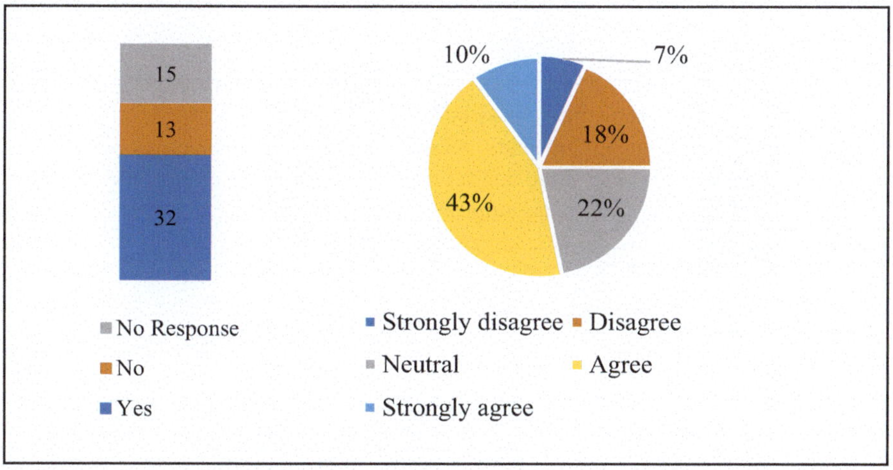

Fig. 13.50 Respondents' perception of Question-3

(i) Use risk-informed land-use planning instead of embankment-centric flood control, limiting the growth of settlements in areas exposed to residual flooding.
(ii) Encourage deliberate and cooperative displacement while guaranteeing service access, tenure security, and livelihood continuity.
(iii) Encourage social protection and diversification of sources of income, especially for informal and dislocated workers.
(iv) Adopt hybrid resilience methods that combine structural protection with ecosystem-based and community-centered measures;

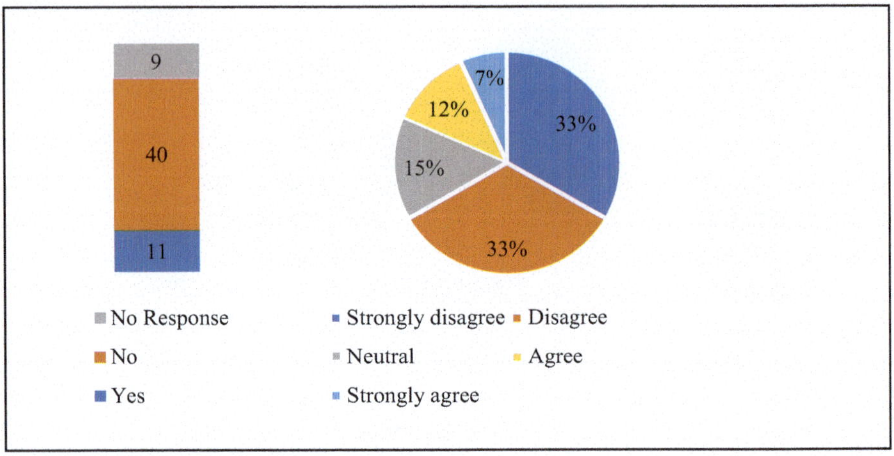

Fig. 13.51 Respondents' perception of Question-4

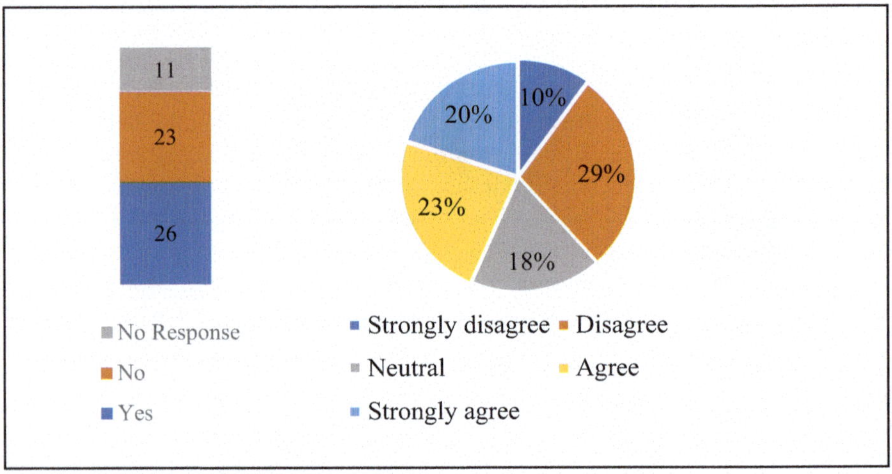

Fig. 13.52 Respondents' perception of Question-5

(v) enhance institutional capacity and governance at the municipal level by incorporating people's perception data into decision-making activities. In support of SDG 1 (No Poverty), SDG 11 (Sustainable Cities and Communities), and SDG 13 (Climate Action), these policies collectively portray displacement as an equitable but adaptive response to the climate crisis. They also directly relate to the book's emphasis on shifting population dynamics and sustainable solutions.

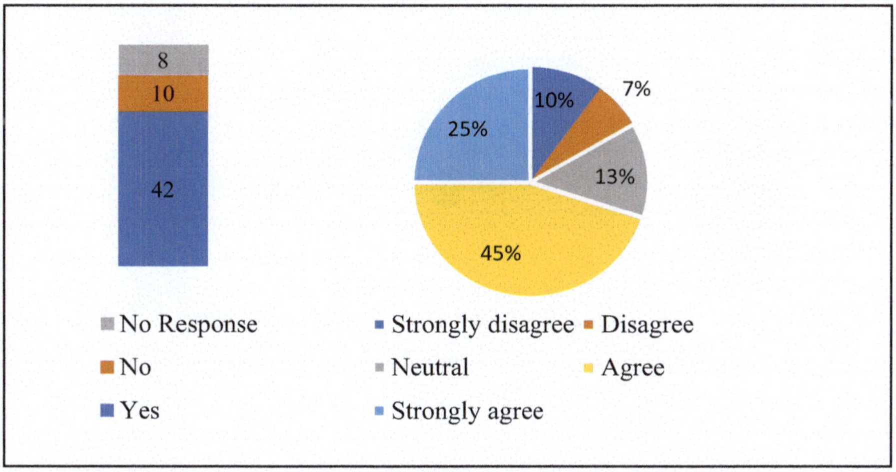

Fig. 13.53 Respondents' perception of Question-6

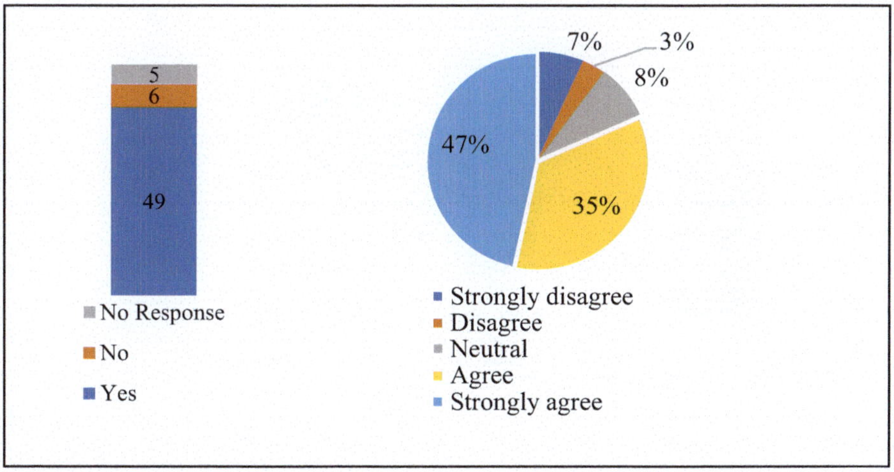

Fig. 13.54 Respondents' perception of Question-7

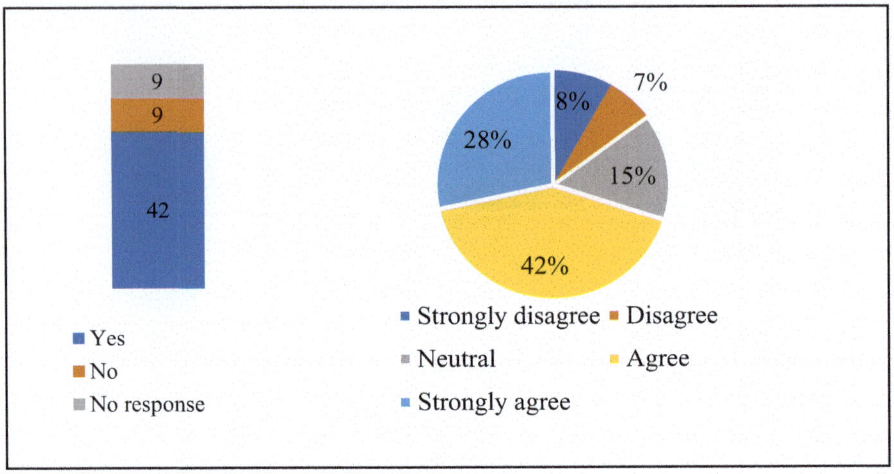

Fig. 13.55 Respondents' perception of Question-8

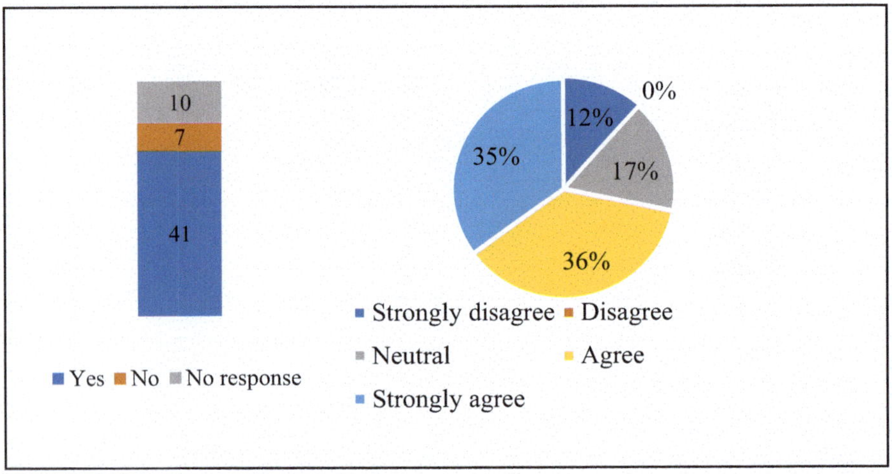

Fig. 13.56 Respondents' perception of Question-9

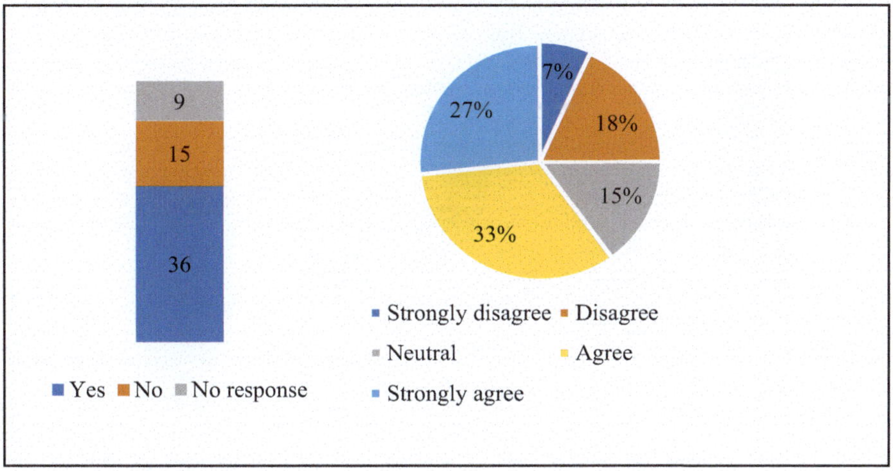

Fig. 13.57 Respondents' perception of Question-10

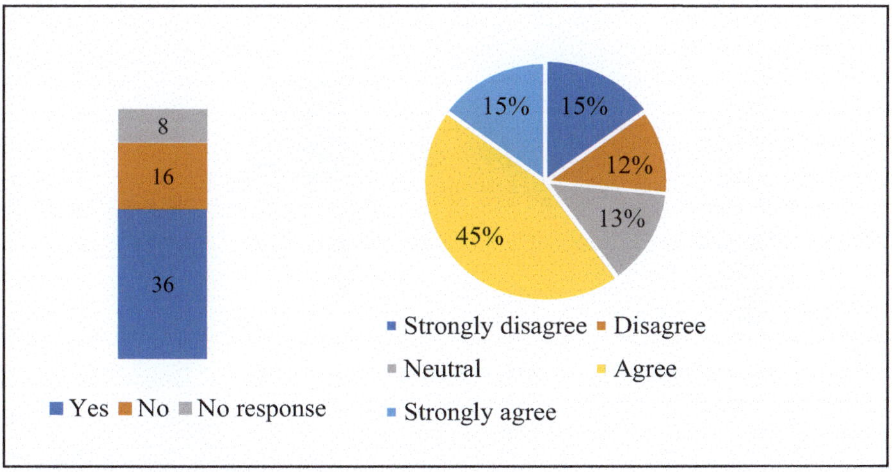

Fig. 13.58 Respondents' perception of Question-11

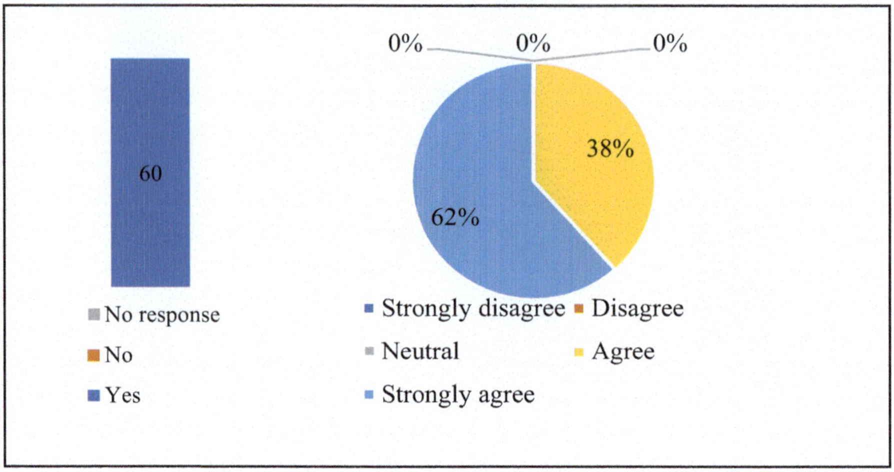

Fig. 13.59 Respondents' perception of Question-12

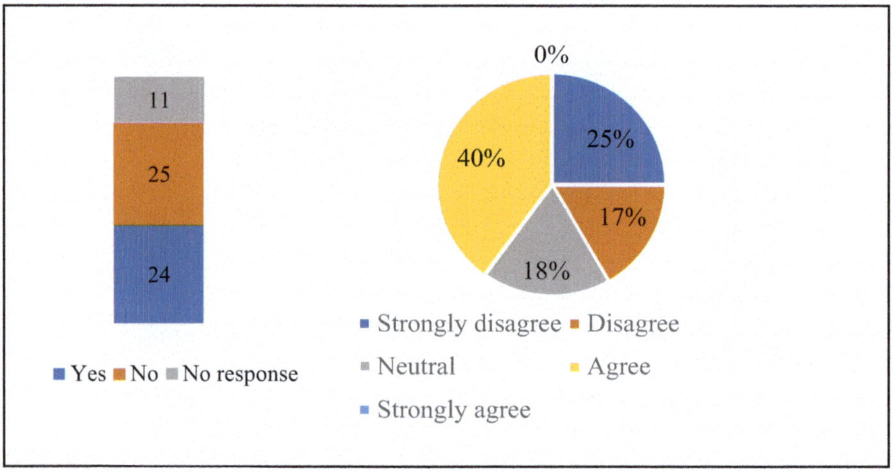

Fig. 13.60 Respondents' perception of Question-13

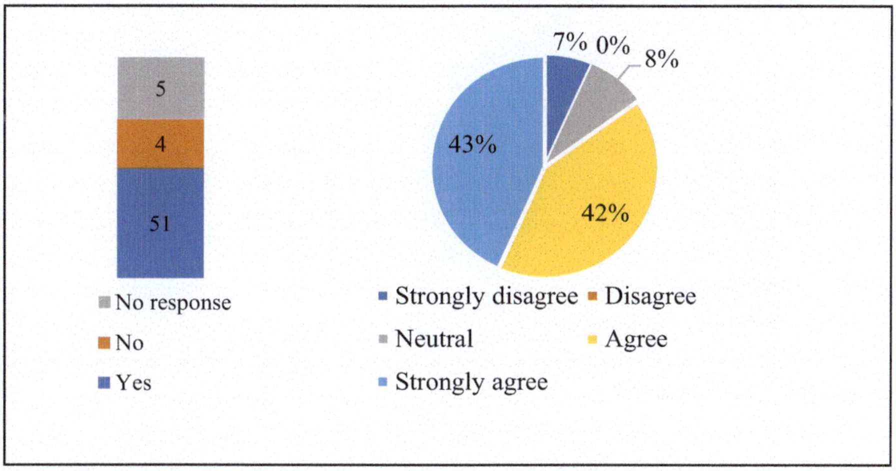

Fig. 13.61 Respondents' perception of Question-14

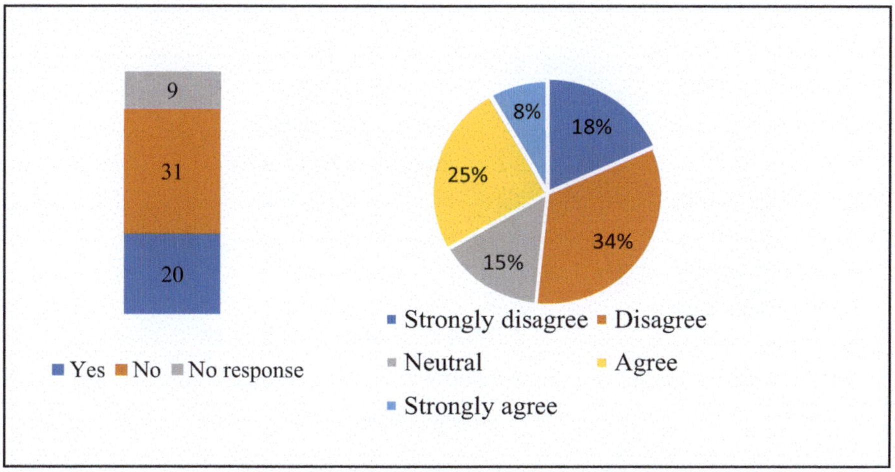

Fig. 13.62 Respondents' perception of Question-15

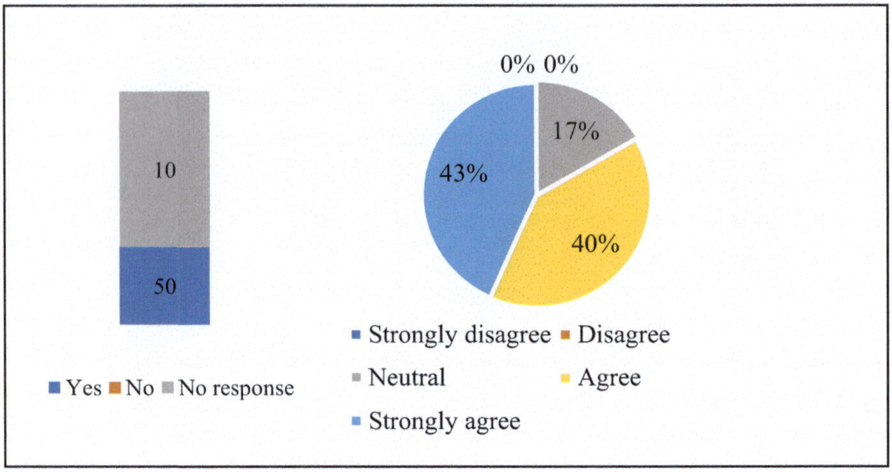

Fig. 13.63 Respondents' perception of Question-16

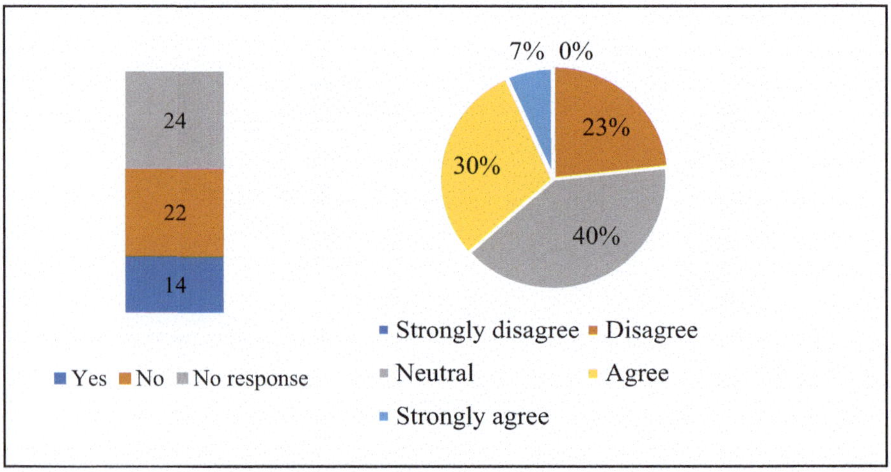

Fig. 13.64 Respondents' perception of Question-17

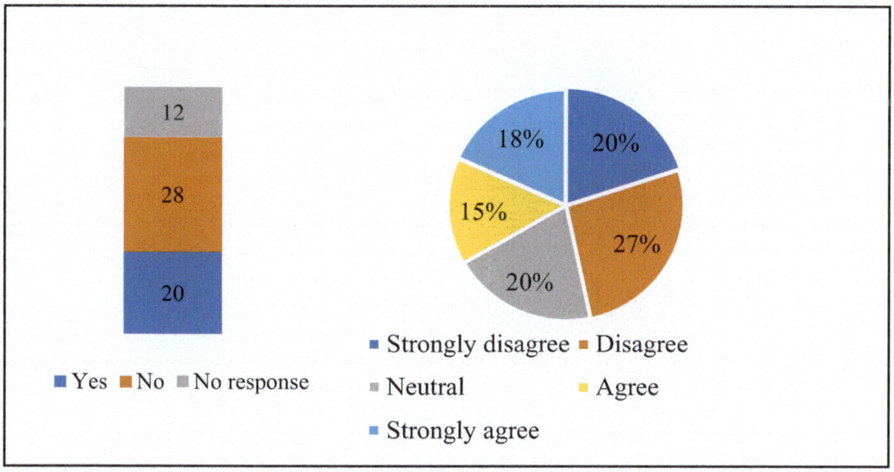

Fig. 13.65 Respondents' perception of Question-18

Table 13.9 Strengths–weaknesses–opportunities–challenges of the study area

Sr. No.	SWOC	Code
	Strengths	
1	Elementary access to education	S1
2	Better transport and communication facilities	S2
3	Better medical facilities	S3
	Weaknesses	
1	No tap water system in the survey area	W1
2	The poor sewage system in the survey area	W2
3	Poor road conditions in the survey area	W3
	Opportunities	
1	Resilient flood prevention and extensive relief distribution	O1
2	Better employment opportunities	O2
3	Better residential privileges	O3
	Challenges	
1	Recent flood occurrence	C1
2	Poverty exists	C2
3	Increase of expenditure	C3

Source Field survey (March 2023)

Table 13.10 SWOC matrix

	S1	S2	S3	W1	W2	W3	O1	O2	O3	C1	C2	C3
S1	–	S1S2	S1S3	S1W1	S1W2	S1W3	S1O1	S1O2	S1O3	S1C1	S1C2	S1C3
S2	S2S1	–	S2S3	S2W1	S2W2	S2W3	S2O1	S2O2	S2O3	S2C1	S2C2	S2C3
S3	S3S1	S3S2	–	S3W1	S3W2	S3W3	S3O1	S3O2	S3O3	S3C1	S3C2	S3C3
W1	W1S1	W1S2	W1S3	–	W1W2	W1W3	W1O1	W1O2	W1O3	W1C1	W1C2	W1C3
W2	W2S1	W2S2	W2S3	W2W1	–	W2W3	W2O1	W2O2	W2O3	W2C1	W2C2	W2C3
W3	W3S1	W3S2	W3S3	W3W1	W3W2	–	W3O1	W3O2	W3O3	W3C1	W3C2	W3C3
O1	O1S1	O1S1	O1S3	O1W1	O1W2	O1W3	–	O1O1	O1O3	O1C1	O1C1	O1C3
O2	O2S1	O2S2	O2S3	O2W1	O2W2	O2W3	O2O1	–	O2O3	O2C1	O2C2	O2C3
O3	O3S1	O3S2	O3S3	O3W1	O3W2	O3W3	O3O1	O3O2	–	O3C1	O3C2	O3C3
C1	C1S1	C1S2	C1S3	C1W1	C1W2	C1W3	C1O1	C1O2	C1O3	–	C1C2	C1C3
C2	C2S1	C2S2	C2S3	C2W1	C2W2	C2W3	C2O1	C2O2	C2O3	C2C1	–	C2C3
C3	C3S1	C3S2	C3S3	C3W1	C3W2	C3W3	C3O1	C3O2	C3O3	C3C1	C3C2	–

Source Authors' calculation based on field survey (March 2023)

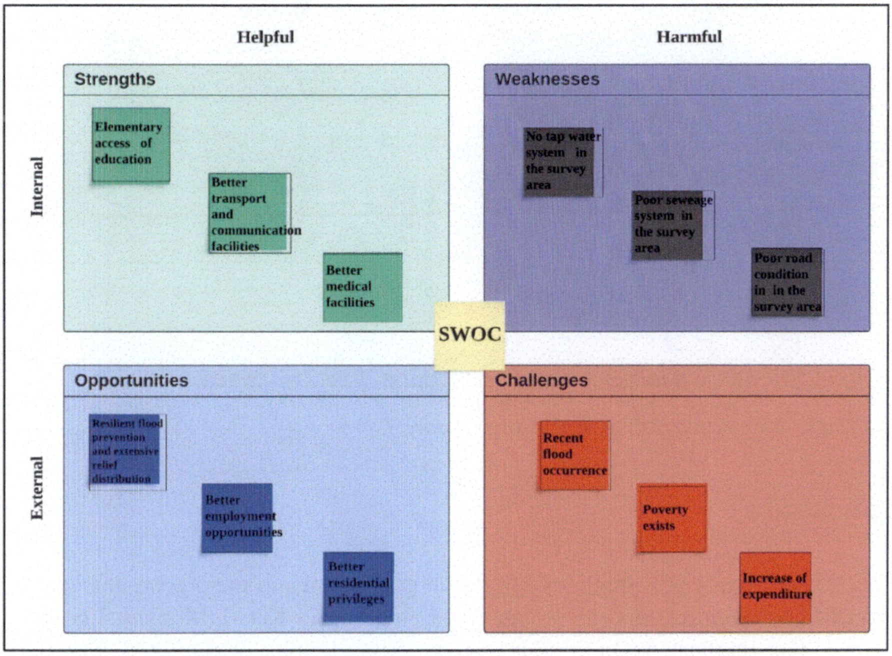

Fig. 13.66 Strengths–weaknesses–opportunities–challenges of the study area

Fig. 13.67 Garbage bins placed in Ward Number 03

Fig. 13.68 Community toilet located in Ward Number 03

Conclusion

This study expands a comprehensive geo-environmental and socio-economic assessment of household displacement caused by flooding in Katwa Municipality, West Bengal. The emphasis is on the Begunkola—Ajay Bandh—Ward No. 3–4 corridor as a case study of a semi-urban riverine system experiencing climate stress. The findings indicate that recurrent monsoonal flooding has acted as a persistent structural stressor rather than a transient threat, progressively modifying land use and cover patterns, livelihoods, settlement stability, and localized population shifts, particularly during the notable events of 1999, 2000, and 2015. Mobility generated by climate in these situations is predominantly limited, repetitive, and socially stratified, rather than permanent or extensive, as demonstrated by nearly 98% of surveyed households experiencing short-distance relocation. Long-term resilience remains obstructed by ongoing infrastructural inadequacies, notably inadequate drainage, limited access to potable water, and sanitation challenges, even though relocating to municipal wards has somewhat aided socio-economic recovery by improving access to education, health care, and non-agricultural employment. The results underscore a principal conclusion of the study: absent further socio-economic and institutional endeavors, structural flood control measures cannot ensure long-term adaptation. This work illustrates how community knowledge may be systematically transformed into robust, policy-relevant evidence by integrating perception-driven, AHP-validated vulnerability assessments with multi-temporal remote sensing-based land-use analysis. This study enhances comprehension of how embankments concurrently mitigate and redistribute flood risk by identifying river proximity, inadequate drainage, and monsoon intensity as primary factors of displacement, while also revealing community perceptions consistent with the established levee effect. The Katwa case offers micro-scale insights applicable to semi-urban environments in South and Southeast Asia, Africa, and global deltaic regions. The study provides a replicable interdisciplinary framework for flood-prone regions facing similar socio-environmental issues, despite constraints related to sample size, cross-sectional design, and the absence of hydrodynamic modeling. It directly promotes SDG 13 (Climate Action), SDG 11 (Sustainable Cities and Communities), and SDG 1 (No Poverty) among the global development agendas. It underscores that sustainable solutions must integrate physical flood mitigation with livelihood diversification, improved basic services, and participatory local government.

Acknowledgements We want to acknowledge Katwa Municipality and Katwa Irrigation Sub-Division Office for providing the necessary maps and information. We extend our gratitude to the survey team, the CBCS Semester-VI (General) students of Katwa College, and the respondents who reside near Ajay Bandh of Ward Number 03 in Katwa Municipality for their active assistance.

Authors Contribution All authors have contributed equally to this work.

Funding The research was not supported by any funding.

Data Availability Statement The datasets during and/or analyzed during the current study available from the corresponding author on reasonable request.

Potential Conflicts of Interest On behalf of all authors, the corresponding author states that there is no conflict of interest.

Informed Consent Informed consent was obtained from all individual participants included in the study.

AI Disclosure We declare that we have used Grammarly (https://app.grammarly.com/) and QuillBot (https://quillbot.com/) (all Basic versions) in online mode for grammar and spelling check, as well as sentence construction error corrections in this chapter.

Abbreviations

AC	Adaptive Capacity
AHP	Analytical Hierarchy Process
AUC	Area Under the Curve
AWiFS	Advanced Wide Field Sensor
BPL	Below Poverty Line
CI	Consistency Index
COVID-19	Coronavirus Disease 2019
CR	Consistency Ratio
DEM	Digital Elevation Model
E	Exposure
GIS	Geographic Information System
IMD	India Meteorological Department
IPCC	Intergovernmental Panel on Climate Change
IRS	Indian Remote Sensing Satellite
LULC	Land Use/Land Cover
MCDM	Multi-Criteria Decision Making

MLC	Maximum Likelihood Classification
NASA	National Aeronautics and Space Administration
NDBI	Normalized Difference Built-up Index
NDMA	National Disaster Management Authority
NGO	Non-Governmental Organization
NRSC	National Remote Sensing Centre
ROC	Receiver Operating Characteristic
S	Sensitivity
SC	Scheduled Caste
SCP	Semi-Automatic Classification Plugin
SFCC	Standard False Colour Composite
SRTM	Shuttle Radar Topography Mission
ST	Scheduled Tribe
SWOC	Strengths–Weaknesses–Opportunities–Challenges
UNDRR	United Nations Office for Disaster Risk Reduction
UN-Habitat	United Nations Human Settlements Programme
USGS	United States Geological Survey
V	Vulnerability

References

Adger WN (2006) Vulnerability. Glob Environ Chang 16(3):268–281. https://doi.org/10.1016/j.gloenvcha.2006.02.006

Adger WN (2016) Place, well-being, and fairness shape priorities for adaptation to climate change. Glob Environ Chang 38:46–55. https://doi.org/10.1016/j.gloenvcha.2016.03.009

Adger WN, Barnett J, Brown K, Marshall N, O'Brien K (2013) Cultural dimensions of climate change impacts and adaptation. Nat Clim Chang 3:112–117. https://doi.org/10.1038/nclimate1666

Alam GMM, Alam K, Mushtaq S (2020) Climate change perceptions and local adaptation strategies of hazard-prone rural households in Bangladesh. Clim Risk Manag 28:100234. https://doi.org/10.1016/j.crm.2020.100234

Banerjee A, Bardhan S, Mukhopadhyay A (2018) Flood risk and embankment failure in the lower Ajay River basin, West Bengal, India. Geogr Rev India 80(3):231–242. https://www.geographicalreviewofindia.org

Bardhan S (2015) Flood hazards, embankments and livelihood vulnerability in the lower Ajay River basin, West Bengal. Indian J Geogr Environ 14:45–58. https://www.ingeo.org

Bezboruah KC, Tripathi S, Das A (2021) Climate-induced migration and peri-urbanization in South Asia. Popul Space Place 27(6):e2441. https://doi.org/10.1002/psp.2441

Bhattacharya S, Das S (2021) Riverbank erosion and flood vulnerability in the lower gangetic plain of West Bengal. J Geol Soc India 97(4):421–432. https://doi.org/10.1007/s12594-021-1705-9

Bhushan N, Rai K (2004) Strategic decision making: applying the analytic hierarchy process. Springer. https://doi.org/10.1007/978-3-662-09515-8

Black R, Bennett SRG, Thomas SM, Beddington JR (2011) Climate change: migration as adaptation. Nature 478:447–449. https://doi.org/10.1038/478447a

Castells-Quintana D, Lopez-Uribe MP, McDermott TKJ (2022) Climate change and human mobility. World Dev 150:105707. https://doi.org/10.1016/j.worlddev.2021.105707

Census of India (2001) District census handbook: Barddhaman, West Bengal (Series 20, Part A & B). Directorate of Census Operations, West Bengal, Government of India

Census of India (2011) District census handbook: Purba Bardhaman. Government of India. https://censusindia.gov.in

Congedo L (2021) Semi-automatic classification plugin documentation. QGIS Association. https://fromgistors.blogspot.com/p/semi-automatic-classification-plugin.html

Coppock JT, Rhind DW (1991) The history of GIS. In: Maguire DJ, Goodchild MF, Rhind DW (eds) Geographical information systems. Wiley, pp 21–43. https://onlinelibrary.wiley.com

Cutter SL, Boruff BJ, Shirley WL (2008) Social vulnerability to environmental hazards. Soc Sci Q 84(2):242–261. https://doi.org/10.1111/1540-6237.8402002

de Sherbinin A, Castro M, Gemenne F, Cernea MM, Adamo S, Fearnside PM, Shi G (2012) Preparing for resettlement associated with climate change. Science 334(6055):456–457. https://doi.org/10.1126/science.1208821

District Disaster Management Plan, Purba Bardhaman (2023) Government of West Bengal. https://wb.gov.in

Dutta S (2010) Flood management and riverbank erosion in eastern India. Econ Polit Wkly 45(38):57–63. https://www.epw.in

Elum ZA, Lawal AI (2022) Climate change adaptation, resilience and vulnerability. Environmental Development 41:100683. https://doi.org/10.1016/j.envdev.2021.100683

Fawcett T (2006) An introduction to ROC analysis. Pattern Recogn Lett 27(8):861–874. https://doi.org/10.1016/j.patrec.2005.10.010

Ghimire R, Ferreira S, Dorfman J (2015) Flood-induced displacement and migration. World Dev 72:112–125. https://doi.org/10.1016/j.worlddev.2015.02.004

Ghosh A (2012) Flood vulnerability and urban expansion in riverine West Bengal. J Regional Geogr 34(2):155–168

Hanley JA, McNeil BJ (1982) The meaning and use of the area under a ROC curve. Radiology 143(1):29–36. https://doi.org/10.1148/radiology.143.1.7063747

Hill T, Westbrook R (1997) SWOT analysis: it's time for a product recall. Long Range Plan 30(1):46–52. https://doi.org/10.1016/S0024-6301(96)00095-7

Hosmer DW, Lemeshow S, Sturdivant RX (2013) Applied logistic regression (3rd ed.). Wiley. https://onlinelibrary.wiley.com

IDMC (2023) Global report on internal displacement 2023. Internal Displacement Monitoring Centre. https://www.internal-displacement.org

India Meteorological Department (2000) Meteorological data (rainfall and temperature). Ministry of Earth Sciences, Government of India. https://mausam.imd.gov.in

India Meteorological Department (2015) Gridded rainfall and temperature datasets. Ministry of Earth Sciences, Government of India. https://mausam.imd.gov.in

India Meteorological Department (2023) Daily gridded rainfall and temperature data products. Ministry of Earth Sciences, Government of India. https://mausam.imd.gov.in

IPCC (2007) Climate change 2007: impacts, adaptation and vulnerability. Cambridge University Press. https://www.ipcc.ch

IPCC (2014) Climate change 2014: impacts, adaptation and vulnerability. Cambridge University Press. https://www.ipcc.ch

IPCC (2021) Climate change 2021: the physical science basis. Cambridge University Press. https://www.ipcc.ch

IPCC (2022) Climate change 2022: impacts, adaptation and vulnerability. Cambridge University Press. https://www.ipcc.ch

Kakinuma K et al (2020) Environmental migration and displacement in Bangladesh. Environ Res Lett 15(11):114046. https://doi.org/10.1088/1748-9326/ab9f03

Kiker GA, Bridges TS, Varghese A, Seager TP, Linkov I (2005) Application of MCDA in environmental decision making. Integr Assess 6(4):95–109

Kothari CR (2004) Research methodology: methods and techniques (2nd ed.). New Age International. https://www.newagepublishers.com

Likert R (1932) A technique for the measurement of attitudes. Archives of Psychology 140:1–55
Lillesand TM, Kiefer RW, Chipman JW (2015) Remote sensing and image interpretation (7th ed.). Wiley. https://onlinelibrary.wiley.com
Malczewski J (2006) GIS-based multicriteria decision analysis. Int J Geogr Inf Sci 20(7):703–726. https://doi.org/10.1080/13658810600661508
Mallick B, Vogt J (2014) Population displacement after cyclone Aila. Nat Hazards 73:1–18. https://doi.org/10.1007/s11069-014-1069-6
McLeman R (2018) Migration and displacement under climate change. Popul Environ 39(4):319–338. https://doi.org/10.1007/s11111-018-0296-4
McLeman R, Gemenne F (2018) Climate-induced migration. Popul Environ 39(4):319–338
Mirza MMQ (2011) Climate change, flooding in South Asia. Reg Environ Change 11:95–107. https://doi.org/10.1007/s10113-010-0184-7
Moore ID, Grayson RB, Ladson AR (1991) Digital terrain modeling: a review of hydrological, geomorphological, and biological applications. Hydrol Process 5:3–30. https://doi.org/10.1002/hyp.3360050103
Mucherera B, Spiegel SJ (2022) Flood risk redistribution and socio-environmental vulnerability. Prog Disaster Sci 14:100227. https://doi.org/10.1016/j.pdisas.2022.100227
Mukhopadhyay A (2009) Flood control, embankments and vulnerability in eastern India. Indian J Geogr 81(2):175–186
Mukhopadhyay A et al (2018) Flood management paradox in embanked rivers. Environ Earth Sci 77:394. https://doi.org/10.1007/s12665-018-7539-1
Nandy S (2013) Drainage congestion and flood hazards in West Bengal. J Indian Geogr Stud 32(1):89–101
NASA & U.S. Geological Survey (2023) Shuttle radar topography mission (SRTM) digital elevation model. EarthExplorer. https://earthexplorer.usgs.gov
NDMA (2008) National flood management guidelines. Government of India. https://ndma.gov.in
NRSC (2012) Bhuvan: Indian geo-platform of ISRO. National Remote Sensing Centre. https://bhuvan.nrsc.gov.in
Patton MQ (1999) Enhancing the quality and credibility of qualitative analysis. Health Serv Res 34(5):1189–1208
Rahman MM (2017) Climate-induced migration and urban vulnerability in eastern India (Unpublished doctoral dissertation). University of Calcutta
Rahman MM et al (2018) Flood vulnerability assessment using AHP. Nat Hazards 91(2):459–479. https://doi.org/10.1007/s11069-017-3139-6
Roy S (2012) Channel migration and flood hazards of the Ajay River. J Geol Soc India 80(2):261–270
Roy S, Misra A (2019) Floods, embankments and vulnerability in eastern India. Nat Hazards 99:1037–1056. https://doi.org/10.1007/s11069-019-03771-7
Saaty TL (1980) The analytic hierarchy process. McGraw-Hill
Sarker MH et al (2019) Rivers, floods and human displacement in Bangladesh. Nat Hazards 98:1011–1033. https://doi.org/10.1007/s11069-019-03735-x
Sen A (2016) Livelihood insecurity and flood displacement in peri-urban West Bengal. Econ Aff 61(4):613–621
Sinha R (2016) River embankments and flood hazards in India. Water Resour Manage 30:2889–2902. https://doi.org/10.1007/s11269-016-1329-9
Tacoli C (2017) Migration, urbanization and climate change. Environ Urban 29(2):1–15. https://doi.org/10.1177/0956247817721963
UN-Habitat (2020) World cities report 2020. United Nations. https://unhabitat.org
UNDRR (2019) Global assessment report on disaster risk reduction. United Nations. https://www.undrr.org
UNDRR (2021) Human cost of disasters. United Nations. https://www.undrr.org
United Nations Office for Disaster Risk Reduction (2015) Global assessment report on disaster risk reduction 2015: making development sustainable—the future of disaster risk management. United Nations. https://www.undrr.org

U.S. Geological Survey (2000) Digital elevation model (DEM). U.S. Geological Survey. https://www.usgs.gov
U.S. Geological Survey (2001) Landsat 5 TM / Landsat 7 ETM+ satellite imagery. Earth Resources Observation and Science (EROS) Center. https://earthexplorer.usgs.gov
U.S. Geological Survey (2023) Landsat collection 2 level-1 and level-2 data products. EROS Center. https://earthexplorer.usgs.gov
Vaidya OS, Kumar S (2006) Analytic hierarchy process: an overview. Eur J Oper Res 169(1):1–29. https://doi.org/10.1016/j.ejor.2004.04.028
Warner K, Head G (2011) Climate change, environmental degradation and migration. Popul Environ 33(1):28–50. https://doi.org/10.1007/s11111-011-0140-2
Wisner B, Blaikie P, Cannon T, Davis I (2004) At risk: natural hazards, people's vulnerability and disasters (2nd ed). Routledge. https://www.routledge.com
Zha Y, Gao J, Ni S (2003) Use of normalized difference built-up index. Int J Remote Sens 24(3):583–594. https://doi.org/10.1080/01431160304987

Chapter 14
Climate-Induced Flooding and Population Displacement Toward Urban Centres: Empirical Insights from Bangladesh

Humaira Tabassum Raiyan and Neegar Sultana

Abstract Flooding and population displacement resulting from climate change are increasing due to extreme climate variability. The study focused on the fact that climate-induced flooding is considered a reason for population displacement, as well as a potential cause of rural–urban migration. Empirical investigations have been conducted to validate the framework which was developed based on the comprehensive review of existing literatures. Afterward, statistical analysis is used to determine the population displacement due to flooding.Additionally, ArcGIS 10.8, and Google Earth Engine platform is used to define rainfall pattern and flood inundation risk for empirical investigations. Results reveal that flooding affects population displacement, with most of the displacement occurring from rural to urban areas in Sunamganj. The research contributes a conceptual framework on the increased impact of the climate crisis flood on population displacement in Sunamganj. This framework can serve as a tool to assess the effects of floods on rural–urban migration and disaster risk reduction for development purposes. Overall, the framework will facilitate the identification of concepts and relations between extreme climate-induced floods and migration, flood-affected rural zones for resettlement and development, and potential mitigation measures to reduce the impact of climate crisis floods.

Keywords Climate change · Flooding intensity · Human displacement · Sunamganj · Bangladesh

Introduction

Climate change is inducing extreme phenomena including storms, cyclones, and floods which are notably resulting in heightened population displacement (Kakinuma et al. 2020). More specifically, climate-induced floods have a severe negative impact

H. T. Raiyan · N. Sultana (✉)
Department of Geography and Environment, Jagannath University, Dhaka, Bangladesh
e-mail: nsapoly@gmail.com; nsapoly@yahoo.com

P. Ghosh et al. (eds.), *Climate Crisis and Changing Population Dynamics*, Springer Climate, https://doi.org/10.1007/978-3-032-24533-5_14

on economic activity, cultural heritage, infrastructure, property, human health, and livelihood compared to other natural hazards (Sarker et al. 2020; Ahmad and Afzal 2021). Repetitive flooding increases the vulnerability of rural populations by declining economic opportunities (Ahmad and Afzal 2020). Such recurrent uncertainties lead to riverbank erosion, flooding, forcing rural households to migrate temporarily or permanently, both within and outside their communities. The phenomenon of flood and population displacement is not a recent event in the hazard's scenario (Alam et al. 2020; Ahmad and Afzal 2021), as flood hazards result in a massive number of land losses, widespread landlessness, and population displacement, leaving a significant proportion of people homeless (Sarker et al. 2020). Each year, worldwide, natural climatic events such as flood hazards compel millions of rural people to migrate from one region to another, exposing them to severe challenges in livelihood, shelter, and safety, along with the loss of homestead land, farmland, and destruction of standing crops (Rahman and Gain 2020; Ahmad and Afzal 2020). In consequence, the Sustainable Development Goal, especially SDG 10, highlight the needs of declining inequalities both within and between nations, strengthening the significance of ensuring safe, orderly, accountable mobility and migration (Target 7) (Almulhim et al. 2024). However, the achievement of this goal was interrupted due to climate-related extreme hazards like water scarcity, floods, storms and so on. A World Bank report from 2018 projected that climate change could compel over 143 million individuals to relocate within their own countries by 2050, within three regions including South Asia, Africa, and Latin America (Semenza and Ebi 2019). Therefore, it is essential to cautiously evaluate the impacts of climate change on migration to focus on SDG target, especially in at-risk area such as Africa, Latin America, Southeast Asia, and the broader pacific region (Almulhim et al. 2024).

Under this context, Asian, African, and South-American countries are highly vulnerable to flood connected with physical, economical, and infrastructural aspect (Brouwer et al. 2007; Bhattacharjee and Behera 2018; Shah et al. 2018), while having better infrastructural, technological, and administrative set-up, European and North-American countries are less vulnerable (Jamshed et al. 2020a). Therefore, low-income and developing regions are exceedingly vulnerable to climate-induced floods, having limited coping and adaptive capacity due to inadequate economic and institutional resources (Feldmeyer et al. 2017; Jamshed et al. 2020a). Additionally, rural areas can be distinguished by poverty, high dependency on agriculture, inadequate access to infrastructure and services, and limited government attention. These factors render rural areas less capable of developing preparedness and coping mechanisms in response to climate change-related events, such as floods (Jamshed et al. 2020a).

Furthermore, over 90% of the global rural population resides in less developed regions, where settlements are often inadequately supplied with resources and remain underdeveloped (Srivastava and Shaw 2016). However, rising rural–urban interdependencies fuelled demographic, economic, and governance transitions in numerous rural regions (Jamshed et al. 2020a). The reciprocal interdependencies arise when urban centres pulled rural communities through furnish markets, agricultural inputs, customer goods and rural regions benefited urban centres with labour, raw materials, timber, food supply, and various ecological services (Gebre and Gebremedhin 2019). Despite this general consensus, there is scarce of empirical study about the impact of extreme weathers or climate events on these intricate interactions, which may affect the rural household vulnerability (Jamshed et al. 2020a). Most of the developing countries in Asia, a massive part of population is constrained to reside in highly disaster-prone zones due to high population density and widespread poverty (Rahman and Gain 2020). Moreover, inadequate disaster management systems result in significant losses in both human lives and economic stability in developing nations, especially in Bangladesh (Hashmi et al. 2012; Mahmood et al. 2021).

Bangladesh is highly experiencing adverse impact of climate change and identified as one of the most disaster-prone nations globally (Rahman and Gain 2020). The significant rural areas including infrastructures, livelihoods, agricultural output, and drinking water access, etc. are currently threatened through climate-induced flood. The combination of ongoing political and socioeconomic issues with increased vulnerability to natural disasters make Bangladesh especially prone to the enduring impacts of climate change including significant population displacement (IDMC 2015). The recurrent floods affect several districts of northeastern Bangladesh each year, including Mymensingh, Sherpur, Moulvibazar, Habiganj, Kishoreganj, Sunamganj, Netrokona, and Brahmaputra, creating severe humanitarian challenges (Tuli et al. 2024). However, the extensive inundation of Sunamganj district, driven by intense monsoon rainfall and transboundary water inflows from India's northeast, has trapped millions of people and triggered a large-scale humanitarian crisis. Although water levels have been gradually and consistently receding, the humanitarian conditions in the flood-affected areas remain critical, with the situation being particularly severe in Sunamganj district (UNICEF 2022). These increasing vulnerabilities and climate change in Sunamganj are causing a massive displacement of the population toward urban centres.

Previous studies identified the basic connection between population displacement and climate change globally (Warner et al. 2010, Black et al. 2011, Black et al. 2013, Beine and Parsons 2015, Kakinuma et al. 2020, Perch-Nielsen et al. 2008, Ivanova et al. 2024). Lifestyle choices, geographical locations and risk exposure mainly drive the climatic events (Teo et al. 2018) which can't be resisted through cultural, economic, social, political, or geographical boundaries (Ahmad and Afzal 2020). However, climate experts considered flood as a greatest threat than other climatic events (Ali and Rahut 2020). In Hazard Scenario, population displacement due to flood is not recent or new (Alam et al. 2020), it always raises the vulnerability of rural population by hampering employment sources and opportunities (Ahmad and Afzal 2020). Therefore, millions of rural people displaced from one place to another, especially towards urban areas due to the devastating climate-induced flood (Ahmad and Afzal 2020).

On the other hand, previous literature also investigated the climate-induced population displacement issues in Bangladesh. For example, Assraf Seddiky et al. (2024) identified that the climate-related incidents have compelled people from high-risk zone to safer regions. Conversely, the relationship between environmental changes and migration for a population residing along the Jamuna River has been examined by Freihardt (2025), highlighting the significant impacts of floods and riverbank erosion in Bangladesh. Additionally, the influencing factors of climate-related migration and support for proactive measures and solution for displacement in coastal areas of Bangladesh (Rahaman et al. 2025). Moreover, numerous research studies have been conducted on population displacement due to floods in the Sunamganj region, suggesting that the government should develop and implement a cohesive and efficient strategy for managing haors and floods to safeguard the lives and livelihoods of those residing in these areas (Tuli et al. 2024; Akter et al. 2023). For instance, during times of flooding, individuals without land or financial resources often resort to informal borrowing from moneylenders (Surendra 2020). This cycle of debt tends to grow over time, particularly with repeated flooding events, resulting in a significant risk of insolvency and financial distress for numerous affected households (Sultana and Rayhan 2012). Rural–urban migration enables households facing challenges to break free from a cycle of debt (Israt Rayhan and Grote 2007). A vicious cycle of debt, resource depletion, and poverty serves as a powerful driver, resulting in enduring hardship and susceptibility that hinders the capacity of Tanguar haor-based communities in Sunamganj to prepare for, respond to, and recover from future floods (Abedin and Khatun 2020). A case study was conducted on an economically disadvantaged family in Sunamganj to assess temporary environmental migration (Bakth and Hasanuzzaman 2023).

Therefore, previous studies have identified that population displacement is linked to environmental push factors. In contrast, some researchers have acknowledged the presence of a pull factor in urban areas that influences population displacement. Despite the relationship between climate crisis, flooding, and population displacement, establishing it becomes complex due to the complex interplay between nature and society. It is essential to develop a conceptual framework to investigate the relationship between climate crisis-induced flooding and population displacement.

Thus, the chapter firstly aims to understand that climate crisis flooding is a contributing factor to population displacement developing a conceptual framework, secondly serve as a potential driver of rural–urban migration including pull and push factor. Finally, focuses on empirical investigation that evident the connection among climate-induced flood and rural–urban migration. This comprehensive framework is useful to evaluate all dimensions of human responses to climate-induced flooding and also facilitate the impactful policies that focused on mitigating population displacement and urban pressure by providing empirical evidence. It is a novel work which will contribute a broad information about the intricate relation of climate crisis flood and population displacement and both climatic and social factors that influences rural–urban population dynamics.

Theoretical Framework: Interconnections Between Climate-Induced Flooding and Population Displacement

Climate Change-Induced Variability

In the last century, the consistent increase in global temperature, caused by anthropogenic greenhouse gas emission has led to significant changes in climate across international, regional, and local levels (IPCC 2014). These involve increased winter temperature, reduced terrestrial water reserves, and diminishing renewable water resources. The alterations negatively impact on agriculture, influencing the viability, growth stages, and yields of various crop species. Elevated evaporation rates and diminished soil moisture intensify water stress, while changing rainfall patterns and interrupt the farming cycles (Mianabadi et al. 2023). Overall, these changes weaken agricultural output, reduce community resilience, triggering economic uncertainty and exacerbate migration especially from rural to urban settings. Furthermore, the occurrence and severity of flood events are on the rise which is linked to the ongoing impact of climate change (Almasi and Soltani 2017). Extreme weather events triggered by climate change notably floods inflict damage on croplands, degrade topsoil, disrupt agricultural infrastructure and intensify land-use trade-offs (Sharifi et al. 2012; Mianabadi et al. 2023).

Flood Risk Amplification

From thousands of years, human societies have sought to protect themselves from the harmful effects of large-scale flooding. Additionally, flood hazard is heightened the vulnerability to human populations due to the increasing anthropogenic factors like rapid population growth, changes in land use practices, and climate change consequences (Malik et al. 2020; Tuli et al. 2024). This section outlines the literatures that develop frameworks which depict factors trigger flooding and its impact on household.

Intense rainfall is the main reason contributing to extreme water discharges and flooding risk that increase the vulnerability to flood (Feng et al. 2021). Vulnerability refers to the social, economic, and environmental factors that heightened a community's susceptibility to the negative impact of hazards (Sultana et al. 2024). However, the term vulnerability typically demonstrates the complex interplay of various factors in disaster illustrating how social and environmental forces result disaster scenario (Bhattacharjee and Behera 2018). In other words, disasters are the outcome of the vulnerability levels which is established by the interplay of social, physical, environmental, economic and political factors (Turner et al. 2003; Rahman and Gain 2020). All factors of vulnerability are fundamentally interconnected to individual's livelihood and poverty situations (Rahman and Gain 2020).

Flooding can devastate crops, accelerate soil erosion, increase sediment deposition, and diminish land fertility (Mianabadi et al. 2023). Significant rainfall resulting severe flooding that have the potential to intensify poverty and food security through declining agricultural output. Devastating flood also reduces economic efficiency and lowers the income levels of rural communities, and forcing population to move towards economic facilitate regions (Mianabadi et al. 2023). Flood also leads to serious health consequences by raising the occurrence of diseases. Contamination of drinking water with sewage can result in outbreaks of cholera and dysentery (Ashbolt 2004, Mora et al. 2018). In contrast, stagnant water creates breeding grounds for insects, which can potentially lead to the spread of diseases such as malaria and yellow fever (White et al. 2002). Such disease outbreaks frequently contribute to elevated mortality rates (Mora et al. 2018). All these contributing phenomena, including social, economic, environmental, and physical dimensions, connected with people's livelihoods and economic activities, increase the vulnerability of floods in rural areas and force people to displace (Semenza and Ebi 2019).

Flood-Induced Population Displacement

Most of the rural population generally refrains from taking immediate and intense decisions in response to adverse conditions, as they believe such circumstances are not permanent. Instead, they try to adapt the situation over time (McLeman 2018). Lee (1966) classified the factors into four types that influence population displacement: push factors related to the origin, pull factors related to the destination, intervening obstacles, and personal factors. Additionally, in migration theory, push factors are effective when people suffer due to limited job opportunities, low wages, and poor living conditions, such as in rural areas (Semenza and Ebi 2019). Conversely, pull factors are associated with the destination, which offers better job facilities, a good quality of life, and opportunities for personal improvement, typically found in urban areas (Lee 1966). Furthermore, factors that encourage people to migrate are classified into three categories: hopes, pull factors, and fear (Mianabadi et al. 2023). When conditions become worst and intolerable, and no alternative means of livelihood remain available, the forced rural population is driven to displace and abandon their homes.

People experience multiple impacts of climate that give rise to serious challenges. The most vulnerable groups often endure extreme poverty, unemployment, food insecurity, and various economic and environmental hardships (Islam and Baten, 2016; Assraf Seddiky et al. 2024). Disaster-affected populations encounter significant difficulties both in their places of origin and in the areas to which they relocate (Sullivan and Townsend 2022). Due to attractive opportunities and enhanced safety certain individuals prefer to depart from their homeland, even when the option to resettle is still available (Mianabadi et al. 2023). In contrast, most individuals choose to remain in their own houses after disaster like flood (Mianabadi et al. 2023). However, when no viable alternatives exist, people are compelled by push factor to move elsewhere in search of proper livelihood (Hugo et al. 2018). During this period, most affected individuals are left with nothing as disasters like floods disrupt their primary income sources, such as farming, agriculture, business, and forestry (Perch-Nielsen et al. 2008). Therefore, it becomes difficult for them to repay the money with interest that they borrow as a loan from a lending institution or a non-governmental organization for rehabilitation (Assraf Seddiky et al. 2024). Suffering from this complex situation, vulnerable people abandon their original places in search of a better income source and opportunities in cities.

Determinants of Rural–Urban Linkage and Flood Vulnerability

The drivers influencing rural–urban linkages and flood vulnerability exhibit considerable similarity. In both domains, these determinants are commonly categorized into social, economic, infrastructural, institutional, spatial, and environmental dimensions (Birkmann 2006, Lianxiao and Morimoto 2019). The determinants influencing both rural–urban connectivity and flood vulnerability are shown in Fig. 14.1.

Social factors, including demographic characteristics such as age, gender, health, and education, significantly influence individuals' ability to access material assets, including tools, technology, and knowledge of various market processes (Nazari et al. 2015). Households that experience financial challenges, have little savings, and possess few valuable assets often face significant risks from climate vulnerability

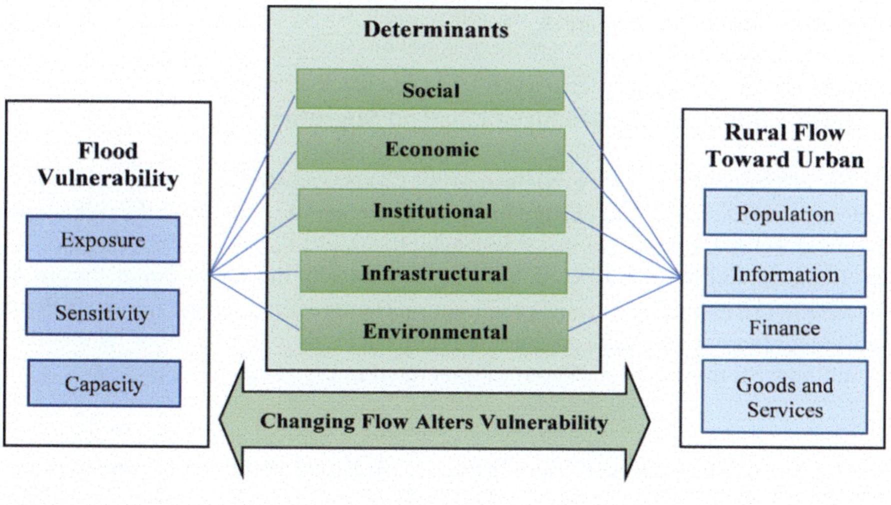

Fig. 14.1 Determinants of rural–urban linkage and flood vulnerability. *Source* Jamshed et al. (2020a)

and socioeconomic changes (Sangpenchan 2011, Zarafshani et al. 2012). Additionally, institutional factors refer to the frameworks and governance of institutions (Bah et al. 2003). Legislative frameworks and their enforcement by appropriate authorities govern development in flood-prone regions, reducing risk exposure (Jamshed et al. 2020a). Further, the availability and accessibility of infrastructure and services significantly affect the coping capacity of households in flood-prone rural regions (Srivastava and Shaw 2016). Environmental factors including presence of river, soil quality, water availability with other ecosystem services significantly influence rural–urban connectivity.

A Conceptual Model of Flood Hazard and Rural Vulnerability

Floods are known to cause population displacement. "A natural hazard forces displacement by damaging land, housing, and material assets" (Haque 1997, Perch-Nielsen et al. 2008). Changing precipitation patterns, such as heavy rainfall, can spoil agricultural activities and create food insecurity (Misra 2014). Natural hazards like floods not only destroy primary economic sources but also disrupt social bonds and family balance, compelling people to migrate. Previous literatures provide evidence that climate change is responsible for changing precipitation patterns that trigger floods and make rural areas too vulnerable, force the population to displace (Perch-Nielsen et al. 2008; Kakinuma et al. 2020; Jamshed et al. 2020a, 2020b; Ivanova et al. 2024; Warner et al. 2010). Frameworks like "influence of climate change on migration through flood," "rural–urban linkages and its influence on rural flood vulnerability" are used to assess the relation between climate-induced flood events and population displacement towards urban areas.

Impact of Climate Change-Induced Flood on Migration

The model proposed by Perch-Nielsen et al. (2008), which has leftmost linkages, illustrates the role of climate change in elevating flood risk. Moving to the right, both direct and indirect effects are described, which can ultimately lead to migration as well as an adaptation strategy. Generally, the influence of this factor diminishes from left to right. For example, extreme precipitation events exert a strong impact on flood occurrence. In contrast, reduced income represents only one of several factors that may influence a migration decision, as illustrated in Fig. 14.2.

Rural–Urban Linkages and Their Influence on Rural Flood Vulnerability

According to Jamshed et al. (2020a), the discourse on rural–urban linkages frames rural areas not as isolated units but as dynamic spaces defined by their interactions

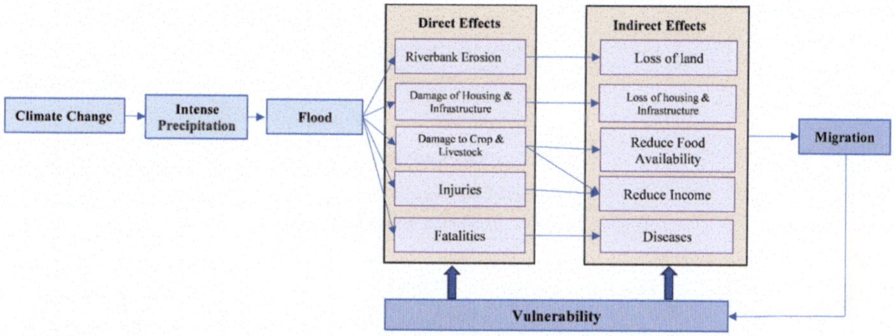

Fig. 14.2 Influence of climate change on migration. *Source* Perch-Nielsen et al. (2008)

with proximate urban centres. The movement of individuals, sharing of information, the transfer of financial resources and the distribution of goods and services demonstrated these interactions and the interconnectedness is crucial for promoting development in improving the livelihoods, well beings and socioeconomic prospects of rural households, presented in Fig. 14.3.

In this framework, floods are defined as phenomena resulting from natural or human-induced factors that can lead to social, economic, physical, and environmental consequences which may be direct or indirect. Vulnerability consists of three components: exposure, susceptibility, and capacity. Here, exposure defined as the severity of flood events assessed by the degree of damage and loss incurred. Conversely, susceptibility refers to the condition of exposed social, economic, and physical systems that make a rural community vulnerable to disaster. Capacity encompasses the ability of rural communities to anticipate, manage, respond to, and adapt to flooding.

In Fig. 14.3, the vulnerability models and factors indicate that population displacement is becoming a concerning issue worldwide, with a changing climate being the triggering factor that increases flood events and their vulnerability in rural areas. Additionally, the increasing rural vulnerability to floods forces people to displace from their original places or their own land, either temporarily or permanently. Ultimately, this flow of displacement occurs mostly in urban or city areas, which is driven by social, economic, institutional, and environmental factors.

The Conceptual Framework

From the previous discussions and frameworks, the current framework has been proposed by integrating the concept of push and pull factors among rural areas (Lee 1966, Martin et al. 2021), flood vulnerability (Jamshed et al. 2020a, Lianxiao and Morimoto 2019), and Climate Change influence on flood (Perch-Nielsen et al. 2008; Mianabadi et al. 2023). The proposed framework describes the effect of pull and

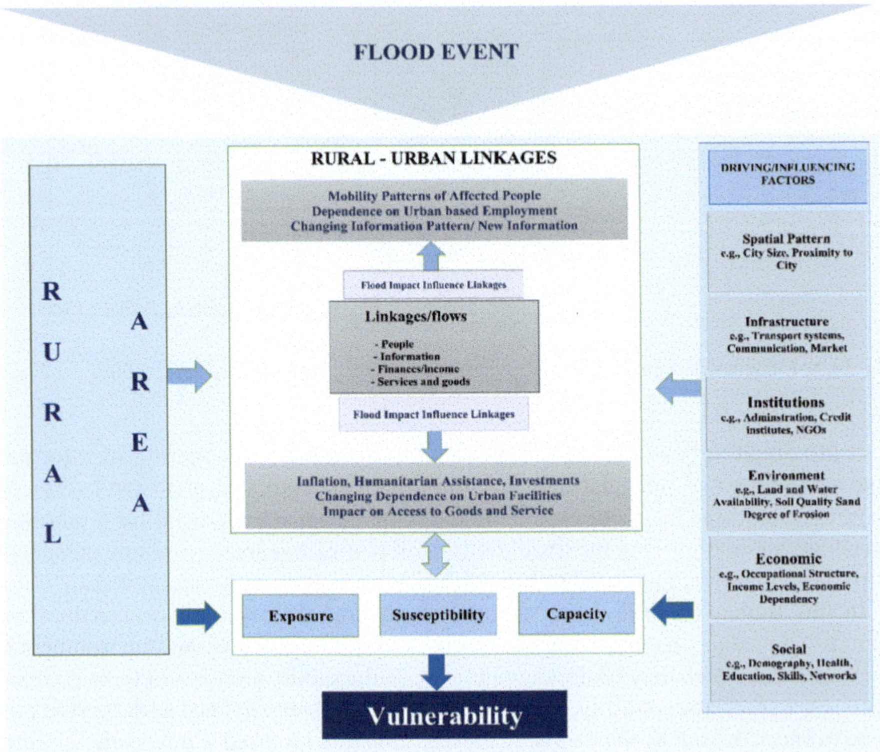

Fig. 14.3 Rural–urban linkages and their influence on rural flood vulnerability. *Source* Jamshed et al. (2020a)

push factors that cause population displacement from rural to urban areas and how climate-induced flood vulnerability works as a push factor in rural areas.

The average annual precipitation in the Sunamganj district of Bangladesh is approximately 5000 mm. Seasonal variations are significant, with approximately 90% of the rainfall concentrated in the four-month monsoon period from June to September. The process of erosion significantly influences siltation and affects the water retention capabilities of rivers. The degradation of the river system caused by siltation is a contributing factor to flooding in Bangladesh (Rahman 2013).

Vulnerability consists of three interrelated components: exposure, sensitivity, and capacity (UNESCO-IHE 2019). Exposure functions as an external factor that influences vulnerability, whereas sensitivity and capacity are the internal factors. Therefore, exposure denotes the susceptibility of the Sunamganj district to flooding based on its geographical location within the sphere of influence. Sensitivity is defined as the extent to which the district is impacted by flooding. At the same time, capacity refers to the Sunamganj district's ability to withstand potential damages or consequences resulting from flooding.

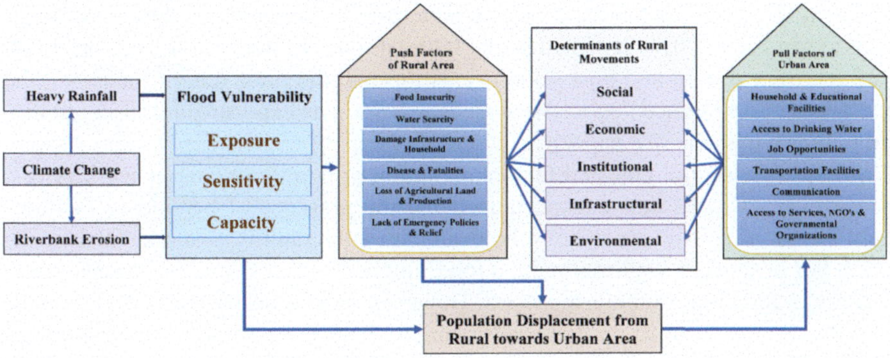

Fig. 14.4 Conceptual Framework to determine the relationship among climate change, flooding, and population displacement toward urban centres. *Source* Authors

Inadequate access to resources, lack of institutional management, unstable social–political structures, and environmental degradation are location-specific factors that include social, economic, institutional, infrastructural, and environmental determinants affecting household livelihood vulnerability to climate crisis events, such as floods. These determinants are connected to both pull and push factors, which capture the feasibility of movement presented in Fig. 14.4.

Social networks are essential for understanding population displacement toward cities. Additionally, the level of exposure to flood vulnerability is influenced by population size (Rana and Routray 2016; Jamshed et al. 2020a). The lower physical capacity of aged individuals and children turns them more vulnerable. Moreover, females are considered as vulnerable due to their lack of social networks (Lianxiao and Morimoto 2019). The health status of individuals within a household encompassing both physical and mental challenges, significantly influences their vulnerability, as it restricts mobility and requires particular attention within flood events (Rana and Routray 2016).

Then, vulnerable financial assets can severely damage due to flood leading to considerable economic challenges for impacted households (Jamshed et al. 2020a). As a result of financial hardship or unpredictable economic conditions, numerous individuals migrated from rural to urban centres (Kakinuma et al. 2020). On the contrary, flood vulnerability may be affected by the economic dependency ratio, since households with higher percentage of income earners have a better ability to cope within flood events (Bhattacharjee and Behera 2018).

In addition, executing evacuation, distributing relief supplies, providing financial assistance, building material seeds and fertilizers, and facilitating the reconstruction of rural infrastructure are collaborated by both formal and informal institutions as emergency response and recovery (Jamshed et al. 2020a). After the flood event, training programs aimed at skill development to enhance livelihood diversification, disaster awareness initiatives, weather forecasting and emerging technologies and extension services on adaptive cropping patterns, etc. are conducted by these institutions (Abid et al. 2016; Jamshed et al. 2020a). Financial institutions support livelihood recovery during and post flood events through the provision of loans and the facilitation of remittance flows. These measures mitigate exposure and susceptibility while improving adaptive capacity.

Moreover, essential physical factors including proximity to emergency services, accessibility of roads and transportation, and the presence of healthcare facilities significantly affect the rural household and its vulnerability to flood (Bah et al. 2003). For instance, a well-maintained road, situated near residences can serve as an effective evacuation route during flood. However, lack of accessibility and damage from flooding to farm to market roads, sanitation system, drinking water supplies and telecommunication services increase vulnerability (Rakib et al. 2017). Further, the occurrence of floods affects natural resources which subsequently influences the variety of functions that rural area can sustain (Das et al. 2025). Flood event can diminish agricultural land, adversely impacting crop cultivation and productivity, while decline in water quality may threaten both human health and crop viability (Jamshed et al. 2020a). Moreover, households located near flood sources, such as rivers, experienced high exposure and risk.

Study Area

The study area located in the southern foothills of the Shillong plateau and characterized as a tectonically uplifted region, surrounded by the Dauki fault to the south and the Oldham fault to the north (Saha et al. 2024). The northeastern Bangladesh contains these wetland regions which is a low-lying basin covers a massive area of 95 haors (Mondal et al. 2021). The vast sedimentary Surma basin has largely formed through the deposit of materials eroded from the Himalayan Mountain. This basin is primarily characterized by alluvial deposits, which significantly influence the geological composition of the area. The presence of Surma River along with its tributaries, the Kushiyara and Jadukata rivers has played a crucial role in shaping the local geography resulting in various fluvial landforms like riverbanks, floodplains, and oxbow lake. These features of the region offer a range of tourist attractions, such as the Hason Raja Museum, Tanguar Haor, Niladri lake, Cotton Garden, Jadukata river, and Barekka tilla (hill).

The area is bounded by Indian state of Meghalaya to the north, by Habiganj and Kishoreganj to the south, to the east by Sylhet, and Netrokona to the west. Sunamganj district contains 12 upazilas. Additionally, the geomorphology of Sunamganj is defined by numerous small waterbodies, river terraces and meandering river channels. The ongoing process of sediment deposition and erosion due to the river system of the region are the reason of its topographical diversity. The region's geological composition consists of sedimentary rocks which are the combination of sand, silt, clay, and gravel (Saha et al. 2024). During the monsoon season (June to September), these geographical features influence significant rainfall and flooding from the rivers

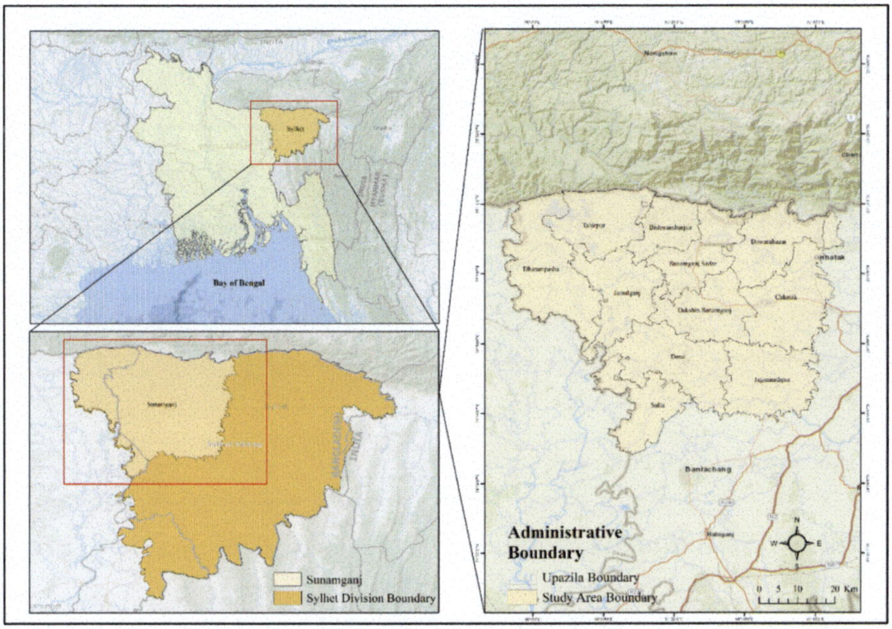

Fig. 14.5 Location of the study area. *Source* Prepared by Authors

and upstream in the Meghalaya, triggering yearly flood events. The prevalence of flood provides a justification for choosing Sunamganj district as the focus of the study, highlighted in Fig. 14.5.

Materials and Methods

Sample Design and Data Collection

The chapter used secondary data for developing indicators and adapted historical maps of study area to generate flood risk amplification. As, rainfall variation triggers climate-induced flood hazards, the rainfall data were collected from the Bangladesh Meteorological Department (BMD) and assessed in ArcGIS 10.8. Additionally, to visualize the flood risk in Sunamganj district, the affected area has been identified through the affected upazila, unions, and infrastructures, and the data were acquired from the District Relief and Rehabilitation Centre (DRRC) of Sunamganj. Also, the population displacement data were gathered from the District Relief and Rehabilitation Centre (DRRC) and analysed utilizing the shapefile of Sunamganj district along with the tools available in the ArcGIS 10.8 platform, represented in Fig. 14.6.

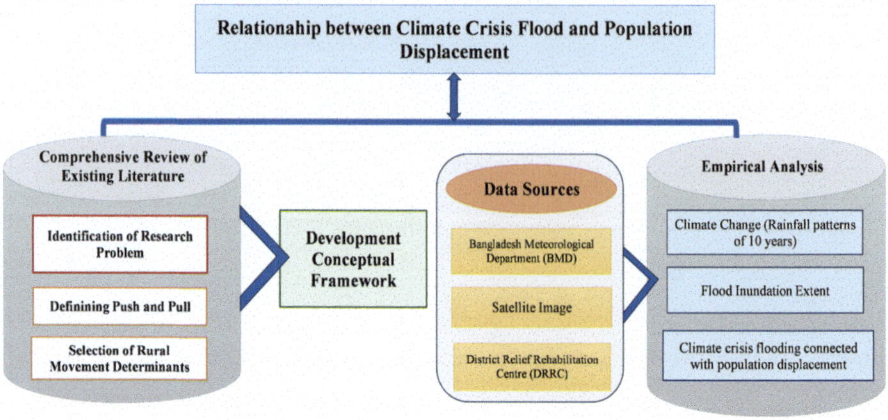

Fig. 14.6 Methodological Framework

Indicators Selection Technique

The indicator selection is the crucial step of developing the framework and empirical analysis. Therefore, indicators have selected reviewing multiple literature including rainfall, flood affected area, total displaced women, men, infants, total population displacement, flood inundation are the variables selected to validate the conceptual framework. The data and their sources have shown in Table 14.1.

Analytical Strategies

Data processing and visualization have been completed using a Geographic Information System (GIS) platform, Google Earth Engine (GEE), and Microsoft Excel. The study area map was generated using the shapefile derived from DIVA GIS.

The 10-year rainfall records were collected and analysed by deriving the mean rainfall of each year in Microsoft Excel, classify them into five classes; very low, low, moderate, high, very high and then visualized in ArcGIS 10.8 platform. Moreover, flood-induced population displacement data were statistically analysed in Microsoft Excel and used as evidence to show the linkage between climate-induced floods and population displacement. Additionally, flood inundation maps were generate using the Eq. 14.1 (Sivanpillai et al. 2021) of NDWI in the 2017, 2022, 2024 satellite image of Sunamganj.

$$\text{NDWI} = \frac{\text{Green} - \text{NIR}}{\text{Green} + \text{NIR}} \tag{14.1}$$

Table 14.1 Description of the datasets and sources used in the study

Datasets	Flood risk and population displacement data	Time period	Source
DIVA GIS	Shapefile	–	DIVA GIS https://diva-gis.org/data.html
Rainfall	Rainfall variation map	2014–2024	Bangladesh Meteorological Department (BMD) https://www.bmd.gov.bd/
Flood inundation	Flood inundation map	2017, 2022, 2024	Google Earth Engine https://earthengine.google.com/
Affected upazila, affected areas	Flood risk of Sunamganj district	2024	District Relief and Rehabilitation Centre (DRRC) https://drro.sunamganj.gov.bd/bn/site/view/officer_list
Population displacement	Flood-induced population displacement	2024	District Relief and Rehabilitation Centre (DRRC) https://drro.sunamganj.gov.bd/bn/site/view/officer_list

The positive value of NDWI indicates flooded area and rest of the area are classified as non-flooded area.

Result: Empirical Evidence Linking Climate-Induced Flood and Population Displacement

Climate Change-Induced Rainfall Variability

The rainfall distribution from 2011 to 2020 has been used as evidence to determine the changing rainfall pattern of Sunamganj district which recognized as one of the climate change stressors, presented in Fig. 14.7.

It can be observed that among the last 10 years, the rainfall distribution was the highest in Doyarabazar, Bishwamvarpur, and Sunamganj Sadar, while Shulla, Jagannathpur, and Derai are the upazilas with the lowest rainfall pattern. However, Tahirpur, Madhyanagar, and Chattak experienced the second-highest amount of rainfall annually. The upazila with a high amount of rainfall distribution every year suffers from the drastic impact of climate change stressors. Therefore, climate change drivers

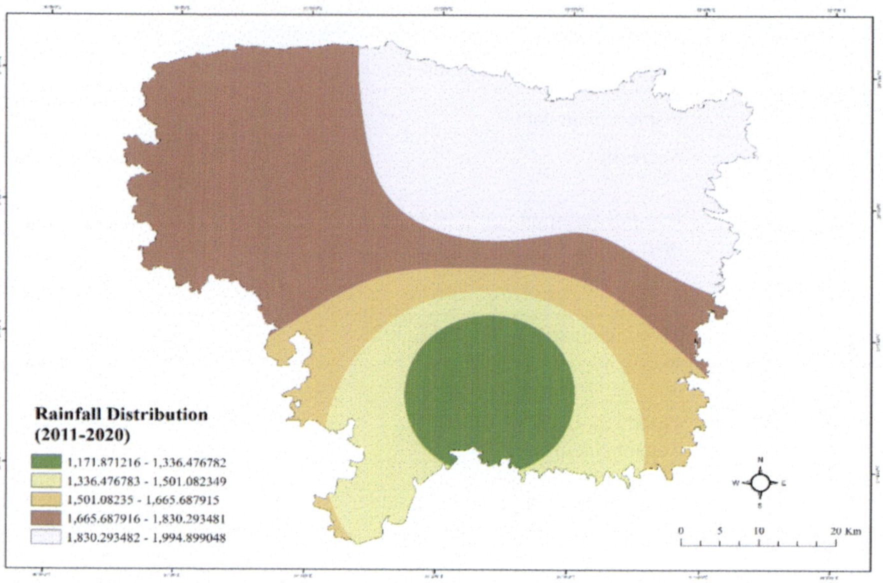

Fig. 14.7 Average annual rainfall map of Sunamganj district (2011–2020)

trigger climate-induced extreme events, such as floods in Sunamganj districts, and cause adverse impacts on properties, as well as human lives and livelihoods.

Flood Inundation Extent

The growth of the population in Sunamganj district is placing a strain on the land resources. The increasing population necessitates additional land for housing, roads, and highways, which frequently results in flooding. Unplanned urbanization and infrastructural expansion in both urban and rural areas increase vulnerability to flooding, particularly during changing rainfall patterns. Moreover, economic, physical, and environmental factors, such as agricultural dependency, weak household and infrastructure, presence of vegetation, soil erosion, distance to the river, and presence of waterbodies, influence flood vulnerability in Sunamganj. The devastating flood events experienced in Sunamganj almost every year, among them the 2017, 2022, and 2024 flood events, recently struck Sunamganj and caused drastic damage, as shown in Fig. 14.8. A vast population suffers from losing their income sources, households, the impact of diseases, injuries, and food and water scarcity in Sunamganj. During the 2017 flood event, it was observed that the flood had inundated almost every region of Sunamganj, except Bishwamvarpur, Doyarabazar, Chattak, and Jagannathpur. Additionally, Tahirpur, Madhyanagar, Derai, Sunamganj Sadar, Sulla, Dharmapasha, and Jamalganj are highly inundated due to the flood event.

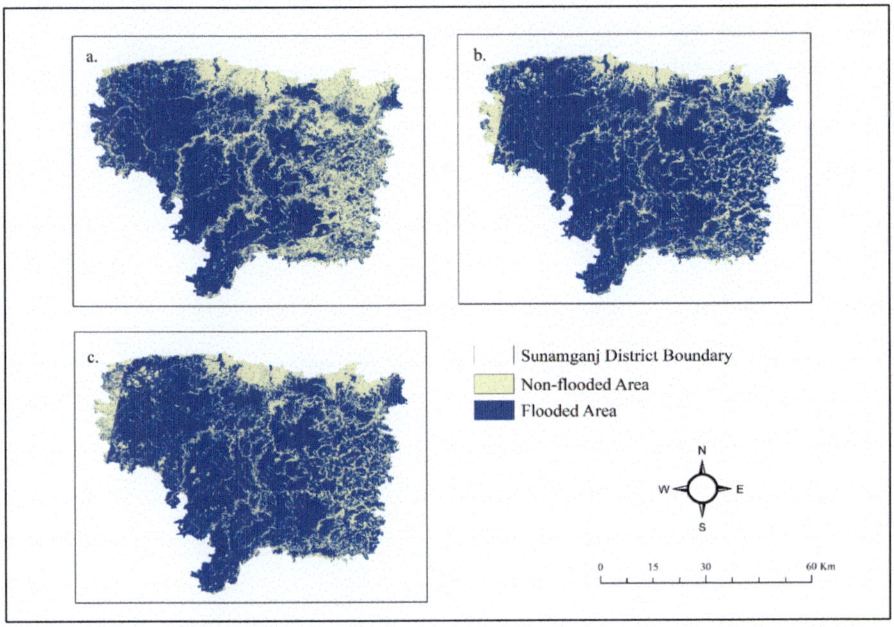

Fig. 14.8 The flood inundation map of Sunamganj District; **a** 2017 Flood event, **b** 2022 Flood event, **c.** 2024 Flood event. *Source* Authors

The 2022 inundation map identifies an increase in the intensity of flood destruction in the Sunamganj district in 2022. Aside from the areas under Bishwamvarpur, Doyarabazar, and Madhyanagar Upazila, nearly all other upazilas were inundated during the 2022 flood event. The flood event posed a significant risk to human lives and had profound consequences for livelihoods, resulting in extensive devastation and damage to agricultural land and infrastructure. Furthermore, Tahirpur, Derai, Sunamganj Sadar, Jagannathpur, Sulla, Dharmapasha, Jamalganj, Chattak, and Jagannathpur experienced significant inundation during the flood event of 2024 in Sunamganj district. The result also identified that almost in every upazila of the study area suffers due to climate-induced flood and every year it is become a constant problem of Sunamganj. The flood risk in Sunamganj district is escalating annually due to notable climate change, intense rainfall, and heightened flood susceptibility. The various types of area affected in different upazilas of Sunamganj during 2024 flood event, demonstrate in Table 14.2.

The result shows that in each upazila, both rural and haor regions are notably affected by flooding. The rural regions of Chattak Upazila, the urban area of Jagannathpur and the haor area of Dharmapasha have experienced the most significant damage, with affected areas measuring 300, 27.21 and 195.11 km^2 respectively. The rural region of Tahirpur, Jagannathpur, Derai, and Doyarabazar have considerable devastation due to flood. The haor regions of Jagannathpur, Shalla, Sunamganj Sadar, and Doyarabazar experienced damage during 2024 flood event, as shown in Fig. 14.9.

Table 14.2 Climate crisis flood affected areas of Sunamganj

Upazila name	Urban area (Km^2)	Rural area (Km^2)	Char area (Km^2)	Hilly area (Km^2)	Haor area (Km^2)
Sunamganj Sadar	10	72.12	0	0	64.68
Jamalganj	6	0	0	0	0
Bishwamvarpur	0	15	0	0	70
Tahirpur	0	192	0	0	1.02
Dharmapasha	5	30	0	0	222
Derai	1	218	0	0	0
Shalla	0	0	0	0	160
Shantiganj	0	70.13	0	0	20
Jagannathpur	27.21	137.14	0	0	195.11
Chattak	3	300	0	0	40
Doyarabazar	4.25	205.6	0	0	72.35
Madhyanagar	0	0	0	0	0

Source District Relief and Rehabilitation Centre (DRRC)

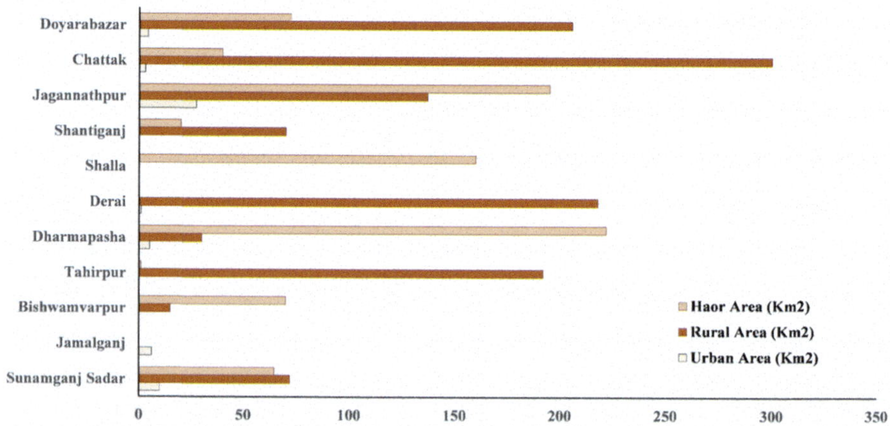

Fig. 14.9 Climate crisis flood affected areas of Sunamganj. *Source* District Relief and Rehabilitation Centre (DRRC) (Prepared by authors)

Thus, Sunamganj district is significantly vulnerable to flood, enduring devastating inundations each year that not only damage properties, infrastructure and agricultural lands but also negatively affect human lives. Furthermore, empirical evidence suggests that rural regions exhibit a higher vulnerability to flooding compared to urban setting.

Climate Crisis Flooding Connected with Population Displacement

Climate change can influence migration choices directly by disrupting ecosystem services or raising flood vulnerability. Similarly, effecting other factors like financial resources or households assets, climate change can have impacts indirectly on population displacement (Freihardt 2025).

Therefore, Table 14.3 exhibits the number of displaced women, men, and infants due to the 2024 climate crisis flood event from different upazilas of Sunamganj district, clearly indicating an extensive variation in the quantity of displaced population in each upazila.

Sunamganj district consists of 12 upazilas, among which Chattak Jagannathpur and Doayarabazar are highly vulnerable to floods, from which the highest number of men, women, and infants have been displaced. For instance, according to DRRC data on the 2024 flood event, about 90,378 women and 95,459 men were displaced from Jagannathpur, which has the highest number of displaced populations among all 12 upazilas. However, in terms of infant displacement, Doyarabazar holds the top position, with a figure of approximately 34,992. Additionally, the number of displaced men and women is approximately 70,380 and 64,516 in Doyarabazar upazila, which occurred due to the 2024 climate crisis flood. Moreover, Chattak is the second flood-affected upazila, dealing with high population displacement, with a total of 89,780 displaced men and 84,200 displaced women, as shown in Fig. 14.10.

Conversely, Madhyanagar upazila consists of the lowest number of displaced populations, including men, women, and infants. The upazila experienced migration

Table 14.3 Population Displacement during the 2024 climate crisis flood event

Upazila name	Displaced women	Displaced men	Displaced infants	Total Displaced Population
Sunamganj Sadar	23,800	24,300	9860	57,960
Jamalganj	4720	3325	832	8877
Bishwamvarpur	55,020	60,250	9200	124,470
Tahirpur	42,000	57,850	17,150	117,000
Dharmapasha	4930	2700	756	8386
Derai	21,420	22,602	7085	51,107
Shalla	11,312	11,900	1259	24,471
Shantiganj	41,590	44,390	14,300	100,280
Jagannathpur	90,378	95,459	34,174	220,011
Chattak	84,200	89,780	25,230	199,210
Doyarabazar	64,516	70,380	34,992	169,888
Madhyanagar	438	480	360	1278

Source District Relief and Rehabilitation Centre (DRRC)

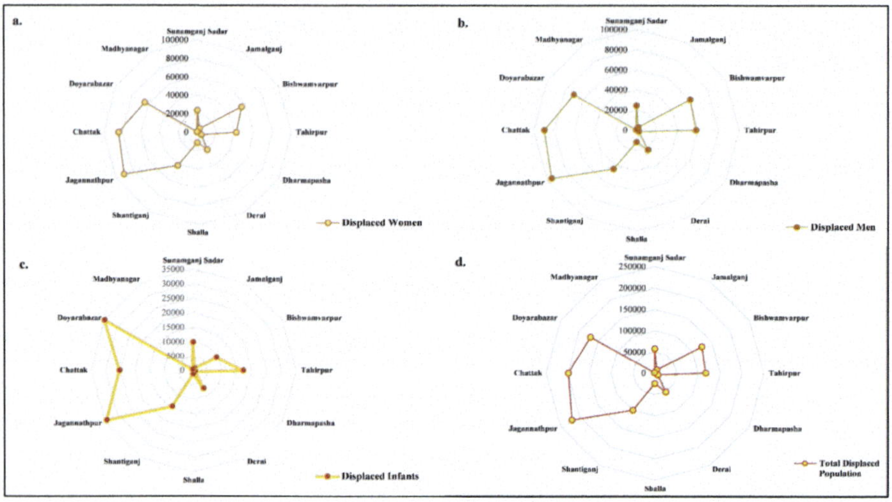

Fig. 14.10 Population displaced due to 2024 climate crisis flood event; **a**. Displaced women; **b**. Displaced men; **c**. Displaced infants and **d**. Total Displaced population

due to the 2024 flood event, with 438 displaced women, 480 displaced men, and 360 displaced infants. The outcomes show that ultimately, climate change is triggering flood events in Sunamganj, damaging properties, economic sources, agricultural lands, and infrastructures, as well as lives, which activate the reasons for population displacement.

Discussion

The result section describes the changed precipitation pattern of last 10 years which define that rainfall has increased in Dowarabazar, Bishwamvarpur, Tahirpur, and Sunamganj Sadar in recent years. Consequently, riverbank erosion, flooding events are heightened in Sunamganj district resulting population displacement. These flooding is evident from flood inundation map which indicates Dowarabazar and Chattak. Sunamganj Sadar, Tahirpur upazila has highly inundated during 2017, 2022, and 2024. Moreover, this extreme flooding forced population to move towards urban areas, as the result shows how flood inundated rural area the most and represents the population displacement that triggered by climate-induced flooding.

The financial implications of climate change amounted to $ 521.37 million from 2009 to 2014, with over 43% of households in Sunamganj experiencing extreme weather events (Rahman et al. 2023). The presence of the Jadukata River and over 95 haors in the region leads to significant riverbank erosion during heavy rainfall, resulting in water overflow and subsequent flooding (Rahman 2013) which is evident in flood risk inundation section. Under this context, the 2017 flood event in Sunamganj resulted in approximately 145 fatalities and affected around 8 million people in the district (Bhuiyan et al. 2024). The 2022 flood impacted 2,690,150 individuals, resulting in the displacement of approximately 650,000 people to shelter centres (Tuli et al. 2024). In 2024, flood water inundated 11 upazilas and the Sunamganj municipality. The extensive devastation resulting from the flooding in 1018 villages and 78 unions has rendered 660,347 individuals homeless.

Alterations in rainfall patterns lead to floods, which are sudden-onset events that can cause significant damage to infrastructure and property, as well as loss of life, and force people to evacuate (Mianabadi et al. 2023). A positive correlation exists between flood risk and migration (Ekoh et al. 2022). Developing countries in Asia, such as Bangladesh, are the regions that frequently experience flooding, and they tend to be inhabited by people facing economic hardships, making them more vulnerable to the impacts of flooding (Walter 2015). The previous literature also identifies that natural hazards, environmental changes caused by climate change, trigger migration, including rural–urban linkages, rural–urban interdependency, riverbank erosion, sea level rise, and increasing temperature, which cause drought and flooding, which cause rural–urban migration (Bhuiyan et al. 2024, Jamshed et al. 2020a, Warner et al. 2010, Gray and Mueller 2012, Rahman 2013, Freihardt 2025, Assraf Seddiky et al. 2024, Bekaert et al. 2021). Therefore, flooding is recognized for causing considerable challenges for impacted households and leading to temporary population displacements, but its effects on long-term population relocation are also noteworthy (Gray and Mueller 2012).

A proved relationship has shared between long-term changing precipitation and migration patterns; however, the influence of socioeconomic factors requires consideration. Moreover, migration also fluctuates in response to increasing precipitation and income sources in Bangladesh. Households reliant on agriculture faced significant crop loss due to riverine flooding amid heightened rainfall, leading to inadequate resources for migration (Call et al. 2017). However, the effect of climate change in Sunamganj district is evident through alterations in rainfall patterns, variations in temperature, a higher occurrence of natural disasters such as floods, and an increase in disease prevalence (Bhuiyan et al. 2024). However, numerous climate crisis flood affected individuals attempting to manage or cope with the significant damage and destruction. People ultimately started seeking relocations when intolerable climate crisis events compel people to leave their homes and belongings as depicted in population displacement analysis.

Migration drivers are commonly seen as elements associated with the labour market (unemployment rate, wage level, etc.), taxation, land and housing conditions, social welfare, political rights and freedoms, and environmental influences (climate, population density, pollution). All these factors can act as both pull and push factors (Lee 1966). A straightforward push logic may be suitable for some irreversible environmental or climatic alterations, such as flooding which is defined through displaced population analysis in the result section. These majority perceive migration not as the optimal solution, since they are bound to their local area or hindered by the economic, social, or emotional expenses associated with relocation (Schewel 2019). Overall, this chapter examines the impact of recurrent floods on demographic trends, livelihoods and social vulnerability, emphasizing displacement as a growing population issue exacerbated by the climate crisis.

Therefore, this chapter is intricately connected to the book through the analysis of climate-induced flooding as a significant factor contributing to population displacement. This chapter examines flood-related migration within the context of broader climate-related migration within the context of broader climate-population interactions, enhancing the understanding of evolving population dynamics. It also addresses adaptive and sustainable responses, including resilience planning, policy interventions and community-based strategies, aimed at mitigating displacement and fostering long-term sustainability.

Limitations and Strengths

This framework acts as an essential tool for evaluating the effects of flooding on rural–urban migration and improving disaster risk reduction for developmental goals. This framework will facilitate the concept identification and relationship among extreme climate-induced flooding and population displacement, flood-affected rural areas suitable for resettlement and development, along with potential mitigation measures. The result ought to aid decision makers, non-profit organizations, communities and diverse development entities encouraging for the integration of disaster risk reduction policy during the planning and implementation of development project.

Using the existing literature, the conceptual framework has introduced which may contain some limitations in terms of its comprehensiveness and accuracy. Moreover, the framework does not seek to include every factor that influence population displacement and related to climate induce flooding. Additionally, there are lack of primary data in the analysis. However, providing more primary data in empirical analysis can enhance and strengthen the aim of this chapter. Also, through identifying more influential factors can improve the conceptual framework.

Conclusion

The chapter revealed how changing climatic phenomena, such as intense precipitation, trigger riverbank erosion that leads to intense flooding in Sunamganj district and affects various determinants, including institutional, social, economic, infrastructural, and environmental aspects, which significantly influence the pull and push factors. These factors significantly influenced the decisions made by individuals and households in the rural areas of Sunamganj district to relocate to urban areas, where they could achieve improved and secure livelihoods. This chapter also determined that most migration trends are from rural to urban regions of Sunamganj and examines the connection between climate crisis flood and population displacement.

Interestingly, with various limitations the methodology of the chapter is flexible to allow the integration of indicators which provides benefit for contextualization and broader applicability of approach. Developing approaches to enhance the resilience and adaptability of communities can minimize the need for migration in response to abrupt and substantial changes. Therefore, overall result remains unaffected, if quantity of indicator has chosen for a particular region. This method can be utilized to create a comparative conceptual framework for any study location.

AI Disclosures

Grammarly software was utilized for grammar checking, language refinement, and stylistic improvement, while QuillBot was used to paraphrase selected sentences and conceptual definitions. These tools were utilized solely to enhance clarity and readability; all intellectual content, analysis, and interpretations are the sole responsibility of the authors.

Acknowledgements We would like to thank the Deputy Commissioner and the District Relief and Rehabilitation Officer of Sunamganj District for providing us with the flood damage data for Sunamganj.

References

Abedin J, Khatun H (2020) Impacts of flash flood on livelihood and adaptation strategies of the Haor Inhabitants: a study in Tanguar Haor of Sunamganj, Bangladesh. Dhaka Univ J Earth Environ Sci 8:41–51

Abid M, Schilling J, Scheffran J, Zulfiqar F (2016) Climate change vulnerability, adaptation and risk perceptions at farm level in Punjab, Pakistan. Sci Total Environ 547:447–460

Ahmad D, Afzal M (2020) Flood hazards and factors influencing household flood perception and mitigation strategies in Pakistan. Environ Sci Pollut Res 27:15375–15387

Ahmad D, Afzal M (2021) Flood hazards, human displacement and food insecurity in rural riverine areas of Punjab, Pakistan: policy implications. Environ Sci Pollut Res 28:10125–10139

Akter N, Islam MR, Karim MA, Miah MG, Rahman MM (2023) Spatiotemporal rainfall variability and its relationship to flash flood risk in Northeastern Sylhet Haor of Bangladesh. J Water Clim Chang 14:3985–3999

Alam GMM, Alam K, Mushtaq S, Sarker MNI, Hossain M (2020) Hazards, food insecurity and human displacement in rural riverine Bangladesh: implications for policy. Int J Disaster Risk Reduct 43:101364

Ali A, Rahut DB (2020) Localized floods, poverty and food security: empirical evidence from rural Pakistan. Hydrology [Online] 7

Almasi P, Soltani S (2017) Assessment of the climate change impacts on flood frequency (case study: Bazoft Basin, Iran). Stoch Env Res Risk Assess 31:1171–1182

Almulhim AI, Alverio GN, Sharifi A, Shaw R, Huq S, Mahmud MJ, Ahmad S, Abubakar IR (2024) Climate-induced migration in the Global South: an in depth analysis. NPJ Climate Action 3:47

Ashbolt NJ (2004) Microbial contamination of drinking water and disease outcomes in developing regions. Toxicology 198:229–238
Assraf Seddiky M, Ara E, Karim A (2024) Climate change-induced hazard risks and migration in Bangladesh: a case study. Curr Res Environ Sustain 7:100253
Bah M, Cissè S, Diyamett B, Diallo G, Lerise F, Okali D, Okpara E, Olawoye J, Tacoli C (2003) Changing rural–urban linkages in Mali, Nigeria and Tanzania. Environ Urban 15:13–24
Bakth N, Hasanuzzaman S (2023) Temporary environmental migration and child truancy: an investigation among hard-to-reach families in Bangladesh. J Soc Econ Dev 25:152–169
Beine M, Parsons C (2015) Climatic factors as determinants of international migration. Scand J Econ 117:723–767
Bekaert E, Ruyssen I, Salomone S (2021) Domestic and international migration intentions in response to environmental stress: a global cross-country analysis. J Demogr Econ 87:383–436
Bhattacharjee K, Behera B (2018) Determinants of household vulnerability and adaptation to floods: empirical evidence from the Indian State of West Bengal. Int J Disaster Risk Reduct 31:758–769
Bhuiyan MS, Paul S, Abdussabur M (2024) Revisiting the causes and effects of recurrent floods in the Haor Region of Sunamganj, Bangladesh: evidence from the 2022 flash flood. Asia Soc Issues 17:e264552
Birkmann J (2006) Measuring vulnerability to promote disaster-resilient societies: conceptual frameworks and definitions. Meas Vulnerability Nat Hazards: Disaster Resilient Soc 1:3–7
Black R, Adger WN, Arnell NW, Dercon S, Geddes A, Thomas D (2011) The effect of environmental change on human migration. Glob Environ Chang 21:S3–S11
Black R, Arnell NW, Adger WN, Thomas D, Geddes A (2013) Migration, immobility and displacement outcomes following extreme events. Environ Sci Policy 27:S32–S43
Brouwer R, Akter S, Brander L, Haque E (2007) Socioeconomic vulnerability and adaptation to environmental risk: a case study of climate change and flooding in Bangladesh. Risk Anal 27:313–326
Call MA, Gray C, Yunus M, Emch M (2017) Disruption, not displacement: environmental variability and temporary migration in Bangladesh. Glob Environ Chang 46:157–165
Das AK, Lee D-S, Woo Y-J, Sultana S, Mahmud A, Yun B-W (2025) The impact of flooding on soil microbial communities and their functions: a review. Stresses [Online], 5.
Ekoh SS, Teron L, Ajibade I, Kristiansen S (2022) Flood risk perceptions and future migration intentions of Lagos residents. Int J Disaster Risk Reduct 83:103399
Feldmeyer D, Birkmann J, Welle T (2017) Development of human vulnerability 2012–2017. J Extrem Events 04:1850005
Feng B, Zhang Y, Bourke R (2021) Urbanization impacts on flood risks based on urban growth data and coupled flood models. Nat Hazards 106:613–627
Freihardt J (2025) Environmental shocks and migration among a climate-vulnerable population in Bangladesh. Popul Environ 47:6
Gebre T, Gebremedhin B (2019) The mutual benefits of promoting rural-urban interdependence through linked ecosystem services. Glob Ecol Conserv 20:e00707
Gray CL, Mueller V (2012) Natural disasters and population mobility in Bangladesh. Proc Natl Acad Sci 109:6000–6005
Haque CE (1997) Hazards in a fickle environment: Bangladesh. Springer Science & Business Media, p 10
Hashmi HN, Siddiqui QTM, Ghumman AR, Kamal MA, Mughal HUR (2012) A critical analysis of 2010 floods in Pakistan. Afr J Agric Res 7:1054–1067
Hugo G, Abbasi-Shavazi MJ, Kraly EP (2018) Introduction: advancing the demography of forced migration and refugees. In: Hugo G, Abbasi-Shavazi MJ, Kraly EP (eds) Demography of refugee and forced migration. Springer International Publishing, Cham
IDMC (2015) Bangladesh: comprehensive response required to complex displacement crisis
IPCC (2014) Climate change 2014: synthesis report. In: Pachauri RK, Meyer LA (eds) Contribution of working groups I, II and III to the fifth assessment report of the intergovernmental panel on climate change. Core Writing Team (Geneva, Switzerland: IPCC), p 151

Islam KM, Baten MA (2016) Exploring dynamics of internal migration to slums of Dhaka city in Bangladesh: a study on drivers and factors. In: Climate change induced migration: the case of Bangladesh. Society for Education & Research Development, Haryana, India, pp 278–292
Israt Rayhan, Grote U (2007) Coping with floods: does rural-urban migration play any role for survival in rural Bangladesh? J Identity Migr Stud 1:82–98.
Ivanova A, Singh D, Rai P, Richards C, Vasconcelos H, Goldsby M (2024) Climate hazards and human migration: literature review. Environ Res: Clim 3:042002
Jamshed A, Birkmann J, Feldmeyer D, Rana IA (2020a) A conceptual framework to understand the dynamics of rural–urban linkages for rural flood vulnerability. Sustainability [Online] 12
Jamshed A, Birkmann J, Mcmillan JM, Rana IA, Lauer H (2020b) The impact of extreme floods on rural communities: evidence from Pakistan. In: Leal Filho W, Nagy GJ, Borga M, Chávez Muñoz PD, Magnuszewski A (eds) Climate change, hazards and adaptation options: handling the impacts of a changing climate. Springer International Publishing, Cham
Kakinuma K, Puma MJ, Hirabayashi Y, Tanoue M, Baptista EA, Kanae S (2020) Flood-induced population displacements in the world. Environ Res Lett 15:124029
Perch-Nielsen SL, Bättig B, Imboden DM (2008) Exploring the link between climate change and migration. Climatic Change 91:375–393
Lee ES (1966) A theory of migration. Demography 3:47–57
Lianxiao, Morimoto T (2019) Spatial analysis of social vulnerability to floods based on the MOVE framework and information entropy method: case study of Katsushika Ward, Tokyo. Sustainability [Online], 11
Mahmood S, Sajjad A, Rahman A-U (2021) Cause and damage analysis of 2010 flood disaster in district Muzaffar Garh, Pakistan. Nat Hazards 107:1681–1692
Malik S, Chandra Pal S, Chowdhuri I, Chakrabortty R, Roy P, Das B (2020) Prediction of highly flood prone areas by GIS based heuristic and statistical model in a monsoon dominated region of Bengal Basin. Remote Sens Appl: Soc Environ 19:100343
Martin S, Singh L, Taylor A, Wahedi L (2021) Dynamic model of displacement. PsyArXiv
Mcleman R (2018) Thresholds in climate migration. Popul Environ 39:319–338
Mianabadi A, Davary K, Mianabadi H, Kolahi M, Mostert E (2023) Toward the development of a conceptual framework for the complex interaction between environmental changes and rural-urban migration. Front Water 5
Misra AK (2014) Climate change and challenges of water and food security. Int J Sustain Built Environ 3:153–165
Mondal S, Akter L, Hiya HJ, Farukh MA (2021) Effects of 2017 early flash flooding on agriculture in Haor areas of Sunamganj. J Environ Sci Nat Resour 12:117–125
Mora C, Spirandelli D, Franklin EC, Lynham J, Kantar MB, Miles W, Smith CZ, Freel K, Moy J, Louis LV, Barba EW, Bettinger K, Frazier AG, Colburn IX JF, Hanasaki N, Hawkins E, Hirabayashi Y, Knorr W, Little CM, Emanuel K, Sheffield J, Patz JA, Hunter CL (2018) Broad threat to humanity from cumulative climate hazards intensified by greenhouse gas emissions. Nat Climate Change 8:1062–1071
Nazari S, Rad GP, Sedighi H, Azadi H (2015) Vulnerability of wheat farmers: toward a conceptual framework. Ecol Ind 52:517–532
Rahaman MA, Saba Z, Akter S, Rownak MRJ (2025) Potential climate-induced migration determinants and decision support of anticipatory actions and climate-displacement solutions in coastal districts of Bangladesh. Int Soc Res Nexus (ISRN) 1:1–19
Rahman MC, Rahaman MS, Biswas JC, Rahman NMF, Islam MA, Sarkar MAR, Islam MS, Maniruzzaman M (2023) Climate change and risk scenario in Bangladesh. Asia-Pac J Regional Sci 7:381–404
Rahman MR (2013) Impact of riverbank erosion hazard in the Jamuna Floodplain Areas in Bangladesh. J Sci Found 8:55–65
Rahman MS, Gain A (2020) Adaptation to river bank erosion induced displacement in Koyra Upazila of Bangladesh. Prog Disaster Sci 5:100055

Rakib MA, Islam S, Nikolaos I, Bodrud-Doza M, Bhuiyan MAH (2017) Flood vulnerability, local perception and gender role judgment using multivariate analysis: a problem-based "participatory action to future skill management" to cope with flood impacts. Weather Clim Extrem 18:29–43
Rana IA, Routray JK (2016) Actual vis-à-vis perceived risk of flood prone urban communities in Pakistan. Int J Disaster Risk Reduct 19:366–378
Saha G, Kabir MN, Islam MS, Khandaker A, Chowdhury P (2024) Flash flood potential risk zonation mapping using GIS-based spatial multi-index model: a case study of Sunamganj District, Bangladesh. Arab J Geosci 17:100
Sangpenchan R (2011) Vulnerability of thai rice production to simultaneous climate and socioeconomic change: a double exposure analysis. In: AGU fall meeting abstracts 2011
Sarker MNI, Wu M, Alam GMM, Shouse RC (2020) Retracted: life in riverine islands in Bangladesh: Local adaptation strategies of climate vulnerable riverine island dwellers for livelihood resilience. Land Use Policy 94:104574
Schewel K (2019) Understanding immobility: moving beyond the mobility bias in migration studies. Int Migr Rev 54:328–355
Semenza JC, Ebi KL (2019) Climate change impact on migration, travel, travel destinations and the tourism industry. J Travel Med 26:taz026
Shah AA, Ye J, Abid M, Khan J, Amir SM (2018) Flood hazards: household vulnerability and resilience in disaster-prone districts of Khyber Pakhtunkhwa province, Pakistan. Nat Hazards 93:147–165
Sharifi F, Samadi SZ, Wilson CAME (2012) Causes and consequences of recent floods in the Golestan catchments and Caspian Sea regions of Iran. Nat Hazards 61:533–550
Sivanpillai R, Jacobs KM, Mattilio CM, Piskorski EV (2021) Rapid flood inundation mapping by differencing water indices from pre- and post-flood landsat images. Front Earth Sci 15:1–11
Srivastava N, Shaw R (2016) 8—Enhancing city resilience through urban-rural linkages. In: Shaw R, Atta Ur R, Surjan A, Parvin GA (eds) Urban disasters and resilience in Asia. Butterworth-Heinemann
Sullivan JP, Townsend K (2022) Climate migration: adding fuel to the ethnocentric fire. Terror Polit Violence 34:914–925
Sultana N, Rayhan MI (2012) Coping strategies with floods in Bangladesh: an empirical study. Nat Hazards 64:1209–1218
Sultana N, Tan S, Hossen MF (2024) Landslide risk assessment by integrating hazards and vulnerability indices in Southeast Bangladesh. Int J Disaster Risk Reduct 114:104991
Surendra V (2020) The moneylender as middleman: formal credit supply and informal loans in Rural India
Teo M, Goonetilleke A, Ahankoob A, Deilami K, Lawie M (2018) Disaster awareness and information seeking behaviour among residents from low socio-economic backgrounds. Int J Disaster Risk Reduct 31:1121–1131
Tuli RD, Rashid KJ, Akter T (2024) Impact analysis of the 2022 flood event in Sylhet and Sunamganj using Google Earth Engine. In: Chandra Pal S, Chatterjee U, Chakrabortty R (eds) Modern cartography series, chap. 3. Academic Press
Turner BL, Kasperson RE, Matson PA, McCarthy JJ, Corell RW, Christensen L, Eckley N, Kasperson JX, Luers A, Martello ML, Polsky C, Pulsipher A, Schiller A (2003) A framework for vulnerability analysis in sustainability science. Proc Natl Acad Sci 100:8074–8079
UNESCO-IHE (2019) Flood vulnerability indices. Institute for Water Education.
UNICEF (2022) Bangladesh humanitarian situation report, January to June 2022. UNICEF, p 11
Walter P (2015) Floods and rural-urban migration in Bangladesh. State Environ Migr 64
Warner K, Hamza M, Oliver-Smith A, Renaud F, Julca A (2010) Climate change, environmental degradation and migration. Nat Hazards 55:689–715
White GF, Bradley DJ, White AU (2002) Drawers of water: domestic water use in East Africa. Bull World Health Organ 80

Zarafshani K, Sharafi L, Azadi H, Hosseininia G, de Maeyer P, Witlox F (2012) Drought vulnerability assessment: the case of wheat farmers in Western Iran. Global Planet Change 98–99:122–130

Chapter 15
Extreme Weather Events, Displacement, and Boundaries: The Anatomy of the 2024 Floods in Rio Grande do Sul (Brazil)

Roberto Rodolfo Georg Uebel

Abstract *Background* Climate change is increasing the intensity of extreme weather events, generating large-scale displacement and exposing gaps in existing protection and governance frameworks; although climate-induced mobility is widely recognised, it remains insufficiently understood how disasters remake borderscapes—the material, administrative, and symbolic boundaries that shape access to territory, rights, and assistance. *Objective* This chapter analyses the 2024 Rio Grande do Sul floods—Brazil's most severe climate disaster—to explain how the event reconfigured borderscapes and what this implies for the institutional recognition and protection of climate-displaced populations. *Methods* The study adopts a multi-scalar, experiential research design combining qualitative field interviews, documentary analysis, and an autoethnographic component based on the author's displacement, interpreted through critical frameworks associated with Saskia Sassen, Milton Santos, and Yanis Varoufakis to link lived disruption to broader territorial and political-economic dynamics. *Results* The floods generated fluid and unequal geographies of mobility, exclusion, and resilience, reshaping administrative jurisdictions and everyday boundary-making practices; governance responses were uneven and often fragmented across levels of government, while media and political actors disputed responsibility and meaning, and grassroots solidarity initiatives emerged as a crucial—though partial—compensatory mechanism for institutional shortcomings. *Conclusion* The chapter argues for institutionalising protection for climate-displaced persons and operationalising integrated, inclusive multi-level governance, recommending that climate justice and social protection be centred in disaster preparedness, emergency response, and post-crisis reconstruction to reduce vulnerability and prevent the normalisation of exclusion after extreme events.

Keywords Climate change · Forced displacement · Borderscapes · Climate refugee · Rio Grande do Sul floods · Territorial governance

---

R. R. G. Uebel (✉)
Department of International Relations, Graduate Program in Administration (PPGA), Escola Superior de Propaganda e Marketing (ESPM), São Paulo, SP, Brazil
e-mail: roberto.uebel@espm.br

P. Ghosh et al. (eds.), *Climate Crisis and Changing Population Dynamics*, Springer Climate, https://doi.org/10.1007/978-3-032-24533-5_15

Introduction

Climate change is increasingly reshaping not only ecosystems but also the political, social, and territorial arrangements through which societies organise space and authority. One under-examined implication of this shift is what it does to borders—not only as juridical lines tied to sovereignty, but also as material infrastructures, administrative competences, and symbolic boundary-making practices that structure everyday life. As extreme events intensify in frequency and severity, borders become less a fixed geography and more a site where vulnerability, mobility, and governance are negotiated in real time.

This chapter examines the redefinition of borders under climate stress through the case of the May 2024 floods in Rio Grande do Sul, southern Brazil. Considered among the most severe disasters in Brazil's history, the floods exposed systemic fragilities across public policy domains—including health, housing, security, logistics, and cross-jurisdictional coordination—while rapidly transforming how territory was accessed, administered, and inhabited. In practical terms, the disaster converted neighbourhoods, municipalities, and transport corridors into shifting "zones" of risk and refuge, where everyday boundaries were submerged, rerouted, or rendered temporarily meaningless.

Rio Grande do Sul offers a particularly instructive case because of the event's scale and its displacement effects. Beyond the immediate human and material losses, the floods triggered patterns of forced migration and internal displacement that reconfigured demographic and territorial dynamics across the state. More broadly, the episode illustrates how disasters intensified by human action can act as catalysts for social and political reorganisation (Cook et al. 2016). Based on primary data reported by Lopes et al. (2024), initial estimates suggest that more than two million people were directly affected and around one million were compelled to relocate—temporarily, provisionally, or permanently—making it the largest recorded internal displacement event in Brazil and among the largest globally associated with climate-related drivers.

While border studies have long centred on sovereignty and state territoriality, they have increasingly moved toward conceptualising borders as dynamic socio-spatial processes. The notion of borderscapes is particularly useful here because it captures borders as multi-layered and contested assemblages—produced through infrastructures (roads, bridges, rivers), administrative regimes (municipal/state/federal competences), and narratives that legitimate inclusion, exclusion, and entitlement. Yet, despite this conceptual advance, two gaps remain salient in current research. First, climate-disaster scholarship often treats displacement primarily as a humanitarian or logistical outcome, without systematically analysing how displacement reconfigures borderscapes across material, administrative, and symbolic registers. Second, work on climate mobility frequently focuses on cross-border migration, leaving large-scale internal displacement and its implications for boundary regimes—and for institutional recognition and protection—comparatively under-theorised, particularly in the Brazilian context.

The 2024 floods demonstrate why these gaps matter. As floodwaters submerged entire cities, including Porto Alegre, infrastructures that ordinarily stabilise territorial order—roads, bridges, municipal delimitations, and service jurisdictions—were disrupted or erased, forcing authorities and communities to improvise new spatial practices of access, triage, and governance. At the same time, political and social boundaries were tested: eligibility for assistance, prioritisation of evacuations, allocation of shelters, and the coordination (or fragmentation) of responses across levels of government revealed how borders are actively produced during crisis conditions. Non-state actors—civil society organisations, local communities, and solidarity networks—also became central to these emergent borderscapes, reflected in grassroots mobilisation and the recurring motto "the people for the people" during the most critical period (Jornal O Diário 2024).

Against this backdrop, the chapter's objective is to explain how an extreme climate event can reorganise borders as lived and governed realities, and what that implies for the recognition and protection of climate-displaced populations. It is guided by the following research question:

How did the May 2024 floods in Rio Grande do Sul reconfigure borderscapes—across material, administrative, and symbolic boundary regimes—through patterns of internal displacement and multi-level governance responses, and what does this reveal about the institutional recognition and protection of climate-displaced populations?

To operationalise this question, the analysis addresses three interrelated sub-questions:

1. Material reconfiguration: How did the disaster disrupt and remap physical infrastructures and spatial delimitations that ordinarily stabilise territorial order?
2. Administrative reconfiguration: How did multi-level governance arrangements enable or constrain crisis management, assistance delivery, and mobility across jurisdictions?
3. Symbolic reconfiguration: How were narratives of displacement, responsibility, and entitlement produced and contested, and with what consequences for who was recognised, protected, or marginalised?

Methodologically, the chapter adopts a qualitative design combining fieldwork (including interviews with affected populations, government authorities, and environmental experts), documentary analysis, and an examination of displacement dynamics before and after the floods to identify emerging patterns and governance challenges. To interpret competing accounts of the disaster and its implications, the chapter mobilises the concept of a "constellation of truths" to analyse how different actors construct, dispute, and institutionalise meanings of crisis, mobility, and boundary-making. Together, these steps support an interdisciplinary contribution to debates on climate migration, territorial governance, and border studies by treating borders not as static lines, but as crisis-sensitive configurations of space, authority, and belonging.

Rationale of the Study: Borders and Climate Change—A New Paradigm

Borders, traditionally understood as demarcation lines separating nation-states, have been the subject of intense debate in the social sciences, particularly in political geography and international relations. Scholars such as Anne-Laure Amilhat Szary (2015) and Anssi Paasi (1998) argue that borders are not merely physical boundaries but also social and political constructs that reflect power relations, collective identities, and historical dynamics.

In the context of climate change and the fragmentation of multipolarity,[1] this perspective gains new relevance. Physical limits, such as rivers, mountains, and coastlines, are subject to significant alterations due to extreme weather events, including floods, droughts, and rising sea levels. These changes challenge the notion of borders as static and immutable entities, highlighting their dynamic and contingent nature. As Paasi (2003) observes, borders are "constantly evolving processes" that reflect not only physical geography but also human interactions and the power structures that shape them.

Moreover, political and social borders are equally affected by climate change. The intensification of natural disasters, such as the 2024 floods in Rio Grande do Sul, can lead to the reconfiguration of territorial jurisdictions, the redefinition of property rights, and the emergence of new social conflicts. In this sense, borders are not just lines on a map but also spaces of contestation and negotiation, where various actors—states, local communities, and international organizations—compete for control and access to scarce resources.

The concept of borderscapes, introduced by Rajaram and Grundy-Warr (2007) and further developed by Krichker (2019), provides a useful lens for understanding the fluidity of borders in times of crisis. Borderscapes refer to the symbolic and material landscapes of borders, encompassing not only their physical and political dimensions but also the narratives, imaginaries, and social practices that constitute them. In the context of climate change, borderscapes become particularly fluid as natural disasters and population displacements challenge established borders and create new geographies of inclusion and exclusion.

For instance, the 2024 floods in Rio Grande do Sul not only altered the physical landscape of the state, particularly in the Taquari and Sinos Valley and the metropolitan region of Porto Alegre, but also reconfigured the societal and political dynamics that define borders. Entire communities were displaced, critical infrastructure was destroyed, and relations between various levels of government—municipal, state, and federal—were strained. In this context, borders became spaces of uncertainty and improvisation, where state responses and local initiatives intersected in complex and sometimes contradictory ways. As Rückert et al. (2024, p. 19)

[1] We begin with the assumption that the current international system is undergoing a process of fragmentation in both multipolarity and the multilateral trade system itself, accelerated by the rise of far-right movements that are anti-system, anti-integration, and anti-globalization, as well as by the broader trend of deglobalization.

poignantly note: "Unfortunately, it is imperative to mention that emotional impacts are considered equivalent to the trauma of war. The population was utterly disoriented and depressed, as the loss of loved ones, homes, neighbourhoods, cemeteries, jobs, and a sense of belonging has been extremely traumatic."

The fluidity of borders in times of crisis is also emphasized by scholars such as Wendy Brown (2010), who argues that contemporary borders are increasingly "porous" and "hybrid." Brown (2010) examines how globalization and climate change are challenging the sovereignty of nation-states and redefining borders as spaces of multiple belongings and loyalties. In the case of the Rio Grande do Sul floods, this porosity is evident in how affected communities sought refuge in neighbouring areas, often crossing municipal and state boundaries in search of safety and resources.

Climate change is redefining not only physical and political borders but also migration patterns and human settlement dynamics. Scholars such as Norman Myers (2002) and François Gemenne (2011) have highlighted the growing role of "climate refugees"—individuals and communities forced to relocate due to natural disasters, environmental degradation, and climate change. Although the term "climate refugee" remains a subject of academic and political debate, it captures the reality of millions of people being displaced by extreme weather events.

In the case of Rio Grande do Sul, the 2024 floods forced approximately one million people to relocate, many of whom had to abandon their homes and communities in search of safer areas. According to data from the platform developed by Possantti et al. (2024), 242,000 homes were affected in the Porto Alegre metropolitan region, 22,000 in the Taquari Valley, and 46,000 in the Sinos Valley. Figure 15.1 illustrates the most affected regions, coinciding with the areas of highest forced displacement.

This displacement not only challenges existing boundaries but also creates new social and political tensions as host communities deal with the influx of migrants, even if temporary, and the competition for scarce resources, such as water. As reported by Araújo (2024): "When the mayor of Porto Alegre, Sebastião Melo (MDB), recommended on Sunday (May 05) that the population of Porto Alegre head to the coast to escape the collapse of electricity and water supplies in the capital due to the worst flood in history, he omitted a crucial detail. Upon arriving at the coast, after hours of traffic congestion, many 'refugees' from the disaster found torrential rain, flooded neighbourhoods, and shortages of gasoline and supplies—conditions like or worse than those they had left behind. Of the 23 municipalities along the 300-km coastal strip between Palmares do Sul and Torres, known as the northern coast, only six (Arroio do Sal, Imbé, Osório, Terra de Areia, Tramandaí, and Três Cachoeiras) were unaffected by the unusual climatic conditions that struck the state of Rio Grande do Sul, according to the state government. In some cases, the causes of the emergency were neither torrential rain nor the overflow of lakes and lagoons." As Gemenne (2011) observes, climate-induced displacements often exacerbate existing inequalities, as the most vulnerable populations—such as the poor, the elderly, and ethnic minorities—are disproportionately affected.

Furthermore, climate change is reshaping human settlement patterns and territorial dynamics. In many regions, coastal areas, and floodplains, traditionally densely

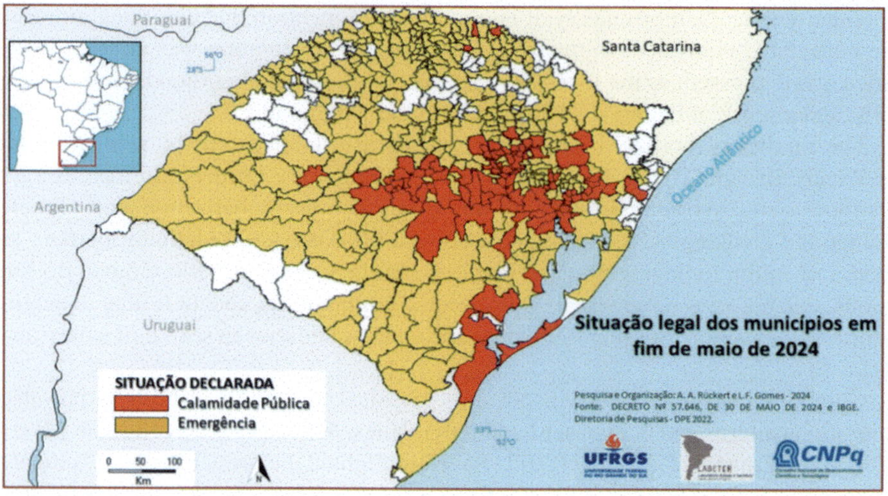

Fig. 15.1 Municipalities in the State of Rio Grande do Sul under declared states of public calamity and emergency as of May 31, 2024. *Legend* Red (public calamity). Orange (emergency). *Source* Rückert et al. (2024, p. 4)

populated, are becoming increasingly vulnerable to floods and other natural disasters. This is leading to a redefinition of urban and rural borders as populations move to safer areas and governments seek to redraw risk and vulnerability maps. As David Harvey (1996) argues, territorial reconfiguration is a deeply political process that reflects the power relations and social inequalities structuring society. In the case of Rio Grande do Sul, this will become evident with the real estate speculation following the floods, where non-flooded areas in Porto Alegre, for example, saw property values increase by up to 67%, ranking among the most valued neighbourhoods in Brazil, as described by Quesada (2024).

The relationship between natural disasters, forced displacement, and state responses is a central theme in the literature on climate change and borders. Scholars such as Alexander Betts (2010) and Susan Martin (2014) have emphasized the need for a more coordinated and inclusive approach to addressing the challenges of climate-induced displacement. This includes not only providing humanitarian assistance and rebuilding infrastructure but also creating governance mechanisms that allow for the participation of affected communities and the integration of local perspectives into public policies.

In the case of the Rio Grande do Sul floods, state responses were marked by a combination of emergency actions and long-term initiatives. At the federal level, the Brazilian government implemented a package of emergency measures, including the release of financial resources, the deployment of rescue teams, and the establishment of temporary shelters. At the state and municipal levels, local authorities worked to coordinate relief and reconstruction efforts, often in partnership with non-governmental organizations and local communities.

However, as Betts (2010) notes, state responses to natural disasters are often fragmented and insufficient, reflecting the limitations of existing governance structures. In the case of Rio Grande do Sul, for example, there were criticisms of the slow and uncoordinated response between various levels of government, as well as the inadequacy of available resources to address the scale of the disaster. Moreover, state responses often prioritize urban areas over rural and peripheral communities, exacerbating existing inequalities. As reported by journalists from Grupo RBS, for example, only 62.2% of public funds allocated for the state's reconstruction had been disbursed a year after the tragedy.

The literature review presented in this section highlights the complexity and multidimensionality of borders in the era of climate change. Physical, political, and social borders are being redefined by extreme weather events, such as the 2024 floods in Rio Grande do Sul, which challenge traditional notions of sovereignty, territoriality, and belonging. The concept of borderscapes offers a useful lens for understanding the fluidity and porosity of borders in times of crisis, while the literature on climate-induced displacement and state responses underscores the need for a more coordinated and inclusive approach to addressing the challenges of climate change.

In this sense, borders are not merely lines on a map but also spaces of contestation and negotiation, where different actors compete for control and access to scarce resources. As the reviewed authors argue, borders are constantly evolving processes that reflect the social, political, and environmental dynamics shaping them. In the context of climate change, this perspective gains new relevance, emphasizing the need to rethink borders as spaces of inclusion, solidarity, and cooperation, rather than mere barriers of exclusion and control. In this vein, the next section will focus on the methodology employed in this chapter, which, combined with these theoretical lenses, will enable the achievement of the results presented in the subsequent sections.

Materials and Methods

Database

Studying mobility during crisis periods—such as an extreme climate disaster—requires methodological adaptation due to (i) disrupted access to field sites and institutions, and (ii) the adverse conditions simultaneously affecting researchers and participants. In the case of the 2024 floods in Rio Grande do Sul, data collection was constrained by flooded roads and streets, the temporary closure of public institutions, and the displacement of potential interviewees. These constraints were particularly visible in the shelter system: in Porto Alegre alone, 193 shelters were recorded between May and September 2024, while the entire state registered 803 official shelters, located in social centres, schools, repurposed public buildings, universities, clubs, and parish halls, among other facilities (Chagas 2024; Bortolon 2024).

Given these limitations, the study draws on a multi-source database combining documentary, administrative, and proxy indicators of displacement and mobility:

1. **Institutional and administrative sources (primary documentary data)**: datasets and official information released by public institutions and entities providing public-interest information.
2. **Public accountability and reconstruction monitoring tools**: including the dashboard developed by Grupo RBS (introduced in the previous section), which tracks the execution of public expenditures allocated to reconstruction.
3. **Media reports as documentary evidence of displacement experiences**: local and national journalism is used as a primary documentary source to capture the challenges faced by displaced groups and to triangulate official estimates. This strategy is consistent with specialised literature on climate change-induced migration, which recognises the value of media documentation in contexts where real-time mobility is difficult to capture through standard instruments (Black et al. 2011).
4. **Mobility proxy indicators (secondary data)**: to estimate displacement magnitude and directionality, we use time-series data on electricity, water, and fuel consumption, as well as vehicle counts derived from highway toll stations linking affected areas to destination/refuge regions. These indicators allow inference about short-run population changes when individual-level registries are incomplete or delayed.

Sample Design

This study examines a defined target group referred to as the temporary migrant population, meaning individuals and households displaced by the floods who sought refuge in other regions and states and, in some cases, crossed into neighbouring Argentina and Uruguay. The sample design is structured around a geographically stratified approach aligned with the spatial distribution of impacts and known displacement corridors.

Fieldwork Regions and Rationale

Post-flood field research focuses on three regions selected for their relevance to the disaster's impacts and displacement dynamics:

- **Metropolitan Region of Porto Alegre (including the capital)**: a major urban concentration, heavily affected, and a primary node for shelters and emergency response coordination.
- **Taquari Valley**: one of the severely impacted areas, with acute disruption and pronounced displacement pressures.
- **Sinos Valley**: an industrial-urban corridor with significant exposure and mobility linkages.

Participant Categories

To ensure analytical coverage of governance and lived experience, interviews are organised across three participant categories:

1. **Affected populations (temporary migrants)**: individuals displaced from flood-affected localities, including those housed in shelters and those who relocated to host municipalities.
2. **Government authorities and operational responders**: actors involved in crisis coordination, assistance provision, and territorial management across municipal and state levels.
3. **Environmental and technical experts**: specialists identified through the literature review and event documentation, contributing interpretive accounts of environmental drivers, risk dynamics, and reconstruction constraints.

Sampling and Recruitment Strategy

Given the crisis context and the mobility of the target population, sampling prioritises feasibility, safety, and triangulation:

- **Purposive sampling** is applied to identify participants who directly experienced displacement and/or were responsible for operational decision-making.
- **Site-based recruitment** (where feasible) is conducted through shelters, host municipalities, and community hubs involved in assistance networks.
- **Snowball sampling** complements purposive selection to reach displaced participants dispersed across temporary housing arrangements and to access networks that formed during the emergency period.

Ethical and Contextual Considerations

Interviews represent a methodological challenge due to the traumatic context affecting both interviewees and the researcher, which can influence recall, narrative framing, and emotional readiness to participate. These constraints are explicitly acknowledged as part of the research design and are treated analytically (rather than as noise), without compromising the study's objectives or validity.

Analytical Strategies

The analysis combines qualitative interpretation with triangulation across documentary and proxy-indicator datasets, using three complementary strategies.

(1) **Narrative and meaning-making analysis: "constellation of truths"**

To interpret competing accounts of responsibility, entitlement, assistance, and mobility decisions, the study employs the concept of a "constellation of truths" (De Geroni 2021). This framework supports the systematic comparison of narratives across affected populations, public authorities, and civil society actors, allowing the manuscript to retain coherence when accounts diverge due to trauma, political contestation, or informational uncertainty.

(2) **Pre–post mobility assessment and baseline reconstruction**

Because the event is unprecedented in Brazil's disaster history, the chapter places particular emphasis on reconstructing mobility baselines and comparing trends before and after the floods. This is necessary to distinguish flood-induced displacement from pre-existing regional mobility patterns.

To do so, the study analyses secondary indicators—electricity and water consumption, fuel consumption, and toll-station vehicle flows—to infer population change in key destination areas and outflows from affected localities.

(3) **Cross-validation and estimation harmonisation using AI-assisted synthesis**

To harmonise secondary datasets and generate consistent estimates, the study uses artificial intelligence tools (ChatGPT 4.0 and DeepSeek) as support instruments for data integration and trend comparison, not as primary evidence. Both platforms produced convergent trends consistent with press reporting and estimates from municipal and state authorities. For example, analyses of electricity and water consumption in affected and destination municipalities align with evidence reported by Dilly (2024), including increased water consumption in Capão da Canoa between May 4 and 14 compared to the same period in the previous year, and an observed population increase in the northern coast region alongside stable electricity supply conditions.

(4) **Modelling displacement pathways**

Finally, the chapter adapts the model proposed by McLeman and Smit (2006) (Fig. 15.2) to outline patterns of temporary migration flows in Rio Grande do Sul during the 2024 floods. This adapted model provides an organising heuristic to relate hazard exposure, vulnerability, decision-making constraints, and destination selection, supporting the subsequent discussion of how these flows interacted with the reconfiguration of internal and external borders.

With this methodological foundation, the next section analyses the Rio Grande do Sul floods through the theoretical lenses and tools outlined above, focusing on their impact on mobility and on the reconfiguration of material, administrative, and symbolic borders.

Results and Discussion: The Case of the Floods in Rio Grande do Sul (2024)

The floods that devastated Rio Grande do Sul in 2024 represent one of the most severe environmental and humanitarian disasters in Brazil's recent history. This extreme event, amplified by climate change, not only exposed the socio-environmental vulnerability of the region but also redefined territorial, social, and political dynamics, challenging governance structures and traditional notions of borders. Below, we describe the disaster, its impacts, the crisis response, patterns of forced migration and internal displacement, and changes in territorial governance and local border regimes, with a particular focus on the concept of borderscapes.

The floods of May 2024 in Rio Grande do Sul resulted from a combination of climatic, geographic, and human action factors, as outlined in initial studies by

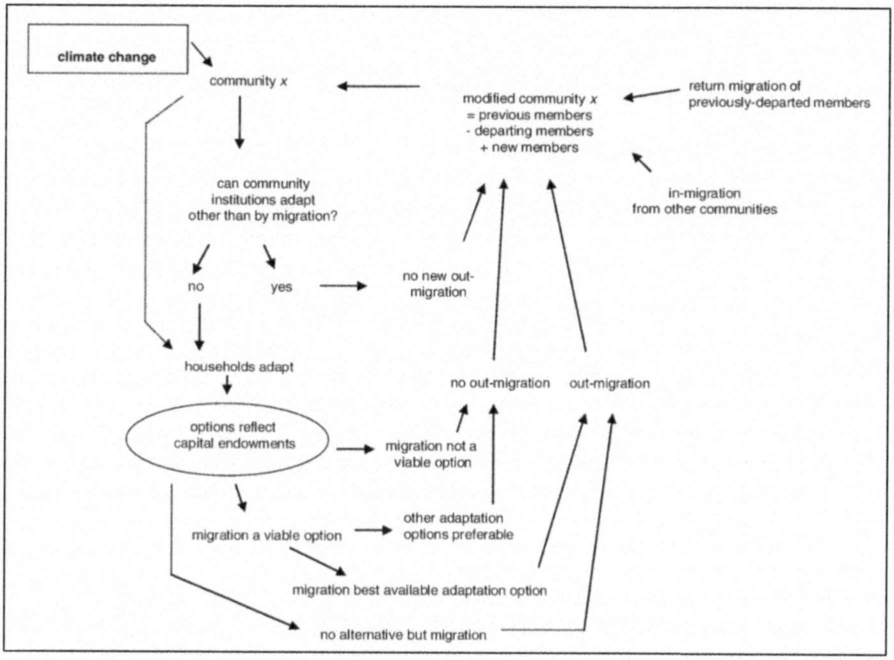

Fig. 15.2 McLeman and Smit's (2006) climate change-induced migration model. *Source* McLeman and Smit (2006, p. 47)

Oliveira et al. (2025). The state, located in southern Brazil, is naturally prone to intense rainfall events due to its proximity to meteorological systems such as the South Atlantic Convergence Zone. However, in May 2024, these patterns were exacerbated by global climate change, which intensified the frequency and magnitude of extreme weather events. Torrential rains lasting weeks exceeded historical levels, causing rivers such as the Sinos, Taquari, and Jacuí, as well as Lake Guaíba in the capital Porto Alegre, to overflow, inundating urban and rural areas across more than one hundred municipalities. In total, 478 of the 497 municipalities in Rio Grande do Sul were directly affected by the May 2024 floods, meaning 96% of the state's territory suffered some form of structural impact from the flooding.

The environmental impacts were devastating. Riverine and coastal ecosystems were severely degraded, with significant biodiversity loss. Permanent preservation areas, such as riparian forests, were eroded, compromising the natural regeneration capacity of rivers. Water contamination by industrial waste, sewage, and chemicals exacerbated the environmental crisis, creating a long-term threat to public health and ecosystem recovery, which could take over a decade, as anticipated by Maliszewski and Faverin (2024) and Hirata (2024).

On the societal and economic front, the disaster affected over two million people, with approximately 500,000 displaced and more than two hundred confirmed deaths. Cities such as Porto Alegre, Canoas, and Eldorado do Sul were partially submerged, with critical infrastructure—including hospitals, schools, and transportation networks—severely damaged. Rural communities, particularly those reliant on family farming, faced total crop and livestock losses, jeopardizing their livelihoods and food security. Social inequality exacerbated the crisis, with poorer and marginalized populations disproportionately affected. The Porto Alegre International Airport, the only major airport in the state, remained closed from early May until late October 2024, disrupting tourist flows crucial to the economies of the Serra Gaúcha, Southern Border, and Northern Coast regions, as well as the movement of goods. The estimated economic impact of the 170-day airport closure was approximately ten billion reais for the industrial sector (Neves 2024) and 1.4 billion reais for the tourism sector (Gallisa et al. 2024).

The crisis response was marked by coordinated efforts but also significant challenges. At the federal level, the Brazilian government declared a state of public calamity, releasing emergency funds and mobilizing national forces, including the Army and Civil Defence, for rescue operations and humanitarian aid. However, delays in resource allocation and bureaucratic hurdles were criticized for hampering immediate response efforts.

At the state level, the Rio Grande do Sul government implemented a contingency plan that included evacuating at-risk areas, distributing food, water, and medicine, and setting up temporary shelters. Despite these efforts, the scale of the disaster overwhelmed the state's response capacity, exposing weaknesses in prevention infrastructure and risk management. The lack of investment in early warning systems and resilient infrastructure was identified as a key factor exacerbating the crisis. Additionally, state authorities were accused of having relaxed hundreds of environmental regulations and laws in the preceding years (Karkow Bones and Araújo 2024), which we consider a contributing factor to the heightened risk and impacts of the May 2024 floods.

Non-governmental organizations (NGOs) and civil society initiatives played a crucial role in the crisis response. Dozens of NGOs and local projects were on the front lines, providing medical, veterinary, psychological, and logistical support. Fundraising campaigns and donations of food and clothing were widely promoted on social media, mobilizing national and international solidarity. However, the lack of coordination between governmental and non-governmental actions led to overlapping efforts and, in some cases, resource mismanagement and diversion, as reported by G1 RS (2024).

The 2024 floods triggered an unprecedented wave of forced migration and internal displacement in Rio Grande do Sul's history. An estimated 500,000 to 1 million people were forced to abandon their homes, many of which were destroyed or rendered uninhabitable due to structural damage and water contamination. Displacement occurred both within the state and to neighbouring regions such as Santa Catarina and Paraná, straining the infrastructure and public services of these areas. The migration flow was even encouraged by government authorities, such as the mayor of Porto Alegre, who stated, "I don't want to sound alarmist, but we have a water problem, and this is a decisive factor. If we lack electricity and water, the city will completely collapse. So, I urge those who have a beach house and can leave Porto

Alegre to do so" (O Antagonista 2024). The city did indeed collapse, as documented by Lopes et al. (2024). The photographic record in Figs. 15.3 and 15.4 illustrates the collapse of Porto Alegre, a scenario repeated in 477 other cities across Rio Grande do Sul.

Fig. 15.3 Downtown Porto Alegre flooded in the second week of the flood. *Source* Agência Brasil (2024)

Fig. 15.4 Porto Alegre International Airport flooded in May 2024. *Source* Metsul Meteorologia (2024)

Emerging migration patterns revealed socio-economic and regional inequalities. Rural populations, particularly small farmers, were disproportionately affected, often lacking the resources to rebuild their lives in new locations. In urban areas, low-income families struggled to access temporary shelters and resettlement programs, often being relegated to already precarious urban peripheries.

Challenges associated with internal displacement included the lack of integrated public policies for resettlement and reconstruction. The absence of a unified registry of displaced persons hindered equitable resource distribution and the identification of specific needs. Additionally, the psychological trauma of losing homes and communities, coupled with uncertainty about the future, created a humanitarian crisis that will take years to overcome, as noted by da Silva et al. (2024).

The 2024 floods also prompted significant changes in territorial governance and local border regimes, redefining the concept of borderscapes—a notion that extends beyond physical and political borders to encompass the symbolic, cultural, and social dimensions of territorial boundaries. The term borderscapes refers to how borders are experienced, represented, and negotiated in everyday life, particularly in contexts of crisis and transformation.

In the case of Rio Grande do Sul, the floods reconfigured borderscapes in multiple ways. First, physical borders were drastically altered. Rivers that once demarcated municipal and state boundaries became impassable barriers, isolating communities and hindering access to essential services, as shown by the flood map in Fig. 15.5. In some areas, river courses shifted entirely, erasing old demarcations and creating new dividing lines. These physical changes challenged existing governance structures, necessitating the redefinition of jurisdictions and the creation of new coordination mechanisms between municipalities and states.

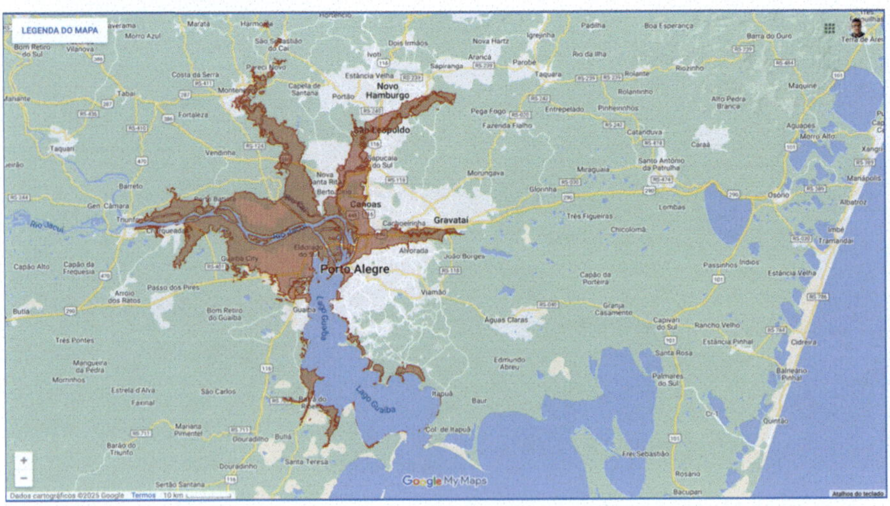

Fig. 15.5 Map of the flood extent in the Porto Alegre Metropolitan Area. *Source* Possantti et al. (2024)

Moreover, the floods redefined social and cultural borderscapes. Historic communities that had lived in certain areas for generations were forced to relocate, severing territorial and identity ties. The resettlement of populations in new areas created symbolic borders between "old" and "new" residents, often exacerbating social and economic tensions. In some regions, new borderscapes of exclusion emerged, where displaced populations were marginalized in urban peripheries or remote rural areas, lacking access to basic services or economic opportunities.

The crisis also highlighted the importance of cross-border borderscapes. The Uruguay River basin, spanning parts of Brazil, Argentina, and Uruguay, required an integrated approach to water resource management and disaster prevention. However, the lack of effective institutional mechanisms for regional cooperation limited the effectiveness of these efforts. The floods underscored the need for cross-border governance that considers not only political boundaries but also the ecological, social, and cultural dynamics defining the region's borderscapes.

Finally, the 2024 floods challenged traditional notions of sovereignty and territorial control. In the context of climate crisis, national and regional borders have become increasingly fluid and permeable. Forced migrations and internal displacements redefined the boundaries of the nation-state, demanding new forms of governance that transcend political borders and incorporate the complexities of borderscapes. This includes the need for public policies that recognize and value the multiple dimensions of borders—physical, symbolic, and cultural—and promote inclusion and social justice in contexts of territorial transformation.

The evidence suggests that the 2024 floods in Rio Grande do Sul were not merely a natural disaster but a catalytic event that redefined physical, social, and political borders, highlighting the relevance of the borderscapes concept. The crisis response, though marked by collective efforts from society and impacted communities, revealed structural flaws in governance and risk management, for which municipal, state, and federal authorities bear responsibility. Forced migration and internal displacement underscored the need for integrated and inclusive public policies. Finally, changes in territorial governance and local border regimes point to a future where climate resilience and social justice must be central priorities in reconstruction and territorial planning, considering the multiple dimensions of borderscapes. This new reality will be further explored in the next section, where narratives from interviews will corroborate this emerging conjuncture.

Challenges and Limitations of the Study: Narratives and Border Disputes in the Era of Climate Extremes

As discussed in previous sections, the construction of discourses on climate displacement and their broader repercussions still face resistance, both from public agents—namely, policymakers—and within academia itself, which has yet to reach a consensus on the concept of the "climate refugee."

In this regard, by employing the term "displaced by extreme climate events," this section, dedicated to the analysis of results, also seeks to question the role of both the state and non-state actors in redefining borders and territorial limits. As previously addressed, these boundaries also underwent a process of transformation during the May 2024 floods, which affected over 90% of Rio Grande do Sul's territory. The impacts of these floods extended beyond national borders, affecting frontier regions of Uruguay and Argentina, as well as the Brazilian state of Santa Catarina.

Regarding climate discourse, in the case of Rio Grande do Sul—a state that has historically been a pioneer in environmental agendas in Brazil, particularly due to the influence of environmentalists such as José Lutzenberger, who was a Brazilian environmentalist from Rio Grande do Sul, known for his strong advocacy against deforestation and industrial pollution; he founded AGAPAN, one of Brazil's first environmental organizations, and played a key role in promoting ecological awareness. As Brazil's Environment Secretary in the early 1990s, he influenced national policies on sustainable development,—the 2024 floods, in our view, represent a turning point. This event has drawn a clear divide between public officials, particularly at the municipal and state levels, and scientists and researchers.

In the interim, the critical role of the local press became evident as a tool for reporting on the unfolding events following the floods. This included coverage of the availability of shelters and reception centres, guidance from civil defence authorities, and designated donation drop-off points. At a later stage, some journalists also positioned themselves as critics of government inaction—particularly at the municipal level—and the lack of disaster preparedness. However, this critique was soon overshadowed by special reports on dramatic rescues, such as the widely covered case of Cavalo Caramelo (The Guardian 2024), and analyses of the financial burden of rebuilding the state. The media agenda quickly shifted, prioritizing narratives on the economic impact of the floods—such as the temporary closure of the state's only international airport and its consequences for tourism—over the accountability of public officials who failed to take preventive measures to mitigate the disaster's severity.

Notably, in the discourse surrounding climate displacement, two distinct political-media narratives emerged. The first, at the height of the floods, was reflected in statements by mayors of the hardest-hit cities, urging residents to evacuate. On the other hand, in destination cities, some local governments—despite being entirely unaffected by the floods—declared a state of emergency due to the influx of temporary residents. This was exemplified by the case of Imbé, a city on the northern coast of Rio Grande do Sul, as illustrated in the following news report: "Located in one of the few regions of Rio Grande do Sul that remained unaffected by the recent heavy rains, Imbé declared a State of Public Calamity on Wednesday night (May 8). According to local authorities, this manoeuvre was intended to secure funding necessary to accommodate displaced residents from other parts of the state seeking refuge in the northern coastal region. Following the declaration, the mayor faced criticism. "Some people will question, 'But Imbé was not directly affected by this climate disaster.' And it is true—we were not directly impacted like cities in the Metropolitan Region or the Taquari Valley. However, in relation to the climate disaster that struck these areas, we were indirectly affected by the migration of people from the Metropolitan Region," stated Mayor Ique Vedovato in a video posted on social media. According to him, Imbé received approximately 5000 displaced residents from other parts of the state over the past five days. The mayor also emphasized that there was no need to establish a public shelter to accommodate those arriving in the city after the floods. Nonetheless, data collected by the Civil Defence and technical analyses conducted by the municipal government justified the decision to declare a State of Calamity to better assist these 'migrants.' "Given the increased demand for healthcare services, we decided to enact this declaration to ensure that we can, in accordance with legal provisions, provide the best possible assistance to these individuals" (Oliveira 2024).

It is evident from the mayor's statement that climate-displaced individuals, even if only temporarily displaced, were perceived as a "problem" that had, in fact, escalated into a public calamity. This type of discourse is well-documented in the literature, wherein migrants are framed as problems, calamities, and risks rather than as individuals who have suffered the impacts of a larger event or phenomenon (Almulhim et al. 2024).

The second category of political-media discourse emerged ex post, in the initial days following the floods, as authorities faced the challenge of maintaining the more than eight hundred shelters established across Rio Grande do Sul. At the same time, climate-displaced individuals not only lost their material possessions to the floodwaters but also saw their jobs and sources of income disappear. Sayad (2004) highlights that employment is often the central priority in times of climate crisis, given the uncertainties surrounding the future. Within this discourse, the predominant narrative revolved around promises of financial aid from government agencies, alongside media scrutiny over the disbursement of these funds. Officials assured that the assistance would arrive within days—a promise that remained unfulfilled.

Initially announced by the federal government during the first week of the floods, the R$5,000 Emergency Assistance payment was only disbursed in the first weeks of June—one month after the disaster. I can personally attest to this, as I was among the recipients, having been forced to stay away from my home for 35 days. My building (Fig. 15.6), located within the flood-affected zone, had its ground floor submerged for three weeks, remained without electricity for the same period, and lacked potable water for two weeks.

In addition to the Emergency Assistance payment—which was granted to only one member per household, a decision we consider flawed, as many families had multiple members affected by income loss—the federal government also authorized the Emergency FGTS[2] Withdrawal. This measure allowed workers to access a portion of their compulsory savings, typically reserved for situations of public calamity. However, it is noteworthy that this withdrawal was only made available nearly two months after the disaster, in mid-July 2024.

[2] The FGTS (Fundo de Garantia do Tempo de Serviço) is a Brazilian workers' severance fund, created to provide financial security in cases of dismissal without cause, retirement, home purchase, or emergencies like natural disasters. Employees accumulate monthly deposits from their employers, and withdrawals (*saques*) are allowed under specific conditions defined by law.

Fig. 15.6 The entrance of my building on the second day of the flood, when we left our home and became climate displaced. A new border logic was emerging on the horizon. May 04, 2025. *Source* Author's collection

What stands out in this context is that Caixa Econômica Federal, the bank responsible for managing the fund, introduced a new category: "public calamity withdrawal due to extreme climate events." This categorization represents, albeit in a limited form, an institutional acknowledgment of the impact of climate change on household incomes.

At the state level, specific income transfer mechanisms were created for targeted groups, including low-income populations, self-employed micro-entrepreneurs, and independent artists—provided they met strict eligibility criteria. This approach drew criticism from large segments of the population who were excluded, particularly those not registered in official social welfare databases.

At the municipal level, the city government of Porto Alegre announced a financial aid program for climate-displaced individuals. However, the initiative never materialized due to the lack of regulatory approval by the City Council.

From the author's personal experience, it became evident that the financial assistance provided by various levels of government functioned more as palliative measures rather than genuine financial safeguards in times of crisis. This was due to the limited amount granted, the delays in disbursement, and the bureaucratic hurdles involved in accessing these funds. In contrast, neighbouring countries that faced similar disasters implemented longer-term programs with broader eligibility criteria, ensuring a more comprehensive safety net for affected populations.

Another key issue that emerged was the impact of induced migration on border control policies following the 2024 floods. Although the core of the disaster occurred several hundred kilometres from the national borders, both the border regions and the internal state boundaries of Rio Grande do Sul were significantly affected.

According to studies by Suertegaray (1998) and Strauss et al. (2021), the combination of rising sea levels, intensifying seasonal rainfall, and the processes of desertification and sandization will progressively degrade Rio Grande do Sul's economic production and circulation zones. This deterioration is expected to force the relocation of cities, industries, and populations in the coming decades, with its peak potentially occurring as early as 2054, as suggested by recent research from the International Antarctic Coastal Circumnavigation Expedition (Peixoto 2025).

In this context, the discourse of climate resilience—promoted primarily by scientists, segments of the specialized press, and a minority of public officials and policymakers—must necessarily transcend societal and national boundaries to create new frameworks for mitigating the impacts of extreme climate events, such as the 2024 floods in Rio Grande do Sul. However, the outlook remains unfavourable.

Of the 827 speeches and statements made by Governor Eduardo Leite between May 2024 and February 2025—that is, from the onset of the floods until the present—94 explicitly mentioned the term "climate change", representing 11.36% of the total. In the case of Porto Alegre's mayor, Sebastião Melo, we analysed his Twitter/X activity, as official speeches were not publicly available for review. The findings are particularly revealing: only eight tweets, all prior to 2024, included the term "climate change", with none appearing after the floods. A particularly emblematic tweet, posted on May 14, 2016, states, "Climate change is in our backyards" (Melo 2016). However, many of the mayor's own political supporters have adopted a different stance—one of climate change denialism—which, in turn, Melo himself was accused of perpetuating following the 2024 floods (Velleda 2024).

Similar patterns emerged in the discourse analysis of other mayors from the affected regions. However, given that many of them were either not re-elected or were ineligible to run for another term in the October 2024 municipal elections, their cases cannot be directly compared to the two political figures. Additionally, a search for terms and phrases related to climate displacement yielded no references in any of their public statements, highlighting a severe misalignment between public policy—particularly long-term planning—and the realities experienced by the affected population.

In this context, when invoking the concept of borderscapes, as discussed in previous sections, this study seeks to analyse the social construction of borders in response to extreme climate events—both as a conceptual response to emerging realities and as a framework for mitigating future risks.

Building upon these discussions, I once again highlight my personal experience in witnessing, in loco, what these borderscapes represent—new meanings attributed to locations, places, geographic spaces, territories, and landscapes. It became strikingly clear to me, my wife, and our six cats on May 4, 2025, that a new border—marking the beginning of a safe zone, distant from the floods—emerged at the boundary between the cities of Porto Alegre and Viamão. This was the only open exit from the state capital and extended through the municipalities of Capivari do Sul and Osório. As illustrated in Fig. 15.7, this route marked the beginning of our journey to Torres, a city on the border with the state of Santa Catarina, where we sought refuge for 35 days.

Fig. 15.7 Our borderscape: entrance to the City of Torres, on the border with the state of Santa Catarina, our refuge for 35 days. Photograph taken upon arrival on May 4, 2025. *Source* Author's collection

In this sense, a borderscape, understood as a constellation of truths—another concept explored in this chapter—does not conform to the fixed and conventional notion of a traditional border as commonly defined in the literature. Instead, it is shaped by the perspectives of those who identify, define, and experience it, as was the case for us. The same landscape became a new border for approximately two million displaced individuals who fled Porto Alegre and its metropolitan region in search of climate refuge in the following weeks. This mass displacement triggered a rapid reconfiguration of the landscape: long traffic jams, carjackings, the temporary suspension of tolls and fines for minor infractions such as unpaid vehicle registration, and the emergence of makeshift roadside businesses selling food and water to those on the move. A new, albeit temporary, migration route was taking shape—along with new borderscapes.

To conclude this section, one recurring theme in conversations with climate-displaced individuals in Torres—and also in other cities along Rio Grande do Sul's northern coast, the coast of Santa Catarina, and border towns in Uruguay and Argentina, regions that were less affected by the floods—was the notion of reconstruction. Even when not explicitly stated, the underlying idea of relocation was evident, whether temporary, provisional, or permanent. Although Rio Grande do

Sul's internal and external territorial boundaries remained intact in official geography, it became clear that the social construction of new borders was underway. This process was driven by the spatial reorganization triggered by severely impacted geological conditions, as well as by uncertainties and fears about the future—ranging from the recurrence of extreme climate events and the costs of rebuilding devastated areas to rising violence in partially abandoned urban peripheries. In other words, new borderscapes were emerging. These developments bring forth critical reflections that will be discussed in the next section.

Policy Implications and Recommendations

The Rio Grande do Sul floods provide clear evidence that climate-related displacement is no longer an exceptional humanitarian episode but a recurring governance challenge with direct implications for territorial management, public service delivery, and border governance. The chapter's findings suggest that extreme events reconfigure **borderscapes** by (i) erasing or displacing material boundaries (infrastructure and access routes), (ii) straining administrative and jurisdictional divisions (municipal–state–federal coordination), and (iii) reshaping symbolic boundaries (who is recognised as protected, entitled, and prioritised). These dynamics translate into actionable policy lessons, particularly for jurisdictions that may face repeated shocks under accelerating climate volatility.

First, institutional recognition is a prerequisite for effective protection. The study shows that a central barrier to timely assistance is the absence of a clear, operational category for climate-displaced persons—especially those displaced internally and temporarily. This conceptual and legal gap delays access to services, complicates eligibility rules, and weakens coordination with humanitarian actors. Policymakers can address this by adopting an institutional definition of climate displacement (with tiers for temporary, provisional, and permanent mobility) linked to streamlined access to emergency benefits, health services, documentation, and legal safeguards. Rather than waiting for full international consensus on "climate refugees," domestic frameworks can create interoperable categories that enable rapid activation of protections during declared climate emergencies.

Second, prevention and preparedness must be treated as measurable public obligations, not discretionary plans. The chapter documents how neglected maintenance and incomplete contingency planning—such as insufficient investment in flood prevention systems, inadequate monitoring capacity, and limited preparedness drills—amplified displacement and prolonged instability. Governments can improve outcomes by embedding risk mitigation requirements into municipal master plans and state regulations, including mandatory maintenance cycles for floodgates and pumps, audit trails for preparedness budgets, and performance indicators for early warning systems. In addition, the systematic integration of universities and local research centres into planning consortia strengthens contextual knowledge and reduces dependence on external consultancies that may lack granular understanding of local hydrology, urban vulnerabilities, and mobility corridors.

Third, crisis governance requires permanent, multi-level coordination structures. The state's shelter landscape—hundreds of improvised facilities across diverse institutions—illustrates the need for standing governance mechanisms capable of rapidly allocating responsibilities across municipalities, state agencies, and federal authorities. A policy priority is therefore the establishment of permanent climate-risk monitoring and response offices, with cross-level protocols for data-sharing, procurement, shelter governance, and population tracking. In border and cross-border regions, these structures should include coordination channels with neighbouring countries and subnational counterparts to anticipate and manage cross-border spillovers.

Fourth, international organisations should be integrated through pre-agreed triggers and domestic leadership. The findings indicate that support from organisations such as IOM and UNHCR was perceived as delayed, and that mobilisation often depended on public pressure rather than predefined operational thresholds. Policymakers can operationalise faster, more predictable engagement by establishing memoranda of understanding that specify activation criteria, division of labour, and integration with domestic actors (armed forces, civil defence, public health systems, and vetted NGOs). This ensures international expertise is mobilised in ways aligned with domestic priorities and tailored to internal displacement realities.

Fifth, information governance is a core instrument of mobility management. The chapter highlights the role of media ecosystems, dashboards, and digital coordination in shaping who moves, when, and how safely. A practical policy implication is to adopt a standardised crisis information framework that integrates civil defence alerts, verified safe-route maps, shelter capacity updates, and service restoration timelines. Partnerships with technology companies and trusted media outlets can reduce misinformation and improve compliance with evacuation orders. In borderscape terms, this strengthens "useful borders"—safe corridors, functional nodes, and administratively legible access points—during periods when normal territorial order collapses.

Sixth, emergency social protection must be automatic, portable, and low-bureaucracy. The study indicates that displacement quickly generates documentation gaps, banking barriers, and interruptions in essential services. Governments can mitigate these harms by pre-authorising simplified procedures in officially declared calamity zones: rapid document reissuance, temporary service reconnections, facilitated access to savings/benefit mechanisms, and automatic eligibility for targeted social benefits regardless of prior registration. Portability matters: assistance should follow people across municipalities, recognising that displacement often redistributes populations into host regions that need immediate fiscal and service reinforcement.

Seventh, policy learning must be institutionalised through continuous monitoring and revision. Because climate extremes are likely to recur, the most effective policies will be those designed for iterative improvement. Governments should implement post-event evaluation cycles that combine administrative data, university-led assessments, and affected-population feedback to revise protocols, update hazard maps, and adjust mobility and shelter strategies. This converts crisis response from ad hoc improvisation into a progressively more resilient governance system.

Taken together, these recommendations translate the chapter's core insight—borders are dynamically remade by climate shocks—into a pragmatic policy agenda: recognise climate displacement institutionally, build permanent coordination capacity, professionalise prevention and preparedness, govern information as critical infrastructure, and make social protection portable and automatic.

Conclusions

This chapter has argued that the May 2024 floods in Rio Grande do Sul illuminate a broader global trend: under intensifying climate extremes, borders cannot be understood as stable territorial lines but as borderscapes—fluid configurations of infrastructure, administration, and meaning that are rapidly reorganised during crisis. By analysing the floods through displacement patterns and multi-level governance responses, the study advances border and migration scholarship in three ways. First, it brings large-scale internal climate displacement to the centre of borderscape analysis, showing how mobility remakes territorial order even without cross-border migration. Second, it links material disruption (infrastructures and access) with administrative strain (jurisdiction and coordination) and symbolic contestation (entitlement, responsibility, and recognition), demonstrating how these layers interact rather than operate separately. Third, it proposes a methodological approach suited to crisis conditions by combining qualitative fieldwork with documentary triangulation and mobility proxies, while interpreting competing accounts through a "constellation of truths."

The wider implication is straightforward: societies are entering an era in which climate events will repeatedly test the institutional capacity to recognise, protect, and govern displaced populations. Future research should extend this framework comparatively—across Brazilian states, across border regions, and across different hazard types—to refine indicators of displacement, evaluate policy performance over time, and better understand how solidarity networks, digital information systems, and institutional architectures shape who is protected and who is left behind. If borders once symbolised primarily sovereignty and control, climate disasters increasingly reveal a different benchmark of territorial authority: not the ability to draw lines on a map, but the capacity to keep people safe when those lines disappear.

Authors Contribution All authors have contributed equally to this work.

Funding The research was not supported by any funding.

Potential conflicts of interest On behalf of all authors, the corresponding author states that there is no conflict of interest.

Informed Consent Informed consent was obtained from all individual participants included in the study.

Data Availability Statement The datasets during and/or analysed during the current study are available from the corresponding author on reasonable request.

AI Disclosure Statement ChatGPT 5.2 was used for grammatical revision and for the translation of the text from Portuguese into English, as well as for the formatting of references. The text was originally written in Portuguese by the author.

References

Agência Brasil (2024) Situação em abrigos públicos do RS preocupa autoridades. Agência Brasil, 8 May. https://agenciabrasil.ebc.com.br/geral/noticia/2024-05/situacao-em-abrigos-publicos-do-rs-preocupa-autoridades

Almulhim AI, Alverio GN, Sharifi A et al (2024) Climate-induced migration in the Global South: an in-depth analysis. npj Clim Action 3:47. https://doi.org/10.1038/s44168-024-00133-1

Amilhat Szary A-L (2015) Qu'est-ce qu'une frontière aujourd'hui? PUF, Paris

Araújo LA (2024) 'Refugiados da chuva' enfrentam inundações também no litoral gaúcho. BBC News Brasil, 13 May. https://www.bbc.com/portuguese/chapters/ce5le8p7yvjo

Betts A (2010) Survival migration: failed governance and the crisis of displacement. Cornell University Press, Ithaca

Black R, Adger WN, Arnell NW, Dercon S, Geddes A, Thomas D (2011) The effect of environmental change on human migration. Glob Environ Chang 21(S1):S3–S11. https://doi.org/10.1016/j.gloenvcha.2011.10.001

Bortolon E (2024) Tristeza e esperança são sentimentos relatados por atingidos que ainda estão em abrigo. Brasil de Fato, 26 June. https://www.brasildefators.com.br/2024/06/26/tristeza-e-esperanca-sao-sentimentos-relatados-por-atingidos-que-ainda-estao-em-abrigo

Brown W (2010) Walled states, waning sovereignty. Zone Books, New York

Chagas G (2024) População em abrigos no RS formaria cidade do tamanho de Venâncio Aires, uma das mais atingidas por cheia; veja dados. G1, 23 May. https://g1.globo.com/rs/rio-grande-do-sul/noticia/2024/05/23/populacao-em-abrigos-no-rs-formaria-cidade-do-tamanho-de-venancio-aires-uma-das-mais-atingidas-por-cheia-veja-dados.ghtml

Cook J, Oreskes N, Doran PT, Anderegg WRL, Verheggen B, Maibach EW, Carlton JS, Lewandowsky S, Skuce AG, Green SA (2016) Consensus on consensus: a synthesis of consensus estimates on human-caused global warming. Environ Res Lett 11(4):048002. https://doi.org/10.1088/1748-9326/11/4/048002

da Silva D, da Silva RKI, Dal Molin E (2024) Intervenções em riscos e desastres: O papel da Psicologia no acolhimento das demandas emocionais da população que sofre com as enchentes no Rio Grande do Sul. Revista Ibero-Americana de Humanidades, Ciências e Educação 10(5):5378–5390, 27 May. https://doi.org/10.51891/rease.v10i5.14265

De Geroni R (2021) Co-existing with water in complexity: the institutional gaze (Master's thesis). Universidade Federal de Santa Catarina. https://repositorio.ufsc.br/handle/123456789/235502

de Oliveira L (2024) Imbé decreta estado de calamidade pública após receber onda de gaúchos de regiões atingidas pela chuva. GZH – GaúchaZH, 8 May. https://gauchazh.clicrbs.com.br/geral/noticia/2024/05/imbe-decreta-estado-de-calamidade-publica-apos-receber-onda-de-gauchos-de-regioes-atingidas-pela-chuva-clvygx6la002i0152vg8uj3v8.html

Dilly B (2024) Litoral Norte se consolida como refúgio emergencial e permanente de gaúchos. GaúchaZH, 1 July (updated 5 July). https://gauchazh.clicrbs.com.br/comportamento/noticia/2024/07/litoral-norte-se-consolida-como-refugio-emergencial-e-permanente-de-gauchos-clxz38j5901in015m2sx346fi.html

G1 RS (2024) MP apura desvio de doações a atingidos por enchentes no RS; secretário e vereador são suspeitos. G1, 4 June. https://g1.globo.com/rs/rio-grande-do-sul/noticia/2024/06/04/mp-apura-desvio-de-doacoes-a-atingidos-por-enchentes-no-rs-secretario-e-vereador-sao-suspeitos.ghtml

Gallisa C, Quadros R, Paz M (2024) Turismo no RS pode ter prejuízo de até R$ 1,4 bilhão caso Aeroporto Salgado Filho fique fechado até o fim do ano, estima governo. G1, 28 June. https://g1.globo.com/rs/rio-grande-do-sul/noticia/2024/06/28/turismo-no-rs-pode-ter-prejuizo-de-ate-r-14-bilhao-caso-aeroporto-salgado-filho-fique-fechado-ate-o-fim-do-ano-estima-governo.ghtml

Gemenne F (2011) Climate-induced population displacements in a 4 °C+ world. Phil Trans R Soc A 369(1934):182–195

Harvey D (1996) Justice, nature and the geography of difference. Blackwell, Malden

Hirata T (2024) Reconstrução no Rio Grande do Sul deve levar uma década, avalia setor de infraestrutura e indústria de base. Valor Econômico, 19 May. https://valor.globo.com/empresas/noticia/2024/05/19/reconstrucao-no-rio-grande-do-sul-deve-levar-uma-decada-avalia-setor-de-infraestrutura-e-industria-de-base.ghtml

Jornal O Diário (2024) 'É o povo pelo povo': Relatos de voluntários que estão na linha de frente demonstram o caos no Vale do Sinos. Jornal O Diário, 9 May. https://odiario.net/noticias/regiao/e-o-povo-pelo-povo-relatos-de-voluntarios-que-estao-na-linha-de-frente-demonstram-o-caos-no-vale-do-sinos/

Karkow Bones N, Araújo LB (2024) 'Não é hora de procurar culpados': O desmonte de políticas ambientais no governo de Eduardo Leite e a calamidade climática no Rio Grande do Sul de maio de 2024. SciELO Preprints, 10 September. https://doi.org/10.1590/SciELOPreprints.9769

Krichker D (2019) Making sense of borderscapes: space, imagination and experience. Geopolitics 26(4):1224–1242

Lopes R, Malinoski A, Gonzatto M (2024) A enchente de 24 – A história da maior tragédia climática de Porto Alegre, 1st edn. BesouroBox, Porto Alegre

Maliszewski E, Faverin V (2024) Solo do Rio Grande do Sul deve levar décadas para se recuperar. Canal Rural, 28 August. https://www.canalrural.com.br/nacional/rio-grande-do-sul/solo-do-rio-grande-do-sul-deve-levar-decadas-para-se-recuperar/

Martin S (2014) Environmental change and migration: what we know. Cambridge University Press, Cambridge

McLeman R, Smit B (2006) Migration as an adaptation to climate change. Clim Change 76(1–2):31–53. https://doi.org/10.1007/s10584-005-9000-7

Melo S (2016) As mudanças climáticas estão nos nossos quintais. Twitter, 14 May. https://x.com/SebastiaoMelo/status/731489706052046848

Metsul Meteorologia (2024) Novas imagens aéreas mostram Aeroporto de Porto Alegre sob as águas. Metsul, 18 May. https://metsul.com/novas-imagens-aereas-mostram-aeroporto-de-porto-alegre-sob-as-aguas/.

Myers N (2002) Environmental refugees: a growing phenomenon of the 21st century. Philos Trans R Soc B 357(1420):609–613

Neves M (2024) Aeroporto de Porto Alegre retoma operação normal em 16 de dezembro, garante concessionária. Agência Câmara Notícias, 4 December. https://www.camara.leg.br/noticias/1117894-aeroporto-de-porto-alegre-retoma-operacao-normal-em-16-de-dezembro-garante-concessionaria/

O Antagonista (2024) Prefeito de Porto Alegre aconselha viagem para o litoral: 'A cidade vai colapsar totalmente'. O Antagonista, 5 May. https://oantagonista.com.br/brasil/prefeito-de-porto-alegre-aconselha-viagem-para-o-litoral-a-cidade-vai-colapsar-totalmente/

Oliveira M, Camardelio AMP, Stedile NLR (2025) As enchentes ocorridas no Rio Grande do Sul: Degradação ambiental, negacionismo e minimização do Estado. RECIMA21 – Revista Científica Multidisciplinar 6(1). https://doi.org/10.47820/recima21.v6i1.6094

Paasi A (1998) Boundaries as social processes: territoriality in the world of flows. Geopolitics 3(1):69–88

Paasi A (2003) Territory. In: Agnew J, Mitchell K, Toal G (eds) A companion to political geography. Blackwell, Malden, pp 109–122

Peixoto J (2025) 'Em 30 anos, devemos repetir maio de 2024', diz pesquisador da UFRGS que participou de expedição à Antártica. GZH – GaúchaZH, 3 February. https://gauchazh.clicrbs.com.br/ambiente/noticia/2025/02/em-30-anos-devemos-repetir-maio-de-2024-diz-pesquisador-da-ufrgs-que-participou-de-expedicao-a-antartica-sobre-enchentes-em-porto-alegre-cm6pj4os2007f01cnjksn7kgy.html

Possantti I, Müller J, Ruhoff A (eds) (2024) Cheias no Rio Grande do Sul: Base de dados e informações geográficas na região hidrográfica do Lago Guaíba e na Lagoa dos Patos em 2024. UFRGS. https://storymaps.arcgis.com/stories/a81d69f4bccf42989609e3fe64d8ef48

Quesada B (2024) Bairros altos de Porto Alegre ficam entre os mais valorizados do país após tragédia no RS. Exame, 21 August. https://exame.com/mercado-imobiliario/bairros-altos-de-porto-alegre-ficam-entre-os-mais-valorizados-do-pais-apos-tragedia-no-rs/

Rajaram PK, Grundy-Warr C (2007) Borderscapes: hidden geographies and politics at territory's edge. University of Minnesota Press, Minneapolis

Rückert AA, Vicente FJ, Gomes LFR (2024) A tragédia climática no Rio Grande do Sul em 2024: Anotações sobre uma catástrofe anunciada. Geografares 4(39). https://doi.org/10.47456/geo.v4i39.46646

Sayad A (2004) The suffering of the immigrant (D. Macey, Trans.; P. Bourdieu, Foreword). Polity Press, Cambridge

Strauss BH, Kulp SA, Rasmussen DJ, Levermann A (2021) Unprecedented threats to cities from multi-century sea level rise. Environ Res Lett 16(11):114015. https://doi.org/10.1088/1748-9326/ac2e6b

Suertegaray DMA (1998) Deserto Grande do Sul: Controvérsias. Editora da Universidade, Porto Alegre

The Guardian (2024) Brazil floods: Horse stranded on roof is rescued as death toll rises to 107 people. The Guardian, 9 May. https://www.theguardian.com/world/2024/may/09/brazil-floods-horse-stranded-rescued

Velleda L (2024) Mesmo após enchente, apoiadores de Sebastião Melo disseminam negacionismo climático. Sul21, 15 August. https://sul21.com.br/geral-1/2024/08/mesmo-apos-enchente-apoiadores-de-sebastiao-melo-disseminam-negacionismo-climatico/

Chapter 16
Living on the Edge: Climate Variability and Migration Patterns Among Char Dwellers in Jamalpur, Bangladesh

Raihan Riaz, Saima Ikbal Safi, Md. Abdul Khaleque, Md. Mohiuddin, Mallik Akram Hossain, and Mumtahina Tisa

Abstract The northern region of Bangladesh frequently experiences floods, riverbank erosion, and is subject to extreme weather events, making life, livelihood, and settlements highly challenging. This study aims to identify the main factors of migration and how climatic stressors, particularly riverbank erosion and flood frequency, affect people's migration decisions. The data was collected through a semi-structured questionnaire from purposively selected 382 most vulnerable households. This study employed descriptive statistics, including frequencies and percentages for all categorical variables, and employed Spearman correlation with 95% confidence interval (CI). The study found that more than half of the respondents migrate primarily as a survival tactic rather than their own choice due to the consequences of frequent occurrences of floods and river bank erosion. The study also demonstrates that increasing the rate of seasonal and temporary migration to cities like Dhaka and Mymensingh, especially among young men. On the contrary, women and children tend to stay in vulnerable Char settlements. The study established a relationship ($P = 0.007$) between livelihood vulnerability and migration decision patterns among community people. Additionally, the study found that the lack of institutional and infrastructural support limits adaptive capacity and increases dependencies on unstable mobility options. The study recommended improving the socio-economic and infrastructural level at the community level, and creating more livelihood and alternative livelihood opportunities for developing climate resilient community. Finally, this study would

R. Riaz (✉) · S. I. Safi · Md. Mohiuddin · M. A. Hossain · M. Tisa
Department of Geography and Environment, Faculty of Life and Earth Sciences, Jagannath University, Dhaka, Bangladesh
e-mail: raihanrtopu@gmail.com

M. A. Hossain
e-mail: mallik.a@geography.jnu.ac.bd

Md. A. Khaleque
Department of Disaster Management & Resilience, Bangladesh University of Professionals (BUP), Dhaka, Bangladesh

P. Ghosh et al. (eds.), *Climate Crisis and Changing Population Dynamics*, Springer Climate, https://doi.org/10.1007/978-3-032-24533-5_16

be helpful for development practitioners, government, non-government organizations, and policymakers to prevent the further marginalization of Char communities' migration decision and enable better basic resource allocation and interventions prioritization.

Keywords Climate variability · Climate change-induced displacement · Migration · Char land · Riverbank erosion · Jamalpur

Introduction

Climatic change is defined as widespread, long-term changes in climatic patterns and their components, such as temperature, precipitation, air pressure, humidity, and rising sea levels, across many parts of the world (Abbass et al. 2022; Riaz and Nasher 2023). It is a global problem that affects all areas of the geopolitical, social, ecological, and environmental spheres (Ali et al. 2023). Climate change alters temperatures, precipitation, sea levels, and the frequency and intensity of extreme events, thereby increasing Bangladesh's vulnerability (Hossain et al. 2022). According to the Global Climate Risk Index 2020, Bangladesh ranks seventh globally in terms of climate-related impacts (Fernández et al. 2024; Sultana et al. 2025). In Bangladesh's Jamalpur District, climate change is substantially affecting agriculture and food security, making it difficult to feed the entire population with existing agricultural output (Anderson et al. 2020). Because of the district's limited economic opportunities and limited livelihood options, many Char residents migrate to larger cities, such as Mymensingh and Dhaka, each year. As a result, it is critical to determine the primary causes and migration patterns among Char dwellers in Bangladesh.

Human activities, such as the use of fossil fuels, deforestation, and industrial processes, are the primary drivers of climate change (CC), which produces long-term changes in global climate patterns (Mondal et al. 2024; Riaz 2024). Bangladesh is among the most vulnerable countries to the impacts of climate change. Besides the consequences of climate change, 40 million people in the country are food insecure, with an additional 11 million facing acute hunger (Das et al. 2024; Riaz et al. 2025). Every year, riverbank erosion and flooding in Jamalpur cause significant harm to the environment, society, and infrastructure (Rahman and Gain 2020). Every year, roughly 10,000 ha of land are lost, displacing families from multiple Upazilas and damaging homes, crops, and embankments (Mullick et al. 2020). These disasters damage livelihoods, increase poverty, and leave thousands of people homeless. Additionally, habitats are degraded, and standing crops and fertile soil are eroded. As a result, people are forced into mass migration, typically to urban slums, where one-quarter of the population suffers from river erosion and flooding (Alam 2017; Han et al. 2024).

Bangladesh's population is exposed to a range of hazards due to its geographical location, particularly riverbank erosion during the rainy season (Paul et al. 2021;

Riaz and Sultana 2025). Bangladesh is known as a disaster-prone country, with riverbank erosion and flooding that substantially affect people's lives and settlements throughout the year, especially during the rainy season (Maharjan et al. 2020). Every year, thousands of people are displaced by ongoing riverbank erosion, which poses challenges for residents in the river region. According to Shamim et al. (2025), an analysis conducted by the Internal Displacement Monitoring Center in 2015 estimated that 4.7 million people in Bangladesh were affected by natural disasters between 2008 and 2014. The Intergovernmental Panel on Climate Change estimates that approximately 50 million people will have migrated by 2050 due to climate change (Ferris 2020). Due to climate change, riverbank erosion is increasing, and people are being forced to relocate from their homes, which also poses serious risks of uprooting communities, destroying infrastructure, impacting agriculture, increasing vulnerability, and exacerbating food insecurity, malnutrition, and poverty (Khatun et al. 2025; Sultana and Riaz 2025). However, these challenges threaten rural households and may serve as a "push" factor for migration. Economic possibilities and social disparities between rural and urban areas also encourage migration. In Bangladesh, roughly two-thirds of permanent migration is from rural to urban areas, with many people returning home within six months; this trend has accelerated significantly over the previous three decades (Barua et al. 2019; Mohiuddin 2022). The permanent rural–urban migration rate per decade was 1.2 per thousand in 1984, but it rose to 22.5 per thousand by the following year, 2010 (Assraf Seddiky et al. 2024). Recently, a quarter of urban slum residents have relocated from areas impacted by riverbank erosion, where they lost their livelihoods and financial assets (Neef et al. 2023). People from the Char land in Jamalpur, near the Jamuna basin, leave their ancestral lands due to landlessness and the loss of livelihoods caused by riverbank erosion (Mohiuddin and Nowreen 2023). According to Mondal et al. (2024), approximately 3000 families experience homelessness annually due to the relentless erosion of the Jamuna River. For this reason, people living along the river are adversely affected by riverbank erosion and quickly fall into poverty after losing their homes, homestead gardens, farmland, and means of subsistence (Morshed et al. 2025). Riverbank erosion directly affects approximately one million people, both within and outside the embankment, and causes changes in occupation, loss of family and community ties, increased vulnerability, and further effects of erosion-induced migration in Jamalpur (Islam 2018; Kartiki 2011). People in Char land are most negatively affected in the Kulkandi Union of Islampur Upazila in Jamalpur District. They lose their livelihoods, including farmland, standing crops, and even change occupations. Property is lost to riverbank erosion, prompting many residents of the Kulkandi Union to relocate to informal settlements in cities, thereby increasing population pressure in urban areas (Riaz et al. 2025). Therefore, it is necessary to examine climate variability and migration patterns among Char dwellers in Jamalpur.

Several recent migration studies that poorly address the variables influencing migration provide a comprehensive review of the relationship between climate change and migration, focusing on the factors that make areas highly vulnerable to significant climatic hazards, such as floods and severe river erosion (Abbass et al. 2022; Habib et al. 2021; Maharjan et al. 2020). Researchers have highlighted that changes in the global environment increase the risk of disasters, threaten livelihoods, and are likely to influence migration patterns in the near future (Ibrahim and Mensah 2022). Another study by Hossain et al. (2022) highlights the factors driving migration in response to climate change and examines the associated socio-economic and socio-demographic characteristics. Some studies examine the coping mechanisms used in riverbank erosion (Barua et al. 2019). In Bangladesh, migration has traditionally been shaped by social and economic factors; however, climate-related migration is rapidly increasing and now accounts for up to 40% of internal migration decisions (Gupta et al. 2025). According to Bernzen et al. (2019), in Bangladesh, economic considerations are the primary drivers, with environmental stress acting as a secondary factor of migration. Research also critiques that environmental disasters have displaced millions of people, and rural-to-urban migration along coastlines has increased rapidly (Ranjan 2023). By 2050, up to 13.3 million people could be affected by climate-related issues, including migrants (Ninawe et al. 2018). According to (Akther et al. 2021; Das et al. 2024; Sajjad et al. 2024) riverbank erosion and floods are significant hazards that displace millions of people and force them to migrate. Additionally, (Chowdhury et al. 2022) demonstrate that riverbank erosion is a significant threat to the Jamalpur District, resulting in economic losses, loss of livelihoods, and asset damage. As a result, most affected individuals are forced to change jobs and migrate to urban areas.

However, the comprehensive review of literature on climate change and migration identifies numerous significant research gaps. Current research often overemphasizes environmental determinism, overlooking the complex interactions among economic, political, and cultural factors that shape migration trends. There is very little long-term research on migrants' adaptability and changes in their livelihoods; most studies are event-specific and short-term. This study used only primary data; secondary data or advanced tools such as GIS, RS, and machine learning models were not employed to map erosional zones and susceptible areas. Large-scale estimates or localized case studies predominate in the methodology, whereas integrative mixed-method approaches would be helpful in this research. Furthermore, there is a lack of studies on planned migration and how to enhance the adaptive capacity. There is no relevant research on the characteristics of the population in Jamalpur that warrants examination. On the contrary, the study primarily relied on primary data, including household survey data, Focus Group Discussions (FGDs), and Key Informant Interviews (KIIs). Accordingly, a questionnaire was designed for the sample survey. The authors identified indicators for this study related to climate variability, livelihood scenarios, vulnerability context, migration decisions, and the adaptive capacity of the study population. Therefore, the main aim of the study is to identify the primary factors of migration, including climate change, floods, and riverbank erosion, and to investigate the relationship between climate variability and migration patterns among Char dwellers in Jamalpur. Additionally, the study examines the relationship between climate variability and migration patterns in these communities, which are severely affected by riverbank erosion. Moreover, the study aims to provide essential insights into effective adaptation techniques for households facing riverbank erosion in Bangladesh's Char land regions by addressing the specific challenges these communities face. The current study will answer the following questions to fulfill the research objectives: (a) What kind of people are being displaced as a result of riverbank erosion? (b) What effects do displaced persons have on their financial situation? (c) What kinds of psychological stress and sociocultural hardships affect displaced persons as a result of river erosion? (d) How have the displaced people coped with, adapted to, or mitigated the effects of river erosion?

Therefore, this study addresses the gap by focusing on riverbank erosion in the Char regions of Jamalpur and by providing insights into how climate change exacerbates its impacts on vulnerable populations. By addressing these gaps, this research offers a more comprehensive view of riverbank erosion risk in Char regions, in contrast to larger studies that often overlook these communities' unique vulnerability. Most existing research focuses on climate-induced migration at the local or national level. However, this study focuses on the livelihood options and experiences of populations whose vulnerability is influenced by both seasonal and long-term climatic changes. This study emphasizes the relationship between migration and climate change in Bangladesh. Furthermore, it promotes the development of adaptation and policy initiatives tailored to Jamalpur's challenges, which are frequently overlooked in migration research. Moreover, the findings may aid NGOs, development partners, and policymakers in developing ways to strengthen the resilience of Char communities. Finally, as illustrated by the UNFCCC and the Global Compact for Migration, this study contributes to the international discourse on climate policy, including loss and damage, migration as an adaptation strategy, climate justice, and the achievement of the SDG goal on Climate Action (13).

Study Area

The study was conducted in Kulkandi Union, Islampur Upazila, in Jamalpur District of Bangladesh. The union covers 28.33 km^2 and is situated between 24.7833° N and 89.9333° E (Riaz and Mohiuddin 2025). This union lies in the eastern floodplain of the Jamuna River, an administrative subdivision bordered by the Garo Hills to the north. The Jamuna River and its tributaries significantly influence the country's hydrological system (BBS 2022). The average annual temperature in Kulkandi is 28.3 °C, and the average yearly rainfall is 2174 mm. The highest and lowest average temperatures recorded are 33.3 °C and 12 °C, respectively (BMD 2024; Riaz and Nasher 2023).

The land cover of the union consists mainly of croplands, with homestead forests and scattered plantations of eucalyptus, mahogany, akashmoni, and koroi having largely replaced the natural vegetation. Agriculture remains the main driver of the local economy, followed by river fishing, labor migration, and informal off-farm businesses (BBS 2022). The study area predominantly produces paddy, jute, and winter vegetables. Seasonally Chars are used for short-term farming but are often lost to erosion. Due to significant negative impacts on farms, homesteads, and physical assets, many households mainly rely on wage labor and seasonal migration to cities to improve their livelihoods (Mohiuddin 2022; Riaz 2023).

The Kulkandi Union is considered fragile due to frequent flooding and riverbank erosion, as well as risks from physical, natural, social, human, and financial factors. Riverbank erosion and seasonal floods occur annually, causing people to migrate and lose their livelihoods in the area studied (Hossain et al. 2022). Because of the ongoing threat of rapid climate change, residents have had to move their homes and lose their farmland, making migration a necessary means to sustain their livelihoods (Riaz 2024). The precise study area was identified through expert opinion, by assessing the risk of climatic hazards to individuals, evaluating livelihood vulnerability resulting from climate variability, and determining the intention to relocate due to climate change and related threats in the region. Therefore, these specific factors helped guide the selection of the research area. The location of the study area is illustrated in Fig. 16.1.

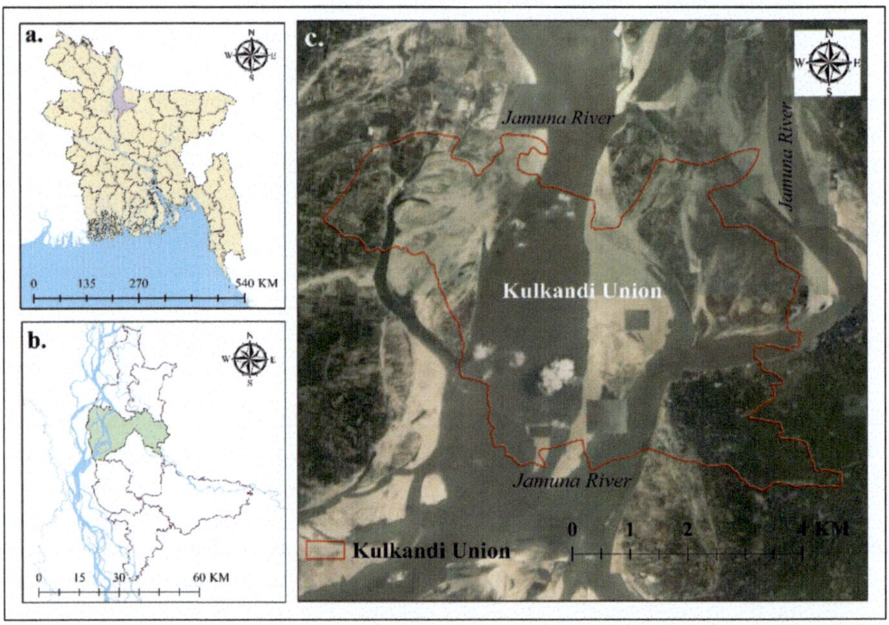

Fig. 16.1 Location of the study area: **a** Location of Jamalpur District in Bangladesh, **b** Location of the Islampur Upazila in Jamalpur District, **c** Location of the Kulkandi Union in the Islampur Upazila

Materials and Methods

This study employs both primary and secondary data to examine migration patterns and related issues in the Islampur Upazila of Jamalpur District. Primary data are primarily collected through household-level questionnaire surveys, whereas secondary data are obtained from published articles, reports, and official government and non-government sources. Additionally, the overall methodology of this study is illustrated in Fig. 16.2.

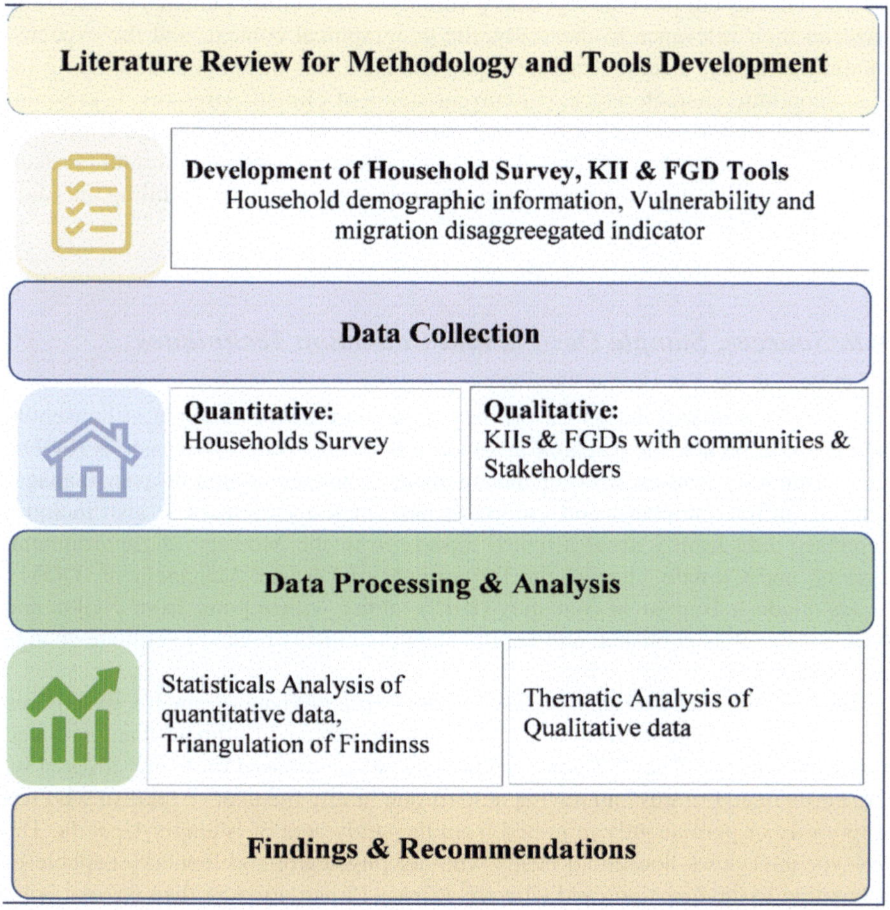

Fig. 16.2 Overall methodology of the study

Tools Preparation Techniques

The study primarily relies on household survey data, and to facilitate data collection, a questionnaire was designed for the sample survey. First, the authors identified indicators related to climate variability, livelihood scenarios, vulnerability context, migration decisions, and the adaptive capacity of the study population. The questionnaire was developed based on previous literature, taking into account the study objectives and geographical location. After drafting the questionnaire, the authors conducted a pilot field visit at the study sites, involving 30 household surveys. During the pilot survey, many variables were either included or excluded based on their relevance to the study, the geographical context, and the expected outcomes. Finally, a semi-structured questionnaire was developed, addressing (a) socio-economic characteristics, (b) current status of climatic stressors, (c) respondents' livelihood assets scenario, (d) vulnerability level of the Kulkandi Union, (e) key indicators of limited livelihood assets, (f) climate change and migration decisions, and (g) adaptive capacity and institutional support aimed at fulfilling the study objectives.

Data Sources, Sample Design, and Collection Techniques

The study is primarily based on primary data sources, for example, a household survey. Additionally, the research involved four Focus Group Discussions (FGDs) with community leaders, disaster management volunteers, union disaster management committee members, and groups of fishermen and farmers. It also includes secondary data from journal articles, reports from the Ministry of Environment, Forestry, and Climate Change, the Department of Disaster Management (DDM), the Bangladesh Bureau of Statistics (BBS), along with reports from NGOs and INGOs, as well as published and unpublished documents related to climate change and migration.

However, while collecting data using a semi-structured questionnaire, the authors recruited five graduate students who had prior experience with climate change-related household surveys. The 382 households were selected based on being affected by climate-induced hazards and having at least one family member or relative who has temporarily or permanently migrated from the study area of Islampur Upazila. The surveyor interviews household heads who are physically and mentally capable of responding to the questions and who are at least 18 and no more than 65 years old. Before starting the conversation with respondents, it is ensured that these interviews are voluntary, responses are anonymous, and that the survey takes 15–25 min to complete. After each individual's survey is completed, the questionnaire is rechecked for inconsistencies, and any gaps are immediately filled by listening to the previously recorded audio. Finally, a study was conducted in 382 households of the Kulkandi Union in Islampur Upazila, Jamalpur District.

Data Analysis Strategies

The study employed statistical analysis in SPSS Version 25.0. Descriptive statistics and frequency analysis were performed with a 95% confidence interval. For the correlation analysis, this study used Spearman's correlation with a 95% confidence interval (CI) to examine the relationship between livelihood asset scores and the likelihood of migration in Islampur Upazila.

Results

The chapter analyzes the indicators that influence households in the Char to migrate due to their current vulnerabilities to climate-induced hazards and extreme weather conditions associated with their livelihoods in Kulkandi Union of Islampur Upazila, Jamalpur District. Their primary strategy for mitigating the adverse effects of climate-related hazards in this area is to migrate in search of a better livelihood. The study found that most respondents were frequently affected by various disasters, particularly twice a year, which makes it difficult for them to manage their livelihoods in the study area. In addition, limited access to livelihoods, structural and institutional vulnerabilities, and a lack of adaptive capacity have deteriorated living conditions, prompting people to migrate elsewhere in search of more suitable livelihood opportunities. The study found that the Kulkandi Union has a limited number of educated individuals who have completed higher secondary education (3.66%). Most people are engaged in day labor and farming activities for their livelihood. Additionally, the respondents' detailed socio-demographic information is presented in Table 16.1.

Current Status of Climatic Stressors

Current climate change has increased the frequency of hazards, including floods, riverbank erosion, and drought, in northern Bangladesh. The study area is prone to various risks due to its spatial characteristics, as the Jamuna River surrounds the region. In this study, 60% of respondents described undergoing floods 1–2 times per year, while 50% stated experiencing riverbank erosion at the same frequency. These results show that flood and riverbank erosion are the most common hazards in this union. Additionally, respondents also observed that total precipitation and the frequency of unprecedented climatic events have increased over the decades. Moreover, 62% of respondents believe that more frequent hazards and extreme weather are attributable to climate change, and that it has had the most significant impact on the community (Table 16.2).

The study's findings indicate that, due to climate change, both slow- and sudden-onset hazards, particularly flooding and riverbank erosion, significantly affect the

Table 16.1 Socio-demographic profile of the respondents

Variables	Categories	Kulkandi Union	
		f	%
Age	18–30	52	13.61
	31–40	128	33.51
	41–50	76	19.9
	51–60	80	20.94
	Above 60	46	12.04
Gender	Male	236	61.78
	Female	146	38.22
Education	Illiterate	262	68.59
	Primary school	66	17.28
	SSC	36	9.42
	HSC	14	3.66
	Graduate	4	1.05
Occupation	Unemployed	24	6.28
	Government services	12	3.14
	Private/NGO job	18	4.71
	Farmer (land owner)	62	16.23
	Farmer (leaser)	76	19.9
	Day labor	142	37.17
	Business	48	12.57

Source Field Survey, 2024

community's livelihood activities. These hazards also affect the community, other sectors, and social services, including critical infrastructure.

Livelihood Assets Status of the Respondent

The section elaborates on the relationship between land use and access to resources, including water, infrastructure, and income opportunities, to understand livelihood patterns in the study area. Most respondents have limited access to land, with only 62% owning less than 0.5 acres. On the other hand, the respondents were also less educated; most had only a primary education. The survey shows that 48% had primary education, 22% were illiterate, and only 18% had access to vocational skills. This limited access to schooling limits the community's options for alternative livelihood. Housing conditions were also poor: 33% of respondents lived in substandard houses, and 31% described their homes as pucca.

Table 16.2 Current status of climatic stressors

Indicator	Response categories	%
Flood frequency (per year)	1–2 times	60
	> 2 times	35
	None	5
Incident of riverbank erosion (per year)	1–2 times	30
	> 2 times	50
	None	20
Changes in rainfall pattern	Yes	76
	No	14
	Not sure	10
Extreme weather frequency	Never	2
	Rarely	15
	Sometimes	38
	Often	35
	Very often	10
Perception of climate change impact	Strongly disagree	3
	Disagree	5
	Neutral	4
	Agree	62
	Strongly agree	26

Source Field Survey, 2024

However, Fig. 16.3 shows that respondents had limited access to credit, leaving them financially excluded, with a minimum monthly income of 5000 BDT. Additionally, regarding social support and interaction through community membership, such as networking with others, only 73% of participants had no community membership, and over half (53%) reported that social support was rarely, never, or sometimes available. A large proportion of participants lived more than 1 km from critical infrastructure.

Vulnerability Level of the Kulkandi Union

The study mainly focused on the existing level of vulnerability of the study area due to climate change. Rapid changes in climate variability (temperature, rainfall, and extreme weather) and frequent climatic hazards (floods, riverbank erosion) increase susceptibility to climate-related hazards. Moreover, the area's vulnerability stems not only from climate change but also from its intensification, which is further exacerbated by structural and institutional factors. Our study found that 62% of respondents

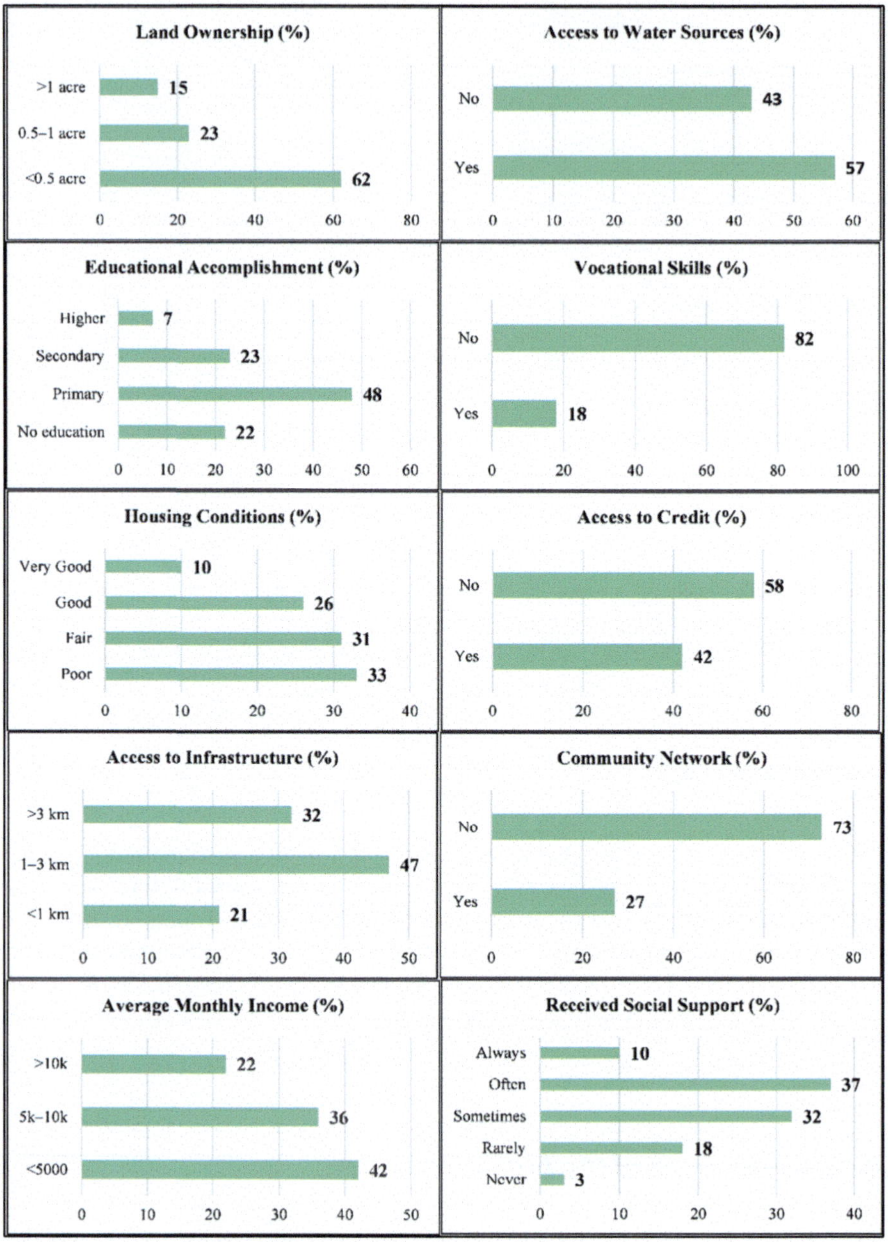

Fig. 16.3 Livelihood assets status of the respondent

Fig. 16.4 Vulnerability level of the Kulkandi Union

reported that public services are limited in the region, indicating they face difficulties accessing basic needs such as healthcare, education, and transportation. On the other hand, the study area is geographically separated from the mainland, which also limits its access to basic services. 56% respondents said that the distance to the mainland is about 5–10 km from their homes (Fig. 16.4).

The study also examines respondents' access to early warning messages; responses indicate that only 18% always receive them, suggesting a limited early warning system for a large portion of respondents. Limited early warning makes people more vulnerable to climatic shocks. Another factor that increases the community's vulnerability is institutional neglect; 72% of respondents reported that these institutions have failed to adequately serve the community. As a result of these vulnerabilities, the community people tend to migrate for betterment. Findings show that 54% respondents have migrated 1–2 times in the last 10 years.

Conceptual Link: Livelihood, Vulnerability, and Migration

The study aimed to establish the correlation among livelihood, vulnerability, and migration. It is well established that livelihood and vulnerability are strongly related to migration. This study's findings, based on both quantitative and qualitative data, reveal the current state of climatic shocks and stress, and the unprecedented changes in climatic variability in the region have significantly affected ways of living and livelihoods. Limited access to livelihood capitals, named natural, human, physical, financial, and social capitals, makes the community more vulnerable to climate change. Regarding natural assets, community members have limited land ownership, indicating a high level of vulnerability (Table 16.3). In addition, the sub-indicator of human capital, vocational skills, and education also shows high vulnerability because of a lack of skilled people, which ultimately reduces the working scope. On the other hand, two major sub-components of physical assets, housing condition and inaccessibility to critical infrastructure, showed high vulnerability. Similarly, lower monthly income and limited access to loans among households indicate their vulnerability with respect to financial assets. Most respondents were not involved in any community organizations, indicating a high level of vulnerability with regard to social assets. In sum, findings from most major components suggest that livelihood assets heighten community vulnerability to climate change and variability (Fig. 16.5).

The study conducted a Spearman's rank correlation analysis, which confirmed a significant negative relationship ($\rho = -0.58$, $p = 0.007$) between livelihood capital scores and the likelihood of migration. The correlation shows that households with fewer livelihood assets are more likely to migrate from the region in search of better livelihood opportunities.

Table 16.3 Key indicators of limited livelihood assets

Asset type	Indicator	Observation
Natural	Land ownership	62% have less than 0.5 acres
Human	Vocational skills, education	82% lack vocational skills
Physical	Poor housing, limited infrastructure access	33% live in poor housing; 47% travel 1–3 km to access services
Financial	Monthly income, credit access	42% earn less than 5000 BDT; 58% lack credit
Social	Community membership, support networks	73% are not involved in community organizations

Source Field Survey, 2024

Livelihood Assets (Low) → Increased Vulnerability → Higher Propensity to Migrate

Fig. 16.5 Conceptual link: livelihood, vulnerability, and migration

Adaptive Capacity and Institutional Support Regarding Vulnerability

Adaptive capacity refers to the ability to manage adverse conditions by drawing on local knowledge and institutional support when a community faces unprecedented climatic events. Most households did not receive sufficient assistance from institutions, including disaster aid, training, participation in community disaster plans, and access to post-disaster support. However, most respondents mentioned the presence of NGOs in their communities. Still, most participants reported not receiving training on climate change and climatic hazards, indicating limited adaptive capacity, particularly in responding to hazards. Similarly, few households have a disaster plan and have received assistance from any institution. The results indicate a significant gap in both adaptive capacity and institutional capacity. Due to a lack of resilience and institutional support, the area is highly vulnerable and drives out migration.

Climate Change-Induced Migration: Decisions and Patterns

The study aimed to examine how climate variability, climatic stress, and shocks affect the livelihoods of the Char people and how these factors influence their decision to migrate. 45% of the respondents demonstrated that climate change and its variability prompt them to migrate. Climatic disasters, including floods, riverbank erosion, droughts, and climate variability itself. People prefer to migrate seasonally, with destinations often in urban areas rather than in rural areas or abroad, because metropolitan areas offer easier access to livelihood opportunities (Mohiuddin 2022). A portion of respondents (65%) reported that the migration strategy yielded positive economic outcomes. However, more than 50% of respondents migrate as a survival mechanism rather than in a collective, planned manner, indicating that climate-related vulnerabilities are significant drivers of migration in the region (Fig. 16.6).

The study suggests that the government should address the community's sustainable livelihoods in the study area. The study found that the region lacks access to natural resources, public services, vocational skills, diversified livelihood training, formal financing, and other basic facilities (Fig. 16.7).

Discussion

The study identifies the primary factors of migration of Char dwellers and establishes a relationship between climate vulnerability and intention to migrate from Kulkandi Union of Islampur Upazila to others betters location (urban centers) for their livelihood. Additionally, this chapter also showed the migration decision and pattern, and the type of migration of the community members. The study found that this study area

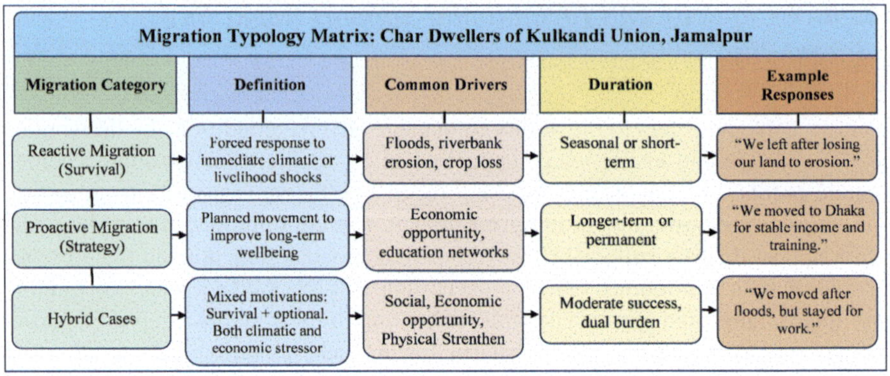

Fig. 16.6 Migration typology matrix: Char dwellers of Kulkandi Union, Jamalpur

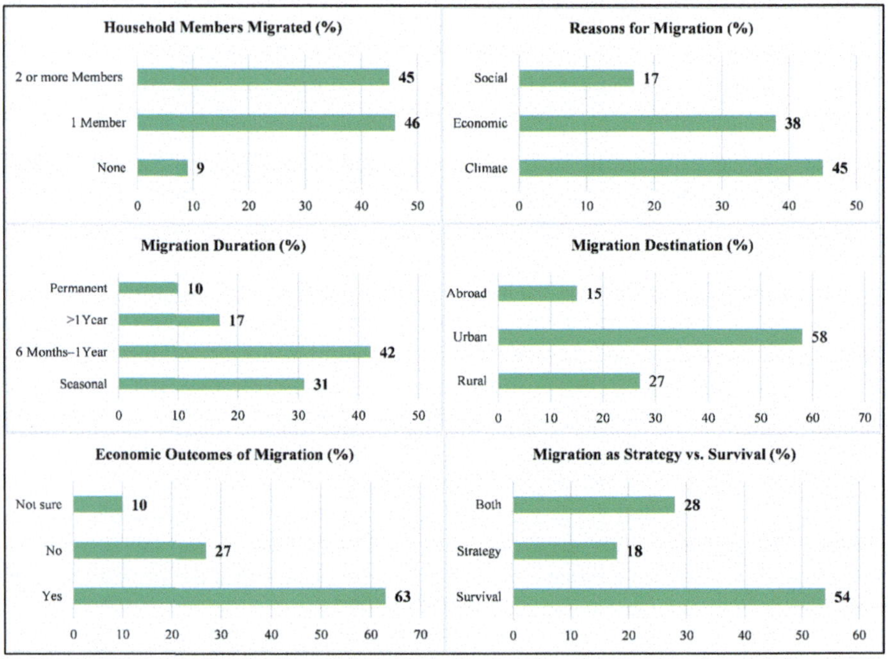

Fig. 16.7 Respondents' migration decisions pattern

is frequently affected by floods, river bank erosion, and different climatic shocks, leading to increasing their vulnerability and preparedness for decision intention to migrate as seasonal, temporary, or permanently displaced from their home.

However, Char dwellers are primarily exposed to climate change and climate-induced hazards, and they respond to rapid shifts in climate patterns amid multiple factors (human, social, physical, financial, and natural) that drive community members to migrate in search of a suitable livelihood. Our study aimed to explore the factors most closely associated with the decision to relocate. People in the Char primarily depend on agriculture for their livelihoods. However, unprecedented changes in climate variability, namely reduced rainfall and temperature fluctuations, have rendered crop harvesting in these areas impossible. In addition, this study found that multiple hazards, primarily flooding and riverbank erosion, occur in the region annually and, occasionally, twice. These phenomena threaten their daily lives and livelihoods. The findings are similar to the previous studies, where authors found that frequent climatic hazard-induced, climatic shocks and stress, as well as climatic variability, have negatively affected communities due to their geographical features, socio-demographic profiles, and fragile adaptation strategies (Assraf Seddiky et al. 2024; Hossain et al. 2022).

Moreover, the findings indicate that livelihood vulnerability depends not only on climate change but also on household characteristics. Most households cite limited access to land and a lack of skilled labor, stemming from inadequate vocational and traditional education, which has exacerbated their vulnerability (Alam et al. 2019; Ali et al. 2023). In addition, the community has low financial stability (low income and limited access to credit) and limited engagement with social or community organizations, which limits its capacity to recognize climate change. Studies suggest that multi-capital deprivation reduces coping strategies in response to climatic shocks and stress, findings that are consistent with the earlier study findings (Azam et al. 2021). Furthermore, the study site is located on an island separate from the mainland. As a result, community members face challenges in accessing critical infrastructure, including healthcare, educational institutions, and transportation. These findings are consistent with a previous study where the author found that community member faces infrastructural and basic services-related challenges (Mohiuddin 2022). Moreover, the study found that the area has extreme unevenness in land tenure, a lack of institutional support, and inadequate disaster preparedness, including insufficient early warning dissemination, leaving the community inextricably prone to climate change.

Therefore, the impact of climate change unevenly disrupted the community's diverse livelihood options, prompting migration. Similar findings reported by Kartiki (2011) indicate that most people affected by climate change migrate primarily due to climate change rather than economic or social factors. Furthermore, Hossain (2020) stated that the lack of adaptation capacity and institutional support leaves the community unable to deal with climate change. However, studies find that livelihood vulnerability mainly stems from resource scarcity and limited livelihood options in the community. This study finds that some people migrate to other places as a survival mechanism to adapt to adverse situations, and both pull and push factors in the study area influence the decision to relocate temporarily (Han et al. 2024).

The study's comprehensive analysis of migration causes, decision pattern, and migration scenarios in response to climate change reveals that in context of socio-economic effects on affected areas of Bangladesh. Additionally, original data from field survey (Households Survey, FGD, KIIs) with affected communities considered one union of Islampur Upazila that may lead to effects on the final concluding results of this study. However, further study considering the distance places households in two or three union study sites, which create the conclusion from this study is more precise, location-specific, and pertinent.

From the study findings, it is recommended that the government should also focus on climate-resilient livelihood creation and make them ensure the opportunity for basic services. Additionally, sector-specific particular enhancement of their basic services regarding education, health, and employment opportunities. This chapter also suggests that facilitated awareness programs, organized training, improved financial services, and non-interest loan management for climate vulnerable community. Finally, this chapter recommended that cross-border cooperation and data collection on community people migration trends, patterns, and decisions will enable more efficient and timely decision ensure that migration or intention to migrate becomes a choice rather than a necessity for climate vulnerable community of Bangladesh.

Conclusion

Bangladesh is particularly vulnerable to the effects of climate change, and if sea levels rise by 80 cm or more by 2100, up to 30 million people may need to relocate. Therefore, the study aims to identify factors that drive the migration of Char residents due to livelihood vulnerabilities caused by climate change hazards and anomalous climate fluctuations in Kulkandi Union, Islampur Upazila, Jamalpur District. The study surveyed 382 households using semi-structured questionnaires and conducted Spearman's rank correlation to examine the relationship between livelihood assets and willingness to migrate from Islampur Upazila.

The study found that 60% of respondents experienced floods 1–2 times a year, while 50% reported riverbank erosion occurring at the same frequency. It also indicates that 20% of respondents were illiterate, whereas 48%, 23%, and 7% had primary, secondary, and higher education levels, respectively. Only 18% had vocational skills, which could limit their options for alternative livelihoods. The study highlights how limited livelihood choices exacerbate vulnerability, leading to increased migration as a negative coping mechanism. Most migration is seasonal and short-term, with people moving to nearby urban centers, such as Mymensingh and Dhaka, in search of better livelihood opportunities.

The study's findings identify significant policy gaps for the study site's community. The qualitative findings suggest that the government and other relevant stakeholders should take immediate action to alleviate people's suffering. The government and NGOs working in the field of climate change should focus on mapping existing resource and structural vulnerabilities. The government should identify the core challenges faced by respondents, such as hospitals, educational institutions, institutional financing, and agricultural incentives, to encourage people to engage in agriculture. This study also suggests that a comprehensive plan that integrates capital-based sustainable livelihoods, including access to natural resources, public services, and skilled labor, and shifts traditional farm-based activities toward diversified employment opportunities. On the other hand, humanitarian support from NGOs and awareness of positive adaptation strategies, such as training on climate change and its impacts, can help people enhance their capacity and reduce forced migration. A comprehensive disaster preparedness plan should be developed to prompt a response to climate change. Finally, the study helps guide the government in recognizing the significant challenges people face with asset ownership, including limited access to institutional financing and other vital services such as healthcare, education, and basic facilities, all of which are essential for developing a climate-resilient society and meeting the SDG 13 (Climate Action) goal.

Limitations of the Study

The study examined the migration patterns of respondents in response to climatic stress and shocks, as well as to livelihood vulnerabilities. However, it did not discover other factors that push respondents to migrate, such as health, psychological well-being, and water-related problems. Further studies should be conducted to evaluate the effectiveness of locally adaptive strategies and the support of the government and other NGOs in eliminating livelihood vulnerability across a larger study area and with a broader sample size.

Acknowledgements The authors would like to express their sincere gratitude to the study participants from the Char area of Kulkandi Union in Islampur Upazila, who generously gave their time for this study. Finally, the authors are indebted to the students from the Department of Geography and Environment at Jagannath University, Dhaka, who assisted with data collection.

Authors' Contribution All authors have contributed equally to this work.

Funding The research was not supported by any funding.

Potential Conflicts of Interest: The corresponding author, on behalf of all authors, states that there are no conflicts of interest.

Informed Consent: Informed consent was obtained from all individual participants included in the study.

Data Availability Statement The datasets used and/or analyzed during the current study are available from the corresponding author on reasonable request.

AI Disclosure Statement The authors confirm that the AI tools contributed zero percent to the core research content, data analysis, and methodology. They were only used to assist in improving the presentation of the manuscript.

Appendix: Household Questionnaire

Assessing the Climate Change-Induced Migration Pattern of Jamalpur, Bangladesh

We are conducting a study to assess the climate change-induced migration patterns in Jamalpur, Bangladesh. By completing this questionnaire, you are consenting to participate in this research. Your responses will be kept confidential and used solely for academic purposes. Participation is voluntary, and you may withdraw at any time without consequence. Your insights will contribute to a better understanding of how climate change affects migration patterns and help inform future policies.

Section 1: Socio-demographic Information of the Respondents

1. What is your age group? ☐ 18–30 ☐ 31–40 ☐ 41–50 ☐ 51–60 ☐ Above 60
2. What is your gender? ☐ Male ☐ Female
3. What is your highest level of education? ☐ Illiterate ☐ Primary school ☐ SSC ☐ HSC ☐ Graduate
4. What is your occupation? ☐ Unemployed ☐ Government Services ☐ Private/ NGO Job ☐ Farmer (Land owner) ☐ Farmer (Leaser) ☐ Day labor ☐ Business

Section 2: Current Status of Climatic Stressor

1. How often do you experience flooding in your area per year? ☐ None ☐ 1–2 times ☐ More than 2 times
2. How often does riverbank erosion occur in your area per year? ☐ None ☐ 1–2 times ☐ More than 2 times
3. Have you observed changes in rainfall patterns in your area? ☐ Yes ☐ No ☐ Not sure
4. How often do you experience extreme weather events (e.g., storms, cyclones) in your area? ☐ Never ☐ Rarely ☐ Sometimes ☐ Often ☐ Very Often
5. To what extent do you agree with the statement: Climate change is having an impact on my community? ☐ Strongly Disagree ☐ Disagree ☐ Neutral ☐ Agree ☐ Strongly Agree

Section 3: Livelihood Assets Status of the Respondent

1. How much land do you own? < ☐ 0.5 acre ☐ 0.5–1 acre > ☐ 1 acre
2. Do you have access to water sources? ☐ Yes ☐ No
3. What is your highest level of educational accomplishment? ☐ No education ☐ Primary ☐ Secondary ☐ Higher
4. Do you have vocational skills? ☐ Yes ☐ No
5. How would you describe your housing conditions? ☐ Poor ☐ Fair ☐ Good ☐ Very Good
6. How far is the nearest infrastructure (e.g., road, market) from your location? < ☐ 1 km ☐ 1–3 km > ☐ 3 km
7. What is your average monthly income? < ☐ 5000 ☐ 5 k–10 k > ☐ 10 k

8. Do you have access to credit? ☐ Yes ☐ No
9. Are you part of a community network? ☐ Yes ☐ No
10. How often do you receive social support? ☐ Never ☐ Rarely ☐ Sometimes ☐ Often ☐ Always

Section 4: Vulnerability Level of the Kulkandi Union

1. How far is your location from the mainland? < ☐ 5 km ☐ 5–10 km > ☐ 10 km
2. What is the availability of public services in your area? ☐ Yes ☐ Limited ☐ No
3. Have you experienced any land tenure disputes? ☐ Yes ☐ No
4. How many times have you experienced displacement in the last 10 years? ☐ None ☐ 1–2 times ☐ More than 2 times
5. How often do you receive early warnings for natural disasters? ☐ Always ☐ Sometimes ☐ Rarely ☐ Never
6. To what extent do you agree with the statement: "Institutional negligence has affected my community"? ☐ Strongly Disagree ☐ Disagree ☐ Neutral ☐ Agree ☐ Strongly Agree

Section 5: Respondents' Migration Decisions Pattern

1. How many household members have migrated? ☐ None ☐ 1 ☐ 2 or more
2. What are the main reasons for migration in your household? ☐ Climate ☐ Economic ☐ Social
3. What is the duration of migration for household members? ☐ Seasonal ☐ 6 months–1 year ☐ More than 1 year ☐ Permanent
4. What is the primary migration destination for your household members? ☐ Rural ☐ Urban ☐ Abroad
5. Have there been economic outcomes due to migration in your household? ☐ Yes ☐ No ☐ Not sure
6. Do you view migration as a strategy or a survival mechanism? ☐ Survival ☐ Strategy ☐ Both

Thank You

References

Abbass K, Qasim MZ, Song H, Murshed M, Mahmood H, Younis I (2022) A review of the global climate change impacts, adaptation, and sustainable mitigation measures. Environ Sci Pollut Res 29:42539–42559. https://doi.org/10.1007/s11356-022-19718-6

Akther S, Shakira US, Bhuiyan MH (2021) Riverbank erosion trend analysis and its impact on socio-economic condition of the inhabitants of Islampur Upazila in Jamalpur District JnU. J Life Earth Sci 6

Alam GMM (2017) Livelihood cycle and vulnerability of rural households to climate change and hazards in Bangladesh. Environ Manage 59:777–791

Alam E, Momtaz S, Bhuiyan HU, Baby SN (2019) Climate change impacts on the coastal zones of Bangladesh: Perspectives on tropical cyclones, sea level rise, and social vulnerability. Bangladesh I: climate change impacts, mitigation and adaptation in developing countries, Springer, pp 145–166

Ali A, Akhtar R, Hussain J (2023) Unveiling high mountain communities' perception of climate change impact on lives and livelihoods in Gilgit-Baltistan: evidence from people-centric approach. Environ Commun 17:602–617. https://doi.org/10.1080/17524032.2023.2229044

Anderson RM, Heesterbeek H, Klinkenberg D, Hollingsworth TD (2020) How will country-based mitigation measures influence the course of the covid-19 epidemic? The Lancet 395:931–934

Assraf Seddiky M, Ara E, Karim A (2024) Climate change-induced hazard risks and migration in Bangladesh: a case study. Curr Res Environ Sustain 7:100253. https://doi.org/10.1016/j.crsust.2024.100253

Azam G, Huda ME, Bhuiyan MAH, Mohinuzzaman M, Bodrud-Doza M, Islam SMD-U (2021) Climate change and natural hazards vulnerability of char land (bar land) communities of Bangladesh: application of the livelihood vulnerability index (LVI). Glob Soc Welfare 8:93–105

Barua P, Rahman SH, Molla MH (2019) Impact of river erosion on livelihood and coping strategies of displaced people in south-eastern Bangladesh. Int J Migr Residential Mobility 2:34–55. https://doi.org/10.1504/IJMRM.2019.103275

BBS (2022) Statistical year book. Bangladesh Bureau of Statistics (BBS), Dhaka, Bangladesh. https://bbs.Gov.Bd/site/page/29855dc1-f2b4-4dc0-9073-f692361112da/statistical-yearbook

Bernzen A, Jenkins JC, Braun B (2019) Climate change-induced migration in coastal Bangladesh? A critical assessment of migration drivers in rural households under economic and environmental stress. Geosciences, p 51

BMD (2024) Climate report-2024. Bangladesh Meteorological Department, Dhaka, Bangladesh

Chowdhury T, Rahman MA, Khan MA, Akter T (2022) Livelihood assets and food consumption status of riverbank erosion hazard people in a selected area of Bangladesh. Arch Agric Environ Sci 7:70–79

Das LC, Mohiul Islam ASM, Aktar S, Sultana F (2024) Climate change challenges in Bangladesh. Int J Big Data Min Glob Warming 06:2430001. https://doi.org/10.1142/S263053482430001X

Fernández S, Arce G, García-Alaminos Á, Cazcarro I, Arto I (2024) Climate change as a veiled driver of migration in Bangladesh and Ghana. Sci Total Environ 922:171210

Ferris E (2020) Research on climate change and migration where are we and where are we going? Migr Stud 8:612–625. https://doi.org/10.1093/migration/mnaa028

Gupta D, Kumar P, Okano N, Sharma M (2025) Climate-induced migration in India and Bangladesh: a systematic review of drivers, impacts, and adaptation mechanisms. Climate, p 81

Habib MA, Alam MR, Ghosh S, Rahman T, Reza S, Mamun S (2021) Impact of knowledge, attitude, and practice on iodized salt consumption at the household level in selected coastal regions of Bangladesh. Heliyon 7

Han Q, Kumar R, Kumar A (2024) Climate change and human migration: perspectives for environmentally sustainable societies. J Geochem Explor 256:107352

Hossain MM (2020) Influential factors of tourists' satisfaction in Bangladesh: evidence from selected tourism destinations at Cumilla region. Can J Bus Inf Stud 2:75–86

Hossain B, Shi G, Ajiang C, Sarker MNI, Sohel MS, Sun Z, Yang Q (2022) Climate change induced human displacement in Bangladesh: implications on the livelihood of displaced riverine island dwellers and their adaptation strategies. Front Psychol 13:964648. https://doi.org/10.3389/fpsyg.2022.964648

Ibrahim B, Mensah H (2022) Rethinking climate migration in sub-Saharan Africa from the perspective of tripartite drivers of climate change. SN Soc Sci 2:87. https://doi.org/10.1007/s43545-022-00383-y

Islam MR (2018) Climate change, natural disasters and socioeconomic livelihood vulnerabilities: migration decision among the char land people in Bangladesh. Soc Indic Res 136:575–593. https://doi.org/10.1007/s11205-017-1563-y

Kartiki K (2011) Climate change and migration: a case study from rural Bangladesh. Gend Dev 19:23–38. https://doi.org/10.1080/13552074.2011.554017

Khatun MR, Amin MR, Pervez AKMK, Shaili SJ, Ahmmed MB, Mahedi M (2025) A global scientometric analysis of climate-smart agricultural innovations adoption in drought-prone areas. Asian Res J Agric 18:159–176. https://doi.org/10.9734/arja/2025/v18i3724

Maharjan A, de Campos RS, Singh C, Das S, Srinivas A, Bhuiyan MRA, Ishaq S, Umar MA, Dilshad T, Shrestha K (2020) Migration and household adaptation in climate-sensitive hotspots in south Asia. Curr Clim Change Rep 6:1–16. https://doi.org/10.1007/s40641-020-00153-z

Mohiuddin M (2022) Impact of riverbank erosion on livelihood of char land people in Jamalpur District of Bangladesh. Jagannath Univ J Life Earth Sci 7:95–108

Mohiuddin M, Nowreen S (2023) Climate change-induced livelihood vulnerabilities in south western coastal region of Bangladesh. Jagannath Univ J Life Earth Sci 9:163–178

Mondal A, Barui T and Paul S (2024) Distress in ecological drivers and its impact on homelessness: case study of Sagar Island. In: Homelessness to hope. Elsevier, pp 221–235

Morshed G, Tortajada C, Hossain MS (2025) The state of climate change adaptation research in Bangladesh: a systematic literature review. Mitig Adapt Strat Glob Change 30:31. https://doi.org/10.1007/s11027-025-10219-8

Mullick MRA, Islam KA, Tanim AH (2020) Shoreline change assessment using geospatial tools: a study on the Ganges deltaic coast of Bangladesh. Earth Sci Inf 13:299–316. https://doi.org/10.1007/s12145-019-00423-x

Neef K, Jones E, Marlowe J (2023) The conflict, climate change, and displacement nexus revisited: the protracted Rohingya refugee crisis in Bangladesh. J Peacebuilding Dev 18:231–247. https://doi.org/10.1177/15423166231190040

Ninawe AS, Indulkar ST, Amin A (2018) Impact of climate change on fisheries. In: Biotechnology for sustainable agriculture. Elsevier, pp 257–280

Paul BK, Rahman MK, Crawford T, Curtis S, Miah MG, Islam R, Islam MS (2021) Coping strategies of people displaced by riverbank erosion in the lower Meghna estuary. In: Zaman M, Alam M (eds) Living on the edge: char dwellers in Bangladesh. Springer International Publishing, Cham, pp 227–239

Rahman MS, Gain A (2020) Adaptation to river bank erosion induced displacement in Koyra Upazila of Bangladesh. Prog Disaster Sci 5:100055. https://doi.org/10.1016/j.pdisas.2019.100055

Ranjan A (2023) Climate change, environmental migration and population displacement. In: Environment, climate change and migration in south Asia, Routledge India, pp 12–30

Riaz R (2024) Building responsiveness through community-based adaptation to climate change: lesson learned from coastal zone of Bangladesh. In: Shaikh MAA (ed) Bangladesh Council of Scientific and Industrial Research (BCSIR) Congress-2023. BCSIR, Dhaka, Bangladesh, pp 367–371

Riaz R, Mohiuddin M (2025) Application of GIS-based multi-criteria decision analysis of hydro-geomorphological factors for flash flood susceptibility mapping in Bangladesh. Water Cycle 6:13–27. https://doi.org/10.1016/j.watcyc.2024.09.003

Riaz R, Nasher N (2023) Urban expansion and land use changes of a fringe area in the southern part of Dhaka city (Bangladesh) using geo-spatial techniques. J Indian Geophys Union 27:462–465

Riaz R, Sultana N (2025) Location-specific heterogeneity in landslide resilience index: a household-based comparative study in Rangamati Hill District, Bangladesh. Ecol Ind 171:113154. https://doi.org/10.1016/j.ecolind.2025.113154

Riaz R, Safi SI, Saba Z, Kabir I (2025) Developing community resilience index for flood (CRIF): evidence from extreme flood-prone Jamalpur District of Bangladesh. World Dev Perspect 40:100752. https://doi.org/10.1016/j.wdp.2025.100752

Riaz R (2023) Assessment of riverbank erosion-deposition trend using geospatial techniques in the Padma river, Bangladesh. In: Islam AS (ed) 9th international conference on water and flood management-ICWFM 2023, Bangladesh University of Engineering and Technology (BUET), Dhaka, Bangladesh

Sajjad H, Bhuyan N, Rahaman MH, Ahmed R (2024) Vulnerability to riverbank erosion in India and Bangladesh: a review for future research framework. Authorea Preprints. https://doi.org/10.22541/au.170670920.07550703/v1

Shamim AS, Islam S, Podder T, Runa MSS (2025) Poor livelihood assets and adaptive strategies of the riverbank erosion induced charland people in Bangladesh: a study on the Teesta riverine ecosystem 4

Sultana N, Riaz R (2025) Knowledge, attitude, and practices (KAP) towards flash floods in Sunamganj, Bangladesh. Nat Hazards 121:20421–20452. https://doi.org/10.1007/s11069-025-07647-9

Sultana N, Hossain A, Hossen MF, Riaz R (2025) Landslide susceptibility mapping using frequency ratio (FR) and random forest (RF) model in the Bandarban Hill District, Bangladesh. Phys Geogr 46:467–515. https://doi.org/10.1080/02723646.2025.2548705

Chapter 17
Adapting to Climate Extremes: Policy Strategies for Protecting Public Health

S. Gayathri

Abstract Climate change has intensified the frequency and severity of extreme weather events, posing escalating risks to public health worldwide. These risks are unevenly distributed, disproportionately affecting vulnerable populations and regions with limited adaptive capacity. While the health impacts of climate extremes are well documented, less attention has been paid to how policy strategies address these challenges across different contexts. This study aims to systematically review global and regional policy strategies designed to protect public health from climate extremes, with a particular focus on effectiveness, equity considerations, and governance contexts. A systematic review was conducted following PRISMA guidelines, covering peer-reviewed literature published between 1975 and 2025. Multiple databases were searched, and eligible studies were screened and synthesized using a thematic policy analysis framework. The review identifies several dominant categories of climate–health policy interventions, including early warning systems, urban planning measures, and health system adaptations. The evidence base is geographically uneven, with a strong concentration in high-income regions and limited policy evaluation in low- and middle-income countries. Equity considerations and anticipatory planning remain inconsistently integrated across policy approaches. This review highlights the need for integrated, equity-oriented, and anticipatory climate–health policies. Strengthening institutional coordination and expanding context-specific research are essential for advancing sustainable public health adaptation to climate extremes.

Keywords Climatic extremes · Health impacts · Policy innovation · Vulnerable populations · Climate adaptation

---

S. Gayathri (✉)
Department of Transport Planning, School of Planning and Architecture, Bhopal, India
e-mail: 2024dr010@spab.ac.in

P. Ghosh et al. (eds.), *Climate Crisis and Changing Population Dynamics*, Springer Climate, https://doi.org/10.1007/978-3-032-24533-5_17

Introduction

Background

The escalating frequency and intensity of climatic extremes, such as heatwaves, floods, droughts, and cyclones, pose profound and unequal threats to global public health (Legg et al. 2025; Patwary et al. 2024; Kafi and Ponrahono 2025; Haubrock 2024). Global land-surface temperatures have risen significantly, with 2023 marking the warmest year since 1850, leading to more prevalent and severe extreme events (IPCC 2023; De Luca and Donat 2023). These phenomena directly cause injury, illness, and death, while indirectly affecting health through damaged infrastructure, water and food contamination, infectious disease spread, and severe mental health impacts like anxiety, depression, and post-traumatic stress disorder (PTSD) (Bell JE 2018; Jiménez et al. 2025; Meadows et al. 2024; Weeda et al. 2024; Ejiohuo et al. 2025; Akram and Mushtaq 2024; Li et al. 2021). The global burden of disease attributable to climate change is substantial, with projections indicating 250,000 additional deaths annually by 2030 and direct healthcare costs between USD 2–4 billion per year (Schmitt et al. 2016).

This crisis disproportionately affects vulnerable populations, including children, the elderly, women, low-income individuals, and those with pre-existing conditions, often living in regions with limited adaptive capacity, such as the Global South (Benevolenza and DeRigne 2019; Houser et al. 2022; Teebken 2024; Weeda et al. 2024; Ejiohuo et al. 2025). For instance, children bear a significant portion of the climate-change health burden (Weeda et al. 2024), and higher temperatures are linked to increased childhood anemia in low- and middle-income countries (Zhang and Wang 2023). The intensification of compound extreme events, like heatwaves and droughts, further amplifies these threats (De Luca and Donat 2023; Ghosh et al. 2025; Li et al. 2025), necessitating urgent, comprehensive policy strategies for public health protection (Legg et al. 2025; Weeda et al. 2024; Ejiohuo et al. 2025).

Despite growing recognition that climate extremes pose serious risks to population health, the field remains characterized by fragmented evidence and sector-specific responses. Public health research has largely focused on documenting health impacts, while climate adaptation policy has tended to prioritize hazard management and infrastructure resilience, often without systematic attention to health equity or differential vulnerability. As a result, policy responses frequently operate in silos, limiting their effectiveness in addressing the compound, socially mediated nature of climate-related health risks. This fragmentation represents a critical research problem, particularly in the context of escalating climate extremes and widening health inequities.

The central research problem addressed in this chapter lies not in the lack of evidence that climate extremes adversely affect health, but in the limited synthesis of how policy interventions respond to these impacts across different social, demographic, and institutional contexts. Existing studies often examine individual hazards, population groups, or policy tools in isolation, providing limited insight into their comparative effectiveness, scalability, and equity implications. This gap is particularly pronounced in low- and middle-income countries, where exposure to climate extremes is high but adaptive capacity and institutional resources remain constrained.

Critically, these threats disproportionately burden vulnerable populations globally (Benevolenza and DeRigne 2019; Weeda et al. 2024). Children bear a significant portion of the climate-change health burden, facing elevated risks of preterm birth, respiratory diseases, and anemia (Weeda et al. 2024; Zhang and Wang 2023). The elderly are highly susceptible due to physiological vulnerabilities and pre-existing conditions (Bell JE 2018; Meadows et al. 2024). Women, low-income individuals, racial/ethnic minorities, and residents of informal settlements or rural areas experience higher exposure, possess limited adaptive capacity, and are often overlooked in policy, particularly in the Global South (Tian et al. 2024; Escorcia Hernández et al. 2023; Dewi et al. 2024; Teebken 2024). This pervasive inequity exacerbates existing health disparities and highlights a critical gap in current intervention strategies. To address this gap, this chapter adopts a systematic review and bibliometric approach to examine the evolution, scope, and effectiveness of climate–health policy strategies between 1975 and 2025, with explicit attention to equity and vulnerability (Table 17.1).

Research Gap

Despite growing recognition of climate extremes' profound health impacts, a significant research gap persists in comprehensively reviewing innovative policy interventions that explicitly bridge public health, environmental governance, and social policy (Ejiohuo et al. 2025; Patwary et al. 2024). Current literature often employs narrow definitions of adaptation, overlooking multi-sectoral solutions and the intricate interplay of social, economic, and environmental factors crucial for equitable and effective strategies (Paudel et al. 2025). There is a notable absence of integrated, transdisciplinary analysis, particularly on the efficacy, scalability, and context-specific implementation of such interventions, especially in highly vulnerable, resource-limited settings (Alcayna et al. 2022; Ejiohuo et al. 2025; Patwary et al. 2024). Accordingly, this chapter is framed as a systematic review that focuses specifically on policy strategies for protecting public health from climate extremes, with particular attention to their effectiveness, equity implications, and relevance across global and regional contexts.

Aims and Objectives

The primary aim of this chapter is to critically review and analyze climate-health strategies implemented between 1975 and 2025. It seeks to evaluate the effectiveness of policy tools such as early warning systems, urban greening, and climate-resilient

Table 17.1 Overview of major health impacts associated with different climate extremes, based on existing literature

Climate extreme	Key socio-environmental mediators	Physical health outcomes	Mental health outcomes	Vulnerable populations
Heatwaves and extreme heat	Urban heat island effect, air pollution, damaged infrastructure	Heat-related illnesses (heat stroke, heat exhaustion), exacerbation of chronic conditions (cardiovascular, respiratory, renal), increased mortality, decreased physical activity	Anxiety, stress, depression, aggression, and irritability, decreased mental health scores in older adults, worsening of pre-existing mood disorders	Elderly (especially those over 75), children, individuals with pre-existing chronic or mental illnesses, low-income populations, urban informal settlement residents
Floods and extreme precipitation	Water contamination, stagnant water, damaged infrastructure, social disruption	Drowning and physical injuries, waterborne diseases (cholera, diarrheal diseases), vector-borne diseases (dengue fever) due to new breeding grounds	Post-traumatic stress disorder (PTSD), anxiety and depression, psychological and psychopathological distress, chronic insecurity	Low-income communities and those in informal urban and rural settlements, women, rural communities with limited healthcare access
Droughts	Water scarcity, crop failure and food insecurity, dust storms	Malnutrition and stunting in children, infectious diseases from poor sanitation, undernutrition, increased childhood anemia	Financial and psychological stress for farmers, despair, depression, and increased suicide rates	Rural communities, farmers, children, low-income and marginalized populations

(continued)

Table 17.1 (continued)

Climate extreme	Key socio-environmental mediators	Physical health outcomes	Mental health outcomes	Vulnerable populations
Wildfires	Air pollution	Respiratory illnesses (asthma, bronchitis), burns and injuries	Stress, anxiety, and PTSD from loss of home and livelihood	Individuals with pre-existing respiratory conditions, displaced communities, those with economic dependence on natural environments
Cyclones, hurricanes and severe storms	Damaged infrastructure, contaminated water supply	Traumatic injuries and death, waterborne diseases, interruption of healthcare services	PTSD and severe anxiety, grief over loss of life and property, displacement-related stress and helplessness	Coastal communities, low-income populations
Compound events (e.g., drought-heatwaves)	Cascading and synergistic risks	Amplified health threats, higher mortality rates, straining of human coping capacities	Compounding psychological trauma	Women, vulnerable communities with limited resources and adaptive capacity

Source Alcayna et al. (2022), Bell JE (2018), Benevolenza and DeRigne (2019)

healthcare in mitigating the health impacts of climate extremes. Special emphasis is placed on assessing the responsiveness of these strategies to demographic vulnerabilities, including age, gender, income, and migration status, and on examining their relevance in resource-constrained settings, particularly in the Global South. By synthesizing global case studies and empirical evidence, the chapter advances the argument for an anticipatory, equity-driven policy paradigm that prioritizes survival, dignity, and mental well-being. The overarching objective is to inform policymakers, researchers, and practitioners in designing inclusive, scalable, and health-protective adaptation strategies. This study is closely aligned with the Sustainable Development Goals, particularly SDG 3 (Good Health and Well-being), SDG 11 (Sustainable Cities and Communities), and SDG 13 (Climate Action), by examining how integrated climate–health policy strategies can reduce health risks, strengthen urban resilience, and support effective adaptation to climate extremes.

Paper Structure

This chapter is organized as follows. The Introduction outlines the research context, establishes the relevance of climate extremes for public health, reviews existing literature, identifies key research gaps, and states the objectives of the study. The Methodology section describes the systematic review approach, including data sources, screening procedures, and analytical framework. The Results section presents the findings of the systematic review, focusing on the distribution, characteristics, and thematic patterns of climate–health policy interventions. The Discussion interprets these findings in relation to existing studies, highlights key insights, and discusses policy implications, strengths, and limitations. Finally, the Conclusion summarizes the main contributions of the study, outlines broader implications, and identifies directions for future research.

This chapter offers a novel contribution by integrating global and regional evidence on climate extremes and public health within a unified, policy-oriented analytical framework. Unlike studies that focus on single hazards, locations, or population groups, it synthesizes five decades of international and regional research to examine how climate–health policy responses have evolved across diverse institutional and socio-economic contexts, with particular attention to low- and middle-income regions. By combining foundational literature with recent empirical and policy studies, the chapter situates global climate–health debates alongside region-specific vulnerabilities and governance challenges, enhancing its relevance for an international readership. The objectives of this chapter are therefore not only to review policy interventions, but to critically assess their equity implications, scalability, and alignment with broader development priorities. In doing so, the chapter directly engages with the focal theme of the book by examining how policy strategies can enhance resilience, adaptation, and public health protection in the face of escalating climate extremes—an emphasis that is carried forward into the results and discussion through comparative analysis, global–regional case illustrations, and synthesis of cross-cutting policy lessons.

Methodology

This study adopts a systematic review and bibliometric analysis to evaluate global climate–health policy strategies between 1975 and 2025. Peer-reviewed literature was retrieved from Scopus, Web of Science, and PubMed, complemented by policy documents from WHO, UNFCCC, and IPCC. Eligibility criteria prioritized studies addressing health impacts of climate extremes including heatwaves, floods, droughts, cyclones, and compound events, while accounting for demographic vulnerabilities such as age, gender, income, and migration (Fig. 17.1).

A PRISMA-based screening protocol was employed to ensure transparency and rigor. Bibliometric mapping identified the top ten most influential sources, visualized

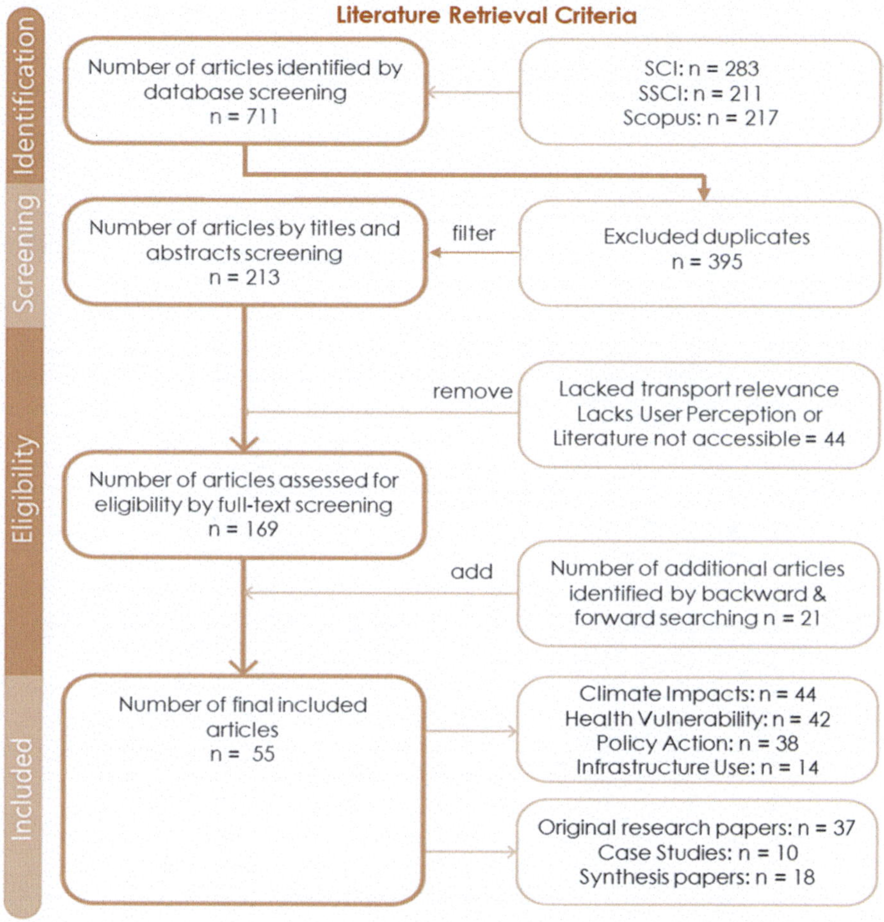

Fig. 17.1 Flow chart of literature identification and selection. Adopted from Page et al. (2021)

as a citation network graph, and thematic clustering generated a word cloud highlighting five dominant clusters: climate, policy, health, equity, and research. Data extraction and coding organized interventions into categories such as early warning systems, heat-health action plans, urban greening, climate-resilient healthcare, and social protection measures (Figs. 17.2 and 17.3).

Case studies from the Global South were emphasized to capture resource constraints and equity dimensions. Comparative thematic synthesis assessed policy effectiveness, scalability, and inclusiveness. Triangulation of empirical, bibliometric, and policy evidence ensured a robust, equity-centered evaluation of adaptation measures across temporal and spatial scales. In line with PRISMA guidelines, the review process involved a structured sequence of identification, screening, eligibility

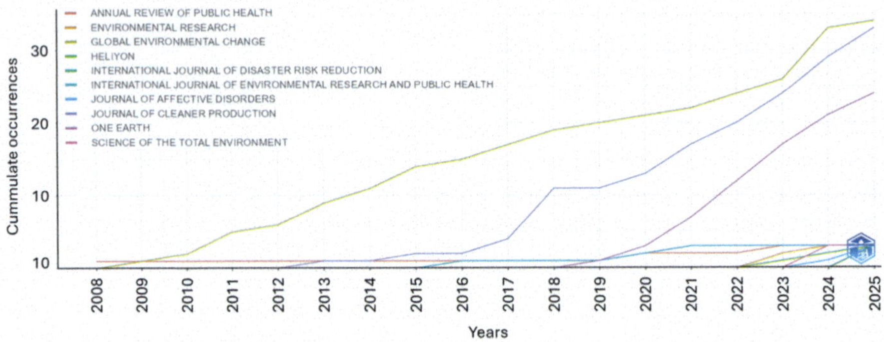

Fig. 17.2 Year wise contribution of top ten sources included in literature review. Generated using: Biblioshiny for R based on the selected literature

Fig. 17.3 Thematic classification of relevant keywords. Generated using: Biblioshiny for R based on the selected literature

assessment, and inclusion. The systematic review constituted the primary methodological approach, while bibliometric analysis was employed as a complementary tool to examine publication trends, influential sources, and thematic clustering within the selected literature. This combined approach enabled both transparent study selection and deeper analytical synthesis of climate–health policy strategies.

Results

The results presented in this section synthesize evidence from the selected studies to identify major categories of climate–health policy interventions, their geographic distribution, and recurring patterns in their design and application. The findings are organized to reflect the systematic review process and analytical framework described in the methodology.

Study Selection and Screening

The database search initially identified **711 records** across Scopus, Web of Science, and PubMed published between 1975 and 2025. After removing **316 duplicate records**, **395 articles** were screened based on titles and abstracts. Of these, **169 full-text articles** were assessed for eligibility. Following the application of inclusion and exclusion criteria, a final set of **55 studies** was retained for systematic review and synthesis. The detailed screening process is presented in the PRISMA flow diagram.

Characteristics of Included Studies

The included studies represent a broad temporal, geographic, and thematic spectrum. Most publications were produced after 2000, reflecting the growing scholarly and policy attention to climate–health interactions in recent decades. Geographically, the evidence base is dominated by studies from high-income regions, particularly Europe and North America, while contributions from low- and middle-income countries remain comparatively limited.

The reviewed studies address a range of climate extremes, including heatwaves, floods, droughts, cyclones, and compound events. Health outcomes examined span physical, mental, and social dimensions, with varying degrees of emphasis across regions. A summary of key study characteristics, including geographic focus, climate hazard type, policy orientation, and key outcomes is presented in Appendix.

Distribution of Climate–Health Policy Interventions

Across the reviewed literature, several recurring categories of climate–health policy interventions were identified. Early warning systems and heat–health action plans emerged as the most frequently reported strategies, particularly in urban and high-income settings. Urban planning and greening interventions were commonly discussed in relation to heat mitigation, while climate-resilient healthcare infrastructure featured prominently in studies focused on emergency preparedness and service continuity.

In contrast, studies from low- and middle-income regions more often highlighted community-based adaptation measures and social protection mechanisms. These interventions were frequently framed in response to resource constraints and informal settlement contexts, reflecting differing institutional capacities and governance structures across regions.

Thematic Synthesis of Policy Strategies

The thematic synthesis of the reviewed studies revealed several cross-cutting patterns in climate–health policy responses. First, policy strategies could be broadly distinguished between anticipatory interventions, which emphasize preparedness and risk reduction, and reactive interventions, which focus on emergency response and recovery. Anticipatory approaches were more commonly documented in high-income contexts, whereas reactive measures dominated in settings with limited forecasting capacity and institutional resources.

Second, differences were observed in the extent to which policies explicitly addressed social vulnerability. Some interventions incorporated targeted measures for high-risk populations, while others adopted universal approaches that assumed uniform exposure and adaptive capacity. These variations were evident across both geographic regions and policy domains.

Third, the role of governance and institutional coordination emerged as a recurring feature across the reviewed studies. Policies were implemented through a range of institutional arrangements, from centralized public health agencies to decentralized, community-led initiatives, highlighting diversity in governance models and implementation pathways.

Regional Patterns and Gaps in the Evidence Base

The synthesis also revealed clear regional imbalances in the existing evidence. Policy interventions documented in high-income regions were more likely to involve long-term planning, integrated governance frameworks, and formal evaluation mechanisms. In contrast, evidence from low- and middle-income countries was more fragmented, with fewer studies examining long-term policy effectiveness or scalability.

Notable gaps were identified in relation to mental health-focused interventions, compound climate events, and rural contexts. These gaps were particularly pronounced in regions experiencing high exposure to climate extremes but limited

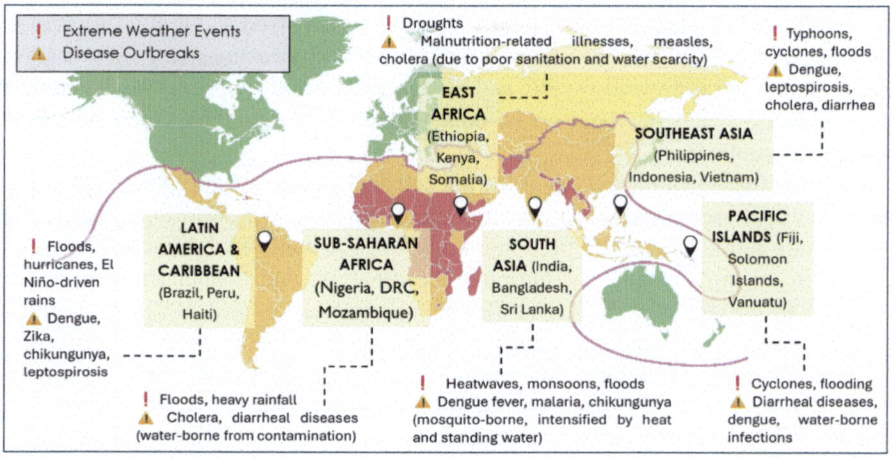

Fig. 17.4 Geographical distribution of extreme weather events and disease outbreaks. *Source* Multiple sources (Appendix)

adaptive capacity, indicating uneven coverage of climate–health policy research across global contexts.

Summary of Systematic Review Findings

Overall, the systematic review identifies a diverse but uneven landscape of climate–health policy responses. While a range of intervention types has been implemented across different regions, their distribution reflects broader disparities in institutional capacity, resources, and research attention. These findings provide the empirical foundation for the subsequent discussion, which examines the implications of these patterns for equity, policy effectiveness, and sustainable public health adaptation (Fig. 17.4).

Discussion

Key Findings

This chapter sets out to systematically examine climate–health policy strategies developed between 1975 and 2025, with particular attention to their effectiveness, equity implications, and relevance across global and regional contexts. By synthesizing evidence from peer-reviewed studies and policy documents, the review clarifies how public health adaptation to climate extremes has evolved and highlights persistent gaps in preparedness, governance, and vulnerability-responsive policy design. Rather than reiterating well-established evidence on health impacts, the discussion focuses on how policy interventions respond to these impacts and why their outcomes vary across contexts.

A central finding of this review is the uneven distribution of anticipatory and reactive policy approaches. Anticipatory strategies, including early warning systems and heat–health action plans, are predominantly documented in high-income regions where forecasting capacity, institutional coordination, and financial resources are relatively strong (Ebi et al. 2021; Bell JE 2018; Weilnhammer et al. 2021). In contrast, many low- and middle-income countries continue to rely on reactive responses that prioritize emergency relief and post-event recovery rather than long-term risk reduction (Alcayna et al. 2022; Patwary et al. 2024; Ejiohuo et al. 2025). Similar patterns have been reported in broader climate adaptation research, which links anticipatory planning to governance stability and institutional capacity (Lesnikowski et al. 2013; Bell L 2024). This imbalance is particularly concerning given evidence that heatwaves and compound extreme events are intensifying globally, with disproportionate impacts in regions least equipped for anticipatory adaptation (Wang et al. 2024; De Luca and Donat 2023; Li et al. 2025).

A second key finding concerns the limited integration of equity considerations within climate–health policy interventions. While some policies explicitly target vulnerable groups such as children, the elderly, and low-income populations, many interventions adopt universal approaches that assume homogeneous exposure and adaptive capacity (Bell JE 2018; Benevolenza and DeRigne 2019; Weeda et al. 2024). The literature consistently demonstrates that climate-related health risks are socially mediated, with vulnerability shaped by age, gender, income, health status, and place of residence (Zhang and Wang 2023; Tian et al. 2024; Escorcia Hernández et al. 2023; Dewi et al. 2024; Teebken 2024). The persistence of equity-blind policy design helps explain why climate extremes continue to produce disproportionate health burdens even where adaptation initiatives exist. This finding aligns with previous research calling for adaptation strategies that move beyond aggregate risk reduction to explicitly address social determinants of health (Ejiohuo et al. 2025; Paudel et al. 2025).

The review also reveals pronounced geographic imbalances in the climate–health evidence base. Most documented policy interventions originate from high-income regions, particularly Europe, North America, and Australia, while empirical evidence from low- and middle-income countries remains comparatively limited and fragmented (Alcayna et al. 2022; Patwary et al. 2024). This imbalance constrains the transferability of policy lessons and limits understanding of how interventions perform under conditions of high exposure and constrained adaptive capacity. Studies from Africa, Asia, and Latin America consistently highlight severe health impacts associated with heatwaves, floods, droughts, and cyclones, particularly among children and marginalized communities, yet policy evaluations from these regions remain sparse (Weeda et al. 2024; Zhang and Wang 2023; Tian et al. 2024). Expanding research coverage in these contexts is therefore essential for developing globally relevant and context-sensitive climate–health policies.

Across regions, institutional coordination and governance capacity emerge as critical determinants of policy effectiveness. Integrated approaches that link public health systems with urban planning, disaster risk reduction, and climate governance frameworks tend to demonstrate greater coherence and continuity than fragmented, sector-specific interventions (Ebi et al. 2021; Hess et al. 2022). Conversely, the review highlights a persistent "adaptation deficit," where growing volumes of climate and health research do not translate into implemented or evaluated policies (Berrang-Ford et al. 2011; Bell L 2024). Many policies that contribute to health resilience are not explicitly framed as climate adaptation, resulting in under-reporting and limited policy learning across contexts (Lesnikowski et al. 2013). This disconnect reinforces the view that climate–health adaptation is not only a technical challenge but also an institutional and political one requiring sustained coordination and accountability.

Policy Implications and Recommendations

The findings of this review point to several policy-relevant implications for strengthening public health protection under escalating climate extremes. First, the persistent dominance of reactive responses highlights the need to prioritize anticipatory adaptation strategies that reduce health risks before extreme events occur. Evidence from high-income regions indicates that early warning systems, heat–health action plans, and preparedness frameworks can significantly reduce morbidity and mortality when supported by effective risk communication and institutional coordination (Ebi et al. 2021; Bell JE 2018; Weilnhammer et al. 2021). Scaling such approaches in low- and middle-income countries requires investment not only in forecasting infrastructure, but also in governance capacity and community-level engagement (Alcayna et al. 2022; Patwary et al. 2024).

Second, the review underscores the necessity of embedding equity as a core principle in climate–health policy design. Universal interventions that fail to account for differential exposure and vulnerability risk reinforcing existing health disparities. Policies must therefore explicitly target populations most at risk, including children, the elderly, women, low-income households, and residents of informal settlements (Benevolenza and DeRigne 2019; Weeda et al. 2024; Tian et al. 2024). Integrating social protection measures with public health and climate adaptation strategies can enhance both effectiveness and fairness, particularly in contexts characterized by high poverty and limited access to healthcare (Ejiohuo et al. 2025; Paudel et al. 2025).

Third, the review highlights the importance of cross-sectoral integration for achieving sustainable adaptation outcomes. Climate–health risks are shaped by interactions between urban design, healthcare systems, environmental management, and social policy. Interventions such as urban greening, climate-resilient healthcare infrastructure, and integrated water management offer co-benefits across multiple domains, but their success depends on coordination across institutional boundaries (Ebi et al. 2021; Hess et al. 2022; Hernández and Martínez 2023). Fragmented governance structures continue to limit scalability and long-term impact, particularly in rapidly urbanizing regions and resource-constrained settings (Dewi et al. 2024; Nhamo et al. 2025).

Finally, the geographic imbalances identified in this review point to the need for stronger international cooperation and knowledge exchange. While lessons from high-income contexts provide valuable insights, they cannot be directly transferred to regions with different socio-economic conditions and governance capacities. At the same time, locally grounded adaptation practices emerging in Africa, Asia, and Latin America remain under-represented in the global evidence base (Alcayna et al. 2022; Weeda et al. 2024). Supporting context-specific research, participatory policy design, and South–South learning is therefore essential for advancing globally relevant climate–health adaptation strategies.

Strengths and Limitations

This study has several limitations that should be acknowledged. As a systematic review, the findings are inherently constrained by the availability, quality, and geographic distribution of published literature. The evidence base remains heavily skewed toward high-income regions, limiting the ability to draw robust conclusions about policy effectiveness in low- and middle-income countries where climate-related health risks are often greatest (Alcayna et al. 2022; Patwary et al. 2024). In addition, variations in study design, definitions of adaptation, and reporting standards complicate direct comparison across contexts. The reliance on secondary sources also restricts the assessment of long-term policy effectiveness and implementation outcomes beyond those reported in the literature.

Despite these limitations, the study offers several important strengths. It provides a comprehensive, long-term synthesis of climate–health policy strategies spanning five decades, using a transparent and systematic review approach grounded in PRISMA principles (Page et al. 2021). By integrating bibliometric analysis with thematic policy synthesis, the chapter bridges fragmented bodies of literature across public health, climate adaptation, and social policy. The explicit focus on equity, vulnerability, and governance enhances the relevance of the findings for both research and practice, contributing to ongoing debates on how to design inclusive, scalable, and sustainable public health adaptation strategies in the face of accelerating climate change (Ejiohuo et al. 2025; Paudel et al. 2025).

Building on these findings and their implications, the following conclusion reflects on how this review advances current understanding of climate–health adaptation and outlines priorities for future research and policy development.

Conclusion

This systematic review synthesized evidence from 55 studies to examine policy strategies aimed at protecting public health from climate extremes across global and regional contexts. By shifting the focus from well-established health impacts to policy responses, the study advances current understanding of how public health adaptation to climate change is being conceptualized and implemented. The findings reveal a diverse but uneven landscape of climate–health policies, with anticipatory and equity-oriented strategies more prevalent in high-income regions, while reactive approaches continue to dominate in many low- and middle-income countries.

The review underscores that effective protection of public health under climate extremes requires integrated, forward-looking, and equity-sensitive policy frameworks. Strengthening institutional coordination across health, climate, and development sectors emerges as a critical determinant of sustainable adaptation outcomes. In this sense, the study contributes to global climate–health scholarship by highlighting not only what types of policies exist, but also where significant gaps persist in preparedness, governance, and inclusivity.

Future research should prioritize context-specific evaluations of policy effectiveness, particularly in regions facing high climate exposure and limited adaptive capacity. Expanding empirical evidence from low- and middle-income countries, alongside longitudinal assessments of policy implementation, will be essential for translating adaptation commitments into measurable health benefits. As climate extremes intensify worldwide, the ability to design and implement inclusive, anticipatory public health policies will increasingly shape global resilience, making climate–health adaptation not a peripheral concern, but a central pillar of sustainable development.

Acknowledgements **Potential Conflicts of Interest** The corresponding author states that there is no conflict of interest.

Data Availability Statement The datasets during and/or analyzed during the current study available from the corresponding author on reasonable request.

AI Disclosure Statement Artificial intelligence tools were used to assist with language editing and clarity. The authors are responsible for the content of the manuscript.

Funding The research was not supported by any funding.

Appendix: Summary of Studies Included in the Systematic Review ($n = 55$)

Author	Year	Region	Climate extreme	Policy type	Key outcome
Ellis	1976	Global	Heatwaves	Clinical guidance and pharmaceutical policy	Established that certain medications interfere with the body's thermoregulatory system, significantly increasing mortality risk during heat events
Steadman	1979	Global	High temperature and humidity (sultriness)	Public health risk assessment	Developed the heat index (apparent temperature) based on human physiological responses and clothing science to assess heat stress risks
Slovic	1987	Global	General environmental hazards	Risk communication and public policy	Analyzed the divergence between scientific and layperson perceptions of risk, which is foundational for developing effective hazard response policies
Von Braun and Webb	1989	West Africa	General climate stress and crop failure	Agricultural technology and labor policy	Demonstrated that the introduction of new crop technologies in climate-stressed settings significantly alters the agricultural division of labor
Grosjean	1991	United States (California)	Atmospheric pollution (peaks)	Air quality monitoring and environmental policy	Established baseline levels for formaldehyde and organic acids in southern California to understand how climate factors influence urban air chemistry

(continued)

(continued)

Author	Year	Region	Climate extreme	Policy type	Key outcome
McKee et al.	1993	Global (framework)	Drought	Drought monitoring and standardized assessment	Developed the standardized precipitation index (SPI), which provides a universal methodology to quantify drought across different time scales
Semenza et al.	1996	United States (Chicago)	1995 heatwave	Emergency response and preventative public health	Identified that social isolation and lack of access to air conditioning were primary drivers of the 700+ deaths during the July 1995 event
Gaffen and Ross	1998	United States	Summertime heat stress	Public health preparedness and climate monitoring	Documented a significant increase in heat stress frequency across the US over several decades
Checkley et al.	2000	Peru	El Niño events	Public health surveillance and gastrointestinal health	Established a direct link between El Niño-induced temperature rises and a substantial increase in pediatric diarrheal admissions
Naughton et al.	2002	United States (Chicago)	1999 heatwave	Emergency response and psychiatric risk monitoring	Identified psychiatric illness (OR 5.7) and social isolation as primary risk factors for heat-related mortality
Cutter et al.	2003	United States	General environmental hazards	Social policy and hazard management	Developed the social vulnerability index (SoVI) to quantify human sensitivity to hazards based on socioeconomic factors

(continued)

(continued)

Author	Year	Region	Climate extreme	Policy type	Key outcome
Meehl and Tebaldi	2004	Global (focused on US/Europe)	Heat waves	Long-term climate policy and mitigation	Projected that heat waves would become significantly more intense, frequent, and longer-lasting throughout the twenty-first century
Few et al.	2004	International (general review)	Floods	Strategic adaptation and health planning	Highlighted the severe post-flood health burden and called for integrated climate-health adaptation strategies
Ahern et al.	2005	Global	Flooding	Epidemiological risk management	Conducted a global review of epidemiologic evidence, linking floods to diverse physical and mental health impacts
McMichael et al.	2006	Global	General climate change risks	International health governance	Provided a landmark synthesis of global health risks, emphasizing that adaptation must address existing health inequities
Smit and Wandel	2006	Global	General climate variability	Adaptation framework development	Refined the scientific understanding of adaptive capacity, providing a foundation for local and national adaptation planning
Gagnon-Lebrun and Agrawala	2007	OECD nations	Climate change impacts	National adaptation policy implementation	Assessed progress in developed nations, noting that while planning was increasing, actual implementation remained limited

(continued)

(continued)

Author	Year	Region	Climate extreme	Policy type	Key outcome
Pitt	2008	United Kingdom	2007 floods	Disaster risk reduction and infrastructure reform	Issued a comprehensive review (the Pitt review) that led to major reforms in UK household insurance and flood prevention policy
Hahn et al.	2009	Mozambique	Floods, droughts, and cyclones	Development planning and risk assessment	Developed the livelihood vulnerability index (LVI) to help planners quantify community-level vulnerability and better target health and development interventions
Johnson and Krishnamurthy	2010	Global	Sudden-onset shocks (floods) and rainfall shortages	Social protection (safety nets, cash transfers)	Argues that social protection mechanisms are essential to facilitate long-term adaptation by reducing the immediate and delayed financial burdens of climate-induced displacement
Vicente-serrano et al.	2010	Global	Drought	Drought monitoring and climatological assessment	Created the standardized precipitation evapotranspiration index (SPEI), a multi-scalar tool sensitive to global warming that improves the accuracy of identifying water imbalances

(continued)

(continued)

Author	Year	Region	Climate extreme	Policy type	Key outcome
Wilhelmi and Hayden	2010	United States (urban)	Extreme heat	Urban planning and social vulnerability management	Proposed a new framework to reduce urban vulnerability by connecting social characteristics of people to the physical characteristics of their place of residence
Berrang-ford et al.	2011	Global	Precipitation changes (floods/ droughts) and variability	Institutional and government-led adaptation	While reporting on adaptation is increasing, it remains limited to 5% of total research; adaptive action is primarily concentrated in high-income nations
Lesnikowski et al.	2013	Annex I parties (36 nations)	Extreme weather and health-related risks	National health adaptation and environmental governance	National wealth and size are insufficient for progress without strong policy commitments to environmental governance
Schmitt et al.	2016	Global (focused on US/Asia)	Extreme temperatures and flooding	Risk reduction and economic appraisal	Extreme events are a pressing public health issue with strong welfare implications, though evidence on the cost-effectiveness of ecosystem-based solutions remains thin
Bell	2018	United States	Heatwaves, droughts, wildfires, flooding, and hurricanes	Public health resilience and adaptation	Pathways connect extremes to diverse health outcomes and economic losses, though local factors complicate predictions of disease burden

(continued)

(continued)

Author	Year	Region	Climate extreme	Policy type	Key outcome
Benevolenza and DeRigne	2019	United States	Hurricanes, heatwaves, and flooding	Emergency preparedness and disaster-relief efforts	Underprivileged groups face heightened risks of trauma and physical harm; advancements in clinical practice and policy are needed to stabilize health outcomes
Ebi et al.	2021	Global	Heatwaves, cyclones, floods, and wildfires	Climate-resilient health system adaptation and disaster risk management	Argues that many health risks are preventable through building resilient infrastructure and improving preparedness and response
Li et al.	2021	China (epidemic regions)	Dengue fever extreme outbreaks	Vector control and public health surveillance	Outbreaks are driven by a combination of habitat suitability, nighttime light intensity, and socioeconomic factors
Weilnhammer et al.	2021	Europe	Heatwaves, floods, droughts, and wildfires	Prevention plans and adaptive management	Southern Europe is the most affected region, with heatwaves identified as the deadliest and floods as the most frequent extremes
Alcayna et al.	2022	Global (focused on LMICs)	Floods, droughts, tropical cyclones, and heatwaves	Anticipatory action and public health preparedness	Identified hydrometeorological triggers for disease outbreaks; found a lack of evidence linking climate extremes to underlying vulnerability
Hess et al.	2022	Global	Extreme heat events (EHEs)	Heat action plans (HAPs) and iterative adaptive management	Immediate investment in climate-resilient health systems can significantly reduce heat-related morbidity and mortality

(continued)

(continued)

Author	Year	Region	Climate extreme	Policy type	Key outcome
Houser et al.	2022	United States (Indiana)	Floods and heatwaves	Local adaptation policy and adaptation taxes	Trust in government efficacy is the most important predictor of public support for local adaptation efforts
De Luca and Donat	2023	Global	Hot, dry, and compound hot-dry extremes	Water resource management and disaster mitigation	Projects that the frequency and intensity of compound heat and drought events will rise significantly, threatening global water security
Escorcia Hernández et al.	2023	Global	Extreme heat and flooding (linked to pandemic shocks)	Resilient urban planning and sustainable mobility	Finds that inclusive urban governance addressing social equity and healthy environments provides co-benefits for environmental stressor resilience
Escorcia Hernández et al.	2023	Global (1439 documents)	Extreme heat, flooding, and air pollution	Inclusive urban governance and sustainable mobility	Post-pandemic recovery must align sustainable development goals (SDGs) and carbon neutrality to support disadvantaged groups
IPCC	2023	Global	Heatwaves, wildfires, droughts, and floods	Transformative adaptation and redistributive policies	Mortality is 15 times higher in regions with high vulnerability; 1.5 °C warming will likely be surpassed by the late 2030s
Zhang et al.	2023	China	Short-term meteorological drought	Machine learning prediction models	Developed high-accuracy forecasting models to enhance drought prediction and provide a scientific foundation for water resource management

(continued)

(continued)

Author	Year	Region	Climate extreme	Policy type	Key outcome
Akram and Mushtaq	2024	Pakistan	Floods	Mental health services and conflict management	Significant displacement due to floods is linked to varying degrees of anxiety, PTSD, and depression in vulnerable populations.
Bell L	2024	Global (covering 43 countries)	Extreme heat and heatwaves	Targeted location and population interventions	Over one third of heat-related deaths are now attributable to human-induced climate change; impacts are inequitable across communities
Dewi et al.	2024	International (31 countries)	Extreme temperatures, floods, droughts, bushfires, and storms	Adaptation and disaster management in rural/remote areas	Highlights a shift in healthcare utilization toward emergency demand; routine care is often disrupted by service and financial barriers
Haubrock	2024	Germany (rhine-main observatory)	Floods, droughts, and extreme heat	Integrated water resource management and conservation	The cumulative impact of events over time is more significant than single event magnitude; resilience is site-specific
Meadows et al.	2024	Global	Extreme heat events	Public health and health system adaptation	Individuals with pre-existing mental illness face significantly increased vulnerability to mortality and morbidity during heatwaves
Patwary et al.	2024	South and Southeast Asia	Cyclones, floods, storm surges, and heat	Health disaster risk management (health-DRM)	Identifies strong links to PTSD, depression, and anxiety, but notes existing health systems lack explicit climate-mental health integration

(continued)

(continued)

Author	Year	Region	Climate extreme	Policy type	Key outcome
Teebken	2024	USA and China	Flooding and extreme heat	Accidental adaptation vs. redistributive policies	Vulnerability is often locked in by external conditions like affordable housing; current adaptation often focuses on "lower-hanging fruits" rather than root causes
Tian et al.	2024	Global	Extreme heat and heatwaves	Global climate governance and contingency plans	Population exposure to heat rose by 41–58% between 2003 and 2020, with high spatial inequalities concentrated in Africa and Asia
Wang et al.	2024	China (Greater Bay Area)	Flood disasters	Whole-life cycle disaster risk management	Highly urbanized and economically developed areas face the highest risk levels, while ecological protection positively reduces this risk
Weeda et al.	2024	Global	Temperature extremes and air pollution	Public health, adaptation, and mitigation	Children carry 88% of the health burden of climate change, with preterm birth risk increasing by an average of 60% due to temperature extremes
Ghosh et al.	2025	India and Bangladesh	Tropical cyclone Remal, floods, landslides, and hot spells	Disaster risk reduction (DRR) and climate risk management (CRM)	Illustrates the complexity of cascading impacts; advocates for integrating earth observation (EO) data into local disaster planning
Jiménez et al.	2025	Global	Anthropogenic environmental stressors	Urban health policy and environmental management	Scoping review indicates that light pollution (ALAN) acts as a global stressor affecting human Well-being and quality of life

(continued)

(continued)

Author	Year	Region	Climate extreme	Policy type	Key outcome
Kafi and Ponrahono	2025	Nigeria	Windstorms and floods	City-wide disaster risk reduction (DRR) and spatial planning	Disaster vulnerability is exacerbated by selective planning and unregulated development in hazard-prone areas.
Legg et al.	2025	Australia	Bushfires, floods, droughts, and storms	Health system adaptation (workforce and governance)	Adaptations improve workforce capability, but critical gaps remain in access to essential medicines and system financing
Li et al.	2025	55 LMICs (low- and middle-income)	Flooding events	Climate-health services and interventions	The chronic effects of flood exposure on child health are significantly larger than direct mortality from natural forces
Nhamo et al.	2025	South Africa	Floods, droughts, heatwaves, and wildfires	Climate adaptation and resilience strategies	There has been a significant increase in event frequency, with the 2011–2020 decade recording the highest number of extremes
Paudel et al.	2025	Nepal	General climate-induced health risks	Public health policy synthesis	Synthesizes Nepal's mental health system through health system building blocks to improve future response capacity

References

Ahern M, Kovats RS, Wilkinson P, Few R, Matthies F (2005) Global health impacts of flooding: epidemiologic evidence. Epidemiol Rev 27(1):36–46

Akram S, Mushtaq S (2024) Environmental change and floods: the long-ignored effects of displacement on mental health. Front Public Health 12:1–9. https://doi.org/10.3389/fpubh.2024.1434123

Alcayna T, Fletcher I, Gibb R, Tremblay L, Funk S, Rao B, Lowe R (2022) Climate-sensitive disease outbreaks in the aftermath of extreme climatic events: a scoping review. One Earth 5(4):336–350. https://doi.org/10.1016/j.oneear.2022.03.011

Bell JE (2018) Health impacts of climate change. Am J Med 131(12):1438–1441. https://doi.org/10.1016/j.amjmed.2018.07.039
Bell L (2024) The mental health impacts of extreme weather events. Environ Health 23(1):1–4. https://doi.org/10.1186/s12940-024-00898-3
Benevolenza MA, DeRigne LA (2019) The impact of climate change and natural disasters on vulnerable populations: a systematic review of literature. J Human Behav Soc Environ 29(2):266–281. https://doi.org/10.1080/10911359.2018.1527739
Berrang-Ford L, Ford JD, Paterson J (2011) Are we adapting to climate change? Glob Environ Change 21(1):25–33. https://doi.org/10.1016/j.gloenvcha.2010.09.012
Checkley W, Epstein LD, Gilman RH, Figueroa D, Cama RI, Patz JA, Black RE (2000) Effects of El Niño and ambient temperature on hospital admissions for diarrhoeal diseases in Peruvian children. Lancet 355(9202):442–450
Cutter SL, Boruff BJ, Shirley WL (2003) Social vulnerability to environmental hazards. Soc Sci Quart 84(2):242–261
De Luca A, Donat MG (2023) Climate change impacts on mental health: a systematic review of the literature. Environ Res Lett 18(8):083002. https://doi.org/10.1088/1748-9326/acd256
Dewi SP, Kasim R, Sutarsa IN, Dykgraaf SH (2024) A scoping review of the impact of extreme weather events on health outcomes and healthcare utilization in rural and remote areas. BMC Health Services Res 24(1):486–497. https://doi.org/10.1186/s12913-024-11695-5
Ebi KL, Vanos J, Baldwin JW, Bell JE, Hondula DM, Errett NA, Hayes K, Reid CE, Saha S, Spector J, Berry P (2021) Extreme weather and climate change: population health and health system implications. Ann Rev Public Health 42:293–315. https://doi.org/10.1146/annurev-publhealth-012420-105026
Ejiohuo O, Onyeaka H, Akinsemolu A, Nwabor OF, Siyanbola KF, Tamasiga P, Al-Sharify ZT (2025) Ensuring water purity: Mitigating environmental risks and safeguarding human health. *Water Biology and Security, 4*(2), 100341.
Ellis FP (1976) Mortality from heat illness and heat-aggravated disease in the United States. Environ Res 5(1):1–58
Escorcia Hernández JR, Torabi Moghadam S, Sharifi A, Lombardi P (2023) Cities in the times of COVID-19: trends, impacts, and challenges for urban sustainability and resilience. J Clean Prod 432. https://doi.org/10.1016/j.jclepro.2023.139735
Few R, Ahern M, Matthies F, Kovats S (2004) Floods, health and climate change: a strategic review. Tyndall Centre for Climate Change Research
Gaffen DJ, Ross RJ (1998) Increased humidity of United States cities, 1961–1995. Nature 396:729–730
Gagnon-Lebrun F, Agrawala S (2007) Progress on adaptation to climate change in developed countries: an analysis of broad trends. OECD Publishing
Ghosh S, Dawn A, Kour S, Mallick A, Chowdhury A, Kundu K, De Sarkar K, Rahman MR, Sharma P, Rajakaruna P, Rahman MM, Nath AJ, Shaw R (2025) Climate extremes walking together: evidence from recent compounding climate hazards after Remal. Int J Disaster Risk Reduct 118. https://doi.org/10.1016/j.ijdrr.2024.104974
Grosjean M (1991) Climate and culture in the Andes. Science 251(4995):102–105
Hahn MB, Riederer AM, Foster SO (2009) The livelihood vulnerability index: a pragmatic approach to assessing risks from climate variability and change—A case study in Mozambique. Glob Environ Change 19(1):74–88. https://doi.org/10.1016/j.gloenvcha.2008.11.002
Haubrock PJ (2024) Site characteristics determine the prevalence of extreme weather events affecting freshwater macroinvertebrate communities. Sci Total Environ 950:175436. https://doi.org/10.1016/j.scitotenv.2024.175436
Hernández D, Martínez M (2023) Climate change and urban health: a systematic review. Int J Environ Res Public Health 20(4):3165. https://doi.org/10.3390/ijerph20043165
Hess JJ, Errett NA, McGregor G, Busch Isaksen T, Wettstein ZS, Wheat SK, Ebi KL (2022) Public health preparedness for extreme heat events. Annu Rev Public Health 44:301–321. https://doi.org/10.1146/annurev-publhealth-071421-025508

Houser M, Gazley B, Reynolds H, Grennan Browning E, Sandweiss E, Shanahan J (2022) Public support for local adaptation policy: The role of social-psychological factors, perceived climatic stimuli, and social structural characteristics. Glob Environ Change 72:102424. https://doi.org/10.1016/j.gloenvcha.2021.102424

IPCC (2023) Climate change 2023: synthesis report. Cambridge University Press

Jiménez D, Neira Arenas L, Hernández Rincón EH, García Céspedes MJ, Jaimes Peñuela CL (2025) Luminous threats: the health impacts of artificial nighttime light on metabolic and mental health: a scoping review. Wellbeing Space Soc 8. https://doi.org/10.1016/j.wss.2025.100260

Johnson CA, Krishnamurthy K (2010) Dealing with displacement: can "social protection" facilitate long-term adaptation to climate change? Glob Environ Change 20(4):648–655. https://doi.org/10.1016/j.gloenvcha.2010.06.002

Kafi KM, Ponrahono Z (2025) Weather and climate extremes in Nigeria: modeling the perceived impact of spatial planning and community practices on windstorm and flood exposure using PLS-SEM and correlogram. Int J Disaster Risk Reduct 124. https://doi.org/10.1016/j.ijdrr.2025.105554

Legg R, Prior J, McIntyre E, Liu E, Tracy M, Tan L, Dawson A, Richmond J, Perry C (2025) Health system adaptation to extreme weather events in Australia: A scoping review. J Clim Change Health 22. https://doi.org/10.1016/j.joclim.2025.100443

Lesnikowski AC, Ford JD, Berrang-Ford L, Barrera M, Berry P, Henderson J, Heymann SJ (2013) National-level factors affecting planned, public adaptation to health impacts of climate change. Glob Environ Change 23(5):1153–1163. https://doi.org/10.1016/j.gloenvcha.2013.04.008

Li C, Wu X, Wang X, Yin J, Zheng A, Yang X (2021) Ecological environment and socioeconomic factors drive long-term transmission and extreme outbreak of dengue fever in epidemic region of China. J Clean Prod 279:123870. https://doi.org/10.1016/j.jclepro.2020.123870

Li P, Tong M, Wu J, Meng X, Fan X, Sun X, Li J, Xue T (2025) Flood exposure contributes to under-five mortality in low- and middle-income countries. One Earth 8(5):101305. https://doi.org/10.1016/j.oneear.2025.101305

McKee, T. B., Doesken, N. J., & Kleist, J. (1993). The relationship of drought frequency and duration to time scales Proceedings of the 8th conference on applied climatology, 17(22, 179–183

McMichael AJ, Woodruff RE, Hales S (2006) Climate change and human health: present and future risks. Lancet 367(9513):859–869

Meadows J, Mansour A, Gatto MR, Li A, Howard A, Bentley R (2024) Mental illness and increased vulnerability to negative health effects from extreme heat events: a systematic review. Psychiatry Res 332:115678. https://doi.org/10.1016/j.psychres.2023.115678

Meehl GA, Tebaldi C (2004) More intense, more frequent, and longer lasting heat waves in the 21st century. Science 305(5686):994–997

Naughton MP, Henderson A, Mirani MC, Hatch DL, Jeong KJ, Nibbelink AS, Osterholm MT (2002) Heat-related mortality during a 1999 heat wave in Chicago. Am J Prevent Med 22(4):221–227

Nhamo G, Chapungu L, Mutanda GW (2025) Trends and impacts of climate-induced extreme weather events in South Africa (1920–2023). Environ Dev 55. https://doi.org/10.1016/j.envdev.2025.101183

Page MJ, McKenzie JE, Bossuyt PM, Boutron I, Hoffmann TC, Mulrow CD, Shamseer L, Tetzlaff JM, Akl EA, Brennan SE, Chou R (2021) The PRISMA 2020 statement: an updated guideline for reporting systematic reviews. BMJ 372

Patwary MM, Bardhan M, Haque MA, Moniruzzaman S, Gustavsson J, Khan MMH, Koivisto J, Salwa M, Mashreky SR, Rahman AKMF, Tasnim A, Islam MR, Alam MA, Hasan M, Harun MAY, Al N, L., & Islam, M. A. (2024) Impact of extreme weather events on mental health in South and Southeast Asia: a two decades of systematic review of observational studies. Environ Res 250. https://doi.org/10.1016/j.envres.2024.118436

Paudel S, Chalise A, Khatri D, Poudel S, Khanal A (2025) Nepal's mental health system from public health perspective: A thematic synthesis based on health system building blocks. Lancet Regional Health Southeast Asia 36:100588. https://doi.org/10.1016/j.lansea.2025.100588

Pitt M (2008) The Pitt review: learning lessons from the 2007 floods. Cabinet Office

Schmitt LHM, Graham HM, White PCL (2016) Economic evaluations of the health impacts of weather-related extreme events: A scoping review. Int J Environ Res Public Health 13(11):1–19. https://doi.org/10.3390/ijerph13111105

Semenza JC, Rubin CH, Falter KH, Selanikio JD, Flanders WD, Howe HL, Wilhelm JL (1996) Heat-related deaths during the July 1995 heat wave in Chicago. New England J Med 335(2):84–90

Slovic P (1987) Perception of risk. Science 236(4799):280–285

Smit B, Wandel J (2006) Adaptation, adaptive capacity and vulnerability. Glob Environ Change 16(3):282–292

Steadman RG (1979) The assessment of sultriness: Part I: A decathelometric index of atmospheric sultriness. J Appl Meteorol Climatol 18(7):861–873

Teebken J (2024) Vulnerability locked in. On the need to engage the outside of the adaptation box. Glob Environ Change 85:102807. https://doi.org/10.1016/j.gloenvcha.2024.102807

Tian P, Zhang F, Yan Y, Wang H, Liu Y, Zhang H, Li J (2024) Inequalities of population exposure and mortality due to heatwaves in China. J Clean Prod 511:145626. https://doi.org/10.1016/j.jclepro.2025.145626

Vicente-Serrano SM, Beguería S, López-Moreno JI (2010) A multiscalar drought index sensitive to global warming: the standardized precipitation evapotranspiration index. J Clim 23(7):1696–1718

Von Braun J, Webb PJ (1989) The impact of new crop technology on the agricultural production of high-risk farmers in West Africa. World Dev 17(4):513–534

Wang Y, Liu C, Wang Y, Liu Y, Liu T (2024) Climate risk assessment and adaption ability in China's coastal urban agglomerations—A case study of Guangdong-Hong Kong-Macao greater bay area. J Clean Prod 452:142036. https://doi.org/10.1016/j.jclepro.2024.142036

Weeda LJZ, Bradshaw CJA, Judge MA, Saraswati CM, Le Souëf PN (2024) How climate change degrades child health: A systematic review and meta-analysis. Sci Total Environ 920. https://doi.org/10.1016/j.scitotenv.2024.170944

Weilnhammer V, Schmid J, Mittermeier I, Schreiber F, Jiang L, Pastuhovic V, Herr C, Heinze S (2021) Extreme weather events in Europe and their health consequences—A systematic review. International Journal of Hygiene and Environmental Health 233:113688. https://doi.org/10.1016/j.ijheh.2021.113688

Wilhelmi OV, Hayden MH (2010) Connecting people and place: A new framework for assessing urban vulnerability to extreme heat. Environ Res Lett 5(1):014021

Zhang D, Wang Y (2023) Climate change and health: a systematic review. Lancet Planet Health 7(12):e986–e994. https://doi.org/10.1016/S2542-5196(23)00216-7

Chapter 18
Climate-Induced Migration in Coastal Regions and Policy: A Review

Puranjani Ghosh, Shruti Thakur, and Tuhin Ghosh

Abstract Climate-induced displacement is emerging as a critical global challenge as increasing temperatures, sea-level rise, and enhanced frequency of extreme weather events are massively shaping human mobility patterns. Climate change has become a significant driver of migration, especially in Small Island Developing States (SIDS), low-lying coastal regions and river deltas. Natural calamities such as coastal erosion, floods, cyclones, prolonged droughts, and ecosystem degradation are forcing millions to relocate within and across national borders. Despite the expanding recognition of climate migration, existing legal and policy frameworks remain fragmented and inefficient to address the scale and further economic devastation caused by climate-driven mobility. This review introspects on the environmental, social, economic, and political dimensions of climate migration through case studies from Tuvalu, the Mekong Delta in Vietnam, Somalia, and the Volta Basin in Ghana, alongside a focused case study of the Indian Sundarban Delta. The chapter analyses climate migration policies at global, regional, and national levels and analyses adaptation strategies such as planned relocation, climate-resilient infrastructure, and cross-border mobility agreements. The study highlights the urgent need for coordinated global action, strengthened legal frameworks, and climate-resilient development pathways to ensure long-term protection and well-being for displaced populations.

Keywords Climate migration · Internally displaced persons (IDPs) · Sea level rise · Adaptation strategies · Policy responses

Introduction

Climate change has become a strong "threat multiplier." It is increasing the frequency of storms, wildfires, floods, droughts, and landslides. This leads to the disruption of livelihoods, governance structures, and socioeconomic systems. The expanding effects of climate variability are altering human mobility patterns and impacting

P. Ghosh (✉) · S. Thakur · T. Ghosh
School of Oceanographic Studies, Jadavpur University, Kolkata, India
e-mail: puranjanig@gmail.com

P. Ghosh et al. (eds.), *Climate Crisis and Changing Population Dynamics*, Springer Climate, https://doi.org/10.1007/978-3-032-24533-5_18

global social, political, and ecological landscapes. These variations show regional weaknesses interlinked with environmental risks, economic instability, and institutional strength. They do not follow any specific pattern, thus decreasing their predictability (Climate Migration Council and Bergmann 2024; Heslin et al. 2018). Communities already facing poverty, landlessness, and limited access to government resources are at much higher risk due to their weak ability to adapt.

Forced migration in low-lying and vulnerable areas is becoming more common. Human-caused warming, erratic rainfall, and rising sea levels are driving this trend (Richards and Bradshaw 2017). Low- and lower-middle-income countries experienced climate-related displacement almost five times more than high-income countries from 2008 to 2016. Without inclusive, proactive adaptation plans, these displacement trends are expected to worsen, leading to long-term humanitarian, ecological, and financial consequences.

Theoretically, the growing scale of climate-related mobility has led to the emergence of multiple terminologies—such as *climate refugees*, *environmentally induced migrants*, and *environmental refugees*—each attempting to capture different aspects of forced movement (Hiraide 2023; Bera 2013; Myers 2002). It can be stated that migration rarely originates from environmental pressures alone. Climate stressors combined with poverty, unemployment, aggressive urbanisation, political instability, and weak governance systems make it difficult to isolate climate-only motives. Recent scholarship further distinguishes *climate mobility*—planned, adaptive, network-supported movement—from *climate displacement*, which is typically short-term, reactive, involuntary, and rooted in the absence of safer alternatives (Heslin et al. 2018; Easton-Calabria and Sturridge 2025).

Recent data highlights the accelerating scale of the crisis. Internally displaced persons (IDPs) have increased by 51% in the past five years, reaching 75.9 million across 116 countries in 2023 (IDMC 2024). Climate pressures are expanding both geographically and socially, threatening food security, freshwater access, and agricultural stability (Jolly et al. 2024). Although the early discourse focused primarily on Bangladesh and Small Island Developing States (SIDS), more recent trends show significant displacement from climate hazards in high-income nations such as Canada, Australia, and the United States (Oakes 2025). For instance, Canada alone recorded 185,000 wildfire-related displacements in 2023—the highest in its history and one of the largest globally (IDMC 2024). Europe, too, has experienced increasing climate-driven displacement, with floods accounting for 44.8% of climate-related movements between 2008 and 2020 (IDMC 2021).

Despite the expanding scope of climate mobility, the central research problem remains the absence of a coherent, multi-scalar policy framework capable of addressing climate-induced displacement. Global governance mechanisms continue to be fragmented, lacking enforceability, clarity, or integration across scales. Many national governments struggle to adopt proactive or rights-based approaches to climate migration, resulting in displaced groups receiving little institutional recognition or support. This governance gap is especially significant because climate-driven mobility now affects both the Global South and Global North, yet disparities in institutional capacity, adaptation financing, and policy preparedness remain stark.

This problem's significance lies in its potential to deepen long-term vulnerability. As climate hazards intensify, millions risk being displaced into situations with inadequate legal protections, insecure land rights, insufficient livelihood alternatives, and limited access to social services. Understanding these governance shortcomings is essential for designing anticipatory, inclusive, and sustainable policy responses.

To address this gap, the present chapter adopts a multi-regional comparative approach, drawing on empirical insights from Tuvalu, the Mekong Delta of Vietnam, Somalia, and Ghana's Volta Basin. The chapter evaluates existing adaptation frameworks—including planned relocation, disaster-resilient infrastructure initiatives, and regional migration agreements—to understand how different states and regions manage climate mobility. Additionally, the chapter includes an in-depth case study from the Indian Sundarban Delta, particularly the south-western part of Gosaba Island, where severe erosion, saline intrusion, and embankment breaches have triggered escalating displacement among economically marginalised communities reliant on agriculture, fishing, and honey collection. As riverbanks rapidly erode and farmland is lost, these households are forced to relocate, often without compensation, institutional recognition, or access to protective legal frameworks. By integrating these diverse cases, the study provides a clearer understanding of how climate mobility manifests across distinct geographical contexts and assesses the adequacy of existing policy responses. This comparative perspective strengthens the methodological foundation of the chapter and supports its broader aim: to highlight the urgent need for coherent, rights-based, and multi-level policy solutions to govern climate-induced displacement.

Materials and Methods

This chapter stems from secondary research and adopts a systematic qualitative review approach to examine climate-induced displacement, adaptation strategies, and policy responses across vulnerable regions, with particular emphasis on SIDS, deltaic systems, and selected national case studies based on coastline changes. The methodological framework evaluates policy analysis through comparative case study synthesis to identify patterns, gaps, and emerging governance responses to climate mobility. The analysis draws exclusively on secondary sources, such as peer-reviewed journal articles published in international academic journals, books and edited volumes on climate change, migration, and displacement, policy documents, national action plans, and legislative frameworks issued by governments, reports from international organisations such as the United Nations Development Programme (UNDP), International Organization for Migration (IOM), Internal Displacement Monitoring Centre (IDMC), Green Climate Fund (GCF), and the World Bank and official statistics and datasets that highlight displacement, climate risk, and adaptation measures. The material sources were selected through targeted searches of academic databases such as Scopus, Web of Science, Google Scholar, and institutional repositories, using keywords including *climate-induced displacement*, *climate migration*, *planned relocation*, *delta vulnerability*, *SIDS adaptation policies*, and *environmental refugees*.

Results and Discussions

Sea Level Rise and Its Impact on the SIDS

Countries like China, India, Bangladesh, Vietnam, Indonesia, Japan, the U.S., the Philippines, Egypt, and Brazil are really feeling the pinch from rising sea levels. But the small island developing states (SIDS) have it even tougher (Strauss et al. 2015). Their coastlines are crowded with people, they're at low elevations, and they have few ways to adapt, which puts them in a really vulnerable spot (Richards and Bradshaw 2017). Take Kiribati, for example—between 2005 and 2015, a staggering 94% of households reported facing natural hazards, and 80% pointed to sea-level rise as a major danger to their lives and livelihoods (Oakes et al. 2016; Heslin et al. 2018). What's absurd is that, even with all these threats, not a lot of folks are actually moving away from these high-risk areas (Oakes 2025). In fact, according to Oakes' study, only 14% of all migration in that decade was due to environmental changes, and most of it was internal rather than international (Oakes et al. 2016; Heslin et al. 2018). This kind of displacement isn't just about moving from one place to another; it also involves significant nonmaterial losses like health, education, and cultural identity—factors that are increasingly being recognized in discussions about climate change (UNICEF UK 2017; Heslin et al. 2018).

Back in 2016, storms forced nearly 12.9 million people out of their homes globally, showing just how intense extreme weather has become because of climate change (IDMC 2017). Cyclone Winston in Fiji was especially harsh, impacting roughly 540,000 folks, which is about 62% of the country's whole population (Ministry of Economy, Republic of Fiji 2016; OCHA 2016; SBS 2017). Many of those who were hit the hardest had already been battling drought and water issues before the cyclone came along. The storm ended up damaging or wiping out over 30,000 homes, displacing more than 55,000 people, and tragically led to 44 deaths, costing nearly a fifth of Fiji's GDP (Wasko and Sharma 2017). The Caribbean also faces a mix of environmental hazards like hurricanes, earthquakes, and coastal flooding, all worsened by high population density in coastal areas, which raises the risk of displacement (International Organization for Migration 2023). At the 43rd CARICOM Conference, UN Secretary-General António Guterres highlighted the urgency of the situation, noting that the Caribbean is at the "epicentre of the global climate emergency" amid climate shocks, economic downturns, and rising debt (Caribbean 'Ground Zero' for Global Climate Emergency 2022). Experts are pointing out that without solid international support for climate adaptation, some regions might face long-term displacement issues (International Organization for Migration 2023). Take Tuvalu, for example; some areas there have become practically uninhabitable due to stronger storm surges and saltwater contaminating fresh water supplies. As a result, people have to move either within the country or elsewhere to find better living conditions, and we're definitely seeing a clear trend in migration (SPREP 2024; UN ESCAP et al. 2007).

Economic Dependence and Climate Migration in SIDS

Many small island developing states (SIDS) possess small and narrowly based economies that depend heavily on external inputs such as imports, tourism, and remittances. According to research by the International Organization for Migration (IOM), remittances accounted for 40.6% of GDP in Tonga, 20.6% in Comoros, and 19.8% in Jamaica in 2023. In 2022 alone, migrants sent over $22 billion to SIDS, providing essential financial flows that support poverty reduction, climate adaptation, and the advancement of sustainable development initiatives. Simultaneously, physical displacement on low-lying islands is no longer a hypothetical scenario but a current reality. In Panama, rising sea levels are actively submerging Gardi Sugdub and 37 other inhabited islands, most of which are situated less than one meter above sea level (The Economist 2024). As part of a "planned retreat" strategy, the Panamanian government relocated 300 families to newly constructed housing on the mainland in June 2024, marking a significant step in government-led adaptation planning (The Economist 2024).

The dominant discourse on climate-induced migration from SIDS is often critiqued for lacking nuance. A growing body of literature suggests that the portrayal of migration as an inevitable and tragic consequence of environmental degradation—emphasized through terms like "evacuation," "fleeing," "abandonment," and "displacement"—does not adequately capture the complexity of mobility decisions. While such terminology may accurately describe certain emergency scenarios, migration in SIDS occurs along a continuum ranging from voluntary to forced movement, shaped by multiple socio-economic, cultural, and political factors beyond environmental change alone (Hugo 1996; Kelman 2018). Historically, mobility has functioned as an adaptive strategy for many island communities, allowing them to respond to shifting environmental and economic conditions without losing their agency or autonomy (Kelman 2018). This more dynamic understanding challenges the one-dimensional "climate refugee" narrative, which tends to frame island populations as passive victims driven solely by external environmental forces (Connell 2016; Yamamoto and Esteban 2014). One illustrative case is that of the Carteret Islands in Papua New Guinea (PNG), where migration is frequently attributed to rising sea levels. However, such interpretations often overlook the contribution of resource constraints and weak governance structures, which hinder the implementation of local adaptation strategies (Kelman 2018). In such contexts, relocation may occur not because people lack alternatives, but because those alternatives are not adequately supported or made viable (Stojanov 2014).

Despite the profound vulnerabilities that SIDS face, their populations have consistently demonstrated agency and adaptability through historical patterns of movement. Migration has long been embedded within the social and economic fabric of island societies, functioning not simply as a reactive response, but as a deliberate and strategic choice that enhances resilience and long-term survival (Kelman 2018). While climate change adds unprecedented pressure to this dynamic, it is important to interpret contemporary migration trends within this longer trajectory of mobility and adaptation. However, this does not imply that migration is uniformly positive or without adverse consequences. For instance, Cape Verde has witnessed substantial emigration, which has led to a "brain drain" effect, undermining local institutional capacity and impeding national development objectives (Atchoaréna et al. 2008). This highlights the dual nature of migration: on the one hand, it presents opportunities for adaptation and risk reduction and on the other, it can also create new vulnerabilities and threaten the sustainability of island communities over the long term.

Climate Migration in Major Deltas

Coastal and deltaic communities in developing economies face extreme risks due to geographic exposure and economic instability (Fernández et al. 2024).

Mississippi Delta

The Mississippi River Delta (MRD), situated in south-eastern Louisiana, has been severely affected by climate change, which has resulted in environmental problems that have forced people to migrate elsewhere. The restoration of ecosystems in the MRD is hindered by wide navigation channels, which need to be included into current pursuits of natural resource extraction, urbanization, agriculture, and fishing (Day et al. 2007; Anderson et al. 2021). Over 25% of wetlands in the MRD were lost in the last century as a result of causes such as rising sea levels, frequent precipitation, and increased river discharge (Day et al. 2024). Due to the region's decreased habitability as a result of these environmental changes, people are moving to more stable locations, which is causing population shrinkage (Day et al. 2024). Evidence showcased by Twilley et al. (2016) depicts a correlation between increased flooding, land loss, and population migration in Terrebonne Bay over the last 78 years. This reduction in delta landscape area has led to a decrease in population due to storm surge risks as the ratio of land to water continues to decline (Twilley et al. 2016). Some regions are currently uninhabitable and inaccessible owing to severe and frequent flooding that is being made worse by climate change (Schilling et al. 2024). Communities like West Alton, Missouri, have experienced large declines in population as a result of recurrent floods; here the ongoing risk of floods has compelled locals to leave their houses in search of safer areas (Schilling et al. 2024). The response of land loss to variations in volumetric sedimental discharge along coastal shorelines has been found to result in proportionate increases in storm-surge risks as wetlands drown, migration proceeds landward, and forested wetlands are lost as a result of human encroachment (Twilley et al. 2016). In contrast to Atchafalaya Bay, where wetlands are preserved by adequate sedimental discharge and the percentage of sediment volume provided by organic production which is naturally declining the risks of storm surges, whereas Terrebonne Bay is massively suffering due to reduction in sedimental discharge (Twilley et al. 2016). According to Foufoula-Georgiou et al. (2011), these problems with deltaic coasts taken together have been identified as a multi-trillion dollar crisis that will affect trade, public safety, and regional riches globally. One example of relocation brought on by climate change in the MRD region is the Isle of Jean Charles in Louisiana where, because of erosion and rising sea levels, the island has lost a large portion of its landmass, forcing its indigenous population to relocate (Simms et al. 2021). This instance demonstrates how migratory patterns within the MRD are directly impacted by climate change.

Nile Delta

Sea-level rise is threatening to submerge large areas of land in the low-lying coastal areas of the Nile Delta, lower Egypt, where estimates suggest that a one-meter rise in sea level could submerge 12–15% of Egypt's agricultural land, potentially uprooting up to 6.7 million people who rely on these areas for an income (Sivakumar et al. 2013). In the Nile Delta area, particularly in Kafr El-Sheikh, one of Egypt's northernmost regions, land is being swiftly consumed by coastal erosion (Earth Journalism Network 2023). Internal displacement is occurring as a result of the farmers leaving behind land that has been rendered infertile due to saltwater intrusion and the consecutive crop failures hampering food security (Earth Journalism Network 2023). The uninhabitable conditions are forcing the residents of the impacted communities to flee their homes and sometimes even permanently abandon them (Earth Journalism Network 2023). The study conducted by Elshinnawy et al. (2017) shows that the risk of coastal flooding in the Nile Delta is greatly increased by sea level rise, especially in low-lying places like Ras El Bar, Port Said, and Damietta, endangering both homes and livelihoods. It has been highlighted in the same study that the residents are being forced to relocate due to increased flooding and land loss, which would further contribute to Egypt's internally displaced population aggravated by climate change (Elshinnawy et al. 2017). In the past, migration has served as a coping strategy for Nile Delta populations dealing with environmental difficulties (Baker Institute 2024). Addressing the effects of climate change in the area at present is thought to require incorporating migration into adaptation plans devised by the governmental beneficiaries (Baker Institute 2024).

Volta Delta

The Volta Delta, which resides on the eastern coast of Ghana, is relatively tiny and rural in comparison to other deltas worldwide, yet it is highly distinctive because of its dynamic character (Appeaning-Addo et al. 2020). Research reveals that environmental hazards, including flooding and shoreline erosion, significantly influence migration intentions (Abu et al. 2022). People who live in high "out-migration" areas—where migration tends to flow outside of the impacted area—and have been affected by these dangers frequently view movement as a means of adapting to climate-related difficulties (Abu et al. 2022). The migration of Ghanaian agricultural labourers are reinforced by fluctuating rainfall and diminishing soil productivity (Atiglo and Codjoe 2023). People relocate as a result of erosion that disrupts coastal operations (Van der Geest et al. 2010; Rain et al. 2011; Atiglo and Codjoe 2023). Massive upstream migration from the Lower Volta was clearly caused by the construction of the Akosombo Dam between 1961 and 1965, which created the Volta Lake; Tsikata (2006) specifically addresses the large-scale exodus from the Lower Volta brought on by the building of the Akosombo Dam (Atiglo and Codjoe 2023). Villages and farmlands were inundated as a result of dam building, which compelled the government to prepare a plan for the relocation of those impacted by the dam's construction (Tsegai 2005; Tsikata 2006). The devastation of the environment that people relied on for their livelihoods prompted the mass exodus; the majority of these were fishermen and farmers (Atiglo and Codjoe 2023). Towns around Volta Lake, where people could engage in farming or fishing, were a popular migratory destination; Accra, Ashaiman and Tema are other locations (Atiglo and Codjoe 2023). While some migrants lost their familial links to their place of origin, others continued to send remittances to sustain their livelihoods (Atiglo and Codjoe 2023).

Mekong Delta

Another highly vulnerable deltaic region in the world to climate change would be the Mekong Delta, which covers a sizable portion of southern Vietnam (Takagi et al. 2016). This area is particularly susceptible to the effects of sea-level rise due to its low-lying features; approximately nine million people live in the interior areas of the delta, where 1.7 million hectares are flooded annually (Hanh and Furukawa 2007; Takagi et al. 2016). Currently, flooding is primarily restricted to the flood season (Takagi et al. 2016). However, in the near future, it might happen at any time of year because of the combined effects of land subsidence and increasing sea levels, which would be extremely dangerous for these low-lying settlements (Takagi et al. 2016). It is crucial to remember that even in the absence of severe sea level rise situations, flooding may become more common and persistent (Takagi et al. 2016). Farmers in An Giang Province, Mekong Delta, have reported major environmental changes, such as rising temperatures, erratic rainfall, and an increase in the frequency of extreme weather events, according to a study conducted by Bayrak et al. (2023), that used an inductive approach to observe smallholder farmers in order to compare the effects of different adaptation strategies on human mobility. According to the same findings by Bayrak et al., (2023), changing irrigation techniques, diversifying revenue streams, and shifting crop calendars are all part of the climate change adaptation plan indigenously developed. Although some households choose local adaptations to sustain their livelihoods, others view migration to metropolitan regions as a coping mechanism for environmental pressures (Bayrak et al. 2023). Effective adaptation is hampered by issues such as a lack of institutional support, restricted financial resources, and poor information availability (Bayrak et al. 2023).

Niger Delta

The Niger Delta, situated directly on the Gulf of Guinea on the Atlantic Ocean in Nigeria, has been undergoing serious environmental difficulties, such as oil spills, gas flaring, and deforestation, all of which damage arable land and disrupt traditional livelihoods (Ighedosa 2019). Communities have been forced to migrate in search of more sustainable living circumstances elsewhere as a result of these situations (Ighedosa 2019). Like all other coastal deltaic regions in the African continent, the Niger Delta is also suffering the impacts of sea level rise and coastal erosion that is submerging farmlands and habitable lands. Saltwater intrusion into freshwater systems due to rising sea levels deteriorates water quality and makes farming and fishing less feasible (Nwilo and Badejo 2006). Local livelihoods have been significantly impacted by this, especially for those who depend on freshwater for drinking and irrigation (Nwilo and Badejo 2006). Homes, farms, and infrastructure are being destroyed by more frequent and severe seasonal and flash floods (Abayomi et al. 2021). In cities like Port Harcourt and Warri, urban flooding is made worse by heavy rainfall and inadequate drainage.

Floods also cause water sources to become contaminated, which can spread illness and have an adverse effect on health (Abayomi et al. 2021). Climate events can also affect oil infrastructure, which could result in greater spills and emissions (Akpodiogaga-a and Odjugo 2010). The crops and aquatic habitats that local populations rely on for sustenance and revenue are destroyed by frequent oil spills that poison farmlands and waterways (Babatunde 2020). Gas flaring has further damaged agricultural and human health by causing acid rain and air pollution (Babatunde 2020). In addition to degrading the land, these environmental risks weaken the local population's ability to adapt to climate change, increasing vulnerabilities and forcing many to leave when their standard of living becomes unsustainable (Akpodiogaga-a and Odjugo 2010). Ten years ago, armed conflict was the main cause of displacement in Nigeria, especially in the northeast (IOM DTM 2023; WFP 2023a, b; Ojo 2023). However, in the northwest states, criminal and communal violence, including conflicts between farmers and herders, was the cause of roughly three-quarters of all displacements in 2023 (IOM DTM 2023; WFP 2023; Ojo 2023). At the end of 2023, there were nearly 3.3 million persons internally displaced due to violence and war, with roughly half of them in Borno state of Nigeria (IOM 2024). Despite the decline from 3.6 million as observed in the year 2022, Nigeria continued to remain one of the top ten countries in the world, home to several IDPs (IOM 2024).

Climate Change Intensifying Disasters and the Rise in Internally Displaced People (IDPs)

According to the Internal Displacement Monitoring Centre (IDMC) (2024), natural calamities lead to the emergence of an additional 26.4 million IDPs in 148 nations and territories. In the past ten years, this is the third-highest number (Internal Displacement Monitoring Centre [IDMC] 2024). Approximately 8.8 million people were displaced, which is one-third of 26.4 million, in China and Turkey alone (Internal Displacement Monitoring Centre [IDMC] 2024). Canada and New Zealand, two countries with a high-income level, recorded their greatest number of displacement ever (IDMC 2024). In 2017, a catastrophic hurricane season struck the Caribbean and southeast United States, where tens of thousands of people were forced into temporary shelters in August after Hurricane Harvey caused major flooding in both Louisiana and Texas (Richards and Bradshaw 2017). Hurricane Irma, which followed Harvey, devastated a number of Caribbean islands, including Barbuda, Saint Martin, and Anguilla, before reaching Florida. Two weeks later, Puerto Rico and the Dominican Republic was devastated by Hurricane Maria, another hurricane that came dangerously close to setting a record (Richards and Bradshaw 2017). Although the number of displacements due to natural hazards was the lowest since 2017 in the East Asian and Pacific zone, they had the greatest number of catastrophe displacements worldwide as recorded by IDMC in 2023 (IDMC 2024). In 2023, storms were responsible for 53% of the 4.8 million documented catastrophe displacements in East Asia and the Pacific and 3.5 million, or 39%, were caused by floods (IDMC 2024). The first few months of 2023 saw the majority of flood-related displacements across Southeast Asia. The Philippines saw 589,000 flood displacements in January, out of a total of one million reported for the year and the areas that were most impacted were Mimaropa, Eastern Visayas, and the Zamboanga Peninsula, which do not often experience flooding at that season (Floodlist Philippines 2023). Similarly, a single incident in January that caused floods in seven regencies in the province of Aceh due to heavy rainfall was responsible for about two-thirds of the 184,000 migrations that were documented in Indonesia in 2023 (BNPB 2023; ASEAN 2023; Floodlist Indonesia 2023). The effects of the floods were probably exacerbated by deforestation and land degradation (Kompas 2023a; b). In 2023, geophysical hazards caused 639,000 internal displacements in East Asia and the Pacific, the highest figure since 2019 (IDMC 2024). The region's location on the Pacific Ring of Fire makes it the most vulnerable to seismic and volcanic activity, with Indonesia, Japan, Papua New Guinea, and the Philippines being the main hotspots (IDMC 2024). The Pacific and the East Asian regions are not only combating floods, cyclones, heavy downpour but also volcanic activity and earthquakes.

In 2023, the Americas saw the greatest number of wildfire displacements worldwide, at 277,000, as in the preceding seven years (IDMC 2024) and Canada saw its most disastrous wildfire season on record, with over seven times as much area burned as the yearly average due to the country's hottest summer in 76 years (Government of Canada 2023). The number of people without houseing will rise sharply when millions more become climate refugees as a result of their homes being destroyed by floods or fires, highlighting the fact that climate change has worsened homelessness over the span of years (Rozsa 2024). 43 million children have been displaced by climate catastrophes in only five years, according to UNICEF (2023a, b).

In Pakistan historic floods caused damage estimated at $5.6 billion in housing, $3.7 billion in agriculture, and $3.3 billion in transportation (WFP 2023; OCHA 2023a, b). People have been displaced as a result of anti-encroachment activities involving settlements across Karachi's natural drains caused by recent record-breaking floods in Pakistan. People from many groups were among them, and they subsequently had to move (Ayub 2023). But compared to Muslim households, Christian and Hindu families have fewer choices. They might have to relocate farther than other groups because of the city's residential segregation (Ayub 2023). The number of earthquake-induced movements reached 491,000, the highest level since 2015 in South Asia (IDMC 2024). More than 75% of these were documented in the western Afghan province of Herat, which housed the greatest number of conflict-related displaced persons in the nation when hundreds of thousands were uprooted from their homes due to a series of earthquakes and subsequent aftershocks (HPG 2022; UNHCR 2023; Islamic Relief 2023; IOM DTM 2023a, b; IDMC 2024). In the western Nepalese region of Karnali, a 6.4 magnitude earthquake and its aftershocks caused substantial damage to infrastructure and forced 107,000 people to relocate (IFRC 2023; UNCT Nepal 2023). Particularly in the hardest-hit districts of Jajarkot and Rukum West, the administration moved swiftly to arrange search and rescue missions and offer financial assistance to those in need; nevertheless, in certain places, landslides made it difficult to distribute aid (Nepal Red Cross 2023; UNICEF 2023a, b). Given Karnali's high seismic risk, which had been increasing since the end of 2022, this was the first time there that displacement was noted by IDMC (2024). In addition, the region has the greatest proportion of poverty in the nation and inadequate infrastructure, which makes residents less resilient to the effects of natural catastrophes and displacement (IDMC 2024). The need for warm clothing, food, and shelter to combat health concerns intensified as winter approached in the mountainous location, where temperatures frequently dropped below zero (ACAPS 2023; CARE 2023). At the end of 2023, at least 40,000 people were still displaced while their homes were being rebuilt due to their stone constructions and lack of reinforcing, many of them could not endure the seismic activity (UN RC/UNCT 2023). Since the Gorkha earthquake in 2015, which caused 2.6 million people to be displaced, the Karnali earthquake caused the second-highest number of related displacements in the nation (IDMC 2024). Other geophysical dangers, such as dry mass movements, which have also caused displacement in the area, are more likely to occur after earthquakes (IDMC 2024). Although they are rarely photographed, reports of similar movements surfaced in early 2023 in the Indian town of Joshimath in the state of Uttarakhand (IDMC

2024). The authorities classified the area a land subsidence zone and moved up to 4,000 residents to temporary relief centres when fractures showed in homes and infrastructure (National Emergency Response Centre 2023; NDTV 2023). By the end of 2023, about 1,000 individuals were still displaced (IDMC 2024). The most severe incident in India, which caused over 91,000 deaths, was in June 2023, across the state of Assam when 20 districts saw intense rainfall that caused several rivers to overflow (IDMC 2024). The effects of the floods were thought to have been exacerbated by soil erosion (Floodlist India 2023; HAI 2023). Assam frequently experiences flooding, and in recent years, there have been reports of considerable relocation (IDMC 2024). As a result, the government has taken a number of actions, such as creating a hazard atlas and a thorough disaster management plan that covers community relocation, evacuation coordination, recovery, and remedies (Cachar DDMA 2023; Government of Assam 2023).

These records shed light on the fact that climate change has not only intensified the disruptive ability of natural disasters but also affected its frequency and patterns, to which neither developed or underdeveloped countries are spared. However, the marginal communities of impoverished countries are the ones who have to undergo the majority of the suffering.

Case Study: Erosion and Internal Displacement in the Indian Sunderban Delta

The Indian Sundarban Delta (ISD) showcases one of the most climate-vulnerable deltaic systems globally, exposed to recurrent storm surges, cyclones, severe coastal erosion, flooding, and frequent embankment failures (Hajra and Ghosh 2018). This vulnerability is aggravated by rapid population growth within an ecologically fragile landscape (Ghosh 2012). The south-western edge of the ISD, along with the islands of Sagar, Ghoramara, and Mousani, has been identified as among the world's most at-risk deltaic zones due to heavy population concentrations and escalating environmental pressures (Hazra et al. 2002; Hazra et al. 2014; Hazra et al. 2016). Over-exploitation of natural resources, saline water inundation from cyclonic surges, and continuous shoreline erosion has dangerously degraded land quality and agricultural productivity, undermining the livelihoods of farming-dependent communities (Hajra and Ghosh 2018).

Empirical evidence collected from areas such as Ghoramara Island has highlighted rapid erosion and repeated embankment failures, enhancing exposure to storm surges and triggering large-scale displacement (Ghosh et al. 2014). Low-income farmers and fishers are often forced to migrate to mainland locations such as Kakdwip and Namkhana or to relatively larger islands like Sagar, frequently without institutional support or formal rehabilitation mechanisms (Ghosh et al. 2014). Extreme climatic events, especially Cyclone Amphan in 2020, have further intensified erosion processes, inundating homes and agricultural land and displacing additional households (Halder et al. 2021). Satellite data from 1975 to 2010 confirm a persistent erosion-intensive trend across the ISD, with several settlements now permanently submerged, resulting in forced migration and irreversible livelihood loss (Ghosh et al. 2014). Migration in the ISD context often functions as an adaptive response aimed at mitigating poverty and constrained access to resources. However, unregulated migration has also produced a growing pool of unskilled laborers absorbed into informal sectors such as construction and daily wage labour (Ghosh et al. 2003; Ghosh et al. 2014; Hajra and Ghosh 2018). Employment within the informal economy is characterized by job insecurity, exploitation, and the absence of wage regulation, further exacerbating socio-economic vulnerability among displaced populations (Hajra and Ghosh 2018).

A vulnerability assessment of the GBM delta based on exposure to climate hazards, sensitivity, and adaptive capacity identifies Gosaba, Basanti, Patharpratima, Sagar, Sandeshkhali-II, and Hingalganj as community development (CD) blocks facing 'high' climate risk (Haq et al. 2015). Persistent riverbank and coastal erosion, coupled with soil salinisation and land submergence, has rendered many residents internally displaced persons (IDPs). Previous erosion-control interventions—including mud bags, bamboo grids, and porcupine structures—have largely failed due to unstable soil conditions and riverbank undercutting, leaving both existing and planned infrastructure increasingly exposed to destruction.

Gosaba, an underdeveloped CD block comprising a cluster of islands in the South 24 Parganas district of West Bengal, exemplifies the lived realities of climate-induced displacement. Characterized by limited accessibility and infrastructural deficits, the region has experienced repeated embankment failures. In the southern part of Gosaba, near Pakhiralay and Tiger More, 42 families once resided along the riverbank. Progressive erosion forced all households to abandon their homes as land, ponds, and agricultural fields were submerged. Despite two consecutive embankment construction efforts, structural failures persisted, offering no long-term protection. Most affected families lacked alternative land ownership and belonged to economically weaker sections, rendering displacement both involuntary and economically devastating. The loss of agricultural land proved particularly catastrophic, directly undermining household food security and income stability.

At present, these displaced households have relocated approximately 200 m inland, where they reside in temporary huts clustered around a large pond. Limited financial resources compel them to remain within hazard-prone zones, exposing them to recurrent floods, cyclones, and ongoing erosion. A lack of awareness regarding legal rights and entitlements has further marginalized these communities, leaving them vulnerable to policy ambiguity and inadequate implementation of welfare schemes intended for economically weaker sections. One erosion-affected resident from southern Gosaba described the prolonged nature of displacement:

> Our house used to be at Dakshin Gheripara. We had quite some land there, but erosion started affecting it, and that land gradually went into the river. There was some land in front of our house, and we built our next home there. It was devastated as well when cyclone 'Aila' struck in 2009. We migrated to the island's interior, around 200 m from the river. Since then, we

> have been living here for the past nine years. We do not have any more land anywhere, and we are on shifting grounds.

These lived experiences reflect prolonged suffering exacerbated by governmental inaction and the absence of comprehensive rehabilitation policies. As climate change disproportionately burdens populations with minimal responsibility for its causes, migration has become an increasingly unavoidable coping strategy against environmental degradation. Displaced individuals in the ISD are predominantly drawn from marginalized, farm-based communities that have lost land, livelihoods, and family members to erosion, floods, and cyclones. For many, displacement is not a choice but an inevitable outcome, while others remain in high-risk zones due to the absence of viable alternatives.

Chronic underdevelopment across much of the ISD further deepens vulnerability by restricting access to education, skill development, and livelihood diversification. Economic opportunities beyond subsistence agriculture remain limited, reinforcing unemployment, low per capita income, and entrenched poverty. In this context, IDPs require targeted financial assistance, institutional support, and skill development initiatives from the state to cope with present climatic impacts and to build long-term adaptive capacity.

Policies in Practice

Global Policy

The protection of IDPs and those impacted by climate change is the focus of numerous international organizations like multiple UN agencies. The United Nations High Commissioner for Refugees (UNHCR) offers guidance on the application of current international human rights and refugee regulations to cross-border displaced people in the context of climate change. This includes the implementation of the 1951 Refugee Convention and relevant regional instruments and their guidelines in such circumstances for ensuring the immediate and long term protection of refugees' fundamental rights protection as per their 'Climate change impacts and cross-border displacement: International refugee law and UNHCR's mandate 12 December 2023' (UNHCR 2024). However, this mandate does not specifically mention the term 'Climate Refugees', 'Environmental Refugees' or 'Climate induced displacement', thus highlighting that environmental refugees constitute a grey area between refugees emerging due to natural disasters and those inflicted by violence and conflict, thereby highlighting their lack of representation in International Laws and legal frameworks. The UNHCR's Strategic Framework for Climate Action (2021) policy recognizes that climate change is a growing factor impacting global displacement, particularly when it intersects with poverty, violence, and weak governance. It therefore seeks to safeguard climate migrants and internally displaced people (UNHCR 2024). However, its approach to protection does not include placing IDPs under a separate legal category. Instead, it emphasizes using existing structures to ensure protection, particularly in cases where environmental degradation coexists with persecution or conflict (UNHCR 2024). Promoting legal frameworks to protect displaced people and bolstering efforts to mitigate and adapt to climate change through partnerships and international coordination with states, non-governmental organizations, and international bodies (such as the United Nations Framework Convention on Climate Change (UNFCCC)) to address displacement risks in climate-vulnerable regions is among their main areas of focus (UNHCR 2024). The International Organization for Migration (IOM) has implemented several policies and strategies to address the challenges faced by individuals displaced due to climate change. The IOM's commitment to environmental migration is outlined in its "Institutional Strategy on Migration, Environment, and Climate Change (2021–2030)" and includes the following goals: assisting and protecting migrants and displaced people impacted by environmental factors; and developing solutions by facilitating safe and regular migration pathways in the context of climate change (IOM 2024). In order to reduce forced migration, the program also seeks to strengthen communities' adaptability and resilience (IOM 2024). By ensuring that migration policies take climate change into account, these goals seek to assist migrants—including displaced people—and their communities (IOM 2024). IOM highlights the significance of adopting precise language and promotes the use of "climate migrants" rather than "climate refugees," since the latter is not acceptable under the law. To safeguard people who have been displaced by environmental circumstances, the organization promotes the application of current legal frameworks, such as humanitarian, human rights, and refugee legislation. The United Nations Development Programme (UNDP) promotes the incorporation of human mobility factors into national climate change planning and policy. In order to guarantee that the needs of displaced populations are met in a comprehensive manner, this strategy seeks to close the gap between the humanitarian, development, climate, and environmental sectors. This integration, especially throughout Latin America and the Caribbean, is highlighted in a policy brief created under UNDP's Climate Promise. The UNDP acknowledges the growing incidence of migration brought on by climate change as well as the current policy shortcomings in this field (International Energy Agency 2014). It highlights the necessity of all-encompassing strategies that include support, safety, and long-term fixes for individuals displaced by climate change. This entails helping voluntary migrants adjust to climate change and controlling climate hazards for those who are left (International Energy Agency 2014).

The UNFCCC first acknowledged the importance of human mobility in relation to climate change when it created the Cancun Adaptation Framework (2010) during the 16th Conference of the Parties (COP16) in Cancun, Mexico. This approach prompted countries to incorporate migration, planned relocation, and displacement brought on by climate change into their adaptation plans. In order to address medium- and long-term adaptation demands, including possible displacement difficulties, it also suggested creating national adaptation plans, or NAPs (Diab and Scissa 2023). The UNFCCC created the Warsaw International Mechanism (WIM) at COP19 in Warsaw, Poland, to redress loss and damage related to the effects of climate change. Migration was acknowledged by the WIM as a crucial topic of concern. However, it has been revealed by policy critiques that despite all these measures, most of the nations that were part of the Cancun Adaptation Framework have not actually implemented the NAPs and therefore the gap in policy lies in implementation.

There are certain International Non-Governmental Organizations (NGOs) like the International Federation of Red Cross and Red Crescent Societies (IFRC) which provides emergency relief and disaster preparedness, the Refugees International which advocates for policies to support refugees and displaced persons, including climate migrants, the Norwegian Refugee Council (NRC) which focuses on displaced populations, including those affected by climate change, the Internal Displacement Monitoring Centre (IDMC) that researches and tracks displacement trends worldwide, CARE International which aims to work on climate resilience and humanitarian aid for displaced communities, the Oxfam International that supports the climate adaptation and displacement responses and the Human Rights Watch (HRW) that documents human rights violations, including those affecting refugees and climate migrants. Some global financial and development institutions like the World Bank have undertaken the Climate Change Action Plan which funds projects that help communities adapt to climate change and reduce displacement, the Green Climate Fund (GCF) that provides financial resources for climate adaptation and mitigation, reducing forced migration. The Asian Development Bank (ADB) that has initiated the Climate Change and Migration Initiative which supports policies to address climate migration in Asia.

Policy in European Union (EU)

The European Union (EU) provides assistance to refugees and displaced people both inside and outside of Europe through its European Civil Protection and Humanitarian Aid Operations (ECHO). The Temporary Protection Directive was created in 2001 with the intention of standardizing EU policies toward displaced people during periods of large-scale immigration (Warner 2010). By streamlining administrative processes and guaranteeing access to social welfare, employment, and residency permits, it provides temporary and urgent protection. Nevertheless, the regulation has been activated infrequently, and its implementation is subject to a formal decision by the EU Council. However, relocation brought on by environmental factors is not specifically addressed (Warner 2010). The European Commission published a staff working document in July 2022 with the title "Addressing displacement and migration related to disasters, climate change, and environmental degradation" (European Commission 2022; DG ECHO 2023). This study identifies future goals in this area and describes current EU policies and practices on displacement caused by disasters and climate change (European Commission 2022; DG ECHO 2023). It shows that the necessity to address the intricate relationship between migration and climate change is becoming increasingly recognized at the EU level (European Commission 2022; DG ECHO 2023). People who do not fit the definition of refugees but require protection because of war, major conflicts, or natural catastrophes are covered by

Sweden's asylum system (Warner 2010). This clause makes it possible to provide protection to those who are unable to return to their native countries due to natural disasters (Warner 2010). In a similar approach, the 'Aliens Act' of Finland allows for the granting of residency permits to people who are unable to return to their homeland because of an environmental catastrophe and this shows that environmental factors are recognized as valid reasons for protection (Warner 2010).

Policy in SIDS and Latin American Nations

The Pacific Islands Forum (PIF) deals with displacement and migration caused by climate change in Pacific Island countries. Table 18.1 provides a comparative overview of climate change-induced displacement policies in selected SIDS and Latin American nations, highlighting national strategies, policy effectiveness, and key limitations, with particular emphasis on the role of regional mechanisms such as the Pacific Islands Forum (PIF).

Policy in African Nations

African nations, including Nigeria, Kenya, and Mozambique, are adopting climate-smart agricultural policies, resettlement initiatives, and early warning systems to address both internal and cross-border displacement. Table 18.2 presents a comparative overview of climate change-induced displacement policies in selected African nations, outlining national adaptation measures, institutional responses, policy effectiveness, and key limitations in addressing climate-driven migration.

Policy in Coastal Nations

Coastal nations like Bangladesh, the Netherlands, and the United States are investing in infrastructure resilience, managed retreat, and floating city designs to mitigate migration pressures. Table 18.3 summarises climate change-induced displacement policies in selected Asian coastal and deltaic regions, highlighting national adaptation strategies, institutional frameworks, policy effectiveness, and key implementation challenges.

National Policies in India

India currently lacks a comprehensive national legal framework specifically addressing planned relocations due to climate change. DECCMA's analysis of national policies formulated for the Indian Bengal Delta region between 1990 and 2016 highlights that out of all these policies, only six exist that had taken the necessity of adaptation to climate change into consideration (Ayeb-Karlsson et al.

Table 18.1 Climate change-induced displacement policies in SIDS and selected Latin American nations

Country	Policies	Effectiveness	Policy implications and limitations
Kiribati	The government of Kiribati acquired 20 square kilometers of land on the second-largest island in Fiji, Vanua Levu in 2014 (Hermann and Kempf 2017). Although this acquisition was formally planned for fish farming and agricultural projects to increase food security, it has been widely perceived as a possible future relocation site for Kiribati's population, who are at risk from rising sea levels (Hermann and Kempf 2017)	This policy is characterized as an anticipatory and dynamic adaptation approach. This differentiates it from other predominantly reactive climate policies prevalent in most SIDS. By procuring land beyond national boundaries, Kiribati is investing in a long-term climate response in contrast to the other SIDS which mostly focus on short term goals like post disaster management, developing 'in-situ adaptation' like building sea walls, embankments, financing mangrove restoration. Although the policy's effectiveness as a planned relocation method can be challenged by the lack of a well defined legal, institutional, and governance arrangements concerning cross-border resettlement. Severe and sensitive issues like citizenship status, land tenure rights, socio-cultural integration, and political representation in the host country still remain unresolved.	The Kiribati–Fiji land acquisition underlines the urgent need for governing bodies to formulate internationally recognized frameworks to support swift planned cross-border relocation as an adaptation strategy. It expresses the limitations of ad hoc or bilateral initiatives in the absence of formal agreements addressing sovereignty, legal protection, and long-term integration of displaced populations. More broadly, this case demonstrates that land-based adaptation measures must be embedded within comprehensive climate mobility governance structures, combining anticipatory planning with enforceable legal safeguards to ensure equitable, sustainable, and rights-based outcomes for climate-displaced communities.

(continued)

Table 18.1 (continued)

Country	Policies	Effectiveness	Policy implications and limitations
Tuvalu	The 'Falepili Union pact', which established a unique human migration corridor for Tuvalu citizens, was signed by Australia and Tuvalu in November 2023 (Schofield and Anggadi 2024). In order to allay Tuvalu's worries about displacement brought on by climate change, this agreement permits up to 280 Tuvaluans to migrate to Australia each year (Schofield and Anggadi 2024)	The Falepili Union Pact observes a significant innovation in climate mobility governance by formally institutionalizing cross-border migration as an adaptive strategy. Its effectiveness underlies in its forward looking recognition of displacement risks and the creation of a legally sanctioned, orderly, and rights-based migration channel. By ensuring gradual and voluntary migration, the pact intends to decrease the likelihood of sudden humanitarian crises interconnected with abrupt displacement. Although, its effectiveness can be challenged by the limited scale of admissions relative to Tuvalu's total population and by uncertainties surrounding crucial matters like long-term residency rights, citizenship pathways, and social integration outcomes for migrants	The pact forms an important pattern for bilateral climate mobility agreements, illustrating how destination countries can rationalize operations through responsibility-sharing in response to climate change. It highlights how migration corridors can be complemented with in-situ adaptation and planned relocation strategies, specifically for economically weaker nations with limited territorial options. Simultaneously, the agreement underscores the urgency for wider regional and multilateral frameworks to ensure scalability, sustainability and equity beyond bilateral arrangements. For climate-vulnerable states, the pact demonstrates both the assurance and limitations of migration-based adaptation, reinforcing the necessity of integrating mobility pathways within comprehensive, long-term climate governance systems

(continued)

Table 18.1 (continued)

Country	Policies	Effectiveness	Policy implications and limitations
Fiji	Fiji has taken the initiative to create prepared relocation plans for communities at risk from the effects of climate change. To aid in the relocation of impacted villages, the Fijian government developed the 'Planned Relocation Guidelines' in 2018. Benge and Neef (2020) prioritized the importance of community involvement and also identified the barriers towards tailor out the plans into practice by the government. Furthermore, Bertana (2020) highlights the complexity of these adaptation techniques by revealing the impact of power dynamics in community engagement throughout relocation operations in Fiji	The guidelines constitute a substantial development in climate adaptation governance by providing a formal framework for planned relocation within national boundaries. Their strong focus on community engagement in the form of consultations enhances social legitimacy and helps mitigate resistance to relocation by taking into account the indigenous traditional knowledge. Although empirical studies have revealed substantial implementation challenges. Benge and Neef (2020) identify issues related to administrative capacity, financial resources, and the translation of participatory principles into operational practice. Bertana (2020) further illustrates that power asymmetries within communities and between state actors and local populations often shape decision-making processes, complicating equitable participation and potentially putting vulnerable groups in precarious positions. These factors limit the consistent effectiveness of the guidelines in practice	This policy implies that formal relocation frameworks is necessary but not sufficient enough to sustain strong institutional capacity, prolonged funding, and mechanisms to address internal power dynamics. It focuses on the the requirement for clearer accountability structures, long-term financing methods which will ensure inclusive participation across gender, class, and social hierarchies. Fiji's Planned Relocation Guidelines provides an important policy model for other SIDS, demonstrating both the prospects and limitations of planned relocation as an adaptation strategy. For such frameworks to be scalable and adaptable, they must be embedded within comprehensive climate governance systems that prioritize equity, transparency, and long-term livelihood security for relocated communities

Source Compiled by the authors based on Pacific Islands Forum Secretariat documents, national policy frameworks, and secondary literature (Hermann and Kempf 2017; Warner 2010; SPREP 2024).

2016). Table 18.4 provides an overview of climate change-induced displacement and

Table 18.2 Climate change-induced displacement policies in selected African nations

Country	Policies	Effectiveness	Policy Implications and Limitations
Egypt	• With assistance from international partners like the UNDP and the Green Climate Fund (GCF), the Egyptian government has put in place a number of adaptation measures to mitigate the effects of climate change, especially land degradation and sea level rise, in the Nile Delta region. • Among the coastal protection infrastructure undertaken by the Egyptian Government include the construction of sea walls, wave breakers, and embankments to shield vulnerable areas from sea-level rise and coastal erosion • The UNDP and GCF funded project of "Enhancing Climate Change Adaptation in the North Coast of Egypt" (2017–2024) aimed at protecting 69 km of coastline using bio-engineering solutions (UNDP 2017) • The Egyptian government further undertook the restoration of coastal dunes and wetlands through mangroves to act as natural buffers against flooding and saltwater intrusion (El Raey 2010) • To protect the IDPs of the Nile Delta region, the government has further instilled climate resilience into 'Egypt's Vision 2030' and 'National Adaptation Plan (NAP)' (United Nations Development Programme 2018a, b; Government of Egypt 2022; International Energy Agency 2022)	Key infrastructure-based measures include the construction of seawalls, wave breakers, and embankments designed to shield exposed areas from coastal erosion and storm surges. Complementing these hard-engineering approaches, the UNDP–GCF funded project *"Enhancing Climate Change Adaptation in the North Coast of Egypt"* (2017–2024) has sought to protect approximately 69 km of coastline through bio-engineering solutions, marking a shift toward more environmentally sustainable adaptation practices (UNDP 2017). The restoration of coastal dunes and wetlands, including mangrove-based natural buffers, has further strengthened ecosystem-based protection against flooding and saltwater intrusion (El Raey 2010). Collectively, these measures have enhanced coastal resilience and delayed land loss in high-risk zones, thereby reducing immediate displacement pressures. However, the effectiveness of these interventions remains contingent on long-term maintenance, financial sustainability, and the accelerating pace of sea-level rise. While protective infrastructure can reduce short- to medium-term risks, it does not fully eliminate the potential for future displacement, particularly in low-lying and subsiding areas of the Nile Delta	The integration of climate resilience into national development frameworks such as *Egypt's Vision 2030* and the *National Adaptation Plan (NAP)* demonstrates a strategic commitment to mainstreaming climate adaptation into long-term planning and safeguarding populations vulnerable to displacement (UNDP 2018a, b; Government of Egypt 2022; International Energy Agency 2022). This policy alignment strengthens institutional coherence and enhances access to international climate finance. Nevertheless, Egypt's experience highlights the limitations of infrastructure-centric adaptation when confronted with intensifying climate hazards. The case underscores the need to complement protective measures with forward-looking displacement planning, livelihood diversification, and social protection mechanisms for internally displaced persons (IDPs). More broadly, the Nile Delta illustrates how adaptation policies can reduce immediate risks while simultaneously revealing the necessity for integrated approaches that link in-situ resilience with long-term human mobility strategies

Source Compiled by the authors based on national policy documents, reports from international organisations including IOM, IDMC, UNDP, and relevant secondary literature.

Table 18.3 Climate change-induced displacement policies in selected Asian coastal and deltaic regions

Country	Policies	Effectiveness	Policy implications and limitations
Bangladesh	The 2009 Climate Change Strategy and Action Plan for Bangladesh (BCCSAP) is a comprehensive plan that emphasizes the protection and resettlement of populations affected by climate impacts while outlining adaptation and mitigation options (Kisinger and Matsui 2021) Bangladesh regularly participates in global conferences to draw attention to the predicament of climate migrants. In March 2023, Foreign Secretary Masud Bin Momen emphasized the necessity for international assistance in carrying out Bangladesh's National Adaptation Plan 2023–2050 and urged for global action and enhanced climate funds to support displaced people (AIIB 2024)	The BCCSAP has been effective in mainstreaming climate adaptation into national development planning and in recognizing displacement as a central climate risk. Its focus on infrastructure resilience, disaster risk reduction, and resettlement planning has contributed to improved preparedness and reduced mortality from extreme weather events. Bangladesh's continued engagement in international climate forums further enhances policy effectiveness by mobilizing external financial and technical support. In March 2023, Bangladesh's Foreign Secretary Masud Bin Momen underscored the country's dependence on international assistance to implement the *National Adaptation Plan (NAP) 2023–2050*, calling for increased climate finance and collective global responsibility to support climate-displaced populations (AIIB 2024). This sustained diplomatic advocacy has positioned Bangladesh as a leading voice on climate migration in global negotiations Nevertheless, implementation gaps persist due to resource constraints, high population density, and the scale of exposure, limiting the reach of resettlement and long-term livelihood rehabilitation programmes	Bangladesh's experience demonstrates that early recognition of climate displacement within national policy frameworks is crucial for anticipatory planning. However, it also highlights the structural limitations of domestic adaptation efforts in the absence of adequate international support. The country's reliance on global climate finance underscores the need for stronger multilateral mechanisms to address loss and damage and to support large-scale displacement governance. Bangladesh's approach illustrates how national strategies can lay the institutional groundwork for displacement management, but long-term effectiveness will depend on sustained international cooperation, predictable funding flows, and the integration of mobility planning into broader development and social protection systems
USA	The United States Department of Housing and Urban Development (HUD) provided $48.3 million in 2016 to relocate people from Isle of Jean Charles, located in Terrebonne Parish, Louisiana, in response to the environmental issues of deltaic erosion and sea level rise (Stankard 2024). This program is the first community-scale, climate-driven resettlement program in the US to get government funding (Stankard 2024). The implementation of the resettlement project has been the subject of various studies, which have highlighted issues including ambiguous federal leadership and the lack of defined procedures for relocations brought on by climate change (Soundararajan 2022). These elements have caused strained community relations and extended migration schedules	The programme's primary effectiveness lies in its symbolic and institutional significance, as it established a precedent for recognizing climate-driven resettlement as a legitimate policy response. The allocation of federal funds enabled the acquisition of relocation land, housing development, and initial community planning, thereby reducing immediate exposure to environmental risks. However, empirical assessments reveal considerable implementation challenges. Studies highlight ambiguous federal leadership, fragmented governance responsibilities, and the absence of clearly defined procedural frameworks for climate-induced relocation (Soundararajan 2022). These governance gaps contributed to prolonged implementation timelines, uncertainty among affected residents, and strained community relations, ultimately undermining the programme's operational effectiveness	The Isle de Jean Charles resettlement underscores the critical importance of clear institutional mandates, coordinated governance structures, and transparent relocation protocols in climate-driven displacement contexts. While the programme demonstrates that high-income countries possess the financial and administrative capacity to support planned relocation, it also reveals that the absence of established legal and procedural frameworks can significantly impede outcomes. The case highlights the need for a nationally coherent climate displacement policy in the United States, including standardized guidelines, community consent mechanisms, and long-term livelihood restoration strategies. More broadly, it illustrates that financial resources alone are insufficient; effective climate resettlement requires robust governance, sustained community engagement, and legal clarity to ensure just and timely relocation outcomes

Source Compiled by the authors based on national adaptation plans, government policy documents, and reports from international organisations including UNDP, IOM, IDMC, and relevant secondary literature relocation policies in India, outlining national and state-level frameworks, policy effectiveness, and major gaps in addressing long-term climate-driven displacement.

Policies in Indian States

Some states have undertaken relocation projects in response to environmental challenges. For instance, the Odisha state government relocated residents from the coastal village of Satabhaya due to severe erosion and sea-level rise. This relocation was implemented under state disaster management policies, namely the Orissa Resettlement and Rehabilitation Policy of 2006, which aims to enhance the quality of life of the displaced population by recognizing them in the first place and providing them with a relocated settlement that includes better infrastructural facilities, highlighting the role of state governments in addressing specific environmental challenges (Government of Odisha 2006).

The policy that has been implemented in the study area of the case study is the Gitanjali Scheme, a West Bengal government initiative, which is a housing scheme that has been designed for families displaced due to environmental hazards. The scheme specifies that the housing facilities are to be provided and constructed by the District Administration, Municipality, or Development Authorities (Government of West Bengal 2014). Unfortunately, the rural aspect of this policy is only for those families who own land. Discovery in the southern Gosaba region revealed that most families who lost their homes, lands, and property to erosion were not eligible for the Gitanjali housing scheme as they did not possess another piece of land. These families come from economically backward communities. Although most have displaced themselves, some continue to live at the river's mercy, which continues to erode into the island's interiors.

Policy Recommendations

Addressing climate-induced displacement demands a comprehensive and forward-looking policy response rooted in justice, inclusion, and long-term sustainability. First, there is a critical need to expand international and national legal frameworks to formally recognise climate-induced displacement and environmental refugees, who currently fall outside the scope of existing refugee conventions. Legal recognition would provide the foundation for consistent protection mechanisms, cross-border cooperation, and accountability in responding to climate-driven mobility.

At the national level, governments—particularly in highly vulnerable countries such as India—should prioritise the formulation of dedicated climate-induced displacement and planned relocation policies. These frameworks must move beyond immediate disaster relief to address slow-onset processes such as sea-level rise, erosion, and salinisation. Integrating climate-induced displacement into existing

Table 18.4 Climate change-induced displacement and relocation policies in India

Policies	Effectiveness	Limitations
The National Action Plan on Climate Change (NAPCC), which was launched in 2008, outlines India's strategy to tackle climate change through eight national missions, focusing on areas like solar energy, energy efficiency, and sustainable agriculture. While it emphasizes adaptation and mitigation, it does not explicitly address planned relocations (Government of India 2008)	The effectiveness of the NAPCC lies in its comprehensive, mission-based approach, which has facilitated measurable progress in renewable energy deployment, climate-resilient agricultural practices, and institutional awareness of climate risks. By integrating adaptation into sector wise missions, the plan has strengthened India's capacity to address climate variability across multiple domains. However, despite its broad scope, the NAPCC does not explicitly recognize climate-induced displacement or planned relocation as adaptation strategies. This omission limits its effectiveness in addressing the needs of populations facing irreversible environmental degradation, particularly in high-risk regions such as coastal deltas, riverine floodplains, and erosion-prone areas	The absence of planned relocation frameworks within the NAPCC highlights a critical policy gap in India's climate governance architecture. As climate impacts intensify and in-situ adaptation becomes insufficient in certain regions, reliance on sector wise adaptation measures alone may exacerbate displacement vulnerabilities. This underscores the need to expand the NAPCC or complement it with dedicated policy instruments that explicitly address climate-induced mobility, internal displacement, and resettlement. Integrating planned relocation into national climate policy would enhance policy coherence, ensure better protection for affected communities, and align India's climate strategy with emerging global norms on climate mobility and displacement governance.

(continued)

Table 18.4 (continued)

Policies	Effectiveness	Limitations
The Disaster Management Act of 2005 establishes a legal framework for disaster risk reduction and management in India. It facilitates the creation of the National Disaster Management Authority (NDMA), responsible for formulating policies and guidelines to ensure effective disaster response and preparedness. The act empowers authorities to plan and implement evacuation and relocation during disasters, but it does not specifically address relocations due to gradual environmental changes like sea-level rise (Government of India 2005)	The DMA has been effective in strengthening India's disaster response architecture by institutionalizing early warning systems, emergency evacuation protocols, and inter-agency coordination mechanisms. It empowers authorities to plan and implement evacuation and temporary relocation during acute disasters such as cyclones, floods, earthquakes, and tsunamis. This has contributed to improved preparedness and reduced loss of life during sudden-onset events. However, the Act is primarily oriented toward emergency response and short-term displacement management. It does not explicitly address relocation or resettlement arising from slow-onset environmental changes such as sea-level rise, chronic coastal erosion, or salinisation, thereby limiting its applicability to long-term climate-induced displacement scenarios	The absence of provisions for gradual environmental displacement within the DMA reveals a structural limitation in India's disaster governance framework. As climate change increasingly manifests through slow-onset processes, reliance on emergency-centric legislation may result in ad hoc and reactive relocation practices. This underscores the need to expand or reinterpret the DMA to incorporate long-term displacement planning, rehabilitation, and livelihood restoration for affected populations. Integrating slow-onset climate risks into disaster management legislation would enhance policy coherence, bridge the gap between disaster response and climate adaptation, and provide stronger legal protection for communities facing permanent environmental loss
The Rehabilitation And Resettlement Policies undertaken by India concern the rehabilitation and resettlement of communities affected by infrastructure projects, such as the Right to Fair Compensation And Transparency in Land Acquisition, Rehabilitation and Resettlement Act, of 2013 The National Rehabilitation and Resettlement Policy of 2007, constructed by the Department of Land Resources of the Ministry of Rural Development, requires acknowledgement of the existence of IDPs borne out of environmental degradation. When IDPs are concerned, the National/State level Governing bodies have mainly focused on the resettlement and rehabilitation policies and schemes (Ministry of Rural Development, Government of India 2007) These policies focus on development-induced displacement and do not specifically cater to climate-induced relocations (Government of India 2013)	India's rehabilitation and resettlement framework, particularly the *National Rehabilitation and Resettlement Policy (NRRP), 2007* and the *Right to Fair Compensation and Transparency in Land Acquisition, Rehabilitation and Resettlement Act (RFCTLARR), 2013*, has been effective in formalizing compensation, resettlement, and livelihood restoration mechanisms for development-induced displacement. The acknowledgement of internally displaced persons (IDPs) arising from environmental degradation represents an important conceptual expansion within displacement governance	Despite this progress, these policies remain primarily oriented toward project-based displacement and do not explicitly address climate-induced relocation or displacement caused by slow-onset environmental change. This reveals a significant policy gap, indicating the need to extend rehabilitation and resettlement frameworks to incorporate climate-induced mobility within India's broader climate adaptation and disaster governance strategies

Source Compiled by the authors based on Government of India and state government policy documents, disaster management frameworks, and secondary academic literature disaster management, land acquisition, and rehabilitation policies would reduce legal ambiguity and ensure continuity of protection for affected communities.

Policy design must also emphasise community-centred planning. Relocation and adaptation strategies should be participatory, respecting local livelihoods, cultural identities, and social networks. Strengthening local governance institutions, investing in climate-resilient infrastructure, and enhancing early warning systems can empower communities to adapt in situ where possible and relocate with dignity where necessary. Dedicated funding mechanisms and sustained international support—particularly for Small Island Developing States and low-income coastal regions—are essential to translate policy intent into effective implementation.

Finally, regional and international cooperation should be strengthened through shared frameworks, data exchange, and financing mechanisms that recognise climate mobility as a collective global challenge. Only through coordinated, rights-based, and anticipatory policy approaches can climate-induced displacement be managed in a manner that safeguards human dignity, reduces vulnerability, and supports sustainable development in an era of accelerating climate change.

Conclusion

Climate-induced migration is no longer a distant or hypothetical concern; it is an ongoing and intensifying reality across vulnerable coastal, island, and deltaic regions worldwide. The disproportionate burden borne by economically and socially marginalised communities underscores the urgency of proactive, inclusive, and rights-based adaptation strategies. Evidence from Small Island Developing States, African and Asian coastal regions, and the Indian Sundarban Delta demonstrates that while policy responses have begun to emerge at national and regional levels, however, they remain fragmented, reactive, and often insufficient to address the scale and complexity of climate-driven displacement.

The analysis reveals persistent governance gaps, particularly the absence of coherent legal frameworks that explicitly recognise and protect populations displaced by gradual environmental change. Existing mechanisms largely prioritise short-term disaster response or development-induced displacement, leaving climate-induced mobility inadequately addressed. As climate hazards intensify, these institutional shortcomings risk deepening long-term vulnerability, exposing displaced populations to insecure land tenure, livelihood instability, and limited access to social protection.

This chapter highlights that climate-induced displacement is not solely an environmental issue but a multidimensional challenge encompassing social justice, economic stability, and human rights. Addressing climate mobility therefore requires coordinated action across global, regional, and national scales, supported by strengthened legal recognition, sustained financing, and climate-resilient development pathways. Without such integrated approaches, millions will continue to face forced displacement with limited institutional support, undermining both human security and sustainable development goals.

AI Discloser The authors declare that generative artificial intelligence (AI) tools were not used in the conceptualization, analysis, interpretation, or substantive writing of this chapter. The content is based entirely on the authors' original intellectual contributions and critical engagement with existing academic literature and policy documents. Any use of digital tools was limited to basic language editing and formatting support and does not affect the originality, accuracy, or integrity of the scholarly work.

References

Abayomi O, Olayemi TE, Ogungbade T (2021) Environmental pollution and its ecological consequences on the Niger Delta: a review of the literature. Afr J Environ Nat Sci Res 4:27–42

Abu M, Atiglo DY, Addoquaye Tagoe C. and Codjoe S.N. (2022) Drivers of migration intentions in the Volta Delta: investigating the effect of climate-related hazards and adaptation strategies. Front Clim 4:975650. https://doi.org/10.3389/fclim.2022.975650

ACAPS (2023) Briefing note: Nepal earthquake, 10 November 2023. https://www.acaps.org. Accessed 13 Sept 2025

Akpodiogaga-a P, Odjugo O (2010) General overview of climate change impacts in Nigeria. J Hum Ecol 29(1):47–55

Anderson CC, Renaud FG, Hagenlocher M, Day JW (2021) Assessing multi-hazard vulnerability and dynamic coastal flood risk in the Mississippi Delta: the global delta risk index as a social-ecological systems approach. Water 13(4):577

Appeaning Addo K, Brempong EK, Jayson-Quashigah PN (2020) Assessment of the dynamics of the Volta river estuary shorelines in Ghana. Geoenviron Disasters 7:19

ASEAN (2023) Weekly disaster update, 16–22 January 2023. https://reliefweb.int. Accessed 13 Sept 2025

Asian Infrastructure Investment Bank (AIIB) (2024) Bangladesh: climate resilient inclusive development program (subprogram 1). AIIB, Beijing. https://www.aiib.org/en/projects/details/2024/approved/Bangladesh-Climate-Resilient-Inclusive-Development-Program-Subprogram-1.html

Atchoaréna D, Dias Da Graça P, Marquez JM (2008) Strategies for post-primary education in small island developing states (SIDS): lessons from Cape Verde. Comp Educ 44(2):167–185

Atiglo Y, Codjoe S (2023) Migration in the Volta Delta: a review of relevant literature

Ayeb-Karlsson, S., Van der Geest, K., Ahmed, I., Huq, S. And Warner, K. (2016). A people-centred perspective on climate change, environmental stress, and livelihood resilience in Bangladesh. Sustain Sci 11(4):679–694

Ayub, A. (2023) Demolished and displaced: how Karachi's anti-encroachment razing has created a climate refugee crisis., https://pulitzercenter.org/stories/demolished-and-displaced-how-karachis-anti-encroachment-razing-has-created-climate-refugee. Accessed 13 Sept 2025

Babatunde EO (2020) Distributive justice in the age of climate change. Can J Law Jurisprud 33(2):263–292

Baker Institute (2024) Sea-level rise in the Nile Delta: promoting adaptation through circular migration. https://www.bakerinstitute.org/research/sea-level-rise-nile-delta-promoting-adaptation-through-circular-migration. Accessed 13 Sept 2025

Bayrak MM, Van Hieu T, Tran TA, Hsu YY, Nien T, Quynh DTT (2023) Climate change adaptation responses and human mobility in the Mekong Delta: local perspectives from rural households in An Giang Province, Vietnam. Human Soc Sci Commun 10(1):1–14

Benge L, Neef A (2020) Planned relocation as a contentious strategy of climate change adaptation in Fiji. In: Climate-induced disasters in the Asia-Pacific region: response, recovery, adaptation, vol 22. Emerald Publishing Limited, pp 193–212

Bera MK (2013). Environmental refugee: a study of involuntary migrants of Sundarban islands. In: Proceedings of the 7th international conference on Asian and Pacific Coasts (APAC 2013) Bali, Indonesia, pp 916–925

Bergmann J (2024) Advocating for change: a compendium on climate mobility. Climate Migration Council, Palo Alto, California

Bertana A (2020) The role of power in community participation: Relocation as climate change adaptation in Fiji. Environ Plann C: Polit Space 38(5):902–919. https://doi.org/10.1177/2399654420909394

BNPB (2023) Pascabanjir dan longsor Kota Manado, warga lakukan pembersihan, 30 January 2023. https://bnpb.go.id. Accessed 13 Sept 2025

Cachar DDMA (2023) Disaster risk reduction roadmap. https://cahcar.gov.in. Accessed 13 Sept 2025

CARE (2023) CARE Nepal launches emergency response to earthquake, 4 November 2023. https://care.org. Accessed 13 Sept 2025

Connell J (2016) Last days in the Carteret Islands? Climate change, livelihoods and migration on coral atolls. Asia Pac Viewpoint 57(1):3–15. https://doi.org/10.1111/apv.12118

Day JW Jr, Boesch DF, Clairain EJ, Kemp GP, Laska SB, Mitsch WJ, Orth K, Mashriqui H, Reed DJ, Shabman L, Simenstad CA (2007) Restoration of the Mississippi Delta: lessons from hurricanes Katrina and Rita. Science 315(5819):1679–1684

Day JW, Xu YJ, Keim BD, Brown VM, Giosan L, Mann ME, Stephens JR (2024) Emerging climate threats to the Mississippi River Delta: Moving from restoration to adaptation. One Earth 7(4):558–571

Diab JL, Scissa C (2023) Gender, migration and environment in the MENA: vulnerabilities, frameworks and ways forward. J Migration Affairs 5(1–2):11–20

Directorate-General for European Civil Protection and Humanitarian Aid Operations (DG ECHO) (2023) Annual activity report 2022: European civil protection and humanitarian aid operations (ECHO). European Commission, Brussels. https://commission.europa.eu/publications/annual-activity-report-2022-european-civil-protection-and-humanitarian-aid-operations-echo_en. Accessed 13 Sept 2025

Earth Journalism Network (2023) Egypt's Nile Delta under threat, Part 2: can the country reclaim what it's lost?. https://earthjournalism.net/stories/egypts-nile-delta-under-threat-part-2-can-the-country-reclaim-what-its-lost. Accessed 07 Sept 2025

Easton-Calabria E, Sturridge C (2025) Mobile displacement or displaced mobility?: Using transition theory to challenge migration-displacement binaries and biases. Migr Stud 13(1):mnae046. https://doi.org/10.1093/migration/mnae046

El Raey M (2010) Impact of sea level rise on the Arab Region. Dissertation. University of Alexandria. Arab Academy of Science, Technology, and Maritime

Elshinnawy I, Foad IE, Salama I, Issa UH (2017) Impacts of sea level rise on coastal flooding of Nile Delta, Egypt. Int J Sci Res Sci Eng Technol 3(2):71–77

European Commission (2022) Addressing displacement and migration related to disasters, climate change and environmental degradation. Commission Staff Working Document, Publications Office of the European Union, Luxembourg. https://doi.org/10.2795/302703. ISBN 978-92-76-59659-2

Fernandez PNJ, Fernandez EGB, Cadondon JG, Subade RF (2024) Enhancing coastal management through the design and development of an in situ water quality monitoring system. J Coastal Res 40(6):1090–1102

Floodlist India (2023) Over 100,000 hit by floods in 20 districts of Assam, 22 June 2023. https://floodlist.com/asia/india-assam-floods-june-2023. Accessed 13 Sept 2025

Floodlist Indonesia (2023) Sumatra Island floods leave 3 dead and 15,000 homes damaged, 24 January 2023. https://floodlist.com/asia/indonesia-sumatra-island-floods-january-2023. Accessed 13 Sept 2025

Floodlist Philippines (2023) Death toll rises after further floods and landslides, 23 January 2023. https://floodlist.com/asia/philippines-floods-landslides-january-2023. Accessed 13 Sept 2025

Foufoula-Georgiou E, Syvitski J, Paola C, Hoanh CT, Tuong P, Vörösmarty C, Kremer H, Brondizio E, Saito Y, Twilley R (2011) International year of deltas 2013: a proposal. Eos Trans Am Geophys Union 92(40):340–341
Ghosh T (2012) Sustainable coastal tourism: problems and management options. J Geogr Geol 4(1):163. https://doi.org/10.5539/jgg.v4n1p163
Ghosh T, Hajra R, Mukhopadhyay A (2014) Island erosion and afflicted population: crisis and policies to handle climate change. In: International perspectives on climate change: Latin America and beyond. Springer, Cham, pp 217–225
Ghosh T, Bhandari G, Hazra S (2003) Application of a 'bio-engineering' technique to protect Ghoramara Island (Bay of Bengal) from severe erosion. J Coast Conserv 9:171–178. https://doi.org/10.1652/1400-0350(2003)009[0171,AOABTT]2.0.CO;2
Government of Assam (2023) Disaster management plan. https://assam.gov.in. Accessed 13 Sept 2025
Government of Canada (2023), Canada's record-breaking wildfires in 2023: A fiery wake-up call. https://www.canada.ca/en/services/environment/news/2023/10/canadas-record-breaking-wildfires-in-2023%2D%2Da-fiery-wake-up-call.html. Accessed 13 Sept 2025
Government of Egypt (2022) Egypt National Climate Change Strategy (NCCS) 2050. Ministry of Environment, Cairo
Government of India (2005) The Disaster Management Act, 2005. Ministry of Law and Justice
Government of India. (2008). National Action Plan on Climate Change (NAPCC). Prime Minister's Council on Climate Change
Government of India (2013). Right to fair compensation and transparency in land acquisition, Rehabilitation and Resettlement Act, 2013
Government of Odisha (2006). Ministry of Law and Justice, Government of Odisha
Government of West Bengal, Housing Department (2014) Gitanjali housing scheme memorandum and progress report. Housing Department, Kolkata
HAI (2023) LOCAL situation report 013/2023, 30 August 2023. https://www.hai.org. Accessed 13 Sept 2025. d
Hajra R, Ghosh T (2018) Agricultural productivity, household poverty and migration in the Indian Sundarban Delta. Elem Sci Anth 6:3
Halder B, Bandyopadhyay J, Banik P (2021) Evaluation of the climate change impact on urban heat island based on land surface temperature and geospatial indicators. Int J Environ Res 15(5):819–835
Hanh PTT, Furukawa M (2007) Impact of sea level rise on coastal zone of Vietnam. Dissertation. University of the Ryukyus
Haq, I., Omar, M.A.T., Zahra, Q.A. and Jahan, I. (2015). Evaluations of adaptation policies in GBM Delta of Bangladesh (No. 107642). DECCMA working paper, Deltas, Vulnerability and climate change: migration and adaptation, IDRC Project
Hazra S, Ghosh T, DasGupta R, Sen G (2002) Sea level and associated changes in the Sundarbans. Sci Culture 68(9–12):309–321. ISSN: 0036–8156
Hazra S, Das I, Samanta K, Bhadra T (2014) Impacts of climate change in Sundarbans area of West Bengal, India. Unpublished report, Caritas India and SOS, Jadavpur University
Hazra S, Mukhopadhyay A, Chanda A, Mondal P, Ghosh T, Mukherjee S, Salehin M (2016) Characterizing the 2D shape complexity dynamics of the islands of Sundarbans, Bangladesh: a fractal dimension approach. Environ Earth Sci 75(20):1367
Hermann E, Kempf W (2017) Climate change and the imagining of migration: Emerging discourses on Kiribati's land purchase in Fiji. Contemp Pac:231–263
Heslin A, Deckard ND, Oakes R, Montero-Colbert A (2018) Displacement and resettlement: understanding the role of climate change in contemporary migration. In: McNamara KE, Gibson C (eds) Loss and damage from climate change: concepts, methods and policy options. Springer, Cham, pp 237–258
Hiraide LA (2023) Climate refugees: a useful concept? Towards an alternative vocabulary of ecological displacement. Politics 43(2):267–282

HPG (2022) Climate change, conflict and internal displacement in Afghanistan, Humanitarian Policy Group, November 2022. https://odihpn.org/publication/climate-change-conflict-and-internal-displacement-in-afghanistan/. Accessed 13 Sept 2025
Hugo G (1996) Environmental concerns and international migration. Int Migr Rev 30(1):105–131. https://doi.org/10.2307/2547462
IDMC (2017) Global report on internal displacement. Internal Displacement Monitoring Centre, Geneva
IDMC (2021) Disaster displacement: a global review, 2008–2021. Internal Displacement Monitoring Centre
IDMC (2024) Global report on internal displacement 2024. Internal Displacement Monitoring Centre, Geneva. Accessed 13 Sept 2025
IFRC (2023) Nepal, Asia Pacific, Karnali earthquake—Emergency Appeal № MDRNP016, 8 November 2023. https://www.ifrc.org. Accessed 13 Sept 2025
Ighedosa SU (2019) Climate change: vulnerability of the Niger Delta Region in Nigeria. Generations 1(2):4–5
International Energy Agency (2014) World energy outlook 2014. International Energy Agency, Paris. https://www.iea.org/reports/world-energy-outlook-2014
International Energy Agency (IEA) (2022) National climate resilience assessment for Egypt. IEA, Paris. https://www.iea.org/reports/national-climate-resilience-assessment-for-egypt
International Organization for Migration (2023) Displacement report 43, February 2023, Nigeria. https://dtm.iom.int/reports/nigeria-displacement-report-43-february-2023. Accessed 13 Sept 2025
International Organization for Migration (2024) World migration report 2024. IOM Publications Platform. https://publications.iom.int/books/world-migration-report-2024. Accessed 13 Sept 2025
IOM DTM (2023a) Afghanistan: Herat Province: baseline mobility and community based needs assessment, Round 16 (September–December 2022), 8 July 2023. https://displacement.iom.int/reports. Accessed 13 Sept 2025
IOM DTM (2023b) Afghanistan—baseline mobility assessment report, Round 16 (September–December 2022), June 2023. https://displacement.iom.int/reports. Accessed 13 Sept 2025
Islamic Relief Worldwide (2023) Four earthquakes and a sandstorm leave traumatised Afghans in need of mental health support, 17 October 2023. https://www.islamic-relief.org. Accessed 13 Sept 2025
Jolly S, Ahmad N, Scott M (eds) (2024) Climate-related human mobility in Asia and the Pacific: interdisciplinary rights-based approaches. Sustainable development goals series. Springer, Singapore. https://doi.org/10.1007/978-981-97-3234-0
Kelman I (2018) Islandness within climate change narratives of small island developing states (SIDS). Island Stud J 13:149–166
Kisinger C, Matsui K (2021) Responding to climate-induced displacement in Bangladesh: A governance perspective. Sustainability 13(14):7788
Kompas (2023a) Floods hit South Aceh and Southeast Aceh, 21 November 2023. https://kompas.com. Accessed 13 Sept 2025
Kompas (2023b) Southeast Aceh floods: impact of forest damage, 24 August 2023. https://kompas.com. Accessed 13 Sept 2025
Ministry of Economy, Republic of Fiji (2016) Disaster recovery framework
Ministry of Rural Development, Government of India. (2007). National rehabilitation and resettlement policy. Department of Land Resources
Myers N (2002) Environmental refugees: a growing phenomenon of the 21st century. Philos Trans Royal Soc Lond Ser B: Biol Sci 357(1420):609–613. https://doi.org/10.1098/rstb.2001.0953. Accessed 13 Sept 2025
National Emergency Response Centre (2023) Brief regarding ground subsidence around Joshimath, District Chamoli, Uttarakhand, 9 January 2023. https://nerc.gov.in. Accessed 13 Sept 2025

NDTV (2023) In "Sinking" Joshimath, 4,000 evacuated after survey via satellites, 10 January 2023. https://ndtv.com. Accessed 13 Sept 2025
Nepal Red Cross (2023) Jajarkot-Rukum Earthquake response 2023—Situation Bulletin 02, 7 November 2023. https://nepalredcross.org. Accessed 13 Sept 2025
Nwilo PC, Badejo OT (2006) Impacts and management of oil spill pollution along the Nigerian coastal areas. Admin Mar Spaces: Int Issues 119:1–15
Oakes R (2025) Science talks: human mobility in the context of climate change. UNU-EHS. https://www.youtube.com/watch?v=vR2r4J8nOCc. Accessed 12 Sept 2025
Oakes R, Milan A, Campbell J (2016) Kiribati: climate change and migration—relationships between household vulnerability, human mobility and climate change. Report No. 20. United Nations University Institute for Environment and Human Security (UNU-EHS), Bonn. http://collections.unu.edu/view/UNU:5903
OCHA (2016) Tropical Cyclone Winston response and flash appeal final summary, 13 June 2016. goo.gl/PgXJ5W. Accessed 12 Sept 2025
OCHA (2023a) Pakistan: 2022 monsoon floods—Situation report No. 19, 12 August 2023. https://www.unocha.org. Accessed 13 Sept 2025
OCHA (2023b) Pakistan: monsoon rains flash update No. 1, 24 July 2023. https://www.unocha.org. Accessed 13 Sept 2025
Ojo G (2023) Global environmental diplomacy and climate change negotiations: implications for Africa's economies. Otuoke J Soc Sci 3(1):10–36
Rain D, Engstrom R, Ludlow C, Antos S (2011) Accra Ghana: a city vulnerable to flooding and drought-induced migration. In: Cities and climate change: global report on human settlements. UN-Habitat, Nairobi, pp 1–21
Richards JA, Bradshaw S (2017) Uprooted by climate change: responding to the growing risk of displacement. Oxfam, Oxford
Rozsa M (2024) Climate change is making homelessness worse—But experts say we can help, Salon. https://www.salon.com/2024/04/15/climate-change-is-making-homelessness-worse-but-experts-say-we-can-help/. Accessed 13 Sept 2025
SBS (2017) A year after Cyclone Winston, Fiji calls for global action on climate change, 20 February 2017. goo.gl/O2CcJb. Accessed 12 Sept 2025
Schilling KE, Anderson ES, Mount J, Suttles K, Gassman PW, Čerkasova N, White MJ, Arnold JG (2024) Evaluation of flood metrics across the Mississippi-Atchafalaya River Basin and their relation to flood damages. PLoS ONE 19(10):e0307486. https://doi.org/10.1371/journal.pone.0307486
Schofield C, Anggadi F (2024) The Australia-Tuvalu Falepili Union treaty: opportunities and controversies. Asia-Pacific Journal of Ocean Law and Policy 9:85–115
Secretariat of the Pacific Regional Environment Programme (SPREP) (2024) SPREP annual report: 2023. SPREP, Apia, Samoa
Simms JR, Waller HL, Brunet C, Jenkins P (2021) The long goodbye on a disappearing, ancestral island: a just retreat from Isle de Jean Charles. J Environ Stud Sci 11(3):316–328
Sivakumar MV, Ruane AC, Camacho J (2013) Climate change in the West Asia and North Africa region. In: Climate change and food security in West Asia and North Africa. Springer, Netherlands, pp 3–26
Soundararajan S (2022) Structured, scalable, and just: Evaluating the implementation of the Isle de Jean Charles resettlement project through risk-mitigation strategies. Thesis. University of Chicago
Stankard K (2024) Climate, culture, and mothers: the effects of relocation on Isle de Jean Charles's Choctaw Nation. Women Leading Change: Case Stud Women Gender Feminism 8:19–42
Stojanov R (2014) Migration as adaptation? Population dynamics in the age of climate variability. Global Change Research Centre, Academy of the Sciences of the Czech Republic, Brno
Strauss BH, Kulp S, Levermann A (2015) Mapping choices: carbon, climate, and rising seas: our global legacy

Takagi H, Esteban M, Mikami T, Fujii D (2016) Projection of coastal floods in 2050 Jakarta. Urban Clim 17:135–145

The Economist (2024) As seas rise, the relocation of Caribbean islanders has begun. Economist. https://www.economist.com/the-americas/2024/06/04/as-seas-rise-the-relocation-of-caribbean-islanders-has-begun. Accessed 13 Sept 2025

Tsegai D (2005) Impacts of migration and remittances on farm and non-farm self-employment incomes in the Volta Basin of Ghana. In: In: 11th EADI general conference on insecurity and development, Bonn

Tsikata D (2006) Living in the shadow of the large dams: long term responses of downstream and lakeside communities of Ghana's Volta River Project, vol 11. Brill, Leiden

Twilley RR, Bentley SJ Sr, Chen Q, Edmonds DA, Hagen SC, Lam NSN, Willson CS, Xu K, Braud D, Hampton Peele R, McCall A (2016) Co-evolution of wetland landscapes, flooding, and human settlement in the Mississippi River Delta Plain. Sustain Sci 11(4):711–731

UN RC/UNCT (2023) Nepal: Western Nepal earthquake 2023—Situation report No. 01, 6 November 2023. https://nepal.un.org. Accessed 13 Sept 2025

UNCT Nepal (2023) UN agencies launch a joint emergency response to earthquake in western Nepal, 6 November 2023. https://nepal.un.org. Accessed 13 Sept 2025

UNHCR (2023) Survivors of deadly earthquakes in Afghanistan struggle to recover, 7 November 2023. https://www.unhcr.org. Accessed 13 Sept 2025

UNHCR (2024) Global trends: forced displacement in 2024. UN Refugee Agency, Geneva, https://www.unhcr.org/global-trends-report-2024. Accessed 13 Sept 2025

UNICEF (2023a) Children displaced in a changing climate: preparing for a future already underway. United Nations Children's Fund. https://www.unicef.org/reports/children-displaced-changing-climate. Accessed 13 Sept 2025

UNICEF (2023b) Nepal humanitarian situation report No. 5, 20 December 2023. https://unicef.org/nepal. Accessed 13 Sept 2025

UNICEF UK (2017) No place to call home: protecting children's rights when the changing climate forces them to flee. UNICEF. https://downloads.unicef.org.uk/wp-content/uploads/2017/04/No-Place-To-Call-Home.pdf?_ga=2.73316244.623674493.1494945170-1964912592.1483624860, London. Accessed 12 Sept 2025

United Nations (2022) The Caribbean is 'ground zero' for the global climate emergency: Guterres. UN News, 28 July 2022. https://news.un.org/en/story/2022/07/1121902. Accessed 12 Sept 2025

United Nations Development Programme (UNDP) (2017) UNDP annual report 2017. United Nations Development Programme, New York. https://www.undp.org/publications/undp-annual-report-2017. Accessed 13 Sept 2025

United Nations Development Programme (UNDP) (2018a) National adaptation plans in focus: lessons from Egypt. UNDP, New York

United Nations Development Programme (UNDP) (2018b) UNDP annual report 2023. United Nations Development Programme, New York. https://www.undp.org/publications/undp-annual-report-2023. Accessed 13 Sept 2025

United Nations Economic and Social Commission for Asia and the Pacific (ESCAP), International Labour Organization (ILO), United Nations Development Programme (UNDP), United Nations University Institute for Environment and Human Security (UNU-EHS), and the University of the South Pacific (USP) (2007) Climate change and migration in the Pacific: links, attitudes, and future scenarios in Nauru, Tuvalu, and Kiribati

Van der Geest K, Vrieling A, Dietz T (2010) Migration and environment in Ghana: a cross-district analysis of human mobility and vegetation dynamics. Environ Urban 22(1):107–123

Warner K (2010) Global environmental change and migration: governance challenges. Glob Environ Change 20:402–413

Wasko C, Sharma A (2017) Global assessment of flood and storm extremes with increased temperatures. Sci Rep 7(1):7945

WFP (2023) Pakistan floods situation report, 29 September 2023. https://www.wfp.org. Accessed 13 Sept 2025

World Food Programme (2023) Annual Country Report. WFP, Nigeria. https://www.wfp.org/operations/annual-country-report?operation_id=NG01&year=2023#/26985. Accessed 13 Sep 2025
Yamamoto L, Esteban M (2014) Atoll island states and international law: climate change displacement and sovereignty. Springer, Berlin. https://doi.org/10.1007/978-3-642-38186-7

The manufacturer's authorised representative in the EU is Springer Nature Customer Service Centre GmbH, Europaplatz 3, 69115 Heidelberg, Germany. If you have any concerns regarding our products, please contact ProductSafety@springernature.com

Printed and bound by CPI Group (UK) Ltd, Croydon, CR0 4YY

21/09/2026

02220013-0012